VOLUME FIVE HUNDRED AND NINE

METHODS IN ENZYMOLOGY

Nanomedicine

Infectious Diseases, Immunotherapy, Diagnostics, Antifibrotics, Toxicology and Gene Medicine

METHODS IN ENZYMOLOGY

Editors-in-Chief

JOHN N. ABELSON AND MELVIN I. SIMON

Division of Biology
California Institute of Technology
Pasadena, California

Founding Editors

SIDNEY P. COLOWICK AND NATHAN O. KAPLAN

VOLUME FIVE HUNDRED AND NINE

METHODS IN ENZYMOLOGY

Nanomedicine

Infectious Diseases, Immunotherapy, Diagnostics, Antifibrotics, Toxicology and Gene Medicine

EDITED BY

NEJAT DÜZGÜNEŞ
Professor of Microbiology
Department of Biomedical Sciences
University of the Pacific
Arthur A. Dugoni School of Dentistry
San Francisco, CA
USA

ELSEVIER

AMSTERDAM • BOSTON • HEIDELBERG • LONDON
NEW YORK • OXFORD • PARIS • SAN DIEGO
SAN FRANCISCO • SINGAPORE • SYDNEY • TOKYO
Academic Press is an imprint of Elsevier

Academic Press is an imprint of Elsevier
525 B Street, Suite 1900, San Diego, CA 92101-4495, USA
225 Wyman Street, Waltham, MA 02451, USA
32 Jamestown Road, London NW1 7BY, UK

First edition 2012

For information on all Academic Press publications
visit our website at elsevierdirect.com

ISBN: 978-0-12-391858-1
ISSN: 0076-6879

Printed and bound in United States of America
12 13 14 10 9 8 7 6 5 4 3 2 1

Contents

Contributors *xiii*
Preface *xxi*
Volumes in Series *xxv*

1. Enhanced Antiviral Activity of Acyclovir Loaded into Nanoparticles 1

Roberta Cavalli, Manuela Donalisio, Agnese Bisazza, Andrea Civra, Elisabetta Ranucci, Paolo Ferruti, and David Lembo

1. Introduction 2
2. Preparation of Acyclovir-Loaded Nanoparticles and Fluorescent Nanoparticles 4
3. Biocompatibility Assays 8
4. Assessment of the Antiviral Activity 14
5. Conclusions 18

Acknowledgements 18
References 18

2. Gold *manno*-Glyconanoparticles for Intervening in HIV gp120 Carbohydrate-Mediated Processes 21

Paolo Di Gianvincenzo, Fabrizio Chiodo, Marco Marradi, and Soledad Penadés

1. Introduction 22
2. Design, Preparation, and Characterization of *manno*-GNPs 23
3. Evaluation of *manno*-GNPs as Inhibitors of Carbohydrate-Mediated gp120 Interactions 28
4. Validation of *manno*-GNPs as Inhibitors of HIV-1/Cell Infection 33

Acknowledgments 37
References 37

3. Nanoparticle-Mediated Targeted Delivery of Antiretrovirals to the Brain **41**

Supriya D. Mahajan, Wing-Cheung Law, Ravikumar Aalinkeel, Jessica Reynolds, Bindukumar B. Nair, Ken-Tye Yong, Indrajit Roy, Paras N. Prasad, and Stanley A. Schwartz

1. Introduction 42
2. Preparation and Characterization of QD 45
3. Physical Characterization of QDs 47
4. Preparation of QD-Amprenavir-Tf Nanoplex 48
5. Assessment of Amprenavir Levels Using HPLC 49
6. Evaluating the Efficacy of Antiretroviral Containing Nanoplex Using an *In Vitro* BBB Model 49
7. Evaluation of the Antiviral Efficacy of the QD-Tf-Amprenavir Nanobioconjugate 54
8. Concluding Remarks 57

Acknowledgments 58
References 58

4. Antibacterial Activity of Doxycycline-Loaded Nanoparticles **61**

Ranjita Misra and Sanjeeb K. Sahoo

1. Introduction 62
2. Polymeric Nanoparticles 65
3. Chitosan Nanoparticles 65
4. Polyalkylcyanoacrylate Nanoparticles 67
5. poly(d,l-lactide-co-glycolide) Nanoparticles 68
6. poly(ε-caprolactone) Nanoparticles 70
7. Doxycycline-Loaded PLGA:PCL Nanoparticles 70
8. Conclusion 81

References 81

5. Antimicrobial Properties of Electrically Formed Elastomeric Polyurethane–Copper Oxide Nanocomposites for Medical and Dental Applications **87**

Z. Ahmad, M. A. Vargas-Reus, R. Bakhshi, F. Ryan, G. G. Ren, F. Oktar, and R. P. Allaker

1. Introduction 88
2. Materials and Methods 90
3. Results and Discussion 91
4. Concluding Remarks and Future Work 97

Acknowledgments 98
References 98

6. Gastrointestinal Delivery of Anti-inflammatory Nanoparticles **101**

Hamed Laroui, Shanthi V. Sitaraman, and Didier Merlin

1. Introduction 102
2. Material According to Drug Application 103
3. Anti-inflammatory Compounds as Encapsulated Drug 104
4. NSAIDs-Loaded NPs 110
5. Biomaterial Choice 111
6. Targeting NPs to the Colon: Hydrogel-Encapsulated NPs 112
7. Conclusion 118

Acknowledgments 121
References 121

7. Chitosan-Based Nanoparticles as a Hepatitis B Antigen Delivery System **127**

Filipa Lebre, Dulce Bento, Sandra Jesus, and Olga Borges

1. Introduction 128
2. Chitosan-Based Particle Preparation 129
3. Physicochemical Characterization of the Particles 132
4. Antigen Adsorption Studies 134
5. *In vitro* Release Studies 135
6. Evaluation of the Bioactivity of the Antigen 137
7. Cell Viability Studies with Spleen Cells 137
8. Studies on Uptake into Peyer's Patches 139
9. Concluding Remarks 140

References 140

8. Targeting Nanoparticles to Dendritic Cells for Immunotherapy **143**

Luis J. Cruz, Paul J. Tacken, Felix Rueda, Joan Carles Domingo, Fernando Albericio, and Carl G. Figdor

1. Introduction 144
2. Passive Targeting 144
3. Active Targeting 146
4. Particulate Vaccines 147
5. Targeting Gold NP Vaccines to DCs 148
6. Experimental Procedure for Targeted AuNP Preparation 150
7. Targeting Liposome-Based Vaccines to DCs 151
8. Targeting Poly(Lactic-*co*-Glycolic Acid)-Based Vaccines to DCs 154
9. Experimental Procedure for the Preparation of Targeted PLGA NP 155
10. Conclusion 157

References 158

9. **Protein–Carbon Nanotube Sensors: Single Platform Integrated Micro Clinical Lab for Monitoring Blood Analytes** **165**

Sowmya Viswanathan, Pingzuo Li, Wonbong Choi, Slawomir Filipek, T. A. Balasubramaniam, and V. Renugopalakrishnan

1. Introduction 166
2. Prototype 167
3. Concept 168
4. Biosensors 169
5. Microfluidics 169
6. Amperometric Sensor for the Detection of Blood Analytes 169
7. Protein Probes for Detection and Monitoring Molecular Components of Serum 171
8. Protein Engineering and Molecular Biology of Probe Proteins 180
9. Control of the Fermentation Process 181
10. Gene Structure and Purification of Overexpressed Protein 181
11. Immobilization of Proteins on SWCNT 183
12. Immobilization of Proteins on Self-Assembled Monolayers 185
13. Characterization of SWCNT Enzyme Adduct 186
14. Mediators 186
15. Surface Characterization 187
16. Fabrication 188
17. Signal Detection 188
18. Experimental Protocol 188
19. Conclusion 190

Acknowledgments 191
References 191

10. **Investigating the Toxic Effects of Iron Oxide Nanoparticles** **195**

Stefaan J. Soenen, Marcel De Cuyper, Stefaan C. De Smedt, and Kevin Braeckmans

1. Introduction 196
2. Materials 197
3. Methods 199
4. Concluding Remarks 221

Acknowledgments 221
References 222

11. Cytotoxicity of Gold Nanoparticles **225**

Yu Pan, Matthias Bartneck, and Willi Jahnen-Dechent

1. Introduction 225
2. Dosage and Quantification of Gold Nanoparticles 227
3. Aggregation State of Nanoparticles in Fluids 228
4. Cell-Based Nanotoxicity Studies 230

References 239

12. Design of Target-Seeking Antifibrotic Compounds **243**

Tero A. H. Järvinen

1. Introduction 244
2. *In Vivo* Phage Display 246
3. Homing Peptides 251
4. Multifunctional Fusion Proteins 254
5. Function of Multifunctional Fusion Protein 257
6. Concluding Remarks 259

Acknowledgments 259
References 260

13. Design and Fabrication of *N*-Alkyl-Polyethylenimine-Stabilized Iron Oxide Nanoclusters for Gene Delivery **263**

Gang Liu, Zhiyong Wang, Seulki Lee, Hua Ai, and Xiaoyuan Chen

1. Introduction 264
2. Materials 265
3. Methods 266
4. Notes 269
5. Anticipated Results 271
6. Summary 274

Acknowledgments 274
References 275

14. Cell-Penetrating Peptide-Based Systems for Nucleic Acid Delivery: A Biological and Biophysical Approach **277**

Sara Trabulo, Ana L. Cardoso, Ana M. S. Cardoso, Nejat Düzgüneş, Amália S. Jurado, and Maria C. Pedroso de Lima

1. Introduction 278
2. Methods for Preparation of CPP-Based Nucleic Acid Complexes 279
3. Methods for Physical Characterization of CPP-Based Nucleic Acid Complexes and Their Interactions with Membranes 283
4. Methods for Evaluation of Biological Activity of CPP-Based Nucleic Acid Complexes 292

5. Methods for Evaluation of Cytotoxicity of CPP-Based Nucleic Acid Complexes 294
6. Concluding Remarks 296
Acknowledgments 297
References 297

15. Multifunctional Envelope-Type Nano Device (MEND) for Organelle Targeting Via a Stepwise Membrane Fusion Process 301

Yuma Yamada, Hidetaka Akita, and Hideyoshi Harashima

1. Programmed Packaging Concept and Construction of R8-MEND 302
2. Screening of Lipid Compositions for Their Ability to Fuse with Nuclear and Mitochondrial Membranes 304
3. Construction of Tetra-Lamellar MEND (T-MEND) 309
4. Nuclear Gene Delivery 314
5. Mitochondrial Bioactive Molecule Delivery Using a Dual Function-MITO-Porter as a MEND for Mitochondrial Delivery 318
6. Conclusions 322
Acknowledgments 322
References 322

16. Lipopolyplexes as Nanomedicines for Therapeutic Gene Delivery 327

Leire García, Koldo Urbiola, Nejat Düzgüneş, and Conchita Tros de Ilarduya

1. Introduction 328
2. Principle of the Method 329
3. Experimental Procedures 329
4. Application of Lipopolyplexes 332
5. Concluding Remarks 336
Acknowledgments 336
References 337

17. Interfering Nanoparticles for Silencing MicroRNAs 339

Huricha Baigude and Tariq M. Rana

1. Introduction 340
2. Delivery of Short Therapeutic RNAs 341
3. Protocols 344
4. Concluding Remarks 350
Acknowledgments 350
References 350

18. Genetic Nanomedicine: Gene Delivery by Targeted Lipoplexes 355

Nejat Düzgüneş and Conchita Tros de Ilarduya

1. Introduction 356
2. Preparation of Protein–Cationic Lipid–DNA Ternary Complexes 357
3. Gene Delivery *In Vitro* 359
4. Cell Viability Following Transfection 361
5. Enhancement of Transfection *In Vitro* 362
6. Targeted Lipoplexes *In Vivo* 364
7. Concluding Remarks 365

Acknowledgments 365
References 366

Author Index 369
Subject Index 399

CONTRIBUTORS

Ravikumar Aalinkeel
Department of Medicine, Division of Allergy, Immunology, and Rheumatology, State University of New York at Buffalo, Innovation Center, Buffalo, New York, USA

Z. Ahmad
School of Pharmacy and Biomedical Sciences, University of Portsmouth, Portsmouth, United Kingdom

Hua Ai
National Engineering Research Center for Biomaterials, Sichuan University, Chengdu, China

Hidetaka Akita
Faculty of Pharmaceutical Sciences, Hokkaido University, Sapporo, Japan

Fernando Albericio
Institute for Research in Biomedicine, Barcelona Science Park, Barcelona, Spain

R. P. Allaker
Queen Mary University of London, Barts and The London School of Medicine and Dentistry, Institute of Dentistry, London, United Kingdom

Huricha Baigude
Program for RNA Biology, Sanford-Burnham Medical Research Institute, La Jolla, California, USA

R. Bakhshi
Department of Mechanical Engineering, University College London, London, United Kingdom

T. A. Balasubramaniam
Biomedical Engineer, 184 Algonquin Trail, Ashland, Massachusetts, USA

Matthias Bartneck
Helmholtz Institute for Biomedical Engineering, RWTH Aachen University, Aachen, Germany

Dulce Bento
Center for Neuroscience and Cell Biology, University of Coimbra, and Faculty of Pharmacy, University of Coimbra, Pólo das Ciências da Saúde Azinhaga de Santa Comba, Coimbra, Portugal

Agnese Bisazza
Dipartimento di Scienza e Tecnologia del Farmaco, Università di Torino, Torino, Italy

Olga Borges
Center for Neuroscience and Cell Biology, University of Coimbra, and Faculty of Pharmacy, University of Coimbra, Pólo das Ciências da Saúde Azinhaga de Santa Comba, Coimbra, Portugal

Kevin Braeckmans
Laboratory of General Biochemistry and Physical Pharmacy, Department of Pharmaceutical Sciences, Ghent University, Ghent, Belgium

Ana L. Cardoso
CNC-Center for Neuroscience and Cell Biology, University of Coimbra, Coimbra, Portugal

Ana M. S. Cardoso
CNC-Center for Neuroscience and Cell Biology, University of Coimbra, Coimbra, Portugal

Roberta Cavalli
Dipartimento di Scienza e Tecnologia del Farmaco, Università di Torino, Torino, Italy

Xiaoyuan Chen
Center for Molecular Imaging and Translational Medicine, School of Public Health, Xiamen University, Xiamen, China, and Laboratory of Molecular Imaging and Nanomedicine, National Institute of Biomedical Imaging and Bioengineering, National Institutes of Health, Bethesda, Maryland, USA

Fabrizio Chiodo
Laboratory of GlycoNanotechnology, Biofunctional Nanomaterials Unit, CIC biomaGUNE, San Sebastián, Spain

Wonbong Choi
Nanomaterials and Device Lab, College of Engineering, Florida International University, Miami, Florida, USA

Andrea Civra
Dipartimento di Scienze Cliniche e Biologiche, Università di Torino, Torino, Italy

Luis J. Cruz
Department of Tumor Immunology, Nijmegen Centre for Molecular Life Sciences, Radboud University Medical Centre, Nijmegen, The Netherlands

Nejat Düzgüneş
Department of Biomedical Sciences, Arthur A. Dugoni School of Dentistry, University of the Pacific, San Francisco, California, USA

Marcel De Cuyper
Laboratory of BioNanoColloids, IRC, Kortrijk, Belgium

Conchita Tros de Ilarduya
Department of Pharmacy and Pharmaceutical Technology, School of Pharmacy, University of Navarra, Pamplona, Spain

Maria C. Pedroso de Lima
CNC-Center for Neuroscience and Cell Biology, University of Coimbra, and Department of Life Sciences, University of Coimbra, Coimbra, Portugal

Stefaan C. De Smedt
Laboratory of General Biochemistry and Physical Pharmacy, Department of Pharmaceutical Sciences, Ghent University, Ghent, Belgium

Paolo Di Gianvincenzo
Laboratory of GlycoNanotechnology, Biofunctional Nanomaterials Unit, CIC biomaGUNE, San Sebastián, Spain

Joan Carles Domingo
Department of Biochemistry and Molecular Biology, University of Barcelona, Barcelona, Spain

Manuela Donalisio
Dipartimento di Scienze Cliniche e Biologiche, Università di Torino, Torino, Italy

Paolo Ferruti
Dipartimento di Chmica Organica e Industriale, Università degli Studi di Milano, Milano, Italy

Carl G. Figdor
Department of Tumor Immunology, Nijmegen Centre for Molecular Life Sciences, Radboud University Medical Centre, Nijmegen, The Netherlands

Slawomir Filipek
Faculty of Chemistry, University of Warsaw, Warsaw, Poland

Leire García
Department of Pharmacy and Pharmaceutical Technology, School of Pharmacy, University of Navarra, Pamplona, Spain

Hideyoshi Harashima
Faculty of Pharmaceutical Sciences, Hokkaido University, Sapporo, Japan

Tero A.H. Järvinen
Vascular Mapping Laboratory, Center for Nanomedicine, Sanford-Burnham Medical Research Institute at UCSB, University of California, Santa Barbara; Cancer Center, Sanford-Burnham Medical Research Institute, La Jolla, California, USA, and Department of Orthopedic Surgery, University of Tampere and Tampere City Hospital, Tampere, Finland

Willi Jahnen-Dechent
Helmholtz Institute for Biomedical Engineering, RWTH Aachen University, Aachen, Germany

Sandra Jesus
CNC-Center for Neuroscience and Cell Biology, University of Coimbra, and Faculty of Pharmacy, University of Coimbra, Pólo das Ciências da Saúde Azinhaga de Santa Comba, Coimbra, Portugal

Amália S. Jurado
CNC-Center for Neuroscience and Cell Biology, University of Coimbra, and Department of Life Sciences, University of Coimbra, Coimbra, Portugal

Hamed Laroui
Department of Biology, Center for Diagnostics and Therapeutics, Georgia State University, Atlanta, Georgia, USA

Wing-Cheung Law
Institute for Lasers Photonics and Biophotonics, Natural Science Complex, State University of New York at Buffalo, Buffalo, New York, USA

Filipa Lebre
Center for Neuroscience and Cell Biology, University of Coimbra, and Faculty of Pharmacy, University of Coimbra, Pólo das Ciências da Saúde Azinhaga de Santa Comba, Coimbra, Portugal

Seulki Lee
Laboratory of Molecular Imaging and Nanomedicine, National Institute of Biomedical Imaging and Bioengineering, National Institutes of Health, Bethesda, Maryland, USA

David Lembo
Dipartimento di Scienze Cliniche e Biologiche, Università di Torino, Torino, Italy

Pingzuo Li
Children's Hospital, Harvard Medical School, Boston, Massachusetts, USA, and Shanghai Research Center of Biotechnology, Chinese Academy of Sciences, Shanghai, PR China

Gang Liu
Center for Molecular Imaging and Translational Medicine, School of Public Health, Xiamen University, Xiamen; Sichuan Key Laboratory of Medical Imaging, North Sichuan Medical College, Nanchong, China, and Laboratory of Molecular Imaging and Nanomedicine, National Institute of Biomedical Imaging and Bioengineering, National Institutes of Health, Bethesda, Maryland, USA

Supriya D. Mahajan
Department of Medicine, Division of Allergy, Immunology, and Rheumatology, State University of New York at Buffalo, Innovation Center, Buffalo, New York, USA

Marco Marradi
Laboratory of GlycoNanotechnology, Biofunctional Nanomaterials Unit, CIC biomaGUNE, and Biomedical Research Networking Center in Bioengineering, Biomaterials and Nanomedicine (CIBER-BBN), San Sebastián, Spain

Didier Merlin
Department of Biology, Center for Diagnostics and Therapeutics, Georgia State University, Atlanta, and Veterans Affairs Medical Center, Decatur, Georgia, USA

Ranjita Misra
Institute of Life Sciences, Chandrasekharpur, Bhubaneswar, Orissa, India

Bindukumar B. Nair
Department of Medicine, Division of Allergy, Immunology, and Rheumatology, State University of New York at Buffalo, Innovation Center, Buffalo, New York, USA

F. Oktar
Nanotechnology and Biomaterials Application & Research Centre, Marmara University, Istanbul, Turkey

Yu Pan
Helmholtz Institute for Biomedical Engineering, RWTH Aachen University, Aachen, Germany

Soledad Penadés
Laboratory of GlycoNanotechnology, Biofunctional Nanomaterials Unit, CIC biomaGUNE, and Biomedical Research Networking Center in Bioengineering, Biomaterials and Nanomedicine (CIBER-BBN), San Sebastián, Spain

Paras N. Prasad
Institute for Lasers Photonics and Biophotonics, Natural Science Complex, State University of New York at Buffalo, Buffalo, New York, USA

Tariq M. Rana
Program for RNA Biology, Sanford-Burnham Medical Research Institute, La Jolla, California, USA

Elisabetta Ranucci
Dipartimento di Chmica Organica e Industriale, Università degli Studi di Milano, Milano, Italy

G. G. Ren
School of Engineering and Technology, University of Hertfordshire, Hatfield, United Kingdom

V. Renugopalakrishnan
Children's Hospital, Harvard Medical School, Boston, Massachusetts, USA

Jessica Reynolds
Department of Medicine, Division of Allergy, Immunology, and Rheumatology, State University of New York at Buffalo, Innovation Center, Buffalo, New York, USA

Indrajit Roy
Department of Chemistry, University of Delhi, Delhi, India

Felix Rueda
Department of Biochemistry and Molecular Biology, University of Barcelona, Barcelona, Spain

F. Ryan
Department of Mechanical Engineering, University College London, London, United Kingdom

Sanjeeb K. Sahoo
Institute of Life Sciences, Chandrasekharpur, Bhubaneswar, Orissa, India

Stanley A. Schwartz
Department of Medicine, Division of Allergy, Immunology, and Rheumatology, State University of New York at Buffalo, Innovation Center, Buffalo, New York, USA

Shanthi V. Sitaraman
Department of Medicine, Division of Digestive Diseases, Emory University, Atlanta, Georgia, USA

Stefaan J. Soenen
Laboratory of General Biochemistry and Physical Pharmacy, Department of Pharmaceutical Sciences, Ghent University, Ghent, Belgium

Paul J. Tacken
Department of Tumor Immunology, Nijmegen Centre for Molecular Life Sciences, Radboud University Medical Centre, Nijmegen, The Netherlands

Sara Trabulo
CNC-Center for Neuroscience and Cell Biology, University of Coimbra, Coimbra, Portugal

Koldo Urbiola
Department of Pharmacy and Pharmaceutical Technology, School of Pharmacy, University of Navarra, Pamplona, Spain

M. A. Vargas-Reus
Queen Mary University of London, Barts and The London School of Medicine and Dentistry, Institute of Dentistry, London, United Kingdom

Sowmya Viswanathan
Newton - Wellesley Hospital/Partners Healthcare System, Newton, Massachusetts, USA

Zhiyong Wang
National Engineering Research Center for Biomaterials, Sichuan University, Chengdu, China

Yuma Yamada
Faculty of Pharmaceutical Sciences, Hokkaido University, Sapporo, Japan

Ken-Tye Yong
School of Electrical and Electronic Engineering, Nanyang Technological University, Singapore

Preface

This volume of *Methods in Enzymology* introduces the applications and methods of nanomedicine, a field that has been emerging in the past decade, in conjunction with advances in nanotechnology and bionanoscience. The chapters cover applications in antiviral, antibacterial, and genetic nanomedicine, immunotherapy, antifibrotics, diagnosis, as well as the methods used to evaluate the toxicity of nanomedicines.

In the section Antiviral Nanomedicines, Roberta Cavalli and colleagues, in Chapter 1, describe the preparation of polymeric nanoparticles based on a β-cyclodextrin-poly(4-acryloylmorpholine) mono-conjugate as carriers of acyclovir, involving the solvent displacement method. They employ fluorescent-labeled nanoparticles for cellular trafficking studies and provide methods for the assessment of the antiviral activity of the nanoparticles. In Chapter 2, Soledad Penadés and coauthors demonstrate that oligomannoside-coated gold nanoparticles (*manno*-GNPs) are able to interfere with HIV high-mannose glycan-mediated processes and present methods for the preparation and characterization of *manno*-GNPs and the experiments using surface plasmon resonance and saturation transfer difference-NMR techniques. Supriya D. Mahajan and colleagues describe the stable incorporation of amprenavir within a transferrin-conjugated quantum dot, the ability of this nanoparticle to transverse an *in vitro* blood–brain barrier model, and its antiviral efficacy in HIV-1-infected monocytes in Chapter 3.

In the section Antibacterial Nanomedicines, Ranjita Misra and Sanjeeb K. Sahoo (Chapter 4) describe the improvement of the entrapment efficiency of doxycycline-loaded poly(lactide-co-glycolide):poly(ε-caprolactone) nanoparticles, evaluate the efficacy of these nanoparticles against *E. coli*, and show that doxycycline-loaded nanoparticles have superior effectiveness compared to native doxycycline, resulting from the sustained release of the antibiotic from the nanoparticles. In Chapter 5, Robert P. Allaker and colleagues write about copper oxide nanoparticles that are embedded into a polyurethane matrix forming thin porous films and demonstrate their activity against methicillin-resistant *Staphylococcus aureus*. This approach may have applications in designer filters, adhesive films instead of sutures, and mechanically supporting structures.

In Chapter 6 of the section Nanomedicines in Immunotherapy, Didier Merlin and coauthors show the development of a technique for specific targeting of nanoparticles to digestive tract regions, including the colon, using a hydrogel based on electrostatic interactions between positive ions and negative polysaccharides. Thus, nanoparticle degradation in the gastrointestinal tract is reduced, and lower doses of drug can be delivered to the colon to reduce colonic inflammation. Olga Borges and collaborators, in Chapter 7, outline the preparation of chitosan-based particles as an antigen-delivery system for mucosal surfaces, in particular for the hepatitis B surface antigen. They also describe the characterization of the particles in terms of size, morphology, and zeta potential; antigen adsorption onto particles; toxicity of the particles; and particle uptake into lymphoid organs. In Chapter 8, Luis J. Cruz and coauthors discuss the development of targeted nanodelivery systems carrying vaccine components, including antigens and adjuvants, to dendritic cells and provide an overview of antigen-delivery vehicles currently under investigation. They then focus on the use of liposomes, poly(lactic-co-glycolic acid) polymers and gold nanoparticles to obtain safe and efficacious vaccines.

In Chapter 9 of the section Diagnostic Nanomedicine, which deals with the application of nanotechnology to diagnostic medicine, Venkatesan Renugopalakrishnan and colleagues cover the design of unique, single-platform, integrated, multichannel sensors based on carbon nanotube-protein adducts specific to each one of the major analytes of blood. Carbon nanotubes enhance the signals derived from the interaction of the enzymes with the different analytes in blood.

Two chapters focus on the toxicity of nanoparticles in the section Nanomedicine Toxicity. In Chapter 10, Stefaan J. Soenen and colleagues describe the measurement of the potential toxicity of iron oxide nanoparticles, involving acute cytotoxicity, induction of reactive oxygen species, cell-associated iron, cell morphology, cell proliferation, cell functionality, as well as pH-induced or intracellular degradation of the nanoparticles. Willi Jahnen-Dechent and coworkers indicate that we are still far from predicting the toxicological properties of new nanoparticles and summarize some basic concept to assess the toxicity of gold nanoparticles, including death pathways, cell cycle arrest, and oxidative stress in Chapter 11.

In an application of the nanomedicine approach to tissue reconstruction, Tero A.H. Järvinen in Chapter 12 of the section Antifibrotic Nanomedicine, describe the development and use of a targeting system for an antifibrotic biotherapeutic, decorin, based on a vascular homing peptide that specifically recognizes angiogenic blood vessels in regenerating and inflammatory tissues.

The section Genetic Nanomedicine starts with Chapter 13 by Xiaoyuan Chen and collaborators who highlight the basic concepts and applications of nonviral gene delivery vehicles based on low-molecular-weight *N*-alkyl polyethylenimine-stabilized iron oxide nanoparticles. Many cell-penetrating peptides have been applied successfully to mediate the intracellular delivery of nucleic acids. In Chapter 14, Maria C. Pedroso de Lima and coauthors provide a description of the experimental procedures for the preparation of cell-penetrating peptide-based nucleic acid complexes, their biophysical characterization, and the evaluation of the biological and cytotoxic effects of the complexes. Hideyoshi Harashima and colleagues, in Chapter 15, describe their experience with the concept of "programmed packaging," leading to the production of a multifunctional envelope-type nanodevice (MEND) as a nonviral gene delivery system. They focus on the construction of a tetra-lamellar MEND, which can be utilized to overcome intracellular membrane barriers, since it involves stepwise membrane fusion. They also describe the delivery of a bioactive molecule targeted to the nucleus and mitochondria in living cells. Conchita Tros de Ilarduya and collaborators, in Chapter 16, describe an efficient, nonviral gene transfer system that employs polyethylenimine (PEI 800, 25, 22 kDa), a cationic lipid and cholesterol at different lipid/DNA molar ratios and five different protocols of formulation. Some of these ternary complexes mediate highly efficient gene delivery even in the presence of a high concentration of serum. Huricha Baigude and Tariq M. Rana describe the preparation and evaluation of interfering nanoparticles for the delivery of anti-miRNA or siRNA, using a lipid-functionalized poly-L-lysine dendrimer, in Chapter 17. They outline methods for the measurement of gene silencing, including Northern blot, PCR, and reporter gene assays. In Chapter 18, Nejat Düzgüneş and Conchita Tros de Ilarduya describe the preparation and evaluation of plain and targeted lipoplexes, using targeting ligands, including epidermal growth factor and transferrin. Ligand-associated lipoplexes may be used to target DNA or other nucleic acid drugs to specific cells, particularly cancer cells that overexpress the receptors. They also outline two animal models in which transferrin-lipoplexes have been used for antitumor therapy.

I would like to thank Zoe Kruze and Shaun Gamble of Elsevier London and Priya Kumaraguruparan of Elsevier Chennai for all their help in preparing the volume. I would also like to thank my family for their support during the editing of this volume.

I dedicate this volume to my dear children, Maxine and Avery.

NEJAT DÜZGÜNEŞ

Methods in Enzymology

Volume I. Preparation and Assay of Enzymes
Edited by Sidney P. Colowick and Nathan O. Kaplan

Volume II. Preparation and Assay of Enzymes
Edited by Sidney P. Colowick and Nathan O. Kaplan

Volume III. Preparation and Assay of Substrates
Edited by Sidney P. Colowick and Nathan O. Kaplan

Volume IV. Special Techniques for the Enzymologist
Edited by Sidney P. Colowick and Nathan O. Kaplan

Volume V. Preparation and Assay of Enzymes
Edited by Sidney P. Colowick and Nathan O. Kaplan

Volume VI. Preparation and Assay of Enzymes *(Continued)*
Preparation and Assay of Substrates
Special Techniques
Edited by Sidney P. Colowick and Nathan O. Kaplan

Volume VII. Cumulative Subject Index
Edited by Sidney P. Colowick and Nathan O. Kaplan

Volume VIII. Complex Carbohydrates
Edited by Elizabeth F. Neufeld and Victor Ginsburg

Volume IX. Carbohydrate Metabolism
Edited by Willis A. Wood

Volume X. Oxidation and Phosphorylation
Edited by Ronald W. Estabrook and Maynard E. Pullman

Volume XI. Enzyme Structure
Edited by C. H. W. Hirs

Volume XII. Nucleic Acids (Parts A and B)
Edited by Lawrence Grossman and Kivie Moldave

Volume XIII. Citric Acid Cycle
Edited by J. M. Lowenstein

Volume XIV. Lipids
Edited by J. M. Lowenstein

Volume XV. Steroids and Terpenoids
Edited by Raymond B. Clayton

Volume XVI. Fast Reactions
Edited by Kenneth Kustin

Volume XVII. Metabolism of Amino Acids and Amines (Parts A and B)
Edited by Herbert Tabor and Celia White Tabor

Volume XVIII. Vitamins and Coenzymes (Parts A, B, and C)
Edited by Donald B. McCormick and Lemuel D. Wright

Volume XIX. Proteolytic Enzymes
Edited by Gertrude E. Perlmann and Laszlo Lorand

Volume XX. Nucleic Acids and Protein Synthesis (Part C)
Edited by Kivie Moldave and Lawrence Grossman

Volume XXI. Nucleic Acids (Part D)
Edited by Lawrence Grossman and Kivie Moldave

Volume XXII. Enzyme Purification and Related Techniques
Edited by William B. Jakoby

Volume XXIII. Photosynthesis (Part A)
Edited by Anthony San Pietro

Volume XXIV. Photosynthesis and Nitrogen Fixation (Part B)
Edited by Anthony San Pietro

Volume XXV. Enzyme Structure (Part B)
Edited by C. H. W. Hirs and Serge N. Timasheff

Volume XXVI. Enzyme Structure (Part C)
Edited by C. H. W. Hirs and Serge N. Timasheff

Volume XXVII. Enzyme Structure (Part D)
Edited by C. H. W. Hirs and Serge N. Timasheff

Volume XXVIII. Complex Carbohydrates (Part B)
Edited by Victor Ginsburg

Volume XXIX. Nucleic Acids and Protein Synthesis (Part E)
Edited by Lawrence Grossman and Kivie Moldave

Volume XXX. Nucleic Acids and Protein Synthesis (Part F)
Edited by Kivie Moldave and Lawrence Grossman

Volume XXXI. Biomembranes (Part A)
Edited by Sidney Fleischer and Lester Packer

Volume XXXII. Biomembranes (Part B)
Edited by Sidney Fleischer and Lester Packer

Volume XXXIII. Cumulative Subject Index Volumes I-XXX
Edited by Martha G. Dennis and Edward A. Dennis

Volume XXXIV. Affinity Techniques (Enzyme Purification: Part B)
Edited by William B. Jakoby and Meir Wilchek

VOLUME XXXV. Lipids (Part B)
Edited by JOHN M. LOWENSTEIN

VOLUME XXXVI. Hormone Action (Part A: Steroid Hormones)
Edited by BERT W. O'MALLEY AND JOEL G. HARDMAN

VOLUME XXXVII. Hormone Action (Part B: Peptide Hormones)
Edited by BERT W. O'MALLEY AND JOEL G. HARDMAN

VOLUME XXXVIII. Hormone Action (Part C: Cyclic Nucleotides)
Edited by JOEL G. HARDMAN AND BERT W. O'MALLEY

VOLUME XXXIX. Hormone Action (Part D: Isolated Cells, Tissues, and Organ Systems)
Edited by JOEL G. HARDMAN AND BERT W. O'MALLEY

VOLUME XL. Hormone Action (Part E: Nuclear Structure and Function)
Edited by BERT W. O'MALLEY AND JOEL G. HARDMAN

VOLUME XLI. Carbohydrate Metabolism (Part B)
Edited by W. A. WOOD

VOLUME XLII. Carbohydrate Metabolism (Part C)
Edited by W. A. WOOD

VOLUME XLIII. Antibiotics
Edited by JOHN H. HASH

VOLUME XLIV. Immobilized Enzymes
Edited by KLAUS MOSBACH

VOLUME XLV. Proteolytic Enzymes (Part B)
Edited by LASZLO LORAND

VOLUME XLVI. Affinity Labeling
Edited by WILLIAM B. JAKOBY AND MEIR WILCHEK

VOLUME XLVII. Enzyme Structure (Part E)
Edited by C. H. W. HIRS AND SERGE N. TIMASHEFF

VOLUME XLVIII. Enzyme Structure (Part F)
Edited by C. H. W. HIRS AND SERGE N. TIMASHEFF

VOLUME XLIX. Enzyme Structure (Part G)
Edited by C. H. W. HIRS AND SERGE N. TIMASHEFF

VOLUME L. Complex Carbohydrates (Part C)
Edited by VICTOR GINSBURG

VOLUME LI. Purine and Pyrimidine Nucleotide Metabolism
Edited by PATRICIA A. HOFFEE AND MARY ELLEN JONES

VOLUME LII. Biomembranes (Part C: Biological Oxidations)
Edited by SIDNEY FLEISCHER AND LESTER PACKER

VOLUME LIII. Biomembranes (Part D: Biological Oxidations)
Edited by SIDNEY FLEISCHER AND LESTER PACKER

VOLUME LIV. Biomembranes (Part E: Biological Oxidations)
Edited by SIDNEY FLEISCHER AND LESTER PACKER

VOLUME LV. Biomembranes (Part F: Bioenergetics)
Edited by SIDNEY FLEISCHER AND LESTER PACKER

VOLUME LVI. Biomembranes (Part G: Bioenergetics)
Edited by SIDNEY FLEISCHER AND LESTER PACKER

VOLUME LVII. Bioluminescence and Chemiluminescence
Edited by MARLENE A. DELUCA

VOLUME LVIII. Cell Culture
Edited by WILLIAM B. JAKOBY AND IRA PASTAN

VOLUME LIX. Nucleic Acids and Protein Synthesis (Part G)
Edited by KIVIE MOLDAVE AND LAWRENCE GROSSMAN

VOLUME LX. Nucleic Acids and Protein Synthesis (Part H)
Edited by KIVIE MOLDAVE AND LAWRENCE GROSSMAN

VOLUME 61. Enzyme Structure (Part H)
Edited by C. H. W. HIRS AND SERGE N. TIMASHEFF

VOLUME 62. Vitamins and Coenzymes (Part D)
Edited by DONALD B. MCCORMICK AND LEMUEL D. WRIGHT

VOLUME 63. Enzyme Kinetics and Mechanism (Part A: Initial Rate and Inhibitor Methods)
Edited by DANIEL L. PURICH

VOLUME 64. Enzyme Kinetics and Mechanism
(Part B: Isotopic Probes and Complex Enzyme Systems)
Edited by DANIEL L. PURICH

VOLUME 65. Nucleic Acids (Part I)
Edited by LAWRENCE GROSSMAN AND KIVIE MOLDAVE

VOLUME 66. Vitamins and Coenzymes (Part E)
Edited by DONALD B. MCCORMICK AND LEMUEL D. WRIGHT

VOLUME 67. Vitamins and Coenzymes (Part F)
Edited by DONALD B. MCCORMICK AND LEMUEL D. WRIGHT

VOLUME 68. Recombinant DNA
Edited by RAY WU

VOLUME 69. Photosynthesis and Nitrogen Fixation (Part C)
Edited by ANTHONY SAN PIETRO

VOLUME 70. Immunochemical Techniques (Part A)
Edited by HELEN VAN VUNAKIS AND JOHN J. LANGONE

VOLUME 71. Lipids (Part C)
Edited by JOHN M. LOWENSTEIN

VOLUME 72. Lipids (Part D)
Edited by JOHN M. LOWENSTEIN

VOLUME 73. Immunochemical Techniques (Part B)
Edited by JOHN J. LANGONE AND HELEN VAN VUNAKIS

VOLUME 74. Immunochemical Techniques (Part C)
Edited by JOHN J. LANGONE AND HELEN VAN VUNAKIS

VOLUME 75. Cumulative Subject Index Volumes XXXI, XXXII, XXXIV–LX
Edited by EDWARD A. DENNIS AND MARTHA G. DENNIS

VOLUME 76. Hemoglobins
Edited by ERALDO ANTONINI, LUIGI ROSSI-BERNARDI, AND EMILIA CHIANCONE

VOLUME 77. Detoxication and Drug Metabolism
Edited by WILLIAM B. JAKOBY

VOLUME 78. Interferons (Part A)
Edited by SIDNEY PESTKA

VOLUME 79. Interferons (Part B)
Edited by SIDNEY PESTKA

VOLUME 80. Proteolytic Enzymes (Part C)
Edited by LASZLO LORAND

VOLUME 81. Biomembranes (Part H: Visual Pigments and Purple Membranes, I)
Edited by LESTER PACKER

VOLUME 82. Structural and Contractile Proteins (Part A: Extracellular Matrix)
Edited by LEON W. CUNNINGHAM AND DIXIE W. FREDERIKSEN

VOLUME 83. Complex Carbohydrates (Part D)
Edited by VICTOR GINSBURG

VOLUME 84. Immunochemical Techniques (Part D: Selected Immunoassays)
Edited by JOHN J. LANGONE AND HELEN VAN VUNAKIS

VOLUME 85. Structural and Contractile Proteins (Part B: The Contractile Apparatus and the Cytoskeleton)
Edited by DIXIE W. FREDERIKSEN AND LEON W. CUNNINGHAM

VOLUME 86. Prostaglandins and Arachidonate Metabolites
Edited by WILLIAM E. M. LANDS AND WILLIAM L. SMITH

VOLUME 87. Enzyme Kinetics and Mechanism (Part C: Intermediates, Stereo-chemistry, and Rate Studies)
Edited by DANIEL L. PURICH

VOLUME 88. Biomembranes (Part I: Visual Pigments and Purple Membranes, II)
Edited by LESTER PACKER

VOLUME 105. Oxygen Radicals in Biological Systems
Edited by LESTER PACKER

VOLUME 106. Posttranslational Modifications (Part A)
Edited by FINN WOLD AND KIVIE MOLDAVE

VOLUME 107. Posttranslational Modifications (Part B)
Edited by FINN WOLD AND KIVIE MOLDAVE

VOLUME 108. Immunochemical Techniques (Part G: Separation and Characterization of Lymphoid Cells)
Edited by GIOVANNI DI SABATO, JOHN J. LANGONE, AND HELEN VAN VUNAKIS

VOLUME 109. Hormone Action (Part I: Peptide Hormones)
Edited by LUTZ BIRNBAUMER AND BERT W. O'MALLEY

VOLUME 110. Steroids and Isoprenoids (Part A)
Edited by JOHN H. LAW AND HANS C. RILLING

VOLUME 111. Steroids and Isoprenoids (Part B)
Edited by JOHN H. LAW AND HANS C. RILLING

VOLUME 112. Drug and Enzyme Targeting (Part A)
Edited by KENNETH J. WIDDER AND RALPH GREEN

VOLUME 113. Glutamate, Glutamine, Glutathione, and Related Compounds
Edited by ALTON MEISTER

VOLUME 114. Diffraction Methods for Biological Macromolecules (Part A)
Edited by HAROLD W. WYCKOFF, C. H. W. HIRS, AND SERGE N. TIMASHEFF

VOLUME 115. Diffraction Methods for Biological Macromolecules (Part B)
Edited by HAROLD W. WYCKOFF, C. H. W. HIRS, AND SERGE N. TIMASHEFF

VOLUME 116. Immunochemical Techniques (Part H: Effectors and Mediators of Lymphoid Cell Functions)
Edited by GIOVANNI DI SABATO, JOHN J. LANGONE, AND HELEN VAN VUNAKIS

VOLUME 117. Enzyme Structure (Part J)
Edited by C. H. W. HIRS AND SERGE N. TIMASHEFF

VOLUME 118. Plant Molecular Biology
Edited by ARTHUR WEISSBACH AND HERBERT WEISSBACH

VOLUME 119. Interferons (Part C)
Edited by SIDNEY PESTKA

VOLUME 120. Cumulative Subject Index Volumes 81–94, 96–101

VOLUME 121. Immunochemical Techniques (Part I: Hybridoma Technology and Monoclonal Antibodies)
Edited by JOHN J. LANGONE AND HELEN VAN VUNAKIS

VOLUME 122. Vitamins and Coenzymes (Part G)
Edited by FRANK CHYTIL AND DONALD B. MCCORMICK

Volume 139. Cellular Regulators (Part A: Calcium- and Calmodulin-Binding Proteins)
Edited by Anthony R. Means and P. Michael Conn

Volume 140. Cumulative Subject Index Volumes 102–119, 121–134

Volume 141. Cellular Regulators (Part B: Calcium and Lipids)
Edited by P. Michael Conn and Anthony R. Means

Volume 142. Metabolism of Aromatic Amino Acids and Amines
Edited by Seymour Kaufman

Volume 143. Sulfur and Sulfur Amino Acids
Edited by William B. Jakoby and Owen Griffith

Volume 144. Structural and Contractile Proteins (Part D: Extracellular Matrix)
Edited by Leon W. Cunningham

Volume 145. Structural and Contractile Proteins (Part E: Extracellular Matrix)
Edited by Leon W. Cunningham

Volume 146. Peptide Growth Factors (Part A)
Edited by David Barnes and David A. Sirbasku

Volume 147. Peptide Growth Factors (Part B)
Edited by David Barnes and David A. Sirbasku

Volume 148. Plant Cell Membranes
Edited by Lester Packer and Roland Douce

Volume 149. Drug and Enzyme Targeting (Part B)
Edited by Ralph Green and Kenneth J. Widder

Volume 150. Immunochemical Techniques (Part K: *In Vitro* Models of B and T Cell Functions and Lymphoid Cell Receptors)
Edited by Giovanni Di Sabato

Volume 151. Molecular Genetics of Mammalian Cells
Edited by Michael M. Gottesman

Volume 152. Guide to Molecular Cloning Techniques
Edited by Shelby L. Berger and Alan R. Kimmel

Volume 153. Recombinant DNA (Part D)
Edited by Ray Wu and Lawrence Grossman

Volume 154. Recombinant DNA (Part E)
Edited by Ray Wu and Lawrence Grossman

Volume 155. Recombinant DNA (Part F)
Edited by Ray Wu

Volume 156. Biomembranes (Part P: ATP-Driven Pumps and Related Transport: The Na, K-Pump)
Edited by Sidney Fleischer and Becca Fleischer

Volume 173. Biomembranes [Part T: Cellular and Subcellular Transport: Eukaryotic (Nonepithelial) Cells]
Edited by Sidney Fleischer and Becca Fleischer

Volume 174. Biomembranes [Part U: Cellular and Subcellular Transport: Eukaryotic (Nonepithelial) Cells]
Edited by Sidney Fleischer and Becca Fleischer

Volume 175. Cumulative Subject Index Volumes 135–139, 141–167

Volume 176. Nuclear Magnetic Resonance (Part A: Spectral Techniques and Dynamics)
Edited by Norman J. Oppenheimer and Thomas L. James

Volume 177. Nuclear Magnetic Resonance (Part B: Structure and Mechanism)
Edited by Norman J. Oppenheimer and Thomas L. James

Volume 178. Antibodies, Antigens, and Molecular Mimicry
Edited by John J. Langone

Volume 179. Complex Carbohydrates (Part F)
Edited by Victor Ginsburg

Volume 180. RNA Processing (Part A: General Methods)
Edited by James E. Dahlberg and John N. Abelson

Volume 181. RNA Processing (Part B: Specific Methods)
Edited by James E. Dahlberg and John N. Abelson

Volume 182. Guide to Protein Purification
Edited by Murray P. Deutscher

Volume 183. Molecular Evolution: Computer Analysis of Protein and Nucleic Acid Sequences
Edited by Russell F. Doolittle

Volume 184. Avidin-Biotin Technology
Edited by Meir Wilchek and Edward A. Bayer

Volume 185. Gene Expression Technology
Edited by David V. Goeddel

Volume 186. Oxygen Radicals in Biological Systems (Part B: Oxygen Radicals and Antioxidants)
Edited by Lester Packer and Alexander N. Glazer

Volume 187. Arachidonate Related Lipid Mediators
Edited by Robert C. Murphy and Frank A. Fitzpatrick

Volume 188. Hydrocarbons and Methylotrophy
Edited by Mary E. Lidstrom

Volume 189. Retinoids (Part A: Molecular and Metabolic Aspects)
Edited by Lester Packer

Volume 206. Cytochrome P450
Edited by Michael R. Waterman and Eric F. Johnson

Volume 207. Ion Channels
Edited by Bernardo Rudy and Linda E. Iverson

Volume 208. Protein–DNA Interactions
Edited by Robert T. Sauer

Volume 209. Phospholipid Biosynthesis
Edited by Edward A. Dennis and Dennis E. Vance

Volume 210. Numerical Computer Methods
Edited by Ludwig Brand and Michael L. Johnson

Volume 211. DNA Structures (Part A: Synthesis and Physical Analysis of DNA)
Edited by David M. J. Lilley and James E. Dahlberg

Volume 212. DNA Structures (Part B: Chemical and Electrophoretic Analysis of DNA)
Edited by David M. J. Lilley and James E. Dahlberg

Volume 213. Carotenoids (Part A: Chemistry, Separation, Quantitation, and Antioxidation)
Edited by Lester Packer

Volume 214. Carotenoids (Part B: Metabolism, Genetics, and Biosynthesis)
Edited by Lester Packer

Volume 215. Platelets: Receptors, Adhesion, Secretion (Part B)
Edited by Jacek J. Hawiger

Volume 216. Recombinant DNA (Part G)
Edited by Ray Wu

Volume 217. Recombinant DNA (Part H)
Edited by Ray Wu

Volume 218. Recombinant DNA (Part I)
Edited by Ray Wu

Volume 219. Reconstitution of Intracellular Transport
Edited by James E. Rothman

Volume 220. Membrane Fusion Techniques (Part A)
Edited by Nejat Düzgüneş

Volume 221. Membrane Fusion Techniques (Part B)
Edited by Nejat Düzgüneş

Volume 222. Proteolytic Enzymes in Coagulation, Fibrinolysis, and Complement Activation (Part A: Mammalian Blood Coagulation Factors and Inhibitors)
Edited by Laszlo Lorand and Kenneth G. Mann

Volume 223. Proteolytic Enzymes in Coagulation, Fibrinolysis, and Complement Activation (Part B: Complement Activation, Fibrinolysis, and Nonmammalian Blood Coagulation Factors)
Edited by Laszlo Lorand and Kenneth G. Mann

Volume 224. Molecular Evolution: Producing the Biochemical Data
Edited by Elizabeth Anne Zimmer, Thomas J. White, Rebecca L. Cann, and Allan C. Wilson

Volume 225. Guide to Techniques in Mouse Development
Edited by Paul M. Wassarman and Melvin L. DePamphilis

Volume 226. Metallobiochemistry (Part C: Spectroscopic and Physical Methods for Probing Metal Ion Environments in Metalloenzymes and Metalloproteins)
Edited by James F. Riordan and Bert L. Vallee

Volume 227. Metallobiochemistry (Part D: Physical and Spectroscopic Methods for Probing Metal Ion Environments in Metalloproteins)
Edited by James F. Riordan and Bert L. Vallee

Volume 228. Aqueous Two-Phase Systems
Edited by Harry Walter and Göte Johansson

Volume 229. Cumulative Subject Index Volumes 195–198, 200–227

Volume 230. Guide to Techniques in Glycobiology
Edited by William J. Lennarz and Gerald W. Hart

Volume 231. Hemoglobins (Part B: Biochemical and Analytical Methods)
Edited by Johannes Everse, Kim D. Vandegriff, and Robert M. Winslow

Volume 232. Hemoglobins (Part C: Biophysical Methods)
Edited by Johannes Everse, Kim D. Vandegriff, and Robert M. Winslow

Volume 233. Oxygen Radicals in Biological Systems (Part C)
Edited by Lester Packer

Volume 234. Oxygen Radicals in Biological Systems (Part D)
Edited by Lester Packer

Volume 235. Bacterial Pathogenesis (Part A: Identification and Regulation of Virulence Factors)
Edited by Virginia L. Clark and Patrik M. Bavoil

Volume 236. Bacterial Pathogenesis (Part B: Integration of Pathogenic Bacteria with Host Cells)
Edited by Virginia L. Clark and Patrik M. Bavoil

Volume 237. Heterotrimeric G Proteins
Edited by Ravi Iyengar

Volume 238. Heterotrimeric G-Protein Effectors
Edited by Ravi Iyengar

Volume 239. Nuclear Magnetic Resonance (Part C)
Edited by Thomas L. James and Norman J. Oppenheimer

Volume 240. Numerical Computer Methods (Part B)
Edited by Michael L. Johnson and Ludwig Brand

Volume 241. Retroviral Proteases
Edited by Lawrence C. Kuo and Jules A. Shafer

Volume 242. Neoglycoconjugates (Part A)
Edited by Y. C. Lee and Reiko T. Lee

Volume 243. Inorganic Microbial Sulfur Metabolism
Edited by Harry D. Peck, Jr., and Jean LeGall

Volume 244. Proteolytic Enzymes: Serine and Cysteine Peptidases
Edited by Alan J. Barrett

Volume 245. Extracellular Matrix Components
Edited by E. Ruoslahti and E. Engvall

Volume 246. Biochemical Spectroscopy
Edited by Kenneth Sauer

Volume 247. Neoglycoconjugates (Part B: Biomedical Applications)
Edited by Y. C. Lee and Reiko T. Lee

Volume 248. Proteolytic Enzymes: Aspartic and Metallo Peptidases
Edited by Alan J. Barrett

Volume 249. Enzyme Kinetics and Mechanism (Part D: Developments in Enzyme Dynamics)
Edited by Daniel L. Purich

Volume 250. Lipid Modifications of Proteins
Edited by Patrick J. Casey and Janice E. Buss

Volume 251. Biothiols (Part A: Monothiols and Dithiols, Protein Thiols, and Thiyl Radicals)
Edited by Lester Packer

Volume 252. Biothiols (Part B: Glutathione and Thioredoxin; Thiols in Signal Transduction and Gene Regulation)
Edited by Lester Packer

Volume 253. Adhesion of Microbial Pathogens
Edited by Ron J. Doyle and Itzhak Ofek

Volume 254. Oncogene Techniques
Edited by Peter K. Vogt and Inder M. Verma

Volume 255. Small GTPases and Their Regulators (Part A: Ras Family)
Edited by W. E. Balch, Channing J. Der, and Alan Hall

Volume 256. Small GTPases and Their Regulators (Part B: Rho Family)
Edited by W. E. Balch, Channing J. Der, and Alan Hall

Volume 257. Small GTPases and Their Regulators (Part C: Proteins Involved in Transport)
Edited by W. E. Balch, Channing J. Der, and Alan Hall

Volume 258. Redox-Active Amino Acids in Biology
Edited by Judith P. Klinman

Volume 259. Energetics of Biological Macromolecules
Edited by Michael L. Johnson and Gary K. Ackers

Volume 260. Mitochondrial Biogenesis and Genetics (Part A)
Edited by Giuseppe M. Attardi and Anne Chomyn

Volume 261. Nuclear Magnetic Resonance and Nucleic Acids
Edited by Thomas L. James

Volume 262. DNA Replication
Edited by Judith L. Campbell

Volume 263. Plasma Lipoproteins (Part C: Quantitation)
Edited by William A. Bradley, Sandra H. Gianturco, and Jere P. Segrest

Volume 264. Mitochondrial Biogenesis and Genetics (Part B)
Edited by Giuseppe M. Attardi and Anne Chomyn

Volume 265. Cumulative Subject Index Volumes 228, 230–262

Volume 266. Computer Methods for Macromolecular Sequence Analysis
Edited by Russell F. Doolittle

Volume 267. Combinatorial Chemistry
Edited by John N. Abelson

Volume 268. Nitric Oxide (Part A: Sources and Detection of NO; NO Synthase)
Edited by Lester Packer

Volume 269. Nitric Oxide (Part B: Physiological and Pathological Processes)
Edited by Lester Packer

Volume 270. High Resolution Separation and Analysis of Biological Macromolecules (Part A: Fundamentals)
Edited by Barry L. Karger and William S. Hancock

Volume 271. High Resolution Separation and Analysis of Biological Macromolecules (Part B: Applications)
Edited by Barry L. Karger and William S. Hancock

Volume 272. Cytochrome P450 (Part B)
Edited by Eric F. Johnson and Michael R. Waterman

Volume 273. RNA Polymerase and Associated Factors (Part A)
Edited by Sankar Adhya

VOLUME 274. RNA Polymerase and Associated Factors (Part B)
Edited by SANKAR ADHYA

VOLUME 275. Viral Polymerases and Related Proteins
Edited by LAWRENCE C. KUO, DAVID B. OLSEN, AND STEVEN S. CARROLL

VOLUME 276. Macromolecular Crystallography (Part A)
Edited by CHARLES W. CARTER, JR., AND ROBERT M. SWEET

VOLUME 277. Macromolecular Crystallography (Part B)
Edited by CHARLES W. CARTER, JR., AND ROBERT M. SWEET

VOLUME 278. Fluorescence Spectroscopy
Edited by LUDWIG BRAND AND MICHAEL L. JOHNSON

VOLUME 279. Vitamins and Coenzymes (Part I)
Edited by DONALD B. MCCORMICK, JOHN W. SUTTIE, AND CONRAD WAGNER

VOLUME 280. Vitamins and Coenzymes (Part J)
Edited by DONALD B. MCCORMICK, JOHN W. SUTTIE, AND CONRAD WAGNER

VOLUME 281. Vitamins and Coenzymes (Part K)
Edited by DONALD B. MCCORMICK, JOHN W. SUTTIE, AND CONRAD WAGNER

VOLUME 282. Vitamins and Coenzymes (Part L)
Edited by DONALD B. MCCORMICK, JOHN W. SUTTIE, AND CONRAD WAGNER

VOLUME 283. Cell Cycle Control
Edited by WILLIAM G. DUNPHY

VOLUME 284. Lipases (Part A: Biotechnology)
Edited by BYRON RUBIN AND EDWARD A. DENNIS

VOLUME 285. Cumulative Subject Index Volumes 263, 264, 266–284, 286–289

VOLUME 286. Lipases (Part B: Enzyme Characterization and Utilization)
Edited by BYRON RUBIN AND EDWARD A. DENNIS

VOLUME 287. Chemokines
Edited by RICHARD HORUK

VOLUME 288. Chemokine Receptors
Edited by RICHARD HORUK

VOLUME 289. Solid Phase Peptide Synthesis
Edited by GREGG B. FIELDS

VOLUME 290. Molecular Chaperones
Edited by GEORGE H. LORIMER AND THOMAS BALDWIN

VOLUME 291. Caged Compounds
Edited by GERARD MARRIOTT

VOLUME 292. ABC Transporters: Biochemical, Cellular, and Molecular Aspects
Edited by SURESH V. AMBUDKAR AND MICHAEL M. GOTTESMAN

VOLUME 311. Sphingolipid Metabolism and Cell Signaling (Part A)
Edited by ALFRED H. MERRILL, JR., AND YUSUF A. HANNUN

VOLUME 312. Sphingolipid Metabolism and Cell Signaling (Part B)
Edited by ALFRED H. MERRILL, JR., AND YUSUF A. HANNUN

VOLUME 313. Antisense Technology
(Part A: General Methods, Methods of Delivery, and RNA Studies)
Edited by M. IAN PHILLIPS

VOLUME 314. Antisense Technology (Part B: Applications)
Edited by M. IAN PHILLIPS

VOLUME 315. Vertebrate Phototransduction and the Visual Cycle (Part A)
Edited by KRZYSZTOF PALCZEWSKI

VOLUME 316. Vertebrate Phototransduction and the Visual Cycle (Part B)
Edited by KRZYSZTOF PALCZEWSKI

VOLUME 317. RNA–Ligand Interactions (Part A: Structural Biology Methods)
Edited by DANIEL W. CELANDER AND JOHN N. ABELSON

VOLUME 318. RNA–Ligand Interactions (Part B: Molecular Biology Methods)
Edited by DANIEL W. CELANDER AND JOHN N. ABELSON

VOLUME 319. Singlet Oxygen, UV-A, and Ozone
Edited by LESTER PACKER AND HELMUT SIES

VOLUME 320. Cumulative Subject Index Volumes 290–319

VOLUME 321. Numerical Computer Methods (Part C)
Edited by MICHAEL L. JOHNSON AND LUDWIG BRAND

VOLUME 322. Apoptosis
Edited by JOHN C. REED

VOLUME 323. Energetics of Biological Macromolecules (Part C)
Edited by MICHAEL L. JOHNSON AND GARY K. ACKERS

VOLUME 324. Branched-Chain Amino Acids (Part B)
Edited by ROBERT A. HARRIS AND JOHN R. SOKATCH

VOLUME 325. Regulators and Effectors of Small GTPases
(Part D: Rho Family)
Edited by W. E. BALCH, CHANNING J. DER, AND ALAN HALL

VOLUME 326. Applications of Chimeric Genes and Hybrid Proteins
(Part A: Gene Expression and Protein Purification)
Edited by JEREMY THORNER, SCOTT D. EMR, AND JOHN N. ABELSON

VOLUME 327. Applications of Chimeric Genes and Hybrid Proteins
(Part B: Cell Biology and Physiology)
Edited by JEREMY THORNER, SCOTT D. EMR, AND JOHN N. ABELSON

Volume 345. G Protein Pathways (Part C: Effector Mechanisms)
Edited by Ravi Iyengar and John D. Hildebrandt

Volume 346. Gene Therapy Methods
Edited by M. Ian Phillips

Volume 347. Protein Sensors and Reactive Oxygen Species (Part A: Selenoproteins and Thioredoxin)
Edited by Helmut Sies and Lester Packer

Volume 348. Protein Sensors and Reactive Oxygen Species (Part B: Thiol Enzymes and Proteins)
Edited by Helmut Sies and Lester Packer

Volume 349. Superoxide Dismutase
Edited by Lester Packer

Volume 350. Guide to Yeast Genetics and Molecular and Cell Biology (Part B)
Edited by Christine Guthrie and Gerald R. Fink

Volume 351. Guide to Yeast Genetics and Molecular and Cell Biology (Part C)
Edited by Christine Guthrie and Gerald R. Fink

Volume 352. Redox Cell Biology and Genetics (Part A)
Edited by Chandan K. Sen and Lester Packer

Volume 353. Redox Cell Biology and Genetics (Part B)
Edited by Chandan K. Sen and Lester Packer

Volume 354. Enzyme Kinetics and Mechanisms (Part F: Detection and Characterization of Enzyme Reaction Intermediates)
Edited by Daniel L. Purich

Volume 355. Cumulative Subject Index Volumes 321–354

Volume 356. Laser Capture Microscopy and Microdissection
Edited by P. Michael Conn

Volume 357. Cytochrome P450, Part C
Edited by Eric F. Johnson and Michael R. Waterman

Volume 358. Bacterial Pathogenesis (Part C: Identification, Regulation, and Function of Virulence Factors)
Edited by Virginia L. Clark and Patrik M. Bavoil

Volume 359. Nitric Oxide (Part D)
Edited by Enrique Cadenas and Lester Packer

Volume 360. Biophotonics (Part A)
Edited by Gerard Marriott and Ian Parker

Volume 361. Biophotonics (Part B)
Edited by Gerard Marriott and Ian Parker

VOLUME 362. Recognition of Carbohydrates in Biological Systems (Part A)
Edited by YUAN C. LEE AND REIKO T. LEE

VOLUME 363. Recognition of Carbohydrates in Biological Systems (Part B)
Edited by YUAN C. LEE AND REIKO T. LEE

VOLUME 364. Nuclear Receptors
Edited by DAVID W. RUSSELL AND DAVID J. MANGELSDORF

VOLUME 365. Differentiation of Embryonic Stem Cells
Edited by PAUL M. WASSAUMAN AND GORDON M. KELLER

VOLUME 366. Protein Phosphatases
Edited by SUSANNE KLUMPP AND JOSEF KRIEGLSTEIN

VOLUME 367. Liposomes (Part A)
Edited by NEJAT DÜZGÜNEŞ

VOLUME 368. Macromolecular Crystallography (Part C)
Edited by CHARLES W. CARTER, JR., AND ROBERT M. SWEET

VOLUME 369. Combinational Chemistry (Part B)
Edited by GUILLERMO A. MORALES AND BARRY A. BUNIN

VOLUME 370. RNA Polymerases and Associated Factors (Part C)
Edited by SANKAR L. ADHYA AND SUSAN GARGES

VOLUME 371. RNA Polymerases and Associated Factors (Part D)
Edited by SANKAR L. ADHYA AND SUSAN GARGES

VOLUME 372. Liposomes (Part B)
Edited by NEJAT DÜZGÜNEŞ

VOLUME 373. Liposomes (Part C)
Edited by NEJAT DÜZGÜNEŞ

VOLUME 374. Macromolecular Crystallography (Part D)
Edited by CHARLES W. CARTER, JR., AND ROBERT W. SWEET

VOLUME 375. Chromatin and Chromatin Remodeling Enzymes (Part A)
Edited by C. DAVID ALLIS AND CARL WU

VOLUME 376. Chromatin and Chromatin Remodeling Enzymes (Part B)
Edited by C. DAVID ALLIS AND CARL WU

VOLUME 377. Chromatin and Chromatin Remodeling Enzymes (Part C)
Edited by C. DAVID ALLIS AND CARL WU

VOLUME 378. Quinones and Quinone Enzymes (Part A)
Edited by HELMUT SIES AND LESTER PACKER

VOLUME 379. Energetics of Biological Macromolecules (Part D)
Edited by JO M. HOLT, MICHAEL L. JOHNSON, AND GARY K. ACKERS

VOLUME 380. Energetics of Biological Macromolecules (Part E)
Edited by JO M. HOLT, MICHAEL L. JOHNSON, AND GARY K. ACKERS

Volume 399. Ubiquitin and Protein Degradation (Part B)
Edited by Raymond J. Deshaies

Volume 400. Phase II Conjugation Enzymes and Transport Systems
Edited by Helmut Sies and Lester Packer

Volume 401. Glutathione Transferases and Gamma Glutamyl Transpeptidases
Edited by Helmut Sies and Lester Packer

Volume 402. Biological Mass Spectrometry
Edited by A. L. Burlingame

Volume 403. GTPases Regulating Membrane Targeting and Fusion
Edited by William E. Balch, Channing J. Der, and Alan Hall

Volume 404. GTPases Regulating Membrane Dynamics
Edited by William E. Balch, Channing J. Der, and Alan Hall

Volume 405. Mass Spectrometry: Modified Proteins and Glycoconjugates
Edited by A. L. Burlingame

Volume 406. Regulators and Effectors of Small GTPases: Rho Family
Edited by William E. Balch, Channing J. Der, and Alan Hall

Volume 407. Regulators and Effectors of Small GTPases: Ras Family
Edited by William E. Balch, Channing J. Der, and Alan Hall

Volume 408. DNA Repair (Part A)
Edited by Judith L. Campbell and Paul Modrich

Volume 409. DNA Repair (Part B)
Edited by Judith L. Campbell and Paul Modrich

Volume 410. DNA Microarrays (Part A: Array Platforms and Web-Bench Protocols)
Edited by Alan Kimmel and Brian Oliver

Volume 411. DNA Microarrays (Part B: Databases and Statistics)
Edited by Alan Kimmel and Brian Oliver

Volume 412. Amyloid, Prions, and Other Protein Aggregates (Part B)
Edited by Indu Kheterpal and Ronald Wetzel

Volume 413. Amyloid, Prions, and Other Protein Aggregates (Part C)
Edited by Indu Kheterpal and Ronald Wetzel

Volume 414. Measuring Biological Responses with Automated Microscopy
Edited by James Inglese

Volume 415. Glycobiology
Edited by Minoru Fukuda

Volume 416. Glycomics
Edited by Minoru Fukuda

Volume 417. Functional Glycomics
Edited by Minoru Fukuda

Volume 418. Embryonic Stem Cells
Edited by Irina Klimanskaya and Robert Lanza

Volume 419. Adult Stem Cells
Edited by Irina Klimanskaya and Robert Lanza

Volume 420. Stem Cell Tools and Other Experimental Protocols
Edited by Irina Klimanskaya and Robert Lanza

Volume 421. Advanced Bacterial Genetics: Use of Transposons and Phage for Genomic Engineering
Edited by Kelly T. Hughes

Volume 422. Two-Component Signaling Systems, Part A
Edited by Melvin I. Simon, Brian R. Crane, and Alexandrine Crane

Volume 423. Two-Component Signaling Systems, Part B
Edited by Melvin I. Simon, Brian R. Crane, and Alexandrine Crane

Volume 424. RNA Editing
Edited by Jonatha M. Gott

Volume 425. RNA Modification
Edited by Jonatha M. Gott

Volume 426. Integrins
Edited by David Cheresh

Volume 427. MicroRNA Methods
Edited by John J. Rossi

Volume 428. Osmosensing and Osmosignaling
Edited by Helmut Sies and Dieter Haussinger

Volume 429. Translation Initiation: Extract Systems and Molecular Genetics
Edited by Jon Lorsch

Volume 430. Translation Initiation: Reconstituted Systems and Biophysical Methods
Edited by Jon Lorsch

Volume 431. Translation Initiation: Cell Biology, High-Throughput and Chemical-Based Approaches
Edited by Jon Lorsch

Volume 432. Lipidomics and Bioactive Lipids: Mass-Spectrometry–Based Lipid Analysis
Edited by H. Alex Brown

VOLUME 433. Lipidomics and Bioactive Lipids: Specialized Analytical Methods and Lipids in Disease
Edited by H. ALEX BROWN

VOLUME 434. Lipidomics and Bioactive Lipids: Lipids and Cell Signaling
Edited by H. ALEX BROWN

VOLUME 435. Oxygen Biology and Hypoxia
Edited by HELMUT SIES AND BERNHARD BRÜNE

VOLUME 436. Globins and Other Nitric Oxide-Reactive Protiens (Part A)
Edited by ROBERT K. POOLE

VOLUME 437. Globins and Other Nitric Oxide-Reactive Protiens (Part B)
Edited by ROBERT K. POOLE

VOLUME 438. Small GTPases in Disease (Part A)
Edited by WILLIAM E. BALCH, CHANNING J. DER, AND ALAN HALL

VOLUME 439. Small GTPases in Disease (Part B)
Edited by WILLIAM E. BALCH, CHANNING J. DER, AND ALAN HALL

VOLUME 440. Nitric Oxide, Part F Oxidative and Nitrosative Stress in Redox Regulation of Cell Signaling
Edited by ENRIQUE CADENAS AND LESTER PACKER

VOLUME 441. Nitric Oxide, Part G Oxidative and Nitrosative Stress in Redox Regulation of Cell Signaling
Edited by ENRIQUE CADENAS AND LESTER PACKER

VOLUME 442. Programmed Cell Death, General Principles for Studying Cell Death (Part A)
Edited by ROYA KHOSRAVI-FAR, ZAHRA ZAKERI, RICHARD A. LOCKSHIN, AND MAURO PIACENTINI

VOLUME 443. Angiogenesis: *In Vitro* Systems
Edited by DAVID A. CHERESH

VOLUME 444. Angiogenesis: *In Vivo* Systems (Part A)
Edited by DAVID A. CHERESH

VOLUME 445. Angiogenesis: *In Vivo* Systems (Part B)
Edited by DAVID A. CHERESH

VOLUME 446. Programmed Cell Death, The Biology and Therapeutic Implications of Cell Death (Part B)
Edited by ROYA KHOSRAVI-FAR, ZAHRA ZAKERI, RICHARD A. LOCKSHIN, AND MAURO PIACENTINI

VOLUME 447. RNA Turnover in Bacteria, Archaea and Organelles
Edited by LYNNE E. MAQUAT AND CECILIA M. ARRAIANO

VOLUME 448. RNA Turnover in Eukaryotes: Nucleases, Pathways and Analysis of mRNA Decay
Edited by LYNNE E. MAQUAT AND MEGERDITCH KILEDJIAN

VOLUME 449. RNA Turnover in Eukaryotes: Analysis of Specialized and Quality Control RNA Decay Pathways
Edited by LYNNE E. MAQUAT AND MEGERDITCH KILEDJIAN

VOLUME 450. Fluorescence Spectroscopy
Edited by LUDWIG BRAND AND MICHAEL L. JOHNSON

VOLUME 451. Autophagy: Lower Eukaryotes and Non-Mammalian Systems (Part A)
Edited by DANIEL J. KLIONSKY

VOLUME 452. Autophagy in Mammalian Systems (Part B)
Edited by DANIEL J. KLIONSKY

VOLUME 453. Autophagy in Disease and Clinical Applications (Part C)
Edited by DANIEL J. KLIONSKY

VOLUME 454. Computer Methods (Part A)
Edited by MICHAEL L. JOHNSON AND LUDWIG BRAND

VOLUME 455. Biothermodynamics (Part A)
Edited by MICHAEL L. JOHNSON, JO M. HOLT, AND GARY K. ACKERS (RETIRED)

VOLUME 456. Mitochondrial Function, Part A: Mitochondrial Electron Transport Complexes and Reactive Oxygen Species
Edited by WILLIAM S. ALLISON AND IMMO E. SCHEFFLER

VOLUME 457. Mitochondrial Function, Part B: Mitochondrial Protein Kinases, Protein Phosphatases and Mitochondrial Diseases
Edited by WILLIAM S. ALLISON AND ANNE N. MURPHY

VOLUME 458. Complex Enzymes in Microbial Natural Product Biosynthesis, Part A: Overview Articles and Peptides
Edited by DAVID A. HOPWOOD

VOLUME 459. Complex Enzymes in Microbial Natural Product Biosynthesis, Part B: Polyketides, Aminocoumarins and Carbohydrates
Edited by DAVID A. HOPWOOD

VOLUME 460. Chemokines, Part A
Edited by TRACY M. HANDEL AND DAMON J. HAMEL

VOLUME 461. Chemokines, Part B
Edited by TRACY M. HANDEL AND DAMON J. HAMEL

VOLUME 462. Non-Natural Amino Acids
Edited by TOM W. MUIR AND JOHN N. ABELSON

VOLUME 463. Guide to Protein Purification, 2nd Edition
Edited by RICHARD R. BURGESS AND MURRAY P. DEUTSCHER

Volume 464. Liposomes, Part F
Edited by Nejat Düzgüneş

Volume 465. Liposomes, Part G
Edited by Nejat Düzgüneş

Volume 466. Biothermodynamics, Part B
Edited by Michael L. Johnson, Gary K. Ackers, and Jo M. Holt

Volume 467. Computer Methods Part B
Edited by Michael L. Johnson and Ludwig Brand

Volume 468. Biophysical, Chemical, and Functional Probes of RNA Structure, Interactions and Folding: Part A
Edited by Daniel Herschlag

Volume 469. Biophysical, Chemical, and Functional Probes of RNA Structure, Interactions and Folding: Part B
Edited by Daniel Herschlag

Volume 470. Guide to Yeast Genetics: Functional Genomics, Proteomics, and Other Systems Analysis, 2nd Edition
Edited by Gerald Fink, Jonathan Weissman, and Christine Guthrie

Volume 471. Two-Component Signaling Systems, Part C
Edited by Melvin I. Simon, Brian R. Crane, and Alexandrine Crane

Volume 472. Single Molecule Tools, Part A: Fluorescence Based Approaches
Edited by Nils G. Walter

Volume 473. Thiol Redox Transitions in Cell Signaling, Part A Chemistry and Biochemistry of Low Molecular Weight and Protein Thiols
Edited by Enrique Cadenas and Lester Packer

Volume 474. Thiol Redox Transitions in Cell Signaling, Part B Cellular Localization and Signaling
Edited by Enrique Cadenas and Lester Packer

Volume 475. Single Molecule Tools, Part B: Super-Resolution, Particle Tracking, Multiparameter, and Force Based Methods
Edited by Nils G. Walter

Volume 476. Guide to Techniques in Mouse Development, Part A Mice, Embryos, and Cells, 2nd Edition
Edited by Paul M. Wassarman and Philippe M. Soriano

Volume 477. Guide to Techniques in Mouse Development, Part B Mouse Molecular Genetics, 2nd Edition
Edited by Paul M. Wassarman and Philippe M. Soriano

Volume 478. Glycomics
Edited by Minoru Fukuda

VOLUME 496. Research on Nitrification and Related Processes, Part B
Edited by MARTIN G. KLOTZ AND LISA Y. STEIN

VOLUME 497. Synthetic Biology, Part A
Methods for Part/Device Characterization and Chassis Engineering
Edited by CHRISTOPHER VOIGT

VOLUME 498. Synthetic Biology, Part B
Computer Aided Design and DNA Assembly
Edited by CHRISTOPHER VOIGT

VOLUME 499. Biology of Serpins
Edited by JAMES C. WHISSTOCK AND PHILLIP I. BIRD

VOLUME 500. Methods in Systems Biology
Edited by DANIEL JAMESON, MALKHEY VERMA, AND HANS V. WESTERHOFF

VOLUME 501. Serpin Structure and Evolution
Edited by JAMES C. WHISSTOCK AND PHILLIP I. BIRD

VOLUME 502. Protein Engineering for Therapeutics, Part A
Edited by K. DANE WITTRUP AND GREGORY L. VERDINE

VOLUME 503. Protein Engineering for Therapeutics, Part B
Edited by K. DANE WITTRUP AND GREGORY L. VERDINE

VOLUME 504. Imaging and Spectroscopic Analysis of Living Cells
Optical and Spectroscopic Techniques
Edited by P. MICHAEL CONN

VOLUME 505. Imaging and Spectroscopic Analysis of Living Cells
Live Cell Imaging of Cellular Elements and Functions
Edited by P. MICHAEL CONN

VOLUME 506. Imaging and Spectroscopic Analysis of Living Cells
Imaging Live Cells in Health and Disease
Edited by P. MICHAEL CONN

VOLUME 507. Gene Transfer Vectors for Clinical Application
Edited by THEODORE FRIEDMANN

VOLUME 508. Nanomedicine
Cancer, Diabetes, and Cardiovascular, Central Nervous System, Pulmonary and Inflammatory Diseases
Edited by NEJAT DÜZGÜNEŞ

VOLUME 509. Nanomedicine
Infectious Diseases, Immunotherapy, Diagnostics, Antifibrotics, Toxicology and Gene Medicine
Edited by NEJAT DÜZGÜNEŞ

CHAPTER ONE

Enhanced Antiviral Activity of Acyclovir Loaded into Nanoparticles

Roberta Cavalli,* Manuela Donalisio,† Agnese Bisazza,* Andrea Civra,† Elisabetta Ranucci,‡ Paolo Ferruti,‡ *and* David Lembo†

Contents

1. Introduction 2
2. Preparation of Acyclovir-Loaded Nanoparticles and Fluorescent Nanoparticles 4
   2.1. Preparation and characterization of β-CD-PACM inclusion complexes 4
   2.2. Preparation of nanoparticles 6
   2.3. Characterization of β-CD-PACM nanoparticles 6
   2.4. Sterilization 8
   2.5. Physical stability of β-CD-PACM nanoparticles 8
3. Biocompatibility Assays 8
   3.1. Cell viability assay 9
   3.2. Evaluation of the hemolytic properties 10
   3.3. Complement activation assay 11
   3.4. Analysis of Intracellular ROS 11
   3.5. Evaluation of skin irritation potential of β-CD-PACM nanoparticles on EpiVaginal Tissue Model 12
4. Assessment of the Antiviral Activity 14
   4.1. HSV-1 infection and treatment of Vero cells 14
   4.2. Virus titration by the plaque assay 14
   4.3. Inhibition of HSV-1 infection in Vero cells by the acyclovir complex with β-CD-PACM nanoparticles 15
   4.4. Determination of acyclovir concentration in Vero cells 15
   4.5. Evaluation of cellular uptake of coumarin 6 β-CD-PACM inclusion complexes 17
5. Conclusions 18
Acknowledgements 18
References 18

* Dipartimento di Scienza e Tecnologia del Farmaco, Università di Torino, Torino, Italy
† Dipartimento di Scienze Cliniche e Biologiche, Università di Torino, Torino, Italy
‡ Dipartimento di Chmica Organica e Industriale, Università degli Studi di Milano, Milano, Italy

Methods in Enzymology, Volume 509
ISSN 0076-6879, DOI: 10.1016/B978-0-12-391858-1.00001-0

Abstract

The activity of antivirals can be enhanced by their incorporation in nanoparticulate delivery systems. Peculiar polymeric nanoparticles, based on a β-cyclodextrin-poly (4-acryloylmorpholine) monoconjugate (β-CD-PACM), are proposed as acyclovir carriers. The experimental procedure necessary to obtain the acyclovir-loaded nanoparticles using the solvent displacement preparation method will be described in this chapter. Fluorescent labeled nanoparticles are prepared using the same method for cellular trafficking studies. The biocompatibility assays necessary to obtain safe nanoparticles are reported. Section 4 of this chapter describes the assessment of the antiviral activity of the acyclovir-loaded nanoparticles.

1. Introduction

Herpes simplex viruses (HSV) type 1 and 2 (HSV-1 and HSV-2) are closely related pathogens of the Herpesviridae family of DNA viruses. Both cause a lifelong latent infection for which there is no cure or available effective vaccine. HSV-1 is usually transmitted via nonsexual contact and is generally clinically associated with orolabial infection, whereas HSV-2 is typically transmitted sexually and infects anogenital sites. However, HSV-1 and HSV-2 are both capable of infecting mucosal sites irrespective of their anatomic localization and can produce clinically indistinguishable lesions. HSV infection causes various forms of disease, from lesions on the lips, eyes, or genitalia to encephalitis or disseminated disease (Roizman *et al.*, 2007).

Acyclovir is the antiviral drug of choice for treating HSV infections because of its efficacy and moderate toxicity. Acyclovir is a nucleoside analogue that inhibits the viral DNA polymerase and is clinically used by intravenous, oral, or topical routes (O'Brien and Campoli-Richards, 1989). Because of its short half-life (about 2 h) and incomplete absorption (bioavailability about 15–30%) (Collins and Bauer, 1977), acyclovir must be taken in its oral dosage form five times daily (up to 1200 mg/day) and the interval of intravenous formulations is 8 h. Moreover, the intravenous dosage of acyclovir should be administered slowly over 1 h to prevent its precipitation in renal tubules. The development of delivery systems for acyclovir administration might improve its efficacy, thus decreasing the need for high and repeated doses, and limit its adverse side effects.

In recent years, nanoparticles have gained increasing attention as drug delivery systems for their several advantages, including controlled drug release, protection of active molecules from degradation, and cell targeting. They have been proposed as carriers of antiviral drugs for increasing their therapeutic index. Nanoparticulate-based systems may change the release kinetics of antivirals, increase their bioavailability, improve their efficacy, restrict adverse drug side effects, and reduce treatment costs. Moreover, they might permit the delivery of antiviral drugs to specific target sites and viral reservoirs in the body. These features are particularly relevant in viral diseases where high drug doses are needed and many active molecules show a low bioavailability (Lembo and Cavalli, 2010).

The preparation method and the process parameters can markedly affect the formation and the sizes of nanoparticles as well as the amount of drug encapsulated. Size, encapsulation efficiency, surface charge of the nanoparticles are important and stringent properties for therapeutic applications, influencing the biopharmaceutical behavior and cellular uptake of nanoparticles.

Consequently the selection of the proper method of preparation plays an important role in the development of drug-loaded nanoparticles and the choice depends mainly on either the active molecule characteristics or the type of matrix (generally polymers or lipids). In addition to the properties of the drug, the design of nanoparticles as drug delivery systems should consider the capacity to obtain sufficient drug encapsulation.

The large number of preparation methods proposed to obtain nanoparticles can be divided in two general groups: top–down and bottom–up approaches (Verma *et al.*, 2009). The top–down processes consists of the reduction of large particles into smaller particles using milling, high pressure homogenization, and microfluidization. These techniques involve high energy, and, generally, considerable heat is generated. Consequently they are not suitable for thermolabile compounds. The bottom–up approaches require that the drug is dissolved in an organic solvent with excipients and then precipitated. This approach includes emulsion-solvent evaporation, emulsion-solvent diffusion, solvent displacement, spray-drying, and supercritical fluid processes. These types of processes can also adversely affect nanoparticles because of the possible formation of hydrates or polymorphs and the presence of residues of the solvents involved in the preparation.

The solvent displacement method, also named nanoprecipitation or solvent injection, is a flexible one-step method firstly developed by Fessi *et al.* (1989). The method is suitable for different drugs and various types of substances for the formation of the matrix of nanoparticles (Cavalli *et al.*, 2007; Quintanar-Guerrero *et al.*, 1998; Schubert and Müller-Goymann, 2003). Moreover, it is fast, reproducible, cost effective, and it is useful to produce particles mainly in the range of 100–500 nm. This technique requires the use of a solvent completely miscible with water, for example, acetone or ethanol or mixtures. The drop-wise addition of a polymer dissolved in acetone to water under stirring forms nanoparticles. The molecular mechanism of particle formation involves complex interfacial phenomena as recently reported (Beck-Broichsitter *et al.*, 2010; Ganachaud and Katz, 2005; Vitale and Katz, 2003). With this method, particle formation depends on precipitation under condition of spontaneous dispersion or self-assembly of macromolecules (Mora-Huertas *et al.*, 2011).

We choose the solvent displacement method to obtain acyclovir nanoparticles for its simplicity, high reproducibility, and absence of toxic solvents or surfactants in the preparation protocol. Moreover, it is possible to use preformed polymers and to obtain high encapsulation efficiency of the drug in small nanoparticles.

This chapter reports a formulation example of this method of preparation which uses a peculiar polymer matrix composed of a β-cyclodextrin-poly(4-acryloylmorpholine) monoconjugate (β-CD-PACM). Previous studies have shown that this conjugate presents some advantages such as stability, biocompatibility, and an high acyclovir complexation capacity (Bencini *et al.*, 2008). The β-CD-PACM conjugates and water are highly miscible in all proportions and the resultant apparent solutions are stable and not as viscous. Based on these results we used a preformed acyclovir β-CD-PACM conjugate to produce the acyclovir-loaded nanoparticles (Cavalli *et al.*, 2009).

The chapter is divided into three sections. Section 2 describes the preparation and characterization of β-CD-PACM nanoparticles for the delivery of acyclovir and the investigation of their cell uptake. Section 3 describes some of the assays that may be used to assess the biocompatibility of the nanoparticulate formulation. Section 4 provides a description of the methods suitable to assess the antiviral potency of acyclovir-loaded nanoparticles. Overall, the chapter provides a general experimental procedure and methods that should be used to develop nanoparticles aiming at enhancing the therapeutic activity of acyclovir.

2. Preparation of Acyclovir-Loaded Nanoparticles and Fluorescent Nanoparticles

To prepare nanoparticles a two-step procedure is used consisting of the incorporation of acyclovir and of a fluorescent marker in the β-cyclodextrin cavity, and then nanoparticle formation.

2.1. Preparation and characterization of β-CD-PACM inclusion complexes

The β-CD-PACM conjugate used is obtained as previously described (Bencini *et al.*, 2008). Briefly, 6-deoxy-6-mercapto-β-ciclodextrin (β-CD-SH) is prepared and used as a chain-transfer agent in the radical polymerization of 4-acryloylmorpholine (Sigma-Aldrich, St. Louis, USA; distilled before use, bp 158 °C at 50 mmHg). The β-CD-PACM w/w ratio was 32.7 and the overall number-average (Mn) and weight average (Mw) molecular weight

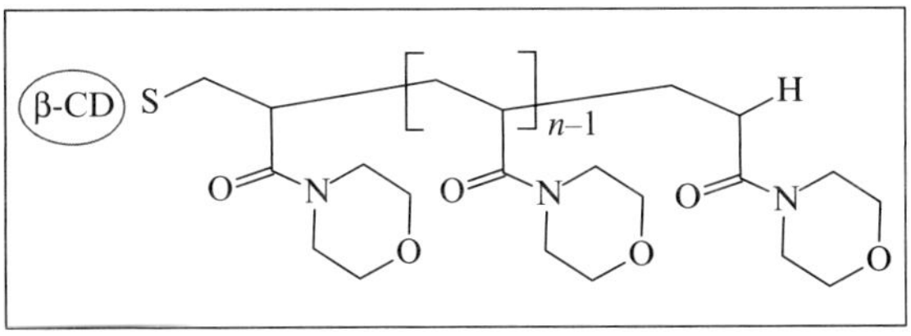

Figure 1.1 Schematic structure of the β-CD-PACM conjugate. (For color version of this figure, the reader is referred to the Web version of this chapter.)

values were 7860 and 13,500, respectively. Figure 1.1 shows the schematic structure of the conjugate.

The inclusion complex between acyclovir (Sigma-Aldrich) and β-CD-PACM conjugate is prepared by adding 10 mg of drug to 3 ml of an aqueous solution of the polymer conjugate (10 mg/ml) in a screw-capped tube; the mixture is left to equilibrate for 3 days at room temperature under moderate magnetic stirring and then centrifuged at 5000 rpm for 10 min (Allegra 64 R centrifuge, Beckman, USA). The supernatant is separated and freeze-dried using an Edwards Modulyo freeze-drier (Edwards, Sanborn, NY, USA; Crawley, UK) to obtain the acyclovir complex in powder form. After characterization, the solid complex is used for the preparation of the nanoparticles. The acyclovir β-CD-PACM complex is characterized by differential scanning calorimetry (DSC) and Fourier transform infrared (FTIR) spectroscopy.

Thermal analysis is carried out using a DSC/7 differential scanning calorimeter (Perkin Elmer, Branford, CT, USA) equipped with a TAC 7/DX instrument controller and the Pyris program. The instrument is calibrated with indium for melting point and heat of fusion before the analyses. A heating rate of 10 °C/min is employed in the 25–300 °C temperature range. Standard aluminum sample pans (Perkin Elmer) are used and about 3 mg of acyclovir complex or free drug are weighed; an empty pan is used as a reference standard. Analyses are carried out under nitrogen purge; triple runs are done on each sample.

For FTIR analysis, the dried acyclovir complex is mixed with spectroscopic grade potassium bromide (KBr) (Merck, Darmstadt, Germany). The acyclovir-KBr mixture is pressed and a KBr pellet (1 cm diameter) is obtained. The pellet contains 20 mg of complex and 100 mg of potassium bromaide. For comparison pellets containing the free drug and β-CD-PACM are also prepared. A System 2000 instrument (Perkin Elmer) is used to record the sample spectra in the range between 4000 and 400 cm^{-1}.

The amount of acyclovir complexed is determined by the HPLC method described below, after dilution in a flask of a weighed amount (2 mg) of the complex with a water:acetonitrile (Carlo Erba, Milan, Italy) solution (50:50 v/v).

The quantitative determination of acyclovir is achieved by HPLC analysis using a Perkin Elmer instrument (L2 Binary Pump, Perkin Elmer) with a Uv–vis spectrophotometer detector (LC 95, Perkin Elmer) with an external standard method. A reverse-phase hypersil ODS column (25 cm × 4.6 mm; Varian, USA) is used with a mobile phase consisting of acetonitrile (Carlo Erba) 20 m*M* ammonium acetate buffer (Sigma-Aldrich) pH 3.5 (12:88, v/v) at a flow rate of 1 ml/min. The detector wavelength is set at 250 nm. The calibration curve is linear in the range 0.5–15 μg/ml.

The inclusion complex between coumarin 6, chosen as a fluorescent marker (ACROS, Geel, Belgium), and CD-PACM is prepared by mixing the fluorescent molecule (2 mg) and 30 mg of the polymer conjugate in 3 ml of distilled water; the suspension is left to equilibrate for 5 days in the dark at room temperature and then centrifuged. The supernatant is separated and freeze-dried with a Modulyo freeze-drier (Edwards, UK) to obtain the complex in powder form.

The coumarin 6 β-CD-PACM complex is characterized by DSC using the method described above and by fluorescent spectroscopy ($\lambda_{ex}=450$ and $\lambda_{em}=490$ nm) using a RF-551 Shimadzu instrument (Shimadzu, Kyoto, Japan).

2.2. Preparation of nanoparticles

The solvent displacement technique is purposely tuned to obtain the β-CD-PACM nanoparticles. An acetone solution containing 13% v/v of water is selected to disperse the acyclovir PACM complex. Two types of polymer nanoparticles are prepared: blank and acyclovir loaded, using the β-CD conjugate as such or preloaded as a complex with acyclovir. Briefly, 3 ml of a polymer or a complex in acetone solution (10 mg/ml) are added drop-wise using a microsyringe to 12 ml of filtered water (MilliQ, Millipore, Billerica, MA, USA) under magnetic stirring at 250 rpm. The aqueous nanoparticle dispersions are then stirred for 5 h at room temperature to eliminate the solvent, washed by diaultrafiltration with a TCF2 Amicon system (Millipore), with a ultrafiltration membrane DIAFLO Amicon cut-off 30,000 Da (Millipore), and stored as aqueous suspension at 4 °C.

Some samples of β-CD-PACM nanoparticles are freeze-dried using a Modulyo freeze-drier (Edwards) to obtain the nanoparticles as a powder.

The same method is used to obtain coumarin 6-loaded nanoparticles.

2.3. Characterization of β-CD-PACM nanoparticles

2.3.1. Sizes and size distribution

The average diameters and polydispersity indices of the nanoparticle colloidal dispersions are determined by photon correlation spectroscopy (PCS) using a 90 Plus instrument (Brookhaven, NY, USA) at a fixed scattering angle of 90° and a temperature of 25 °C. Each reported value is the average of 10 measurements of three different nanoparticle batches. The polydispersity index is the size distribution of the nanoparticle population. The particle size is obtained from the Stokes–Einstein equation

$$D = nKT/(3\pi\eta d),$$

where d is the diameter of thc nanoparticles, D is the translational diffusion coefficient, K is the Boltzman constant, T is the temperature, and η is the viscosities of the medium.

2.3.2. Morphology

Transmission electron microscopy (TEM) analysis is performed using a Philips CM10 (Eindoven, NL) instrument. One hundred microliters of nanoparticles are dispersed in 3 ml of filtered water, and this nanoparticle suspension is sprayed on Formwar-coated copper grid and air-dried before observation.

2.3.3. Surface charge

The electrophoretic mobility and zeta potential of the two types of nanoparticles are determined using a 90 Plus instrument (Brookhaven). For zeta potential determination, samples of the three formulations are diluted with 0.1 m*M* KCl (Sigma-Aldrich) and placed in the electrophoretic cell, where an electric field of about 15 V/cm is applied. Each sample is analyzed at least in triplicate. The electrophoretic mobility measured is converted into zeta potential using the Smoluchowsky equation.

2.3.4. Drug loading and encapsulation efficiency

The amount of acyclovir in the nanoparticles is determined by the HPLC method described above. A weighed amount of freeze-dried nanoparticles is dissolved in methanol, diluted, and injected into the HPLC system. The loading and encapsulation efficiency are calculated as

$$\%\ \text{Encapsulation efficiency} = \frac{\text{actual amount of drug loaded in nanoparticles}}{\text{theory amount of drug loaded in nanoparticles}} \times 100$$

$$\%\ \text{Loading amount of the drug} = \frac{\text{amount of drug in nanoparticles}}{\text{amount of drug loaded nanoparticles}} \times 100$$

2.3.5. Determination of the coumarin content in the fluorescent nanoparticles

The amount of coumarin 6 in the nanoparticles is determined by spectrofluorimetric analysis, using a RF-551 Shimadzu instrument. A weighed amount of freeze-dried nanoparticles is dissolved in methanol and diluted before the sample is measured by fluorescent spectroscopy (λ_{ex} = 450 and λ_{em} = 490 nm) using a RF-551 Shimadzu instrument (Shimadzu).

2.3.6. *In vitro* release kinetics

The *in vitro* release experiments are carried out in phosphate buffer (Carlo Erba) at pH 7.4 at 37 °C using the dialysis bag diffusion technique. About 10 mg of acyclovir-loaded nanoparticles is dispersed in 3 ml of 0.05 *M* phosphate buffer at pH 7.4 and then placed in a dialysis bag using a cellulose dialysis membrane Spectra/Por (Spectrum, CA, USA) with cut-off 3500 Da.

The dialysis bag is then placed in 25 ml of phosphate buffer and incubated at 37 °C under stirring. At fixed times within a period of 2 h, buffer samples are collected, replaced with fresh buffer and the concentration of released acyclovir determined by HPLC analysis in the withdrawn samples using the method previously described. The release of coumarin 6 is carried out at two different pH values to mimic the physiological and intracellular conditions. For the release determination about 2 mg of coumarin 6-loaded nanoparticles are dispersed in 3 ml of 0.05 *M* phosphate buffer at pH 7.4 or 0.1 *M* Hepes buffer (Sigma-Aldrich) pH 5.0 and then placed in a dialysis bag as previously described. The withdrawn samples collected at fixed time are freeze-dried reconstituted with methanol and analyzed by fluorimetric spectroscopy, using an RF-551 Shimadzu instrument (Shimadzu).

2.4. Sterilization

For cell experiments sterile formulations are required. For the preparation of sterile nanoparticles all the solution are filtered through a 0.22-μm membrane (Millipore) or autoclaved. The nanoparticles are then obtained under aseptic conditions.

2.5. Physical stability of β-CD-PACM nanoparticles

The stability of blank and drug-loaded β-CD-PACM nanoparticle aqueous dispersions is evaluated over time. For this purpose, the two types of nanoparticles are maintained at 4 °C. Their average diameter, polydispersity index, and zeta potential values are determined after 24, 48 h and after 1, 3, 6, 12 months. The nanoparticle aqueous dispersions are stable over time without aggregation phenomena.

3. Biocompatibility Assays

Nanoparticles should undergo rigorous processing and characterization as well as safety testing before their application as drug carriers. Safety concerns the evaluation of the biocompatibility and toxic potential of the materials. Toxicity describes a change in form or function of cells while biocompatibility describes the interaction of a material with the host tissues. Once injected in the bloodstream, nanocarriers may interact with erythrocytes and endothelial cells affecting their viability and inducing the production of reactive oxygen species (ROS). Moreover, as any foreign body, they can be taken up from the circulation by phagocytes, including those of the mononuclear phagocytic system. Recognition and elimination of nanoparticles may occur through the absorption of specific proteins, for example, the C3 protein of the complement system (Sahu and Lambris, 2001). Upon binding, the C3 protein will change conformation, expose a reactive site, and release a signaling molecule that triggers the chain of biochemical events called the complement activation cascade. This will lead to the elimination of the particles by the macrophages, thus preventing the distribution of nanoparticles to other tissues and anatomic districts (Vonarbourg *et al.*, 2006). Therefore, in order to develop safe nanocarriers for medical application and to optimize the pharmacokinetic properties of the agent, it is of paramount importance to select those which are devoid of cytotoxic and complement activation effects and do not induce ROS production. If nanoparticles are designed for topical administration, human cell derived, organotypic *in vitro* 3D tissue equivalents can be used to assess cytotoxicity and induction of proinflammatory cytokines.

3.1. Cell viability assay

First of all, it is important to determine the cytotoxic potential of the formulation on the cell line used in the antiviral assays. This allows the calculation of the selectivity index defined as the ratio of the 50% cytotoxic concentration (CC_{50}) to the 50% antiviral concentration (IC_{50}).

The *Vero cells* (ATCC CCL-81) are derived from the kidney of an African green monkey, and are one of the more commonly used mammalian continuous cell lines to propagate HSV and to test anti-HSV compounds.

To test the effect of the acyclovir complex with β-CD-PACM nanoparticles on the viability of Vero cells, cells are seeded at a density of 6×10^4 well^{-1} in 24-well plates. After 24 h, they are incubated with the compounds or left untreated. After 72 h treatment, cell viability is determined by the CellTiter 96 Proliferation Assay Kit (Promega, Madison, WI, USA) according to the manufacturer's instructions. Notably, standard protocol for the MTS test is modified to enable proper evaluation of toxic effect of nanoparticles in cell culture conditions. Preliminary experiments revealed that the optical and chemical properties of nanoparticles might cause errors in the assessment of toxicity using cytotoxicity tests base on optical density measurements. Some of the nanoparticles are able to interfere with the substrate (MTS) increasing amount of formazan formed, even in the absence of living cells. To overcome this limitation, the test must be modified by addition of a second plate where nanoparticles are incubated with MTS in the absence of cells. Optical densities from this plate are subtracted from the optical densities obtained from the cell seeded plate.

The other limitation of the optical density based test is direct influence of nanoparticles on the optical characteristics of the well: to eliminate this limitation the plates are centrifuged and supernatant (free of nanoparticles) was transferred into the second plate (on which the OD measurements were performed). Obtained optical densities are subtracted.

The effect on cell viability of the formulation at different concentrations is expressed as a percentage, by comparing treated cells with cells incubated with culture medium alone.

The nanoparticulate formulation should be also tested on endothelial cells. The human umbilical vein endothelial cells are suitable for cell viability assays. They can be purchased from different sources or, alternatively, can be isolated from normal human umbilical vein. In this case, aseptically taken umbilical cord, preferably abort 20–30 cm length, is clamped and dipped in 70% ethanol for 30 s. Umbilical cord (vein) is washed with PBS to remove blood. Collagenase/dispase solution (Collagenase (type I—1 mg/ml; 242 U/ml)) and dispase (2 mg/ml (about 1.4 U/ml)) solution in Dulbecco's modified Eagle's medium (DMEM—Gibco) containing antibiotic/antimicotic is run into vessel until it appears at bottom end. Bottom end is clamped and more solution is added to the lumen of vessel. Umbilical cord is incubated at 37 °C for 30 min (for time to time the cord is rotated). The cell suspension is collected in 50 ml centrifuge tube. Lumen of vessel is rinsed with PBS and added to cell suspension. Pooled digest is centrifuged at 200 g for 10 min. Cells are washed twice and centrifuged. Final pellet is resuspended in culture medium (DMEM:F-12 (1:1) with Glutamax I, antibiotics/antimicotics, 20% FBS, 18 U/ml heparin, 10 ng/ml EGF) and seeded into flasks. Cells are seeded on collagen coated plastic at conc. 10^4 per well in 96-well plates. After 24 h of adaptation, cells are exposed to the range of concentrations of nanoparticles. Cell viability is determined by the CellTiter 96 Proliferation Assay Kit (Promega) as reported previously at different times postexposure. No cytotoxicity is detectable at the tested concentrations.

3.2. Evaluation of the hemolytic properties

Two hundred microliters of nanoparticle suspension are incubated at 37 °C for 90 min with 1 ml of diluted human blood (1:5 v/v). After incubation, blood containing suspensions are centrifuged at 1000 rpm for 10 min to separate plasma. The amount of hemoglobin released due to hemolysis is measured spectrophotometrically at 543 nm using a Beckman DU 730 spectrophotometer. The hemolytic activity is calculated with reference to blank and completely hemolyzed samples (induced by addition of ammonium sulfate (Carlo Erba) 20% w/v).

3.3. Complement activation assay

Human blood for the experiments is obtained from healthy volunteers. In the second step, plasma is separated by centrifugation (400 g, 10 min) of heparinised (10 U/ml) (S-Monoovette Sarstedt, Nümbrecht, Germany) freshly obtained blood: the separated plasma is then centrifuged once again (800 g, 10 min) and later aliquoted in 2 ml polypropylene tubes and stored in −70 °C. The processing time must never be longer than 1 h.

For the assessment of complement activation potential of nanoparticles, human serum is exposed to the tested nanoparticle at a concentration of 1 mg/ml in 2 ml polypropylene vials for 60 min in a standard CO_2 incubator (5% CO_2, 37 °C); the sample is tested in triplicate. After exposure, sample is cooled on ice and mixed with the cocktail of enzyme inhibitors (Sample Stabilizing Solution) to prevent future activation.

Finally, the sample is stored at −70 °C for future analysis. Generation of the complement activation products is measured using three different ELISA systems (iC3b, iC4b, Bb, SC5-9; QUIDEL® San, Diego, CA, USA), following the manufacturer's instructions. As an indicator of complement activation iC3b fragment release is chosen.

Briefly, 100 μl of iC3b Specimen Diluent (blank), Standards, Controls, and diluted Specimens are pipetted into the antibody-coated wells. The plate is incubated for 30 min at room temperature, then wells are washed for five times with the wash buffer provided in the kit and are added to each well 50 μl of iC3b conjugate antibody. The plate is incubated for 30 min at room temperature, then the wells are washed five times. Finally, 100 μl of Substrate solution are added to each well and the plate is incubated for 15–30 min at room temperature. After this incubation time, 50 μl of Stop solution are added to each well and absorbance is measured at 405 nm using an ELISA mircoplate reader.

Methodology is based on ISO procedure used for biocompatibility assessment of biomaterials in contact with blood. Obtained results indicate relatively low capacity of tested nanoparticle to activate complement system.

3.4. Analysis of Intracellular ROS

The generation of ROS in treated cells is determined by 2,7-dichlorofluoresceindiacetate (DCFH-DA; Sigma-Aldrich) staining. DCFH-DA is nonfluorescent and can diffuse into the cell through the plasma membrane where it is hydrolyzed to DCFH.

Nonfluorescent DCFH is finally converted to green fluorescent dichlorofluorescein (DCF) upon intracellular oxidation. For this assay, Vero cells are seeded in a six-well plate (2×10^5 cells/well) and are grown for 24 h.

After this time, cells are treated with different concentrations of nanoparticles, corresponding to ½ IC_{50}, IC_{50}, and 2 IC_{50}, for 3 h. Then, cells are harvested and washed twice with PBS.

Finally, the cells are resuspended in 1 ml of MEM with 5 μ*M* DCFH-DA and incubated for 10 min at 37 °C. Stock (1 mM) solution of DCFH-DA is prepared in ethanol and stored under liquid nitrogen vapor.

Immediately after the incubation, the samples are analyzed for DCF fluorescence in a flow cytometer (FacsCalibur, BD Biosciences, NJ, USA) at an excitation wavelength of 488 nm and emission wavelengths of 530. The fluorescence data are recorded with the CellQuest program (BD Biosciences) for 20,000 cells in each sample. Flow cytometric data are analyzed using WinMDI software, and the ROS generation is expressed in terms of percentage of cells with DCF (green) fluorescence.

Results indicated that no significant increase of intracellular ROS is detectable at the tested concentrations.

3.5. Evaluation of skin irritation potential of β-CD-PACM nanoparticles on EpiVaginal Tissue Model

To evaluate the skin irritation potential of β-CD-PACM nanoparticles on mucous membranes we analyse the toxicity and inflammatory response on a human cell derived, organotypic *in vitro* 3D tissue equivalent. As a tissue model, the EpiVaginal System purchased from MatTek Corporation (Ashland, MA, USA) is used; it consists of Human 3D Vaginal-Ectocervical Tissues, that is, cultured to form a multilayered and highly differentiated tissues closely resembling that of epithelial tissue *in vivo*. On the base of the foreseen use of the formulation, the study can be extended to other human cell derived, organotypic *in vitro* 3D tissue equivalents, as the EpiAirway and EpiDerm Systems (AIR-100-MM and EPI-200, MatTek Corp.), the differentiated models of the human epidermis and the epithelial tissue of the respiratory tract, respectively. According to the supplier's instructions, the cultures are transferred to six-well plates (containing 0.9 ml of MatTek assay medium per well)—with the apical surface remaining exposed to air—and incubated at 37 °C in 5% CO_2 overnight. The cytotoxicity of nanoparticles on mucous membranes is evaluated using two complementary systems: the MTT ET-50 Tissue Viability Assay, to study the metabolic activity of living cells and LDH release, to measure the accumulation of dead cells.

The MTT ET-50 Tissue Viability Assay is a colorimetric assay system that measures the reduction of a yellow tetrazolium component (MTT) into an insoluble purple formazan product by the mitochondria of viable cells. For this purpose, nanoparticles (100 μ*M*) are applied to the apical surface at the air-tissue interface of quadruplicate tissues for 30 min, 1, 4, and 18 h at 37 °C. When exposure of the samples to the compound is complete, any liquid atop the tissue is decanted and inserts are gently rinsed with PBS to remove any residual material. Then tissues are placed in the MTT solution (300 μl) containing 24-well plate and incubated for 3 h at 37 °C according the MTT kit's protocol (MatTek Corporation). After the incubation period is complete, the inserts are immersed into a extractant solution to lyse the cells and solubilize the colored crystals for 2 h at room temperature on an orbital shaker in the dark. Samples are read using an ELISA plate reader at a wavelength of 570 nm. The amount of color produced is directly proportional to the number of viable cells. The percentage of viability was determined at each of the dose concentrations using the following formula: % viability = 100 × (OD(sample)/OD(negative control)), where OD is the optical density. As negative control tissues are incubated with ultrapure water, 1.0% Triton X-100 is used as positive control. We demonstrate that nanoparticles are not cytotoxic at exposure times and the time required to reduce tissue viability to 50% (ET_{50}) is major of 18 h (data not shown).

A second standard method for quantification of cellular viability is based on the measurement of cytoplasmic enzyme activity released from damaged cells. LDH is a stable cytoplasmic enzyme which is present in most cells and it is released into the cell culture supernatant upon damage of the cytoplasmic membrane. Released LDH in culture medium of treated tissues is measured with the LDH Cytotoxicity Detection Kit (TAKARA Bio Inc., Shiga, Japan). The assay is a colorimetric assay system that measures the reduction of tetrazolium salt INT to red formazan, catalyzed by the LDH-catalyzed conversion of lactate to pyruvate. Briefly, 100 μl of culture medium of treated cells is removed and trasfered into corresponding wells of an optically clear 96-well flat bottom microtiter plate. Next, 100 μl of reaction mixture are added to each well and the plates are incubated for 30 min at room temperature in the dark to avoid photobleaching. Absorbance values for the samples are read using an ELISA plate reader at a wavelength of 490 nm.

The following three controls have to be performed in each experimental setup in order to calculate percent cytotoxicity: a background control, to measure the LDH activity contained in the assay medium, a low control, to measure the spontaneous LDH release, that is, the LDH activity released from the untreated normal cells, and a high control to measure the maximum releasable LDH activity in the cells, that is, the maximum LDH release induced by the addition of Triton X-100.

No difference for the cytoplasmic enzyme LDH release is observed between treated and untreated tissues (data not shown).

The inflammatory response is evaluated monitoring the release of the cytokine IL-1α in the culture medium of treated EpiVaginal tissues at different exposure times of 30 min, 1, 4, and 18 h, as reported previously. After incubation, the concentration of IL-1α in culture medium is measured according to the instructions provided by the manufacturer using the enzyme-linked immunoassay (kit IL-1 alpha ELISA (Bender Medsystem)). The kit included a microplate coated with a antibody specific for IL-1α. One hundred microliters of diluent and 50 μl of sample medium is added to each well in the provided plate, followed by 50 μl of biotin conjugate then covered with a plate cover and incubated for 2 h at room temperature. After washing each well, 100 μl of streptavidin-HRP is added to wells, covered and incubated for 1 h at room temperature. The wells are washed to remove unbound conjugate solution and 100 μl of TMB Substrate solution is added to each well. The plate then incubated in the dark for 20 min. Finally, 50 μl of stop solution is added to each well and absorbance values are measured within 30 min at 450 nm. The concentration of IL-1α is calculated by interpolation of a standard calibration curve. Our results indicate that samples exposed to nanoparticles do not exhibit significant differences in levels of inflammatory cytokine (data not shown).

These data demonstrated that β-CD-PACM nanoparticles not exerted toxicity and inflammatory response on a human-reconstructed epithelial tissue and therefore they can be classified as "not irritant" compound.

4. Assessment of the Antiviral Activity

4.1. HSV-1 infection and treatment of Vero cells

The effect of acyclovir and of the acyclovir-loaded formulation on the production of infectious viruses of HSV-1 is assessed through a yield reduction assay. Vero cells are grown as monolayers in Eagle's minimal essential medium (MEM, Gibco-BRL) supplemented with 10% of heat inactivated fetal bovine serum (FBS, Gibco-BRL) and antibiotics (Zell Shield, Minerva biolabs). The cells are seeded at a density of 6×10^4 well^{-1} in 24-well plates and after 24 h are infected with HSV at a multiplicity of infection (MOI) of 0.01 pfu/cell and then exposed to the drug or to the nanoparticulate formulation for 72 h. After a 1 h adsorption, the virus inoculum is removed and cultures are exposed in duplicate to serial dilutions of the test compounds. Supernatants are pooled as appropriate when a complete cytopathic effect occurs in the untreated control (normally 48–72 h postinfection) and cell-free virus infectivity titers are determined in duplicate by the plaque assay in Vero cell monolayers.

4.2. Virus titration by the plaque assay

Vero cells are seeded at a density of 1×10^4 well^{-1} in 96-well plates and after 24 h are inoculated with increasing dilutions of virus inoculum prepared in chilled maintenance medium (MEM, with 2% serum). After 1 h adsorption at

37 °C, the virus inoculum is removed, and cells are overlaid with 1.2% methylcellulose and incubated for 72 h at 37 °C. Plates are then fixed and colored with 0.1% of crystal violet for 30 min and then gently washed with water. The virus titer is estimated as plaque forming units per ml (pfu/ml) by counting the number of plaques at an appropriate dilution. The percent inhibition of virus infectivity is determined by comparing the virus titer in treated wells to the percent in untreated control wells.

The end-points of the virus yield reduction assay are the inhibitory concentrations of drug which reduced virus yield by 50% (IC_{50}) and by 99% (IC_{99}) versus the untreated virus control. The percent inhibition of virus infectivity is plotted as a function of drug concentration and the IC_{50} and IC_{99} values for inhibition curves are calculated by using the program PRISM 4 (GraphPad Software, San Diego, California, USA) to fit a variable slope sigmoidal dose–response curve.

4.3. Inhibition of HSV-1 infection in Vero cells by the acyclovir complex with β-CD-PACM nanoparticles

The virus yield reduction assay is performed to compare the antiviral activity of plain acyclovir and acyclovir β-CD-PACM nanoparticles. The assay provides a stringent test, which allows multiple cycles of viral replication to occur before measuring the production of infectious viruses.

The dose–response curves shown in Fig. 1.2 and the corresponding IC_{50} and IC_{99} values reported in Table 1.1 demonstrate that the antiviral potency of the acyclovir β-CD-PACM nanoparticles against two HSV-1 isolates is higher than that of free acyclovir. By contrast, the unloaded carriers exhibit no antiviral activity *per se*.

4.4. Determination of acyclovir concentration in Vero cells

The concentration of acyclovir in Vero cells is investigated as a measure of the intracellular accumulation of the drug. After the incubation with loaded nanoparticles, the cells are washed, lysed with a saturated solution of ammonium sulfate, and centrifuged at 4 °C for 10 min. The cellular lysate obtained can be frozen until analysis. Just before analysis they are thawed and centrifuged at 5000 rpm for 10 min at 10 °C. The supernatants are diluted with the mobile phase and injected in the system to estimate acyclovir concentration. The amount of acyclovir taken up inside the cells is calculated from the standard curve obtained in mobile phase with blank cellular lysate added to varying amounts of drug stock solution. A different reversed-phase HPLC method for the determination of acyclovir accumulated in cells is developed. A Spherisorb column (250 mm × 4.6 mm, Waters, Milford, MA, USA) is used, with water (adjusted to pH 2.5 with orthophosphoric acid:methanol (Carlo Erba) (92:8, v:v)) as mobile phase. The detector is set to 252 nm and the flow rate is 1 ml/min. The calibration curve is found

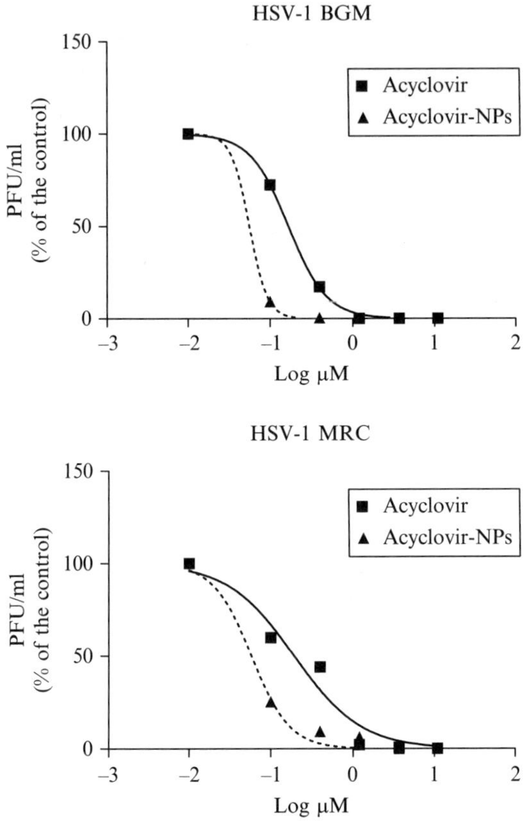

Figure 1.2 Antiviral activity of acyclovir and β-CD-PACM nanoparticles on two clinical isolates of HSV-1 (BGM and MRC). Vero cells are infected at a MOI of 0.01 and then exposed for 72 h to different drug concentrations. Supernatants of cell suspension are assayed for their infectivity by standard plaque reduction assay. Values are the means of three separate determinations.

Table 1.1 Antiviral potency of acyclovir formulations on Vero cells

	HSV-1 BGM		HSV-1 MRC	
	IC_{50}[a]	IC_{99}	IC_{50}	IC_{99}
Acyclovir	0.16	1.96	0.19	18.8
Acyclovir-loaded β-CD-PACM nanoparticles	0.05	0.15	0.05	0.59
β-CD-PACM nanoparticles	>100	>100	>100	>100

[a] Values are given as μ*M*.

to be linear in the range 0.04–10 μg/ml. Uptake enhancement from the formulations is expressed as % uptake versus that of plain acyclovir.

The intracellular drug concentration is considerably higher when the cells are incubated with acyclovir-loaded β-CD-PACM nanoparticles than cells incubated with acyclovir.

4.5. Evaluation of cellular uptake of coumarin 6 β-CD-PACM inclusion complexes

The uptake of fluorescently labeled β-CD-PACM nanoparticles is evaluated by confocal laser scanning microscopy. Exponentially growing Vero cells are plated and cultured overnight in 24-well plates on glass coverslips. The cell monolayers are then incubated with 10 μg/ml of fluorescent labeled nanoparticles for the times indicated and extensively washed with PBS (Carlo Erba) to observe the living cells. The assay is carried out on living unfixed cells to avoid misleading due to the cell fixation protocols. Confocal sections are taken on an inverted Zeiss LSM510 fluorescence microscope. As reported in Fig. 1.3, the images indicate that both compounds are internalized soon after exposure to the cells and that they remain within the cell for at least 24 h, with a cytoplasmic distribution. Interestingly, 1 h after treatment the cells incubated with β-CD-PACM nanoparticles exhibit a thicker layer of bright fluorescence in the perinuclear compartment. Fluorescence is not detected in control cells that have not been exposed to the labeled compounds (data not shown). It may be noticed that the perinuclear accumulation of nanoparticles after internalization in cells seems to be a common feature, as reported in several other studies on particulate delivery systems (Chavanpatil *et al.*, 2007; Chawla and Amiji, 2002;

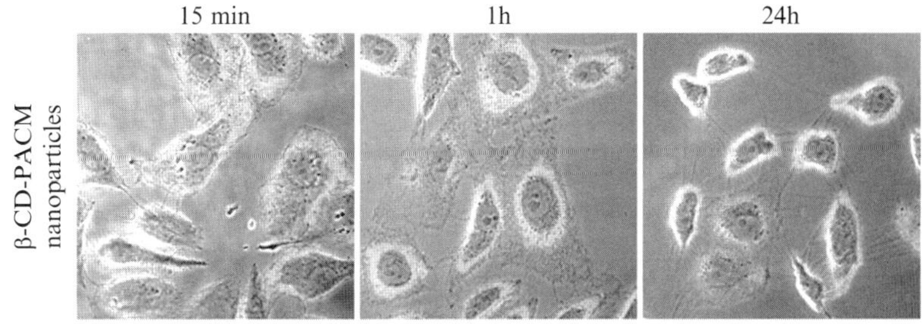

Figure 1.3 Cell uptake of β-CD-PACM nanoparticles. Vero cells are incubated with the compound for the times indicated and then analyzed by confocal laser scanning microscopy without fixation. Merged phase-contrast and immunofluorescence images are shown. β-CD-PACM nanoparticles appear to accumulate in a perinuclear compartment 1 h post-exposure. (See Color Insert.)

Desai *et al.*, 1997; Harush-Frenkel *et al.*, 2007; Lai *et al.*, 2007). Based on these results of the cell uptake assay, it is tempting to speculate that the higher antiviral activity of the acyclovir β-CD-PACM nanoparticle complex is due to the internalization and perinuclear accumulation of the nanoparticles, providing sustained drug delivery in the vicinity of the nucleus, that is the cellular compartment where acyclovir exerts its antiviral activity.

5. Conclusions

The methods presented in this chapter indicate that acyclovir-loaded nanoparticles with sizes lower than 200 nm and spherical shape are prepared. The antiviral activity of acyclovir in nanoparticles is notably superior compared to the free drug. The nanoparticle formulation may be proposed for the intracellular delivery of antiviral drugs.

ACKNOWLEDGEMENTS

Some of the experimental work presented in this chapter was supported by Turin University and Milan University research funds and Regione Piemonte (Ricerca Finalizzata 2008).

REFERENCES

Beck-Broichsitter, M., Rytting, E., Lebhardt, T., Wang, X., and Kissel, T. (2010). Preparation of nanoparticles by solvent displacement for drug delivery: A shift in the "ouzo region" upon drug loading. *Eur. J. Pharm. Sci.* **41,** 244–253.

Bencini, M., Ranucci, E., Ferruti, P., Manfredi, A., Trotta, F., and Cavalli, R. (2008). Poly (4-acryloylmorpholine) oligomers carrying a ß- cyclodextrin residue at one terminus. *J. Polym. Sci. A* **46,** 1607–1617.

Cavalli, R., Trotta, F., Carlotti, M. E., Possetti, B., and Trotta, M. (2007). Nanoparticles derived from amphiphilic cyclodextrins. *J. Incl. Phenom. Macrocycl. Chem.* **57,** 657–661.

Cavalli, R., Donalisio, M., Civra, A., Ferruti, P., Ranucci, E., Trotta, F., and Lembo, D. (2009). Enhanced antiviral activity of acyclovir loaded into ß-cyclodextrin-poly(4-acryloylmorpholine)conjugate nanoparticles. *J. Control. Rel.* **137,** 116–122.

Chavanpatil, M. D., Khdair, A., Gerard, B., Bachmeier, C., Miller, D. W., Shekhar, M. P., and Panyam, J. (2007). Surfactant-polymer nanoparticles overcome P-glycoprotein-mediated drug efflux. *Mol. Pharmacol.* **4,** 730–738.

Chawla, J. S., and Amiji, M. M. (2002). Biodegradable poly(epsilon-caprolactone) nanoparticles for tumor-targeted delivery of tamoxifen. *Int. J. Pharm.* **249,** 127–138.

Collins, P., and Bauer, D. J. (1977). Relative potencies of anti-herpes compounds. *Ann. N. Y. Acad. Sci.* **284,** 49–59.

Desai, M. P., Labhasetawar, V., Walter, E., Levy, R. J., and Amidon, G. L. (1997). The mechanism of uptake of biodegradable microparticles in Caco-2 cells is size dependent. *Pharm. Res.* **14,** 1568–1573.

Fessi, H., Puisieux, F., Devissaguet, J. Ph., Ammoury, N., and Benita, S. (1989). Nanocapsule formation by interfacial polymer deposition following solvent displacement. *Int. J. Pharm.* **55,** R1–R4.
Ganachaud, F., and Katz, J. L. (2005). Nanoparticles and nanocapsules created using the ouzo effect: Spontaneous emulsification as an alternative to ultrasonic and high-shear devices. *Chem. Phys. Chem.* **6,** 209–216.
Harush-Frenkel, O., Debotton, N., Benita, S., and Altschuler, Y. (2007). Targeting of nanoparticles to the clathrin-mediated endocytic pathway. *Biochem. Biophys. Res. Commun.* **353,** 26–32.
Lai, S. K., Hida, K., Man, S. T., Chen, C., Machamer, C., Schroer, T. A., and Hanes, J. (2007). Privileged delivery of polymer nanoparticles to the perinuclear region of live cells via a nonclathrin, non-degradative pathway. *Biomaterials* **28,** 2876–2884.
Lembo, D., and Cavalli, R. (2010). Nanoparticulate delivery systems for antiviral drugs. *Antivir. Chem. Chemother.* **21,** 53–70.
Mora-Huertas, C. E., Fessi, H., and Elaissari, A. (2011). Influence of process and formulation parameters on the formation of submicron particles by solvent displacement and emulsification–diffusion methods: Critical comparison. *Adv. Colloid Interface Sci.* **163,** 90–122.
O'Brien, J. J., and Campoli-Richards, D. M. (1989). Acyclovir. An updated review of its antiviral activity, pharmacokinetic properties and therapeutic efficacy. *Drugs* **37,** 233–309.
Quintanar-Guerrero, D., Allémann, E., Fessi, H., and Doelker, E. (1998). Preparation techniques and mechanisms of formation of biodegradable nanoparticles from preformed polymers. *Drug Dev. Ind. Pharm.* **24,** 1113–1128.
Roizman, B., Knipe, D. M., and Whitley, R. J. (2007). Herpes simplex viruses. *In* Fields Virology 5th edn. pp. 2501–2601. Lippincott Williams & Wilkins, Philadelphia.
Sahu, A., and Lambris, J. D. (2001). Structure and biology of complement protein C3, a connecting link between innate and acquired immunity. *Immunol. Rev.* **180,** 35–48.
Schubert, M. A., and Müller-Goymann, C. C. (2003). Solvent injection as a new approach for manufacturing lipid nanoparticles evaluation of the method and process parameters. *Eur. J. Pharm. Biopharm.* **55,** 125–131.
Verma, S., Gokhale, R., and Burgess, D. J. (2009). A comparative study of top-down and bottom up approaches for the preparation of micro-nanosuspensions. *Int. J. Pharm.* **380,** 216–222.
Vitale, S. A., and Katz, J. L. (2003). Liquid droplet dispersions formed by homogeneous liquid-liquid nucleation: The Ouzo effect. *Langmuir* **19,** 4105–4110.
Vonarbourg, A., Passirani, C., Saulnier, P., and Benoit, J. P. (2006). Parameters influencing the stealthiness of colloidal drug delivery systems. *Biomaterials* **27,** 4356–4373.

CHAPTER TWO

Gold *manno*-Glyconanoparticles for Intervening in HIV gp120 Carbohydrate-Mediated Processes

Paolo Di Gianvincenzo,* Fabrizio Chiodo,* Marco Marradi,*,† *and* Soledad Penadés*,†

Contents

1. Introduction 22
2. Design, Preparation, and Characterization of *manno*-GNPs 23
   2.1. Preparation of *manno*-GNPs 25
   2.2. Characterization of *manno*-GNPs 25
3. Evaluation of *manno*-GNPs as Inhibitors of Carbohydrate-Mediated gp120 Interactions 28
   3.1. Binding study of *manno*-GNPs to 2G12 by SPR 28
   3.2. Study of the 2G12/GNP interaction by STD-NMR 30
4. Validation of *manno*-GNPs as Inhibitors of HIV-1/Cell Infection 33
   4.1. Effect of *manno*-GNPs on HIV-1 neutralization by 2G12 in cellular models 33
   4.2. Inhibition of DC-SIGN-mediated HIV-1 trans-infection of human T cells by *manno*-GNPs 35
Acknowledgments 37
References 37

Abstract

After nearly three decades since the discovery of human immunodeficiency virus (HIV) (1983), no effective vaccine or microbicide is available, and the virus continues to infect millions of people worldwide each year. HIV antiretroviral drugs reduce the death rate and improve the quality of life in infected patients, but they are not able to completely remove HIV from the body. The glycoprotein gp120, part of the envelope glycoprotein (Env) of HIV, is responsible for virus entry and infection of host cells. High-mannose type glycans that decorate gp120

* Laboratory of GlycoNanotechnology, Biofunctional Nanomaterials Unit, CIC biomaGUNE, San Sebastián, Spain
† Biomedical Research Networking Center in Bioengineering, Biomaterials and Nanomedicine (CIBER-BBN), San Sebastián, Spain

Methods in Enzymology, Volume 509
ISSN 0076-6879, DOI: 10.1016/B978-0-12-391858-1.00002-2

are involved in different carbohydrate-mediated HIV binding. We have demonstrated that oligomannoside-coated gold nanoparticles (*manno*-GNPs) are able to interfere with HIV high-mannose glycan-mediated processes. In this chapter, we describe the methods for the preparation and characterization of *manno*-GNPs and the experiments performed by means of SPR and STD-NMR techniques to evaluate the ability of *manno*-GNPs to inhibit 2G12 antibody binding to gp120. The antibody 2G12-mediated HIV neutralization and the lectin DC-SIGN-mediated HIV trans-infection in cellular systems are also described.

1. Introduction

In the fight against the transmission of the human immunodeficiency virus (HIV), strategies are being developed to prevent (vaccines and microbicides) or to eliminate infection (antiretrovirals, therapeutic vaccines). None of these strategies have yet eradicated HIV infection or prevented transmission. A renewed effort is needed to understand the molecular basis of the complex mechanism of virus infection. The recent development of biofunctional nanoparticles and their applications in biomedicine has opened new opportunities to address this problem.

A remarkable feature of HIV is the dense carbohydrate coating of the envelope glycoprotein gp120 (Allan *et al.*, 1985; Barin *et al.*, 1985). This protein is heavily glycosylated with *N*-linked high-mannose type glycans (Feizi and Larkin, 1990; Geyer *et al.*, 1988; Mizuochi *et al.*, 1988). The glycans account for about 50% of the mass of this viral protein, the densest array of carbohydrates observed among human glycoproteins. The carbohydrate coating promotes HIV infection by its interaction with the dendritic cell-specific ICAM 3-grabing nonintegrin (DC-SIGN) expressed in dendritic cells (DCs). The C-type lectin DC-SIGN plays a crucial role in the entry and dissemination of HIV-1 from the mucosal site to T-cell areas in lymphoid tissues (Engering *et al.*, 2002; Fenouillet *et al.*, 1994; Geijtenbeek *et al.*, 2000). The glycan coat also protects the virus from potential immunogenic response. In spite of this role, the broadly neutralizing human antibody 2G12 that binds oligomannoside clusters of gp120 with nanomolar affinity was isolated from infected patients (Sanders *et al.*, 2002; Scanlan *et al.*, 2002; Trkola *et al.*, 1996). The crystal structure of the Fab fragment of 2G12 with oligosaccharide $Man_9GlcNAc_2$ and disaccharide Manα1-2Man was determined (Calarese *et al.*, 2003). The X-ray structure, supported by electron microscopic data, showed that the 2G12 Fab arms are locked together through the variable heavy domain, forming a previously uncharacterized dimer interface region for multivalent interaction with oligomannosides on gp120.

The multivalent presentation of the high-mannose glycans on the gp120 is fundamental for establishing both gp120/DC-SIGN binding and gp120/2G12 interactions. Mimicking the cluster presentation of gp120 high-mannose oligosaccharides is a strategy for designing carbohydrate-based antiviral agents and vaccines. Many different scaffolds were used to multimerize mannose oligosaccharides: Cholic acid (Li and Wang, 2004), galactose (Wang *et al.*, 2004), peptides (Dudkin *et al.*, 2004; Krauss *et al.*, 2007; Wang *et al.*, 2007), dendrons (Kabanova *et al.*, 2010; Wang *et al.*, 2008), and icosahedral virus capsids (Astronomo *et al.*, 2010). Our laboratory has developed a methodology based on gold nanoclusters as a scaffold to present carbohydrates in a multivalent display (glyconanoparticles) (de la Fuente *et al.*, 2001). The carbohydrate ligands are linked to the metallic nucleus by a stable Au—S bound. The carbohydrate molecules confer water solubility and biological functionality to the inorganic core of the nanomaterial. The nanometric size (1–3 nm) of these biofunctional materials is in the same order of magnitude of biological macromolecules such as proteins and nucleic acids, and allows the study, at the molecular level, of biorecognition processes where carbohydrates are involved (de la Fuente and Penadés, 2006; García *et al.*, 2010).

To interfere with gp120 carbohydrate-mediated processes, gold glyconanoparticles presenting different structural motifs of the *N*-linked high-mannose glycans of gp120 (oligomannoside-coated gold nanoparticle, *manno*-GNP) were designed and prepared (Fig. 2.1; Martinez-Ávila *et al.*, 2009a). The *manno*-GNPs were designed to mimic the carbohydrate cluster of gp120. The calcium-dependent lectin DC-SIGN and the antibody 2G12 that bind gp120 through mannose oligosaccharides were chosen as molecular targets to study the efficacy of *manno*-GNPs. We found that oligomannoside-functionalized gold glyconanoparticles are able to inhibit the HIV-1 DC-SIGN-mediated trans-infection of T cells at nanomolar concentrations (Martinez-Ávila *et al.*, 2009b), which validates *manno*-GNPs as an antiadhesive barrier at an early stage of HIV-1 infection.

In this chapter, we describe the design, preparation, and characterization of a set of *manno*-GNPs and methods to study their efficacy as inhibitors of gp120 binding to DC-SIGN or 2G12. We also describe their use as inhibitors of antibody 2G12-mediated HIV neutralization and DC-SIGN-mediated HIV trans-infection in a cellular system that mimics the natural route of infection of T-lymphocytes by DCs.

2. Design, Preparation, and Characterization of *Manno*-GNPs

There are essentially two strategies for the preparation of gold glyconanoparticles: direct gold salt reduction in the presence of thiol-ending ligands in aqueous solution, and ligand exchange on preformed gold nanoclusters (Marradi *et al.*, 2010).

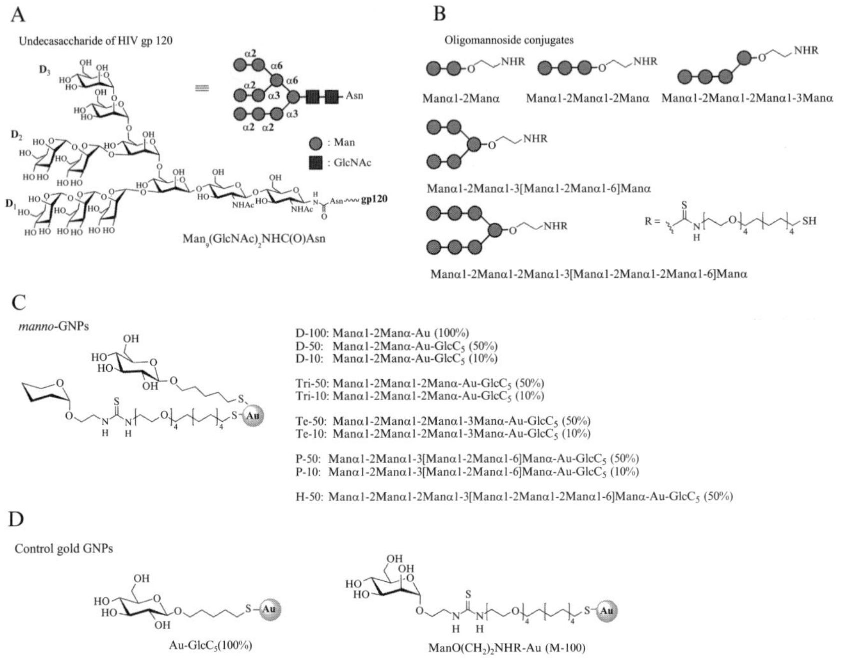

Figure 2.1 (A) gp120 *N*-glycan undecasaccharide; (B) thiol-ending oligomannoside conjugates used for *manno*-GNP preparation; (C) *manno*-GNPs. D, T, Te, P, and H stand for di- tri-, tetra-, penta-, and heptamannoside conjugates, respectively; the numbers indicate the percentages of oligomannosides on GNP, the rest being the 5-(mercapto)pentyl β-D-glucopyranoside ($GlcC_5$) component; (D) control GNPs bearing glucose and mannose conjugates. (See Color Insert.)

manno-GNPs are prepared by direct synthesis (Martinez-Ávila *et al.*, 2009a) using a modified protocol of the Brust–Schiffrin method (Brust *et al.*, 1994). Preparation of *manno*-GNPs require conjugation of the oligomannosides to a spacer ending in a mercapto group. The selected oligosaccharides are structural motifs of the gp120 undecasaccharide $Man_9(GlcNAc)_2$, except for a heptasaccharide that results from adding two mannose residues to the pentasaccharide (Fig. 2.1). To prepare the glycoconjugates, diverse types of spacers have been used with a hydrophobic and/or hydrophilic nature: aliphatic chains (five carbon atoms) to provide rigidity to the sugars on GNP or an amphiphilic mixed aliphatic/polyethyleneglycol linker to provide flexibility and solubility to the nanoparticle. The detailed protocols for the synthesis of the thiol-ending glycoconjugates have been described (Martinez-Ávila *et al.*, 2009a). *manno*-GNPs coated with variable density of oligomannosides (10%, 50%, and 100%) are prepared to assess the effect of presentation on the cluster in their interactions. Gold clusters of 2 nm size can bear up to 100 units of oligomannosides (Barrientos *et al.*, 2003). The mannoside density on the gold surface can be controlled by incorporating a second thiol-ending ligand on the surface.

A glucose conjugate ($GlcC_5S$) is used as a stealth component to control the density of oligomannosides on the *manno*-GNPs. GNPs fully covered by glucose or mannose thiol-ending conjugate are prepared and used as GNP controls. The oligomannoside conjugates and *manno*-GNPs are shown in Fig. 2.1.

2.1. Preparation of *manno*-GNPs

The first step for the direct *manno*-GNPs formation is the preparation of a solution of the thiol-ending neoglycoconjugates that will be linked to the gold surface. The ^{1}H-nuclear magnetic resonance (NMR) spectrum of the neoglycoconjugates mixture is registered to confirm the ligand ratio. Methanol and water are used as solvents for the direct GNP preparation. Usually, peer-shaped flasks are used. However, 2.5 mL Eppendorf tubes are preferably used for thiol-ending ligand amount below 10 μmol.

A typical *manno*-GNP preparation with tetramannoside conjugate and glucose conjugate is carried out as follows: To a methanolic solution of thiol-ending glycoconjugates in the desired ratio (total concentration 0.012 *M*; 3 eq), $HAuCl_4$ (0.025 *M* in H_2O; 1 eq) is added.

A freshly prepared water solution of $NaBH_4$ (1 *M*; 21 eq) is added in four times to the mixture under gentle shaking. It is convenient to perform some holes in the top of the reaction vessel, to avoid overpressure caused by the exothermic reaction of $NaBH_4$ with H_2O that produces H_2. Mixture is left 2 h under shaking at room temperature and, after centrifugation (5 min, 10000 rpm), the dark solid is washed four to five times with methanol (or ethanol). Supernatants containing unreacted ligands are collected and purified on Sephadex LH-20 ($MeOH/H_2O$ 9/1 as eluent). GNPs are dissolved in the minimum volume of NANOPURE water (18.2 MΩ-cm obtained by a Thermo Scientific Barnstead NANOpure DIamond Water System), and purified by dialysis. Briefly, the GNP solution is loaded into 5–10 cm segments of SnakeSkin pleated dialysis tubing (Pierce, 3500 MWCO) and placed in a 3-L beaker full of H_2O under gentle stirring. Water is changed seven to eight times in 72 h; afterward the solution is freeze-dried. The dry dark solid can be stored under Argon, in a dark dry place at 5 °C. It is better to avoid the use of gloves handling the dry solid GNPs (weight them with a Teflon spatula) because of electrostatic charges.

2.2. Characterization of *manno*-GNPs

The gold core of *manno*-GNPs is characterized by UV/Vis spectroscopy and, transmission electron microscopy (TEM). The organic ligands are determined by ^{1}H-NMR and elemental analysis as previously reported (Barrientos *et al.*, 2003; Carvalho de Souza and Kamerling, 2006). The UV/Vis spectra gave an indication of the GNPs' dimensions: small GNPs, with a diameter below 2 nm, do not show the plasmon absorption band at

520 nm at 0.1 mg mL^{-1} in water. UV/Vis spectra are registered in a Beckman Coulter DU 800 spectrophotometer. Infrared spectra (IR) on KBr pellet is a qualitative technique that helps especially when small amounts of ligands are attached on the gold surface. A GNPs spatula tip is enough to run IR spectra. IR were recorded from 4000 to 400 cm^{-1} with a Nicolet 6700 FT-IR spectrometer (Thermo Spectra-Tech).

In order to determine the diameter of the gold nucleus, TEM analysis is performed. A drop of a GNPs solution (10 μL; 0.1 mg mL^{-1} in milliQ water) was placed onto a copper grid coated with a carbon film (Electron Microscopy Sciences) and left drying on air over night at room temperature. TEM analyses are carried out in a Philips JEOL JEM-2100F microscope working at 200 kV. Statistical determination of gold dimension is performed over 200 GNPs using ImageJ program (Java-based image processing program developed at the National Institutes of Health, NIH) (Fig. 2.2). The diameter of the gold core can be correlated to the number of ligands present on the GNP (Hostetler *et al.*, 1998). The total amount of organic ligands is determined by elemental analysis. The ratio of different neoglycoconjugates on the same GNP can be deduced from ^{1}H-NMR spectrum. ^{1}H-NMR of the mixture of unreacted ligands after purification compared to ^{1}H-NMR of reactants before the reaction gave a good qualitative indication about the ratio of different neoglycoconjugates on the GNPs. In the case of *manno*-GNPs the ratio between mannose and glucose ligands in the GNPs is deduced by comparing integration of the mannoside anomeric protons signals with respect to those of glucoside, as shown in Fig. 2.3.

By combining together TEM, elemental analysis and ^{1}H-NMR data, it is possible to calculate an average molecular formula for each *manno*-GNP. As an example, we show the characterization steps of Te-10 GNP. From TEM micrographs the diameter found is 1.4 ± 0.7 nm that corresponds to a gold nanocluster of 116 gold atoms which theoretically can arrange around 53 ligands (Hostetler *et al.*, 1998). From NMR, the ligands ratio of tetrasaccharide and glucose (before and after the preparation of GNPs) is 1:9. From these data,

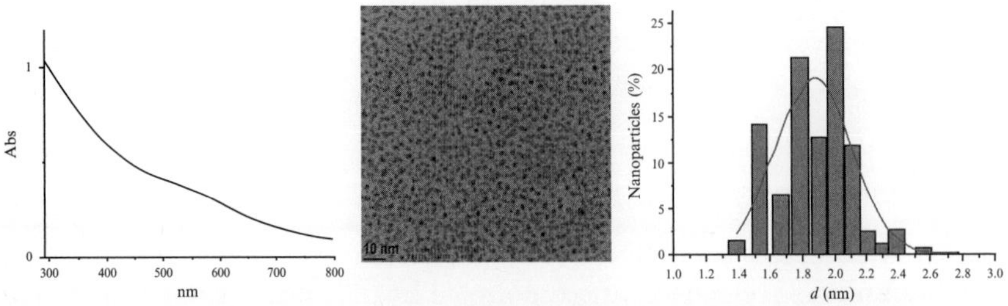

Figure 2.2 Characterization of *manno*-glyconanoparticles: UV/Vis spectrum, TEM micrograph in H_2O, and size-distribution histogram of Te-50. (For the color version of this figure, the reader is referred to the Web version of this chapter.)

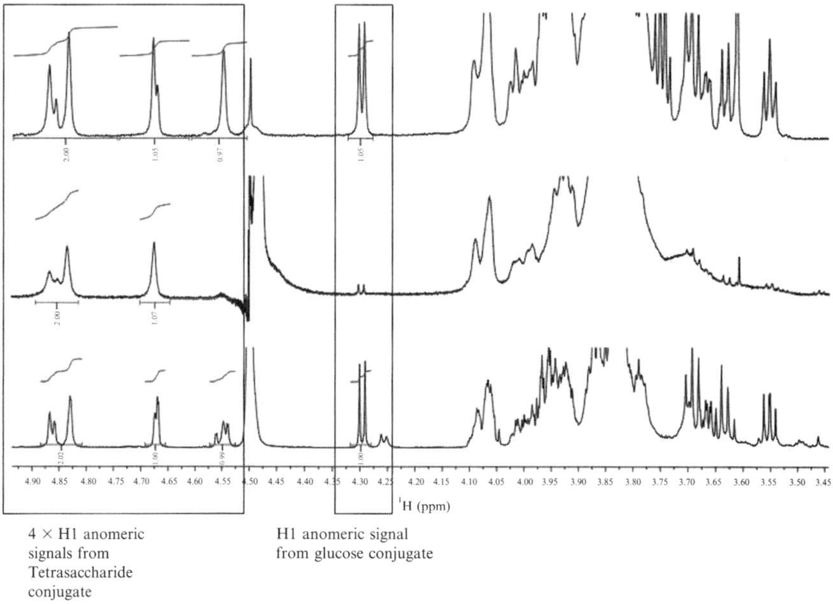

Figure 2.3 ^{1}H-NMR of neoglycoconjugates ratio before and after GNP preparation. *Top panel*: ^{1}H-NMR (500 MHz, CD_3OD) of the mixture of tetrasaccharide/glucose conjugates used to prepare GNP. Integration of selected signals show that the ratio between tetrasaccharide and glucose conjugate is about 1:1. *Middle panel*: ^{1}H-NMR (500 MHz, D_2O, water suppression) of Te-50 GNP. The selected signals show the presence of the two components in the same nanoparticle. *Bottom panel*: ^{1}H-NMR (500 MHz, CD_3OD) of the mixture used to prepare GNP after GNPs preparation. Integration of selected signals show that the ratio between tetrasaccharide and glucose conjugates is 1:1. (For the color version of this figure, the reader is referred to the Web version of this chapter.)

the calculated molecular formula was $(C_{46}H_{85}N_2O_{25}S_2)_7(C_{11}H_{21}O_6S)_{59}Au_{116}$, which correspond to an average molecular weight of 47 KDa and a calculated elemental composition of C 24.63%, H 3.90%, N 0.41%, and S 4.94%. The elemental analysis obtained for the Te-10 GNP was: C 24.40%, H 4.36%, N 0.77%, and S 4.53% that fitted well with the calculated one. From the obtained molecular formula, an average yield can be calculated. The yield of GNPs reaction calculated based on "active" gold atoms (surface gold atoms) is around 35%. The yield based on consumed neoglycoconjugate is 11%, but the unreacted glycoconjugates can be recovered.

The use of enzymes to determine the amount of specific ligands attached to the gold surface is a destructive technique that have been used for the quantification of sialyl Lewis X attached to iron oxide nanoparticles (van Kasteren *et al.*, 2009). However, it has been observed that degradation of lactose GNPs by β-galactosidase does not take place completely due probably to the ligand presentation on the gold surface (Barrientos *et al.*, 2009).

3. Evaluation of *manno*-GNPs as Inhibitors of Carbohydrate-Mediated gp120 Interactions

The potency of the multimerized oligomannoside as inhibitors of the gp120 binding to 2G12 is evaluated by biosensor and magnetic resonance techniques. Biosensors with surface plasmon resonance (SPR) detectors have become an established method of measuring molecular interactions (Rich and Miszka, 2000). Many SPR studies are carried out in order to understand the molecular basis of HIV viral life cycle, HIV drug discovery, and anti-HIV mAb characterization (Rich and Miszka, 2003). *manno*-GNPs were already evaluated by means of SPR in competition experiments as inhibitors of gp120 binding to DC-SIGN (Martinez-Ávila *et al.*, 2009a). *manno*-GNPs bearing disaccharide Manα1-2Manα inhibited gp120/DC-SIGN binding 20,000-fold more efficiently than the free oligosaccharide.

NMR has been widely used to study the interactions between biological macromolecules (usually proteins and antibodies) and their small ligands in solution. Transferred nuclear Overhauser effect enhancement experiments provide dynamic information about a macromolecule–ligand interaction (Jiménez-Barbero and Peters, 2003). An advance in this technique has been achieved by saturation transfer difference (STD) experiments (Berger and Braun, 2004; Mayer and Meyer, 1999; Meyer and Peters, 2003), where selective irradiation of proton signals of protein side chains in a spectral region free of small ligands proton signals causes a magnetization transfer to the protons of the ligands that make closer contacts with the protein. These protons can be identified because they yield more intense signals than remote ones in the final difference spectrum and quantitative information of the binding can be obtained. We found that *manno*-GNPs are synthetic mimics of the 2G12 epitope using SPR technology and STD-NMR spectroscopy. We demonstrated the ability of *manno*-GNPs to bind 2G12 and to inhibit 2G12/gp120 binding, proving that the oligomannosides maintain their functionality and enhance their ability to interact with this antibody when clustered onto gold nanoparticles (Marradi *et al.*, 2011). We describe now SPR and STD experiments which have been carried out to evaluate the direct binding of *manno*-GNPs to 2G12 and the ability to inhibit gp120 binding to 2G12.

3.1. Binding study of *manno*-GNPs to 2G12 by SPR

3.1.1. Direct binding of *manno*-GNPs to 2G12

Interaction measurements are carried out using ProteOn XPR36 biosensor with research-grade GLC sensor chips. Due to its unique technological features, ProteOn is able to processes 36 binding event simultaneously

(Bravman *et al.*, 2006). Antibody 2G12 was kindly supplied by Dr D. Katinger (Polymun Scientific, Vienna, Austria). Recombinant gp120 from HIV-1 CN54 clone (repository reference ARP683) was obtained from the Centre for AIDS Reagents, NIBSC HPA UK (by Prof. Ian Jones, Reading University, UK). ProteOn GLC sensor chips are washed with a short pulse of NaOH (50 m*M*), HCl (100 m*M*), and SDS (0.5%) before using. Channel activation is performed at 25 °C using phosphate-buffered saline (PBS)–Tween 20 as running buffer (PBS, 10 m*M* Na_3PO_4 and 150 m*M* NaCl, pH 7.4) with 0.005% of the surfactant Tween 20. Sensor chip channels are activated with 1-ethyl-3-(3-dimethylaminopropyl) carbodiimide (EDC, 16 m*M*) and *N*-hydroxysulfosuccinimide (sulfo-NHS, 4 m*M*). Thirty microliters of a 1:1 mixture of EDC/sulfo-NHS are injected (contact time: 60 s; flow rate: 30 μL min^{-1}). In channel 1, after activation, 120 μL of a 2G12 solution in acetate buffer pH 5.5 (50 μg mL^{-1}, 10 m*M*) are injected (240 s, 30 μL min^{-1}) until 2600 RU. Channel 2 is used as control and only PBS–Tween 20 buffer is injected. Both channels are then saturated with 100 μL of 1 *M* ethanolamine HCl (200 s, 30 μL min^{-1}). Binding experiments are performed at 25 °C using *tris*(hydroxymethyl)amino methane buffer (Tris-buffered saline: 10 m*M* Tris and 150 m*M* NaCl, pH 7.4) containing 0.005% Tween 20, as running buffer. This buffer is also used to prepare *manno*-GNPs solutions at different concentrations. Six dilutions for each analyte are prepared (1, 0.5, 0.25, 0.125, 0.06, and 0.03 μg mL^{-1}) and injected simultaneously. Each analyte (*manno*-GNPs and control GNPs) is injected under the same conditions (flow rate: 30 μL min^{-1}; contact time: 300 s; dissociation: 300 s). Sensorgrams are obtained by automatic subtraction of the reference channel signal from the 2G12 channel signal. After every injection, channels are regenerated with a short pulse of 3.5 *M* $MgCl_2$ (flow rate: 100 μL min^{-1}; contact time: 30 s). The affinity of gp120 to 2G12 is also evaluated to verify the binding features of the antibody after immobilization on the sensor chip. 2G12 activity slightly decreased after the regeneration conditions due to the instability of the antibody for long periods. To compare the interactions of *manno*-GNPs to 2G12, sensorgrams corresponding to the highest concentration tested (1 μg mL^{-1}) are used. Data shown in Table 2.1 indicate that at this concentration the response of most of the *manno*-GNPs is similar to the gp120 used.

When comparing at the same oligomannoside concentration (0.1 μ*M*), the highest and similar binding affinities for 2G12 correspond to Te-10 and Te-50 (Marradi *et al.*, 2011).

3.1.2. Competition experiments

gp120 is immobilized on a GLC sensor chip (around 8000 RU) using the same methodology described above for 2G12. Antibody 2G12 (final concentration, 100 n*M*) is incubated with five different concentrations of *manno*-GNPs for 10 min at 25 °C in Tris-buffered saline. 2G12/*manno*-GNP complexes and 2G12 (control) are injected simultaneously onto

Table 2.1 Direct binding of GNPs (1 μg mL^{-1}) and gp120 (1.6 μg mL^{-1}) to 2G12

manno-GNP	Binding [Ru][a]	Mannoside chains	GNP conc. [μ*M*][b]	Mannoside conc. [μ*M*][c]
D-10	<2	9	0.014	0.13
D-50	77	22	0.025	0.55
D-100	26	59	0.016	0.94
Tri-10	<5	13	0.012	0.16
Tri-50	66	62	0.010	0.62
Te-10	85	7	0.021	0.15
Te-50	113	56	0.008	0.45
P-10	<5	5	0.017	0.09
P-50	39	28	0.010	0.28
gp120	20[d]	–	0.011	–
M-100	0	40	0.024	0.96
Au-GlcC$_5$	0	0	0.027	0

[a] Taken at 150-s in the association phase of the sensorgrams.
[b] Calculated using the average molecular formulas.
[c] Calculated from the number of oligomannosides per GNP.
[d] The lower SPR response of gp120 with respect to *manno*-GNPs can be due to the gold colloid SPR-enhancing effects.

gp120 and reference channels (flow rate: 30 μL min^{-1}; contact time: 300 s; dissociation: 300 s). The sensor surface between runs is regenerated with a short pulse of 0.1 *M* HCl. In repeated experiments, gp120 shows a slight loss of activity due to HCl regeneration phases. Te-10 and Te-50 GNPs showed the best binding activity to 2G12 (Marradi *et al.*, 2011). The other *manno*-GNPs showed slight effects (D-50 and Tri-50) to moderate effects (D-100 and P-50) at the tested concentrations.

3.2. Study of the 2G12/GNP interaction by STD-NMR

The dissociation constants (K_D) of monovalent 1-aminoethyl oligomannoside/2G12 complex have been previously determined by STD-NMR (Enríquez-Navas *et al.*, 2011) following an improved protocol for single-ligand STD-NMR titrations based on the initial slopes of the build-up curves of the STD amplification factor against ligand concentration (Angulo *et al.*, 2010). The study of the direct interaction between *manno*-GNPs and 2G12 in STD experiments is hampered by the fact that both the antibody and the nanoparticles are macromolecules. We overcame this impasse by carrying out indirect competition experiments in which 2G12-oligomannoside mixtures were titrated with *manno*-GNPs. The displacement of a monovalent oligomannoside from the 2G12 binding sites resulted in a significant reduction of its STD signal. This effect has been used to obtain qualitative and quantitative information on the affinity of *manno*-GNPs for antibody 2G12, as it will be shown in detail in the following section.

3.2.1. Titration of 2G12-oligomannoside mixtures with *manno*-GNPs

For the STD-NMR titration experiments in isotropic solution, the aminoethyl oligomannosides and *manno*-GNPs are lyophilized twice against D_2O (99.9% purity, Sigma-Aldrich). Antibody 2G12 (400 μL at 11.69 mg mL^{-1}, in 2 m*M* acetic acid, 10% maltose-sterile, no preservatives) is dialyzed (10 × 10 mL in the reservoir, 30 min each) against 10 m*M* phosphate buffer (pH 6.7) using a membrane with 100,000 MWCO (Float-A-Lyzer of 5 mm diameter and 500 μL volume, Spectrum Laboratories, Inc.) following the manufacturer's instructions. Unlike SPR experiments, this step is necessary to remove the excess of maltose. To verify whether 2G12 remained inside the membrane, the dialysate is subjected to a Bradford-type protein assay (Bradford, 1976).

3.2.2. 2G12 deuteration

After 2G12 dialysis with 10 m*M* phosphate buffer (pH 6.7), three additional dialyses (3 × 10 mL, 30 min each) are performed with deuterated buffer. To deuterate the buffer 35 mL of the phosphate buffer is freeze-dried and then subjected to three solubilization–lyophilization cycles against 10 mL of D_2O (99% purity, Sigma-Aldrich) before final dissolution in 35 mL of D_2O (99% purity, Sigma-Aldrich). After dialysis with deuterated buffer, 2G12 is carefully recovered from the membrane and lyophilized twice against D_2O (99.9% purity, Sigma-Aldrich).

3.2.3. Competition experiments

For the inhibition experiments, we selected as monovalent ligands the aminoethyl tetramannoside and aminoethyl trimannoside because they have a high affinity for 2G12 ($K_D = 400$ μ*M*, Enríquez-Navas, *et al.,* 2011). STD-NMR titration experiments are recorded at 25 °C in a Bruker DRX 500-MHz spectrometer with a broadband inverse probe using 1 K scans and 32 dummies scans without suppression of the residual HDO signal. The broad signals of the antibody are eliminated by adding a $T_{1\rho}$ filter to the pulse sequence (stddiff3, Bruker). Selective irradiation of the aromatic side chains of 2G12 is achieved by using a typical train of 50 ms Gaussian pulses (Mayer and Mayer, 1999), each one with a total saturation time of 2 s. The absence of aromatic protons in the *manno*-GNPs and in the aminoethyl oligomannosides allowed the selective excitation of antibody aromatic protons without affecting the signals of the ligands.

In a typical experiment, special NMR tubes (3 mm × 100 mm, Bruker) adaptable to a Bruker *MATCH*™ holder are charged with 180 μL containing

2G12 (25 μ*M*) and aminoethyl tetramannoside (8 m*M*) in 10 m*M* deuterated potassium phosphate buffer (pH 6.7). The high ligand/antibody ratio is set up so that more than 90% of the 2G12 binding sites are occupied, fulfilling the condition of competitive inhibition titrations (Cheng and Prusoff, 1973). Increasing volumes (0.6, 2.46, 6.33, 7.85, 9.0, and 18.0 μL) of solution of Te-50 GNP (733 μ*M* in tetramannoside) in D_2O are added to the 180 μL solution of 2G12/aminoethyl tetramannoside mixture. The decrease in the STD signals intensity of the anomeric proton at the nonreducing mannose of aminoethyl tetramannoside is monitored. The STD signal intensity of this proton decreased up to 60% of its original value (in the absence of GNPs) in a dose-dependent way, as shown in Fig. 2.4. This indicates that Te-50 GNP efficiently displaces the free tetramannoside ligand from the binding site of 2G12. Similar results are found with Te-10 GNP, proving that this tetramannoside clustered onto gold nanoparticles enhances its ability to interact with 2G12.

3.2.4. Calculation of the dissociation constant

To calculate the dissociation constant K_D of the oligomannosides multimerised onto the *manno*-GNPs, the STD data obtained from the titrations experiments with Te-10 and Te-50 GNPs are mathematically fitted to Eq. (2.1) (Benie *et al.*, 2003). Equation (1) describes a competitive model and derives from the seminal Cheng–Prusoff equation (Cheng and Prusoff, 1973).

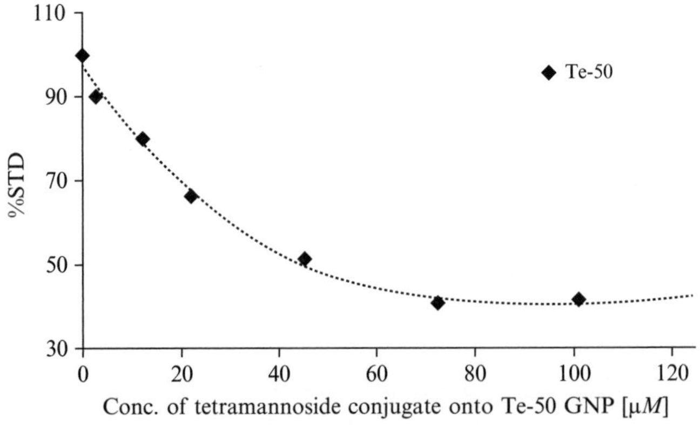

Figure 2.4 Competitive titration of 2G12/aminoethyl tetramannoside complex with Te-50 GNP. The decrease in the STD intensity of the anomeric proton H1 of the mannose ring at the nonreducing end of aminoethyl tetramannoside was monitored as a function of the GNP concentration expressed in terms of tetramannoside. The concentration of the monovalent aminoethyl tetramannoside was set to 8 m*M*. Symbols correspond to experimental data. Lines represent a three-order polynomial fitting (for visualization purposes).

$$I_{\mathrm{STD}} = 100\left(1 - \frac{[K_{\mathrm{D}}/L_0][I_0/K_{\mathrm{I}}]}{1 + [K_{\mathrm{D}}/L_0][1 + (I_0/K_{\mathrm{I}})]}\right). \quad (2.1)$$

In Eq. (2.1), I_{STD} is the monitored STD ^{1}H intensity of the monovalent ligand, K_{D} is the dissociation constant of the monovalent ligand, L_0 is the concentration of the monovalent ligand in the sample, I_0 is the concentration of the inhibitor (in these cases, the oligomannosides onto the *manno*-GNPs), and K_{I} is the dissociation constant of the oligomannoside onto *manno*-GNPs. The calculated inhibition constants of Te-10 and Te-50 GNPs are 4.2 ± 0.5 and 3.6 ± 0.6 μM (expressed in terms of tetramannoside), respectively. This indicates that the affinity of the monovalent aminoethyl tetramannoside for 2G12 ($K_{\mathrm{D}} = 400$ μM) was enhanced ~100-fold by the multimerization of tetramannoside onto gold nanoparticles (Marradi *et al.*, 2011).

4. Validation of *Manno*-GNPs as Inhibitors of HIV-1/Cell Infection

4.1. Effect of *manno*-GNPs on HIV-1 neutralization by 2G12 in cellular models

The inhibitory effect of Te-10 and Te-50 GNPs on the 2G12-mediated neutralization of a replication-competent HIV infection of TZM-bl cells was also demonstrated by competition neutralization assay under conditions that resemble the ones in which normal serum prevents infection of the target cells (Marradi *et al.*, 2011).

4.1.1. Determination of 2G12 neutralizing concentration

Before testing the *manno*-GNPs, experiments to find the conditions in which antibody 2G12 neutralizes the HIV-1 infection of TZM-bl cells is performed. TZM-bl is a luciferase reporter HeLa cell line (Derdeyn *et al.*, 2000; Platt *et al.*, 1998; Wei *et al.*, 2002) that stably expresses CD4, CCR5, and CXCR4 (viral receptor and coreceptors) and also contains separate integrated copies of the luciferase and β-galactosidase genes under the control of the HIV-1 promoter. We used a neutralization assay (see below for further details) that has been previously validated by comparison with the current standardized pseudotype assays and a good agreement is found (Fenyo *et al.*, 2009; Garcia-Perez *et al.*, 2007). Different concentrations of 2G12 are tested against an HIV-1 strain NL4-3 laboratory isolate (Fang *et al.*, 1999; Salminen *et al.*, 1995) to find the optimal conditions for inhibiting the HIV infection of TZM-bl cells (Fig. 2.5A). At 29 μg mL^{-1}, 2G12 could efficiently inhibit the infectivity of the NL4-3

A

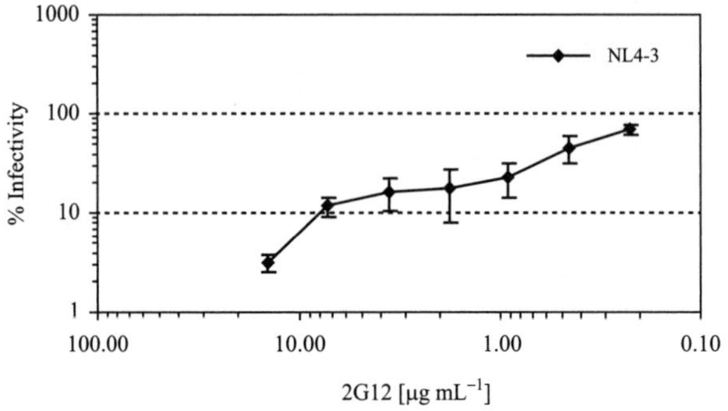

B

Envelope /subtype	Tropism	Sensitivity to 2G12 (IC_{50})	Sensitivity to 2G12 (IC_{50})
AC10 /B	R5	Resistant	Resistant
NL4-3 /B	X4	0.65 μg mL^{-1} (4.3 nM)[a]	0.7 μg mL^{-1} (5 nM)[b]

Figure 2.5 (A) Sensitivity of HIV-1 NL4-3 to 2G12-mediated neutralization using TZM-bl cells. 2G12 was used at 29.2, 14.6, 7.3, 3.65, 1.83, 0.91, 0.46, and 0.23 μg mL^{-1}. (B) IC_{50} of 2G12-mediated neutralization of HIV-1 infection of TZM-bl cells, in the absence of *manno*-GNPs ([a]Four independent measurements; [b]Published by Binley *et al.*, 2004 using a different assay, but the same viral envelopes).

HIV-1 so that this concentration is used in the inhibition assays with the *manno*-GNPs. The data reported in Fig. 2.5A are also used to calculate the concentration of antibody 2G12 which is required to reduce by 50% the HIV-1 infection of TMZ-bl cells in the absence of *manno*-GNPs (IC_{50}, Fig. 2.5B). The obtained IC_{50} (0.65 μg mL^{-1}) is in agreement with the literature (Binley *et al.*, 2004).

4.1.2. Inhibition of *manno*-GNPs of 2G12-mediated HIV-1 neutralization

Serial dilutions of GNPs Te-10 and Te-50 GNPs (the highest concentration being 125 μg mL^{-1}) are preincubated with antibody 2G12 (58 μg mL^{-1}) to a final volume of 50 μL in culture media, in triplicate, for 30 min at 37 °C. The GNP–2G12 solution is then added (1:1, by volume) to the HIV-1 virus (4 ng of p24, NL4-3 strain) and the mixture is incubated for 1 h at 37 °C. At this point, the concentration of 2G12 (29 μg mL^{-1}) is able to neutralize the virus (see Fig. 2.5). The virus–GNP–antibody mixture is added (1:1, by volume) to 10,000 TZM-bl cells (final volume 200 μL). The plate is then placed in a humidified chamber within a CO_2 incubator at 37 °C. Luciferase activity is measured from cell lysates when levels are sufficiently over background to give reliable measurements (at least 10-fold) using the Luciferase Assay System (Promega) and following the manufacturer's recommendations. Virus equivalent to 4 ng of the p24 capsid protein (quantified by an antigen-capture assay; Innogenetics, Belgium) of the NL4-3 strain of HIV-1 is chosen as the lowest level of viral input sufficient to give a clear luciferase signal within the linear range on day 3 postinfection. Neutralization activity is measured in triplicate and reported as the percentage of luciferase activity compared to the luciferase activity corresponding to the wells with virus and no antibody. The 2G12 concentration required to inhibit 50% of viral infectivity (IC_{50}) is determined for each GNP at different concentrations. Competition is observed when the addition of GNPs resulted in a decrease in the neutralizing capacity of the antibody (higher IC_{50}).

It is found that Te-10 and Te-50 GNPs caused a reproducible inhibition of 2G12 neutralization within the micromolar range in experiments using TZM-bl cells. Te-10 GNPs could efficiently inhibit the 2G12-mediated neutralization of the NL4-3 strain at 5.5 μ*M* (37 μ*M* of tetrasaccharide). At this concentration of Te-10 GNP, the amount of 2G12 required to reduce 50% of viral infectivity (IC_{50}) is three times higher (13.3 n*M*) than the one required in the absence of GNPs (4.3 n*M*; Fig. 2.5). When Te-50 GNP is used, similar results were observed (Marradi *et al.*, 2011).

4.2. Inhibition of DC-SIGN-mediated HIV-1 trans-infection of human T cells by *manno*-GNPs

DC-SIGN mediates interactions between DCs and resting T cells, by binding ICAM-3 (Geijtenbeek *et al.*, 2000). DCs in mucosal tissues capture HIV-1 through DC-SIGN–gp120 interactions. After migration to lymphoid organs, DCs promote efficient trans-infection of T cells through DC-SIGN, resulting in a vigorous viral replication.

The *manno*-GNPs are designed to target DC-SIGN receptors present on DCs by mimicking the clustered carbohydrate display of gp120. We found that oligomannoside-functionalized gold glyconanoparticles are able to inhibit the HIV-1 DC-SIGN-mediated trans-infection of T cells at nanomolar concentrations and they could be an antiadhesive barrier at an early stage of HIV-1 infection, preventing viral attachment to DC-SIGN-expressing cells (Martinez-Ávila *et al.*, 2009b).

For trans-infection experiments, Raji DC-SIGN transfected lymphoblastoid B cells are used. *manno*-GNPs are nontoxic to Raji DC-SIGN+ and to human-activated PBMCs at concentrations of 100 mg mL^{-1}, as determined by the CellTiter cell viability assay. The activities of *manno*-GNPs against R5 or X4 HIV-1 are evaluated through an original DC-SIGN transfer assay in which inhibition of HIV-1 infection by GNPs is assessed by use of recombinant viruses carrying the Renilla reporter genes in their genomes (Garcia-Perez *et al.*, 2007). Inhibition of viral replication is proportional to Renilla-luciferase activity in cell lysates.

4.2.1. Cell culture and preparation of PBMCs from blood

Raji cells were kindly provided by Dr. Alfredo Toraño (Instituto de Salud Carlos III, Madrid, Spain), and Raji DC-SIGN+ cells are kindly provided by Dr. Fernando Arenzana-Seisdedos (Institut Pasteur, Paris, France). Both cell lines are cultured in RPMI 1640 medium containing fetal bovine serum (10% v/v), l-glutamine (2 m*M*), penicillin (50 $IUmL^{-1}$), and streptomycin (50 mg mL^{-1}; all from Whittaker M.A. Bio-Products, Walkerville, MD, USA). The 293T cells (used for the production of recombinant viruses) are cultured in Dulbecco's modified Eagle's medium (DMEM) containing fetal bovine serum (10% v/v), l-glutamine (2 m*M*), penicillin (50 $IUmL^{-1}$), and streptomycin (50 mg mL^{-1}). The 293T cells are cultured at 37 °C in a humidified atmosphere with CO_2 (5%) and split twice a week. PBMCs are obtained from buffy coats from healthy donors. PBMCs are harvested from buffy coats by centrifugation over Lymphoprep (Sigma-Aldrich) gradient by standard procedures, stimulated with interleukin-2 (300 $IUmL^{-1}$; Chiron) and phytohaemagglutinin (5 mg mL^{-1}), and incubated at 37 °C under humidified CO_2 (5%) for 48 h.

4.2.2. Trans-infection assay

Raji or Raji DC-SIGN+ cells (10^5 cells per well) are incubated with GNPs for 1 h prior to addition of either R5 or X4 tropic recombinant viruses (JR-Renilla or NL4.3-Renilla, respectively; both 200 ng p24 per well) and left for 2 h at 37 °C for efficient adsorption. Cells are washed extensively with PBS and preactivated PBMCs (10^5 per well) are added. Viral replication is followed by measurement of Relative luminescence unit (RLU) activity in cell lysates. Briefly, cells are harvested and lysed after 48 h and sample activity is measured with the Renilla-luciferase assay system (Promega) according to the manufacturer's protocol. RLUs are obtained by using a luminometer (Berthold Detection Systems, Pforzheim, Germany) after the addition of substrate to cells extracts. All the experiments are performed in parallel with Raji cells as control. IC_{50} values are calculated with GraphPad Prism Software (Sigmoidal dose-response analysis). The results are representative of at least three independent experiments and are shown in Fig. 2.6.

A

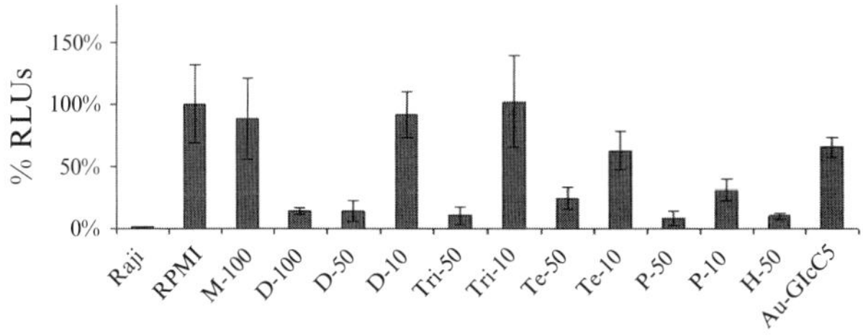

B

GNP	IC_{50} (ng/mL)
D-50	80.5 (~2.0 n*M*)
Tri-50	166 (~1.6 n*M*)
Te-50	41.6 (~0.3 n*M*)
P-50	58.5 (~0.6 n*M*)
H-50	9.2 (~0.5 n*M*)

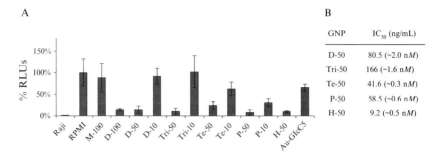

Figure 2.6 (A) Anti-HIV evaluation of *manno*-GNPs at 1 mg mL^{-1} in DC-SIGN-mediated trans-infection of human T cells. HIV-1 recombinant viruses JR-Renilla R5 was used. Raji cells not expressing DC-SIGN (Raji $DC\text{-}SIGN^{-}$) were used as control. Results are expressed as percentages of infection related to untreated control. (B) IC_{50} of selected *manno*-GNPs in $ngmL^{-1}$ and n*M* concentration of oligommannosides. (For color version of this figure, the reader is referred to the Web version of this chapter.)

ACKNOWLEDGMENTS

We thank the Spanish Ministry of Science and Innovation MICINN (grant CTQ2008-04638/BQU), the European Union (CHAARM grant Health-F3-2009-242135, EMPRO LSHP-CT2003-503558, Glycogold MRTN-CT2004-005645), and the Department of Industry of the Basque Country (grant ETORTEK2009) for financial support. Antibody 2G12 was kindly supplied by Dr D. Katinger (Polymun Scientific, Vienna, Austria). Recombinant gp120 from HIV-1 CN54 clone (repository reference ARP683) was obtained from the Program EVA Centre for AIDS Reagents, NIBSC HPA UK, supported by the EC FP6/7 Europrise Network of Excellence, and NGIN consortia, and the Bill and Melinda Gates GHRC-CAVD Project and was donated by Prof. I. Jones (Reading University, UK). TZM-bl was obtained from Dr. J. C. Kappes, Dr. X. Wu, and Tranzyme, Inc., through the National Institutes of Health AIDS Research and Reference Reagent Program, Division of AIDS, National Institute of Allergy and Infectious Diseases. We thank all the members of the group and collaborators who have contributed to the work described here. Dr. P. M. Enríquez-Navas and Dr. E. Yuste are acknowledged for revising Sections 3.2 and 4.1, respectively.

REFERENCES

Allan, J. S., Coligan, J. E., Barin, F., McLane, M. F., Sodroski, J. G., Rosen, C. A., Haseltine, W. A., Lee, T. H., and Essex, M. (1985). Major glycoprotein antigens that induce antibodies in AIDS patients are encoded by HTLV-III. *Science* **228,** 1091–1094.

Angulo, J., Enríquez-Navas, P. M., and Nieto, P. M. (2010). Ligand–receptor binding affinities from saturation transfer difference (STD) NMR spectroscopy: The binding isotherm of STD initial growth rates. *Chem. Eur. J.* **16,** 7803–7812.

Astronomo, R. D., Kaltgrad, E., Udit, A. K., Wang, S. K., Doores, K. J., Huang, C. Y., Pantophlet, R., Paulson, J. C., Wong, C. H., Finn, M. G., and Burton, D. R. (2010). Defining criteria for oligomannose immunogens for HIV using icosahedral virus capsid scaffolds. *Chem. Biol.* **17,** 357–370.

Barin, F., McLane, M. F., Allan, J. S., Lee, T. H., Groopman, J. E., and Essex, M. (1985). Virus envelope protein of HTLV-III represents major target antigen for antibodies in AIDS patients. *Science* **228,** 1094–1096.

Barrientos, A. G., de la Fuente, J. M., Rojas, T. C., Fernandez, A., and Penadés, S. (2003). Gold glyconanoparticles: Synthetic polyvalent ligands mimicking glycocalyx-like surfaces as tools for glycobiological studies. *Chem. Eur. J.* **9,** 1909–1921.

Barrientos, A. G., de la Fuente, J. M., Jiménez, M., Solís, D., Cañada, F. J., Martín-Lomas, M., and Penadés, S. (2009). Modulating glycosidase degradation and lectin recognition of gold glyconanoparticles. *Carbohydr. Res.* **344,** 1474–1478.

Benie, A. J., Moser, R., Bäuml, E., Blaas, D., and Peters, T. (2003). Virus–ligand interactions: Identification and characterization of ligand binding by NMR spectroscopy. *J. Am. Chem. Soc.* **125,** 14–15.

Berger, S., and Braun, S. (2004). 200 and More NMR Experiments. Wiley-VCH, Verlag GmbH & Co. KGaA, Weinheim Ch. 8, Exp. 8.13, pp. 298–301.

Binley, J. M., Wrin, T., Korber, B., Zwick, M. B., Wang, M., Chappey, C., Stiegler, G., Kunert, R., Zolla-Pazner, S., Katinger, H., Petropoulos, C. J., and Burton, D. R. (2004). Comprehensive cross-clade neutralization analysis of a panel of anti-human immunodeficiency virus type 1 monoclonal antibodies. *J. Virol.* **78,** 13232–13252.

Bradford, M. M. (1976). Rapid and sensitive method for the quantitation of microgram quantities of protein utilizing the principle of protein-dye binding. *Anal. Biochem.* **72,** 248–254.

Bravman, T., Bronner, V., Lavie, K., Notcovich, A., Papalia, G. A., and Miszka, D. G. (2006). Exploring "one-shot" kinetics and small molecule analysis using the ProteOn XPR36 array biosensor. *Anal. Biochem.* **358,** 281–288.

Brust, M., Walker, M., Bethell, D., Schiffrin, D. J., and Whyman, R. (1994). Synthesis of thiol derivatised gold nanoparticles in a two-phase liquid/liquid system. *J. Chem. Soc. Chem. Commun.* **7,** 801–802.

Calarese, D. A., Scanlan, C. N., Zwick, M. B., Deechongkit, S., Mimura, Y., Kunert, R., Zhu, P., Wormald, M. R., Stanfield, R. L., Roux, K. H., Kelly, J. W., Rudd, P. M., *et al.* (2003). Antibody domain exchange is an immunological solution to carbohydrate cluster recognition. *Science* **300,** 2065–2071.

Carvalho de Souza, A., and Kamerling, J. P. (2006). Analysis of carbohydrate-carbohydrate interactions using gold glyconanoparticles and oligosaccharide self-assembling monolayers. *Methods Enzymol.* **417,** 221–243.

Cheng, Y. C., and Prusoff, W. H. (1973). Relationship between the inhibition constant (Ki) and the concentration of inhibitor which causes 50 per cent inhibition (I50) of an enzymatic reaction. *Biochem. Pharmacol.* **22,** 3099–3108.

de la Fuente, J. M., and Penadés, S. (2006). Glyconanoparticles: Types, synthesis and applications in glycoscience, biomedicine and material science. *BBA* **1760,** 636–651.

de la Fuente, J. M., Barrientos, A. G., Rojas, T. C., Rojo, J., Canada, J., Fernandez, A., and Penadés, S. (2001). Gold glyconanoparticles as water-soluble polyvalent models to study carbohydrate interactions. *Angew. Chem. Int. Ed.* **40,** 2258–2261.

Derdeyn, C. A., Decker, J. M., Sfakianos, J. N., Wu, X., O'Brien, W. A., Ratner, L., Kappes, J. C., Shaw, G. M., and Hunter, E. (2000). Sensitivity of human immunodeficiency virus type 1 to the fusion inhibitor T-20 is modulated by coreceptor specificity defined by the V3 loop of gp120. *J. Virol.* **74,** 8358–8367.

Dudkin, V. Y., Orlova, M., Geng, X., Mandal, M., Olson, W. C., and Danishefsky, S. J. (2004). Toward fully synthetic carbohydrate-based HIV antigen design: On the critical role of bivalency. *J. Am. Chem. Soc.* **126,** 9560–9562.

Engering, A., van Vliet, S., Geijtenbeek, T. B., and van Kooyk, Y. (2002). Subset of DC-SIGN dendritic cells in human blood transmits HIV-1 to T lymphocytes. *Immunobiology* **100,** 1780–1786.

Enríquez-Navas, P. M., Marradi, M., Padro, D., Angulo, J., and Penadés, S. (2011). A solution NMR study of the interactions of oligomannosides and the anti-HIV-1 2G12 antibody reveals distinct binding modes for branched ligands. *Chem. Eur. J.* **17,** 1369–1707.

Fang, G., Weiser, B., Visosky, A., Moran, T., and Burger, H. (1999). PCR-mediated recombination: A general method applied to construct chimeric infectious molecular clones of plasma- derived HIV-1 RNA. *Nat. Med.* **5,** 239–242.

Feizi, T., and Larkin, M. (1990). AIDS and glycosylation. *Glycobiology* **1,** 17–23.

Fenouillet, E., Gluckman, J. C., and Jones, I. M. (1994). Functions of HIV envelope glycans. *Trends Biochem. Sci.* **19,** 65–70.

Fenyo, E. M., Heath, A., Dispinseri, S., Holmes, H., Lusso, P., Zolla-Pazner, S., Donners, H., Heyndrickx, L., Alcami, J., Bongertz, V., Jassoy, C., Malnati, M., *et al.* (2009). International network for comparison of HIV neutralization assays: The NeutNet report. *PLoS One* **4,** e4505.

García, I., Marradi, M., and Penadés, S. (2010). Glyconanoparticles: Multifunctional nanomaterials for biomedical applications. *Nanomedicine* **5,** 777–792.

Garcia-Perez, J., Sanchez-Palomino, S., Perez-Olmeda, M., Fernandez, B., and Alcami, J. (2007). A new strategy based on recombinant viruses as a tool for assessing drug susceptibility of human immunodeficiency virus type 1. *J. Med. Virol.* **79,** 127–137.

Geijtenbeek, T. B., Kwon, D. S., Torensma, R., van Vliet, S. G., van Duijnhoven, G. C., Middel, J., Cornelissen, I. L., Nottet, H. S., KewalRamani, V. N., Littman, D. R., Figdor, C. G., and van Kooyk, Y. (2000). DC-SIGN, a dendritic cell-specific HIV-1-binding protein that enhances trans-infection of T cells. *Cell* **100,** 587–597.

Geyer, H., Holschbach, C., Hunsmann, G., and Schneider, J. (1988). Carbohydrates of human immunodeficiency virus-structure of oligosaccharides linked to the envelope glycoprotein- 120. *J. Biol. Chem.* **263,** 11760–11767.

Hostetler, M. J., Wingate, J. E., Zhong, C.-J., Harris, J. E., Vachet, R. W., Clark, M. R., Londono, J. D., Green, S. J., Stokes, J. J., Wignall, G. D., Glish, G. L., Porter, M. D., *et al.* (1998). Alkanethiolate gold cluster molecules with core diameters from 1.5 to 5.2 nm: Core and monolayer properties as a function of core size. *Langmuir* **14,** 17–30.

Jiménez-Barbero, J., and Peters, T. (2003). TR-NOE experiments to study carbohydrate-protein interactions. *In* "NMR Spectroscopy of Glycoconjugates," (J. Jiménez-Barbero and T. Peters, eds.), .Wiley-VCH, Verlag GmbH & Co. KGaA, Weinheim.

Kabanova, A., Adamo, R., Proietti, D., Berti, F., Tontini, M., Rappuoli, R., and Costantino, P. (2010). Preparation, characterization and immunogenicity of HIV-1 related high-mannose oligosaccharides–CRM197 glycoconjugates. *Glycoconj. J.* **27,** 501–513.

Krauss, I. J., Joyce, J. G., Finnefrock, A. C., Song, H. C., Dudkin, V. Y., Geng, X., Warren, J. D., Chastain, M., Shiver, J. W., and Danishefsky, S. J. (2007). Fully synthetic carbohydrate HIV antigens designed on the logic of the 2G12 antibody. *J. Am. Chem. Soc.* **129,** 11042–11044.

Li, H., and Wang, L. X. (2004). Design and synthesis of a template-assembled oligomannose cluster as an epitope mimic for human HIV-neutralizing antibody 2G12. *Org. Biomol. Chem.* **2,** 483–488.

Marradi, M., Martin-Lomas, M., and Penadés, S. (2010). Glyconanoparticles: Polyvalent tools to study carbohydrate-based interactions. *Adv. Carbohydr. Chem. Biochem.* **64,** 212–270.

Marradi, M., Di Gianvincenzo, P., Enriquez-Navas, P. M., Martinez-Ávila, O. M., Chiodo, F., Yuste, E., Angulo, J., and Penadés, S. (2011). Gold nanoparticles coated with oligomannosides of HIV-1 glycoprotein gp120 mimic the carbohydrate epitope of antibody 2G12. *J. Mol. Biol.* **410,** 798–810.

Martinez-Ávila, O., Hijazi, K., Marradi, M., Clavel, C., Champion, C., Kelly, C., and Penadés, S. (2009a). Gold *manno*-glyconanoparticles: Multivalent system to block HIV-1 gp120 binding to the lectin DC-SIGN. *Chem. Eur. J.* **15,** 8974–9888.

Martinez-Ávila, O., Bedoya, Luis M., Marradi, M., Clavel, C., Alcamí, J., and Penadés, S. (2009b). Multivalent manno-glyconanoparticles inhibit DC-SIGN mediated HIV-1 trans-infection of human T cells. *Chembiochem* **10,** 1806–1809.

Mayer, M., and Meyer, B. (1999). Characterization of ligand binding by saturation transfer difference NMR spectra. *Angew. Chem. Int. Ed.* **38,** 1784–1788.

Meyer, B., and Peters, T. (2003). NMR spectroscopy techniques for screening and identifying ligand binding to protein receptors. *Angew. Chem. Int. Ed.* **42,** 864–890.

Mizuochi, T., Spellman, M. W., Larkin, M., Solomon, J., Basa, L. J., and Feizi, T. (1988). Carbohydrate structures of the human-immunodeficiency-virus (HIV) recombinant envelope glycoprotein gp120 produced in chinese ovary cells. *Biochem. J.* **254,** 599–603.

Platt, E. J., Wehrly, K., Kuhmann, S. E., Chesebro, B., and Kabat, D. (1998). Effects of CCR5 and CD4 cell surface concentrations on infections by macrophagetropic isolates of human immunodeficiency virus type 1. *J. Virol.* **72,** 2855–2864.

Rich, R. L., and Miszka, D. G. (2000). Advances in surface plasmon resonance biosensor analysis. *Curr. Opin. Biotechnol.* **11,** 54–61.

Rich, R. L., and Miszka, D. G. (2003). Spying on HIV with SPR. *Trends Microbiol.* **11,** 124–133.

Salminen, M. O., Koch, C., Sanders-Buell, E., Ehrenberg, P. K., Michael, N. L., Carr, J. K., Burke, D. S., and McCutchan, F. E. (1995). Recovery of virtually full-length HIV-1 provirus of diverse subtypes from primary virus cultures using the polymerase chain reaction. *Virology* **213,** 80–86.

Sanders, R. W., Venturi, M., Schiffner, L., Kalyanaraman, R., Katinger, H., Lloyd, K. O., Kwong, P. D., and Moore, J. P. (2002). The mannose-dependent epitope for neutralizing antibody 2G12 on human immunodeficiency virus type 1 glycoprotein gp120. *J. Virol.* **76,** 7293–7305.

Scanlan, C. N., Pantophlet, R., Wormald, M. R., Saphire, E. O., Stanfield, R., Wilson, I. A., Katinger, H., Dwek, R. A., Rudd, P. M., and Burton, D. R. (2002). The broadly neutralizing anti-HIV-1 antibody 2G12 recognizes a cluster of 132 mannose residues on the outer face of gp120. *J. Virol.* **76,** 7306–7321.

Trkola, A., Purtscher, M., Muster, T., Ballaun, C., Buchacher, A., Sullivan, N., Srinivasan, K., Sodroski, J., Moore, J. P., and Katinger, H. (1996). Human monoclonal antibody 2G12 defines a distinctive neutralization epitope on the gp120 glycoprotein of human immunodeficiency virus type 1. *J. Virol.* **70,** 1100–1108.

van Kasteren, S. I., Campbell, S. J., Serres, S., Anthony, D. C., Sibson, N. R., and Davis, B. G. (2009). Glyconanoparticles allow pre-symptomatic in vivo imaging of brain disease. *Proc. Natl. Acad. Sci. USA* **106,** 18–23.

Wang, L. X., Ni, J., Singh, S., and Li, H. (2004). Binding of high-mannose-type oligosaccharides and synthetic oligomannose clusters to human antibody 2G12: Implications for HIV-1 vaccine design. *Chem. Biol.* **11,** 127–134.

Wang, J., Li, H., Zou, G., and Wang, L. X. (2007). Novel template-assembled oligosaccharide clusters as epitope mimics for HIV-neutralizing antibody 2G12. Design, synthesis, and antibody binding study. *Org. Biomol. Chem.* **5,** 1529–1540.

Wang, S. K., Liang, P. H., Astronomo, R. D., Hsu, T. L., Hsieh, S. L., Burton, D. R., and Wong, C. H. (2008). Targeting the carbohydrates on HIV-1: Interaction of oligomannose dendrons with human monoclonal antibody 2G12 and DC-SIGN. *Proc. Natl. Acad. Sci. USA* **105,** 3690–3695.

Wei, X., Decker, J. M., Liu, H., Zhang, Z., Arani, R. B., Kilby, J. M., Saag, M. S., Wu, X., Shaw, G. M., and Kappes, J. C. (2002). Emergence of resistant human immunodeficiency virus type 1 in patients receiving fusion inhibitor (T-20) monotherapy. *Antimicrob. Agents Chemother.* **46,** 1896–1905.

CHAPTER THREE

Nanoparticle-Mediated Targeted Delivery of Antiretrovirals to the Brain

Supriya D. Mahajan,* Wing-Cheung Law,† Ravikumar Aalinkeel,* Jessica Reynolds,* Bindukumar B. Nair,* Ken-Tye Yong,‡ Indrajit Roy,§ Paras N. Prasad,† *and* Stanley A. Schwartz*

Contents

1. Introduction 42
2. Preparation and Characterization of QD 45
   2.1. Quantum dots 45
   2.2. Synthesis of double-shelled CdSe/CdS/ZnS QDs 46
   2.3. Aqueous dispersion of the QDs terminated with carboxyl groups 46
3. Physical Characterization of QDs 47
   3.1. High-resolution transmission electron microscopy 47
   3.2. Dynamic light scattering 47
   3.3. Spectrophotometry and spectrofluorometry 47
4. Preparation of QD-Amprenavir-Tf Nanoplex 48
5. Assessment of Amprenavir Levels Using HPLC 49
6. Evaluating the Efficacy of Antiretroviral Containing Nanoplex Using an *In Vitro* BBB Model 49
   6.1. *In vitro* model of the human BBB 49
   6.2. Measurement of TEER across a membrane 50
   6.3. Cell viability measurement using an MTT (3-(4,5-dimethylthiazol-2-yl)-2,5-diphenyltetrazolium bromide, a tetrazole) assay 51
   6.4. Transfer of the Tf-QD-Amprenavir nanoparticles across the BBB and uptake of the nanoplex by HIV-1-infected monocytes 51

* Department of Medicine, Division of Allergy, Immunology, and Rheumatology, State University of New York at Buffalo, Innovation Center, Buffalo, New York, USA
† Institute for Lasers Photonics and Biophotonics, Natural Science Complex, State University of New York at Buffalo, Buffalo, New York, USA
‡ School of Electrical and Electronic Engineering, Nanyang Technological University, Singapore
§ Department of Chemistry, University of Delhi, Delhi, India

Methods in Enzymology, Volume 509
ISSN 0076-6879, DOI: 10.1016/B978-0-12-391858-1.00003-4

6.5. Monocyte isolation from PBMCs) 52
6.6. HIV-1 infection of monocytes 54
7. Evaluation of the Antiviral Efficacy of the QD-Tf-Amprenavir Nanobioconjugate 54
7.1. HIV-1 p24 quantitation using a commercially available p24 ELISA 54
7.2. Quantification of HIV-1 LTR-R/U5 gene expression using real-time quantitative PCR 55
8. Concluding Remarks 57
Acknowledgments 58
References 58

Abstract

Nanotechnology offers a new platform for therapeutic delivery of antiretrovirals to the central nervous system (CNS) where human immunodeficiency virus (HIV-1) is sequestered in patients with HIV-1-associated neurological disorders (HAND). HAND is a spectrum of neurocognitive disorders that continue to persist in HIV-1-infected patients in spite of successful highly active antiretroviral therapy (HAART). Nanoformulated antiretroviral drugs offer multifunctionality, that is, the ability to package multiple diagnostic and therapeutic agents within the same nanocomposite, along with the added provisions of site-directed delivery, delivery across the blood–brain barrier (BBB), and controlled release of therapeutics. We have stably incorporated the antiretroviral drug, Amprenavir, within a transferrin (Tf)-conjugated quantum dot (QD), and evaluated the transversing ability of this Tf-QD-Amprenavir nanoplex across an *in vitro* BBB model and analyzed its antiviral efficacy in HIV-1-infected monocytes. We describe methods for synthesis of the Tf-QD-Amprenavir nanoplex and approaches to evaluate both its BBB transversing capability and antiviral efficacy.

1. Introduction

HIV-1 can enter the central nervous system (CNS) at an early stage of infection and can cause mild, moderate, or severe neurological disorders, classified as HIV-1-associated neurocognitive disorders (HAND; Letendre *et al.*, 2009; Shapshak *et al.*, 2011). HIV-1-infected monocytes can readily cross the blood–brain barrier (BBB) and accumulate in perivascular macrophages and microglia, leading to HAND (Singer *et al.*, 2010). The use of antiretroviral therapies, including HIV-1 protease inhibitors, and nucleoside, nucleotide and nonnucleoside reverse transcriptase inhibitors, has reduced the morbidity and mortality associated with HIV-1 infection, but has failed to eliminate HAND because of the incomplete transport of antiretroviral drugs across the BBB. HAND continue to develop in individuals despite treatment with the new highly active antiretroviral treatments

(HAART). The CNS is a site where the virus can be sequestered for a prolonged period of time, and is not eradicated by HAART (DeLuca *et al.*, 2002; Enting *et al.*, 1998; Letendre *et al.*, 2009).

Systemic delivery of antiretroviral drugs in the brain is hampered severely by the presence of the BBB. The BBB is a complex physiological checkpoint that inhibits the free diffusion of circulating molecules from the blood into the brain. Therefore, fabrication of novel macromolecular carriers that would enhance significantly the delivery of drugs across the BBB holds the key for the treatment of neuro-AIDS and other neurological diseases. HAART commonly requires complex dosing schedules and leads to the emergence of viral resistance and treatment failures. Resistance to particular combinations of drugs may develop even with good drug adherence due to the high genetic diversity of HIV-1 and continuous viral mutations. Development of nanoparticle-based ART regimens could preclude such limitations and result in improved clinical outcomes. Nanoparticle-based formulations have several advantages that can enable potent drug delivery across the BBB, while avoiding any damage to the BBB. These include biocompatibility, nonantigenicity, capability for targeted and controlled drug delivery, multimodality, and the ability to monitor BBB permeation in real time (Kreuter *et al.*, 1995; Pathak *et al.*, 2006; Schroeder *et al.*, 1998; Silva, 2007)

Transport of nanoformulated ARV drugs across the BBB allows for targeting the virus in sequestered sites such as the CNS, and the eradication of the virus from such reservoirs is critical to the effective long-term treatment of HIV/AIDS patients. Therefore, developing nontoxic treatment modalities that provide more sustained dosing coverage may be able to effectively eradicate the virus from the CNS.

Nanotechnology is still in an early stage of innovation; however, it could potentially transform personalized medicine. The translational impact of nanoformulated ARV drugs is not just the delivery of the ARV across the BBB but also the modification of the surface chemistry of nanoparticles to carry high concentrations of therapeutic drugs and/or molecules for tissue-specific recognition and sustained release of the therapeutic drug from the nanoparticles once it reaches its target tissue. Nanoparticles can cross the BBB by passive diffusion and receptor-mediated endocytosis. Site-directed brain delivery of nanoparticles may be possible by use of high-affinity nanoparticle surface ligands to native BBB transporters. Once the nanoparticles targeted to the BBB transporter are in the brain, the encapsulated drug can be released slowly in the CNS tissues, avoiding other organs, and thus reducing peripheral or systemic toxicity. This is the major therapeutic advantage and therefore has tremendous translational potential in the pharmacology industry.

The current chapter is focused on semiconductor nanoparticles called quantum dots (QDs) that have unique optical properties, such as high photostability and emission tunability spanning the visible and near-infrared (NIR) range (Yong *et al.*, 2009a,b). Additionally, owing to their broad surface area and rich surface chemistry, they are ideally suited to be developed as a multimodal nanoplatform upon which other therapeutic and/or biorecognition agents can be attached. Therefore, QDs are expected to play a significant role in the delivery of neurotherapeutic payloads across the BBB via their interaction with specific endogenous receptors that are present on the capillary epithelia comprising the BBB (e.g., transferrin receptors, TfRs) (Bhaskar *et al.*, 2010; Pathak *et al.*, 2006). Transferrin (Tf) binds Fe with high affinity, and it is the primary carrier of Fe across the BBB via the interaction between Tf and TfRs. Tf is present in blood plasma and brain extracellular fluids, and the TfRs are present on brain capillary endothelial cells, neurons, and glial cells (Gaillard *et al.*, 2005; Hallmann *et al.*, 2000; Patel *et al.*, 2009; Wang *et al.*, 2009). Transport across the BBB may involve passive diffusion, transcytosis, endocytosis, or a combination of all three mechanisms. Tf transport from the blood to the brain may be via a receptor-mediated process, as well as by other nonselective mechanisms. Tf is present in the brain interstitial fluid, and it is generally assumed that Fe that transverses the BBB is rapidly bound by brain Tf and can then be taken up by receptor-mediated endocytosis in brain cells (Patel *et al.*, 2009; Wang *et al.*, 2009).

Our current studies suggest that the Tf-QD-Amprenavir nanoplexes cross the BBB via receptor-mediated endocytosis and are degraded to release the Amprenavir. The activity of Amprenavir remains unaltered when conjugated to the QD or the transporter molecule, Tf. The high photostability of QDs emitting in the NIR spectral region facilitates the monitoring of this transport process in "real-time" *in vitro* and *in vivo* via high-resolution optical imaging (Michalet *et al.*, 2005; Prasad, 2003, 2004; Qian *et al.*, 2007; Yong *et al.*, 2009a,b). A QD-based nanoplatform thus facilitates not only the delivery of an antiretroviral drug such as Amprenavir across the BBB via interaction with endogenous TfRs on the BBB but also allows the real-time monitoring of this transendothelial migration process via optical imaging.

Amprenavir is a protease inhibitor that inhibits the production of mature HIV-1 virions. Proteases are enzymes that cleave protein molecules into smaller fragments. HIV-1 protease is vital for both viral replication within the cell and the release of mature viral particles from an infected cell. Amprenavir acts by inhibiting the HIV-1 protease, resulting in the production of noninfectious virions. Amprenavir is used in combination with other antiretroviral drugs, such as zidovudine and lamivudine. Studies show that the 50% inhibitory concentration (IC_{50}) of Amprenavir ranged from 0.012 to 0.08 μM in acutely infected cells and was 0.41 μM in chronically infected cells (Aquaro *et al.*, 2004). Amprenavir exhibits synergistic anti-HIV-1 activity in combination with abacavir, zidovudine, didanosine, or saquinavir and additive anti-HIV-1 activity in combination with indinavir, nelfinavir, and ritonavir *in vitro* (Conway and Shafran, 2000). Although the volume of distribution or bioavailability of Amprenavir is greater than that of saquinavir, it is not significantly better than saquinavir in crossing the BBB. Concentrations of Amprenavir in cerebrospinal fluid are less than 1% of the plasma concentration (Sparidans *et al.*, 2000). Therefore, increasing its bioavailability via nanoformulations has significant benefits as it maintains sufficient Amprenavir blood plasma levels to efficiently suppress the replication of HIV-1 over an extended period of time. A nanoformulation of an ARV drug such as Amprenavir will improve the bioavailability, pharmacology, cytotoxicity, and interval dosing of this antiretroviral drug.

In this chapter, we describe methodologies for the synthesis and characterization of QDs, synthesis of the Tf-QD-Amprenavir bioconjugate, and evaluation of the efficacy of this nanobioconjugate using a well-validated *in vitro* BBB model in HIV-1 infected monocytes by monitoring HIV-1 p24 antigen levels and HIV-1-LTR/RU5 gene expression levels.

2. Preparation and Characterization of QD

2.1. Quantum dots

QDs are semiconductor nanocrystals (in the size range of 1–10 nm) with size-tunable optical and electrical properties. QDs have generated tremendous interest in the scientific community because of their unique optical properties, including broad excitation spectra, narrow, tunable, and symmetric emission profile, and excellent photostability. The use of semiconductor nanocrystals as luminescence probes for numerous biological and biomedical applications has become an area of intense research focus over the past few years. QDs offer several advantages over organic dyes, including increased brightness, stability against photobleaching, broad emission range using a single excitation source, and a large surface to volume ratio. The luminescent nanoparticles whose surfaces have been functionalized with biomolecules have the potential to dramatically outperform conventional organic dyes in imaging of cellular and subcellular structures and in a variety of bioassays. For instance, bioconjugated QDs have been used in cell labeling, tissue imaging, photosensitization for photodynamic therapy, *in vivo* tumor targeting, and drug and gene delivery (Bonoiu *et al.*, 2009; Ding *et al.*, 2011; Samia *et al.*, 2003; Yong *et al.*, 2010). With proper surface coatings, such as silica, polymer, phospholipid, and short thiol ligand, QDs can be well dispersed in aqueous media (e.g., biological buffer) with minimal toxicity (Erogbogbo *et al.*, 2011; Hu *et al.*, 2010; Law *et al.*, 2009; Yong *et al.*, 2009a,b). The new generations of QD have far reaching potential for the study of intracellular processes at the single-molecule level, high-resolution cellular imaging, long-term *in vivo* observation of cell trafficking, tumor targeting, and diagnostics (Michalet *et al.*, 2006). In addition to target specificity, they can also be tailored to include additional functionalities within them for other imaging modalities such as magnetic resonance imaging and positron emission tomography (PET), which will make possible multimodal imaging using a single nanocomposite (Erogbogbo *et al.*, 2010; Law *et al.*, 2009; Liu *et al.*, 2011; Tu *et al.*, 2011).

2.2. Synthesis of double-shelled CdSe/CdS/ZnS QDs

The double-shelled CdSe/CdS/ZnS QDs are synthesized in organic media as reported in the literature (Li *et al.*, 2003; Manna *et al.*, 2002; Yong *et al.*, 2007). The CdSe core is prepared by dissolving 1.6 mmol of cadmium oxide, 6.4 mmol of stearic acid, 12 mL of octadecene, 10 g of octyldecylamine, and 4 g of TOPO into a 100-mL three-necked flask. The reaction mixture is heated slowly under an argon atmosphere to 150–160 °C for 2 h. The temperature is then set to 300 °C for 5 min, producing a colorless and homogeneous mixture. At this temperature, a selenium solution prepared by dissolving 3.2 mmol of Se powder in 6.4 mL of trioctylphosphine (TOP) is rapidly injected. The nucleation of QDs starts immediately and the color of solution turns from colorless to red. The size (emission wavelength) of QDs can be controlled by the duration of aging. At desire wavelength, the reaction is stopped by removing the heating mantle and addition of toluene. The QDs are purified from the surfactants solution by the addition of ethanol and centrifugation.

The synthesis of CdS/ZnS-graded shell on CdSe QD is done by dissolving 2 mmol of cadmium oxide, 4 mmol of zinc acetate, 5.5 g of TOPO, 2 g of stearic acid, and 10 mL of oleic acid into a 100-mL three-necked flask. The reaction mixture is heated to 120 °C for 30 min under an argon flow, and then the QD solution is injected slowly under stirring into the hot reaction mixture. The reaction mixture is held at 120 °C, with a needle outlet that allows the solvent to evaporate. After 15 min of heating, the needle is removed, and the reaction temperature is raised to 210 °C. When the desired temperature is reached, a sulfur solution prepared by dissolving 2 mmol of sulfur powder in 2 mL of TOP is added dropwise into the reaction mixture. The reaction mixture is then held at 210 °C for 10–15 min. The QDs are purified by the addition of ethanol and centrifugation.

2.3. Aqueous dispersion of the QDs terminated with carboxyl groups

The CdSe/CdS/ZnS QDs are transferred in aqueous media by ligand exchange with short thiol chain, mercaptosuccinic acid (MSA). Three millimole of MSA is dissolved in 10 mL of chloroform under vigorous stirring. After stirring for 10–15 min, 4 mL of concentrated (~20 mg/mL chloroform)

QD dispersion is added into this mixture. This solution is stirred overnight at room temperature. The QDs are separated from the surfactant solution by addition of ethanol and centrifugation. The reddish precipitate is redispersed in 10 mL high pressure liquid chromatography (HPLC) water and the solution is further filtered using a syringe filter with a pore diameter of 0.45 μm. The carboxyl functional of MSA not only enhances the colloidal stability of QD but also renders the capability for bioconjugation. This stock solution is kept in the refrigerator at 4 °C for further use.

3. Physical Characterization of QDs

The physical properties of the nanoformulation are established using transmission electron microscopy, dynamic light scattering (DLS), spectrophotometry, and spectrofluorometry.

3.1. High-resolution transmission electron microscopy

High-resolution transmission electron microscopy images are obtained using a JEOL model JEM 2010 microscope at an acceleration voltage of 200 kV. The specimens are prepared by drop-coating the sample dispersion onto an amorphous carbon-coated 300-mesh copper grid, which is placed on filter paper to absorb excess solvent.

3.2. Dynamic light scattering

The size distribution of water-dispersed QD is determined by DLS measurement with a Brookhaven Instruments 90Plus particle size analyzer, with a scattering angle of 90°. The samples are put in a cuvette and the size-distribution profiles are acquired using computer-controlled software (Brookhaven Instruments, Holtsville, NY).

3.3. Spectrophotometry and spectrofluorometry

The absorption spectra of QDs are collected using a Shimadzu model 3101PC UV-vis-NIR scanning spectrophotometer over a wavelength range from 450 to 725 nm. The samples are measured against water as the reference. The QD emission spectra are collected using a Fluorolog-3 spectrofluorometer (Jobin Yvon; fluorescence spectra). The QD solution is filtered with a syringe filter and loaded into a cuvette for measurements over a wavelength range from 450 to 725 nm.

4. Preparation of QD-Amprenavir-Tf Nanoplex

Site-specific targeted delivery of nanoparticles uses a targeting ligand, like a protein/peptide or a monoclonal antibody, which will help the nanoparticles to "home-in" on their specific target. Controlling the number of targeting molecule and drug moiety is critical for successful targeted drug delivery; too low amount of the targeting ligand will result in inefficient targeting and too high amount will trigger immunogenic reactions. For the purpose of targeting across the BBB, we used the molecule Tf as TfRs are overexpressed on the BBB. TfR is an iron-transporting protein-receptor which is present on the BBB in elevated amounts.

For the purpose of traversing across the BBB, we bioconjugate the QDs with Tf, whose corresponding receptors (TfRs) are overexpressed on the BBB. Following the synthesis of the QDs terminated with carboxyl groups, they are covalently bounded with Amprenavir and Tf molecules using simple chemical strategy (Scheme 3.1). Amprenavir is dissolved in a methanol/water mixture, and a 1-μM concentration solution of Amprenavir is used for bioconjugation. The carboxylated QDs, that are stable colloids in aqueous media, facilitate the bioconjugation of Amprenavir and Tf molecules using carbodiimide chemistry. Six hundred microlitres of QD stock solution is mixed with 10 μL of 0.05 M of 1-ethyl-3-(3-dimethylaminopropyl) carbodiimide solution and gently stirred for 30 min to activate the carboxyl groups to amine-reactive intermediates. A mixture of 7 μL of Tf (0.25 mg/mL) and 10 mL of highly polar Amprenavir solution is then added to allow the formation of covalent bonds between the amino terminated Amprenavir and Tf with the intermediates. The reaction is left overnight at

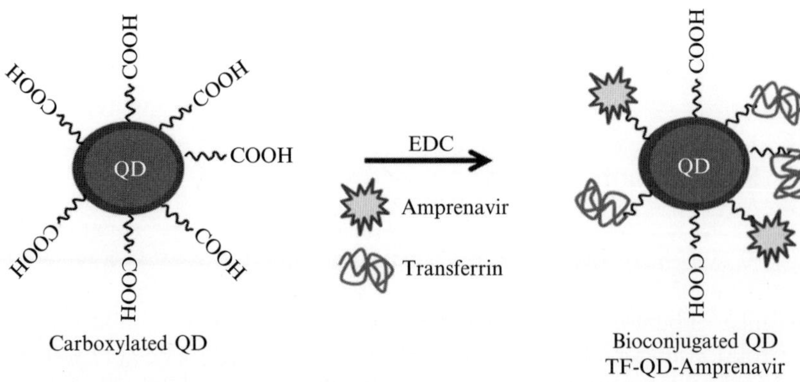

Scheme 3.1 Formation of Tf-QD-Amprenavir nanoplex. (See Color Insert.)

4 °C followed by dialysis and centrifugation for removing the unreacted substances. The amount of Amprenavir in the nanoplex is quantitated using HPLC and is determined to be approximately 0.65 μ*M*.

5. Assessment of Amprenavir Levels Using HPLC

To measure Amprenavir levels by HPLC, a C18 solid phase extraction (SPE) cartridge is used for sample preparation. One milliliter of the culture supernatant from the lower chamber of the *in vitro* BBB model is loaded on to a SPE cartridge that is conditioned prior to loading with 3 mL of 50% methanol. The SPE cartridge is eluted with 2.6 mL of methanol. The eluate is evaporated to dryness and the residue reconstituted with 50:50 solution of acetonitrile and 40 m*M* disodium hydrogen phosphate containing 4% octasulfonic acid.

HPLC conditions: The HPLC column used is 250 × 4.65 mm symmetry C18 column. Additionally, a GuardPak μ-BondaPak C18 Guard column is used. Column temperature is maintained at 37 °C and flow rate is 1.3 mL. Injection volume of the sample is 100 μL, and detection includes monitoring absorbance of the eluate at UV 215 nm (Dailly *et al.*, 2001; Dickinson *et al.*, 2005).

6. Evaluating the Efficacy of Antiretroviral Containing Nanoplex Using an *In Vitro* BBB Model

The BBB is a critical interface and acts as a physical and metabolic barrier between the CNS and the peripheral circulation that serves to regulate and protect the microenvironment of the brain. The primary function of the normal BBB is to establish and maintain homeostasis in the CNS (Bradbury, 1993). The BBB is not rigid and comprises dynamic vessels that are capable of responding to rapid changes in the brain or blood. The BBB is composed of specialized brain capillary endothelial cells and astrocytic endfeet that enhance the differentiation of the BBB endothelium.

6.1. *In vitro* model of the human BBB

Several *in vitro* tissue culture systems have been developed to reproduce the physical and biochemical properties of the intact BBB. A good *in vitro* BBB model used to study neuro-AIDS must reproduce the salient features of the *in situ* BBB and also must allow for manipulations to enable the researcher to mimic neuropathogenic process. A transwell coculture model, such as the Persidsky model (Persidsky and Gendelman, 1997; Persidsky *et al.*, 1997) which uses primary normal human astrocytes (NHAs) and brain microvascular endothelial cells (BMVECs), both cell types that constitute the *in vivo* BBB, grown to confluence on an polyethylene terephthalate membrane insert have been extensively used by researchers because it can be reproducibly formed in large quantities and can provide detailed information about the cellular and molecular mechanisms of a wide variety of neurological disease conditions. We have validated the Persidsky model in our laboratory and believe that is a good model to analyze transport of nanoconjugates across the BBB and evaluate the mechanisms that may be involved in the neuropathogenesis process.

This *in vitro* BBB model uses primary cultures of both BMVECs (Cat # ACBRI-376) and NHAs (Cat # ACBRI-371), which are obtained from Applied Cell Biology Research Institute (ACBRI), Kirkland, WA. Characterization of BMVECs demonstrated that >95% cells were positive for cytoplasmic von Willebrand's factor/Factor VIII. BMVECs are cultured in CS-C complete serum-free medium (ABCRI, Cat # SF-4Z0-500) with attachment factors (ABCRI, Cat # 4Z0-210) and Passage Reagent Group™ (ABCRI, Cat # 4Z0-800). NHAs are cultured in the CS-C medium, supplemented with 10 μg/mL human epidermal growth factor, 10 mg/mL insulin, 25 μg/mL progesterone, 50 mg/mL Tf, 50 mg/mL gentamicin, 50 μg /mL amphotericin-B, and 10% FBS. NHAs are characterized on the basis of >99% of these cells being positive for glial fibrillary acidic protein (GFAP). Both BMVECs and NHAs are obtained at early passage typically passage 2 for each experiment and are used for all experiments between 2 and 8 passages, within the 6–27 cumulative population doublings.

The BBB model used consists of two-compartment wells in a six-well culture plate, with the upper compartment separated from the lower by a 3-μ*M*PET (polyethylene terephthalate) insert (surface area = 4.67 cm^2). The BMVECs are grown to confluency on the upper side of the insert, while a confluent layer of NHAs are grown on the underside. The formation of a functional and intact BBB takes a minimum of 5 days, which can be confirmed by determining the transendothelial electrical resistance (TEER) value as described below (Mahajan *et al.*, 2008, 2010). Using the *in vitro* BBB model, we examine BBB permeability, transendothelial migration, and the efficacy of drug delivery of the QD-Tf-Amprenavir nanoformulation.

6.2. Measurement of TEER across a membrane

TEER across the *in vitro* BBB is measured using an ohm meter Millicell ERS system (Millipore, Bedford, MA Cat # MERS 000 01). Electrodes are sterilized using 95% alcohol and rinsed in distilled water prior to measurement. A constant distance of 0.6 cm is maintained between the electrodes at all times during TEER measurement.

6.3. Cell viability measurement using an MTT (3-(4,5-dimethylthiazol-2-yl)-2,5-diphenyltetrazolium bromide, a tetrazole) assay

The MTT assay is done to evaluate the cell viability of all cells in culture, namely, the peripheral blood mononuclear cells (PBMCs)-derived monocytes, BMVECs, and NHAs. Cell viability is also tested in the monocytes prior to being infected by HIV-1 as well as at 7 days postinfection. Monocyte viability is also measured before and after treatment of the *in vitro* BBB with the QD-Amprenavir-Tf nanoconjugate. No significant difference in cell viability was observed for all three cell types above when treated with Tf-QD-Amprenavir nanobioconjugate, free Amprenavir alone, and Amprenavir-QD bioconjugate as compared to the untreated control, over time points ranging from 6 to 48 h posttreatment. MTT cell-proliferation assay measures the reduction of a tetrazolium component (MTT) into an insoluble formazan product by the mitochondria of viable cells. The MTT assay is a quantitative and sensitive detection of cell proliferation as it measures the growth rate of cells by virtue of a linear relationship between cell activity and absorbance. Typically 10,000 cells suspended in 100 μL of media are incubated with 10 μL of MTT reagent (Cat # 30-1010K; ATCC) for approximately 3 h, followed by addition of a detergent solution to lyse the cells and solubilize the colored crystals. Colorimetric detection is done at a wavelength of 570 nm. The amount of color produced is directly proportional to the number of viable cells. Figure 3.1 shows dose- and time-dependent cytotoxicity of the free Amprenavir, Amprenavir-QD, and QD-Tf-Amprenavir nanoformulations on BMVEC over a 6- to 48-h time period using the MTT assay.

6.4. Transfer of the Tf-QD-Amprenavir nanoparticles across the BBB and uptake of the nanoplex by HIV-1-infected monocytes

The cellular uptake of the Tf-QD-Amprenavir is evaluated using confocal microscopy. A Nikon Eclipse TE2000 microscope equipped with the Nuance GNIR imaging system (Cambridge Research & Instrumentation Inc., Cambridge, MA) is employed, which is capable of multispectral (wavelength-resolved) imaging in the range of 500–950 nm. Custom-designed filter cubes, with corresponding dichroic and emission filters acquired from Omega Optical, are used to cut off the excitation light and obtain high-contrast fluorescence images. Confocal microscopic analysis shows QD staining on both the upper and lower sides of the PET membrane following treatment with Tf-QR-Amprenavir nanoplex. Confocal imaging data shown in Fig. 3.2 shows evidence for the successful traversing of the Tf-QD nanoplex across the *in vitro* BBB. Figure 3.2a and c show

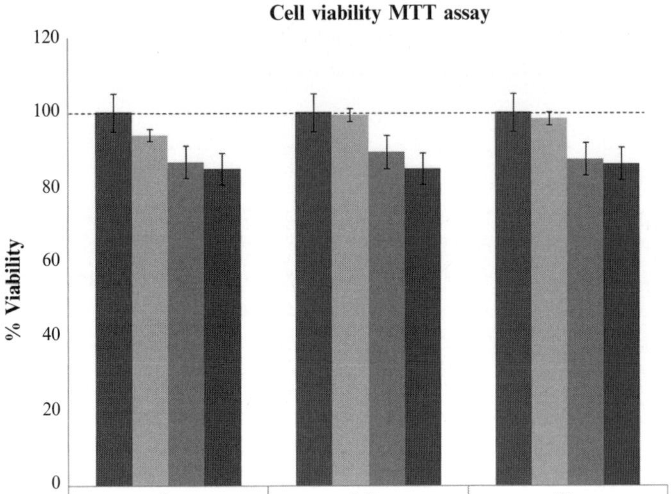

	6 h	24 h	48 h
Untreated control	100	100	100
Amprenavir alone	94.011	99.41	98.39
Amprenavir-QD conjugate	86.82	89.53	87.54
Tf-QD-Amprenavir	84.88	84.88	86.32

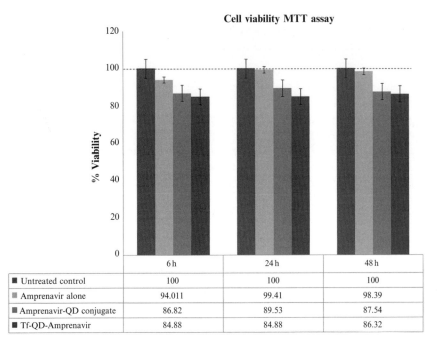

Figure 3.1 Effect of free Amprenavir, Amprenavir-QD, and Tf-QD-Amprenavir on cell toxicity. Dose- and time-dependent cytotoxicity of free Amprenavir, Amprenavir-QD, and QD-Tf-Amprenavir nanoformulations on BMVEC over a 6- to 48-h time period using the MTT assay. Our results show no significant toxicity or no significant difference in cell viability with respect to the untreated control for the above nanoformulations. The results shown are the mean $\pm$ SD of three separate experiments done in duplicate. (For the color version of this figure, the reader is referred to the Web version of this chapter.)

significant uptake of the Tf-QD nanoplex by both upper and lower side of the BBB; however, when the *in vitro* BBB is treated with the nonbioconjugated QD's alone, QD staining is only observed in the upper side (Fig. 3.2b), and not in the lower side (Fig. 3.2d), of the PET membrane. These data support the premise that bioconjugation of transferring to the QD nanoparticles can facilitate transfer across the BBB as shown using our *in vitro* BBB model.

6.5. Monocyte isolation from PBMCs)

PBMCs are isolated from leukocyte-depletion filters obtained from the blood bank at Upstate New York Transplant Services, Buffalo, NY. Leukocytes are eluted aseptically from these filters using standard protocols to yield the so-called Filter buffy coats (Meyer *et al.*, 2005). Filters are flushed

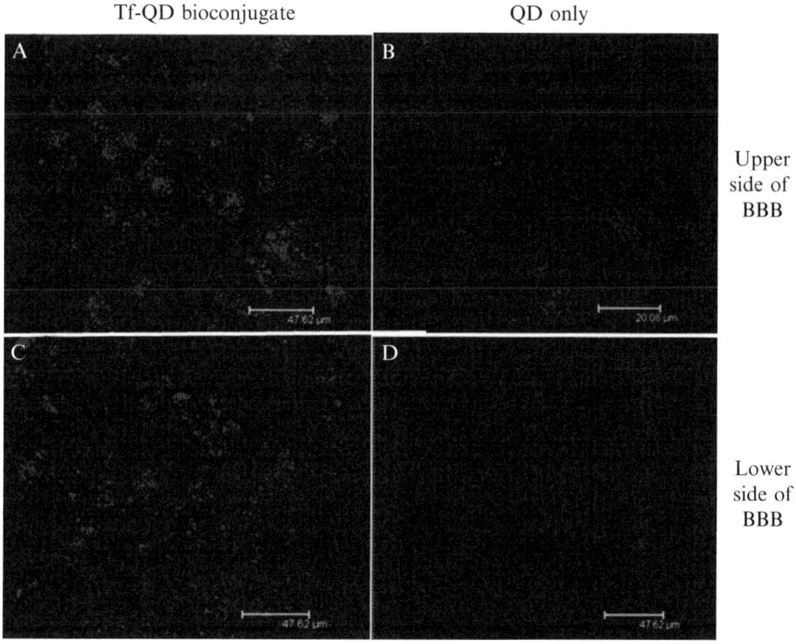

Figure 3.2 Transmigration of nanoplexes across the BBB. Confocal Images of the BBB, (A) upper side of the *in vitro* BBB model following treatment with (a) QD-Tf and (b) QD alone. (B) Lower side of the *in vitro* BBB model following treatment with (c) QD-Tf and (d) QD alone. (See Color Insert.)

at room temperature with sterile filter elution medium (Dulbecco's PBS without $MgCl_2$ and $CaCl_2$, containing 5 mM Na_2-EDTA and 2.5% w/v sucrose) using a sterile 60 mL syringe. A total volume of 200 mL elution medium is collected using gentle pressure to avoid cell disruption or filter leakage. These filter buffy coats are then overlaid on 15 mL Ficoll-Paque® Plus (Amersham-Pharmacia, Piscataway, NJ Cat # 17-1440-03), in a 50-mL culture tube. Samples are centrifuged for 20 min at 700 × g and 20 °C. The filter–PBMC interface is carefully removed by pipetting and washed twice with PBS/EDTA and resuspended in complete RPMI media. Then, the total number of cells is counted using a hemocytometer. Cells are resuspended at a concentration of 2.5×10^6 cells/mL in RPMI + 1% heat-inactivated human serum AB. Cells are then incubated overnight at 37 °C, 5% CO_2 followed by removal of nonadherent cells. Adherent cells are washed three times with warm PBS. Cells are then resuspended in 10 mL complete media (RPMI with 10% FBS). Typically, more than 97% of the recovered cells are monocytes as determined by cell morphology; additionally, CD14 positivity is confirmed by flow cytometry analysis using a FASCan/Cell Quest software (BD Pharmingen, Mountain View, CA).

6.6. HIV-1 infection of monocytes

Monocytes (1×10^5 cells/mL) are infected for 3 h with HIV-1_{IIIB} (NIH AIDS Research and Reference Reagent Program) at a concentration of $10^{3.0}$ $TCID_{50}$/mL cells, equivalent to 10 ng viral isolate/mL of culture media. Following that, the infected cells are washed with Hanks buffered saline, reconstituted in RPMI media (fortified with 10% FBS) and incubated at 37 °C/5% CO_2 for 7 days. Levels of p24 in the culture supernatants are measured using a commercially available p24 ELISA kit (Zeptometrix, Buffalo, NY) 7 days postinfection. These infected monocytes are then washed and reconstituted in fresh culture medium and used for evaluating the anti-HIV-1 efficacy of the nanobioconjugate using the *in vitro* BBB model.

7. Evaluation of the Antiviral Efficacy of the QD-Tf-Amprenavir Nanobioconjugate

Tf-QD-Amprenavir nanobioconjugate, free Amprenavir alone, and Amprenavir-QD bioconjugate are reconstituted in 100 μL of media and are added to the upper chamber of the *in vitro* BBB. 1×10^5 HIV-1-infected monocytes suspended in 1 mL of RPMI complete media are plated in the lower chamber (basolateral end) of the *in vitro* BBB model. The *in vitro* BBB cell culture chambers are incubated at 37 °C and 5% CO_2 for a period of 24–48 h after addition of the Tf-QD-Amprenavir nanobioconjugate, free Amprenavir alone, and Amprenavir-QD bioconjugate to the upper chambers (apical end). At the end of this incubation period, monocytes are harvested from the lower chamber, washed, and then analyzed for antiviral efficacy of the nanobioconjugates by measurement of (a) HIV-1 p24 antigen levels using a commercial p24 ELISA assay and (b) HIV-1 LTR-R/U5 gene expression levels using a quantitative real-time PCR. Triplicate wells are used for each condition tested.

7.1. HIV-1 p24 quantitation using a commercially available p24 ELISA

RETRO-TEK HIV-1 p24 Antigen ELISA (ZeptoMetrix Corporation, Buffalo, NY; Cat # ZMC Catalog #: 0801111) is an enzyme-linked immunoassay used to detect HIV-1 p24 antigen cell culture media. Microwells are coated with a monoclonal antibody specific for the p24 *gag* gene product of HIV-1. Viral antigen in the specimen is specifically captured onto the immobilized antibody during specimen incubation. The captured antigen is then reacted with a high-titered human anti-HIV-1 antibody conjugated with biotin. Following a subsequent incubation with streptavidin-peroxidase, color develops as the bound enzyme reacts with the substrate. The resultant optical density is proportional to the amount of HIV-1 p24 antigen present in

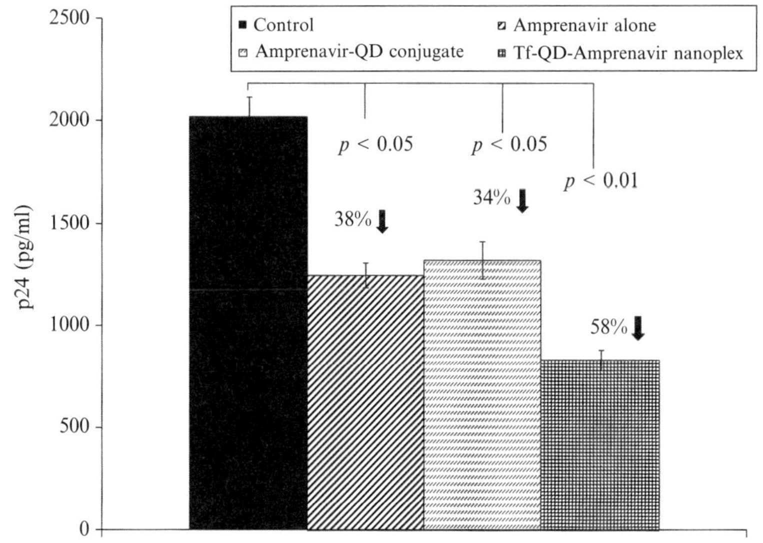

Figure 3.3 Effect of free Amprenavir, Amprenavir-QD, and Tf-QD-Amprenavir nanoplexes, on p24 production in HIV-1-infected monocytes. Culture media from the lower chamber containing the HIV-1-infected monocytes (basolateral end of the *in vitro* BBB model) is harvested 48 h after the free Amprenavir, Amprenavir-QD, and Tf-QD-Amprenavir nanoplexes were added to the apical chamber of the *in vitro* BBB model. p24 Levels in these supernatants are measured using a commercially available ELISA kit. Our results show a 38%, 34%, and 58% decrease in p24 production in the culture supernatants of HIV-1-infected monocytes harvested from the basolateral end of the culture wells treated with free Amprenavir, Amprenavir-QD, and Tf-QD-Amprenavir nanoplexes compared to the untreated control culture well. The results shown are the mean ± SD of three separate experiments done in duplicate.

the specimen. Our results (Fig. 3.3) showed a 38%, 34%, and 58% decrease in p24 production in HIV-1-infected monocytes harvested from lower chamber of the *in vitro* BBB which was treated with free Amprenavir, Amprenavir-QD, and QD-Tf-Amprenavir nanoplexes, respectively. Statistical significance was determined using ANOVA based on comparisons between free Amprenavir, Amprenavir-QD, and Tf-QD-Amprenavir treated wells versus the untreated control. Comparison among the means was performed with the *post-hoc* Bonferroni analysis test using the PRISM statistical analysis software (GraphPad Software Inc., La Jolla, CA). Statistical significance was considered at $p < 0.05$.

7.2. Quantification of HIV-1 LTR-R/U5 gene expression using real-time quantitative PCR

Cytoplasmic RNA is extracted by an acid guanidinium–thiocyanate–phenol–chloroform method using Trizol reagent (Invitrogen, Carlsbad, CA) (Chomczynski and Saachi, 1987). The amount of RNA is quantified using a Nano-Drop ND-1000 spectrophotometer (Nano-Drop™, Wilmington, DE) and the isolated RNA is stored at −80 °C until used. The LTR-R/U5 region represents early stages of reverse transcription of HIV-1. Following conversion of RNA to cDNA using reverse transcription, the relative abundance of mRNA species is quantified by real-time quantitative PCR using the LTR/RU5-specific primers and the Brilliant® SYBR® green QPCR master mix (Stratagene Inc., La Jolla, CA; Cat # 600548-51). The following are the primer sequences used for LTR/RU5 (forward primer 5′-TCTCTCTGGTTAGACCAGATCTG-3′ and reverse primer 5′-ACTGCTAGAGATTTTCCACACTG-3′). Relative expression of mRNA species is calculated using the comparative C_T method (Bustin, 2002). To provide precise quantification of initial target in each PCR reaction, the amplification plot is examined at a point during the early log phase of product accumulation. This is accomplished by assigning a fluorescence threshold above background and determining the time point at which each sample's amplification plot reaches the threshold (defined as the threshold cycle number or C_T). Differences in threshold cycle number are used to quantify the relative amount of PCR target contained within each tube. All data are controlled for quantity of RNA input by performing measurements on an endogenous reference gene, β-actin. Results on RNA from treated samples are normalized to results obtained on RNA from the control sample. The analysis is performed as follows: for each sample, a difference in C_T values (ΔC_T) is calculated for each mRNA by taking the mean C_T of duplicate tubes and subtracting the mean C_T of the duplicate tubes for the reference RNA (β-actin) measured on an aliquot from the same RT reaction. The ΔC_T for the treated sample is then subtracted from the ΔC_T for the control sample to generate a $\Delta\Delta C_T$. The mean of these $\Delta\Delta C_T$ measurements is then used to calculate the expression of the test gene relative to the reference gene and normalized to the control as follows: relative expression/transcript accumulation index $= 2^{-\Delta\Delta C_T}$ (Schefe *et al.*, 2006). This calculation assumes that all PCR reactions are working with 100% efficiency. In our laboratory, all PCR efficiencies were found to be >95%; therefore, this assumption introduces minimal error into the calculations.

Results of our gene expression studies showed that the HIV-1 LTR/RU5 gene expression levels were significantly decreased by 64%, 62%, and 91% in HIV-1-infected monocytes harvested from the lower chamber of the *in vitro* BBB which were treated with free Amprenavir, Amprenavir-QD, and Tf-QD-Amprenavir nanoplexes, respectively. These HIV-1 LTR/RU5 gene expression levels were significantly decreased when compared to the untreated control (Fig. 3.4). Data are the mean ± SD of three separate experiments done in duplicate. Statistical significance was determined using ANOVA based on comparisons between free Amprenavir, Amprenavir-QD, and Tf-QD-Amprenavir treated wells versus the untreated control. Comparison among the means was performed with the *post-hoc*

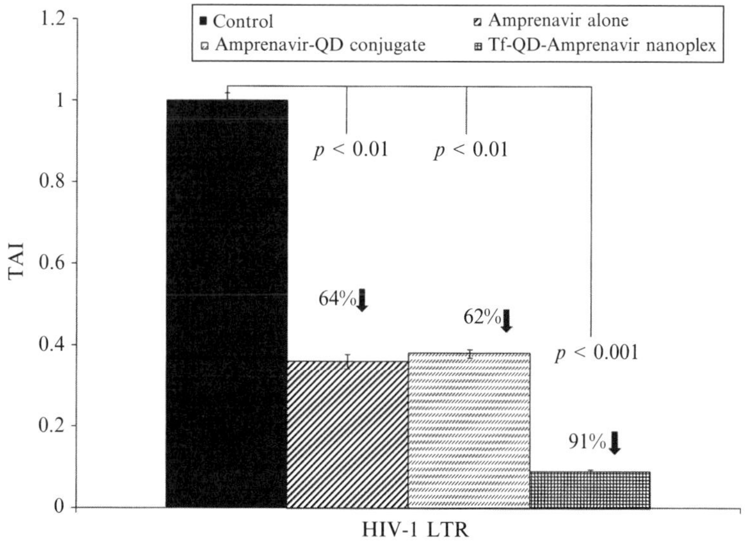

Figure 3.4 Effect of free Amprenavir, Amprenavir-QD, and Tf-QD-Amprenavir nanoplexes on LTR/RU5 gene expression in HIV-1-infected monocytes. HIV-1-infected monocytes are harvested from the lower chamber, 48 h after free Amprenavir, Amprenavir-QD, and Tf-QD-Amprenavir nanoplexes were added to the apical chamber of the *in vitro* BBB model. RNA is extracted, reverse transcribed, and the LTR/RU5 gene expression is quantitated from these HIV-1-infected monocytes using Q-PCR. Our results show a significant decrease in LTR/RU5 gene expression in HIV-1-infected monocytes harvested from the basolateral end of the culture well which were treated with free Amprenavir, Amprenavir-QD, and Tf-QD-Amprenavir nanoplexes as compared to untreated control culture well. The results shown are the mean ± SD of three separate experiments done in duplicate.

Bonferroni analysis test using the PRISM statistical analysis software (GraphPad Software). Statistical significance was considered at $p < 0.05$.

8. Concluding Remarks

The application of nanotechnology provides unprecedented opportunities for addressing many of the gaps in the diagnosis and therapy of diseases. Nanoparticle technology offers a significant advancement in the ability to increase drug translocation across the BBB. The development of BBB permeable, multifunctional drug-loaded nanoparticles will provide an advancement toward the therapy of neurological disorders associated with HIV-1; further, these nanoparticle systems will integrate high-resolution imaging capability in addition to therapeutic modalities. We have demonstrated the ability of a Tf-QD-Amprenavir nanoplex to transverse the BBB and significantly inhibit HIV-1 replication in HIV-1-infected monocytes, demonstrating their anti-HIV-1 efficacy in the brain. The use of such nanotechnology platforms for delivery of antiretroviral drugs will revolutionize the treatment of neuro-AIDS.

ACKNOWLEDGMENTS

Amprenavir was obtained through the AIDS Research and Reference Reagent Program, Division of AIDS, NIAID, NIH. This study was supported by grants from the National Institute of Health ARRA Grant # NIAID-RO1LM009726-01 (S. A. S.), NIDA K01 DA024577 (J. R.), NIDA-1R21DA030108-01 (S. M.); Pfizer Inc. # GA 400IN3 (S. M.); and the Kaleida Health Foundations.

REFERENCES

Aquaro, S., Guenci, T., Di Santo, F., Francesconi, M., Caliò, R., and Perno, C. F. (2004). Potent antiviral activity of amprenavir in primary macrophages infected by human immunodeficiency virus. *Antiviral Res.* **61,** 133–137.

Bhaskar, S., Tian, F., Stoeger, T., Kreyling, W., de la Fuente, J. M., Grazú, V., Borm, P., Estrada, G., Ntziachristos, V., and Razansky, D. (2010). Multifunctional nanocarriers for diagnostics, drug delivery and targeted treatment across blood-brain barrier: Perspectives on tracking and neuroimaging. *Part. Fibre Toxicol.* **7,** 3.

Bonoiu, A., Mahajan, S. D., Ye, L., Kumar, R., Ding, H., Yong, K. T., Roy, I., Aalinkeel, R., Nair, B., Reynolds, J. L., Sykes, D. E., Imperiale, M. A., *et al.* (2009). MMP-9 gene silencing by a quantum dot-siRNA nanoplex delivery to maintain the integrity of the blood brain barrier. *Brain Res.* **1282,** 142–155.

Bradbury, M. W. (1993). The blood-brain barrier. *Exp. Physiol.* **78,** 453–472.

Bustin, S. A. (2002). Quantification of mRNA using real-time reverse transcription PCR (RT-PCR): Trends and problems. *J. Mol. Endocrinol.* **29,** 23–39.

Chomczynski, P., and Sacchi, N. (1987). Single step method of RNA isolation by acid guanidinium thiocyanate-phenol-chloroform extraction. *Anal. Biochem.* **162,** 156–159.

Conway, B., and Shafran, S. D. (2000). Pharmacology and clinical experience with amprenavir. *Expert Opin. Investig. Drugs* **9,** 371–382.

Dailly, E., Thomas, L., Kergueris, M. F., Jolliet, P., and Bourin, M. (2001). High-performance liquid chromatographic assay to determine the plasma levels of HIV-protease inhibitors (amprenavir, indinavir, nelfinavir, ritonavir and saquinavir) and the non-nucleoside reverse transcriptase inhibitor (nevirapine) after liquid-liquid extraction. *J. Chromatogr. B* **758,** 129–135.

DeLuca, A., Ciancio, B., Larussa, D., Murri, R., Cingolani, A., Rizzo, M. G., Giancola, M. L., Ammassari, A., and Ortona, L. (2002). Correlates of independent HIV-1 replication in the CNS and of its control by antiretrovirals. *Neurology* **59,** 342–347.

Dickinson, L., Robinson, L., Tjia, J., Khoo, S., and Back, D. (2005). Simultaneous determination of HIV protease inhibitors amprenavir, atazanavir, indinavir, lopinavir, nelfinavir, ritonavir and saquinavir in human plasma by high-performance liquid chromatography-tandem mass spectrometry. *J. Chromatogr. B Analyt. Technol. Biomed. Life Sci.* **829,** 82–90.

Ding, H., Yong, K. T., Law, W. C., Roy, I., Hu, R., Wu, F., Zhao, W., Huang, K., Erogbogbo, F., Bergey, E. J., and Prasad, P. N. (2011). Non-invasive tumor detection in small animals using novel functional Pluronic nanomicelles conjugated with anti-mesothelin antibody. *Nanoscale* **3,** 1813–1822.

Enting, R. H., Hoetelmans, R. M., Lange, J. M., Burger, D. M., Beijnen, J. H., and Portegies, P. (1998). Antiretroviral drugs and the central nervous system. *AIDS* **12,** 1941–1955.

Erogbogbo, F., Yong, K. T., Hu, R., Law, W. C., Ding, H., Chang, C. W., Prasad, P. N., and Swihart, M. T. (2010). Biocompatible magnetofluorescent probes: Luminescent silicon quantum dots coupled with superparamagnetic iron(III) oxide. *ACS Nano* **4,** 5131–5138.

Erogbogbo, F., Yong, K., Roy, I., Hu, R., Law, W. C., Zhao, W., Ding, H., Wu, F., Kumar, R., Swihart, M. T., and Prasad, P. N. (2011). In vivo targeted cancer imaging, sentinel lymph node mapping and multi-channel imaging with biocompatible silicon nanocrystals. *ACS Nano* **5,** 413–423.

Gaillard, P. J., Visser, C. C., and de Boer, A. G. (2005). Targeted delivery across the blood-brain barrier. *Expert Opin. Drug Deliv.* **2,** 299–309.

Hallmann, R., Savigni, D. L., Morgan, E. H., and Baker, E. (2000). Characterization of iron uptake from transferrin by murine endothelial cells. *Endothelium* **7,** 135–147.

Hu, R., Yong, K. T., Roy, I., Ding, H., Law, W. C., Cai, H., Zhang, X., Vathy, L. A., Bergey, E. J., and Prasad, P. N. (2010). Functionalized near-infrared quantum dots for in vivo tumor vasculature imaging. *Nanotechnology* **21,** 145105.

Kreuter, J., Alyautdin, R. N., Kharkevich, D. A., and Ivanov, A. A. (1995). Passage of peptides through the blood-brain barrier with colloidal polymer particles (nanoparticles). *Brain Res.* **674,** 171–174.

Law, W. C., Yong, K. T., Roy, I., Ding, H., Hu, R., Zhao, W., and Prasad, P. N. (2009). Aqueous-phase synthesis of highly luminescent CdTe/ZnTe core/shell quantum dots optimized for targeted bioimaging. *Small* **5**(11), 1302–1310.

Letendre, S. L., Ellis, R. J., Everall, I., Ances, B., Bharti, A., and McCutchan, J. A. (2009). Neurologic complications of HIV disease and their treatment. *Top. HIV Med.* **17,** 46–56.

Li, J. J., Wang, Y. A., Guo, W., Keay, J. C., Mishima, T. D., Johnson, M. B., and Peng, X. (2003). Large-scale synthesis of nearly monodisperse CdSe/CdS core/shell nanocrystals using air-stable reagents via successive ion layer adsorption and reaction. *J. Am. Chem. Soc.* **125,** 12567–12575.

Liu, L., Law, W. C., Yong, K. T., Roy, I., Ding, H., Erogbogbo, F., Zhang, X., and Prasad, P. N. (2011). Multimodal imaging probes based on Gd-DOTA conjugated quantum dot nanomicelles. *Analyst* **136,** 1881–1886.

Mahajan, S. D., Aalinkeel, R., Sykes, D. E., Reynolds, J. L., Bindukumar, B., Fernandez, S. F., Chawda, R., Shanahan, T. C., and Schwartz, S. A. (2008). Tight junction regulation by morphine and HIV-1 tat modulates blood-brain barrier permeability. *J. Clin. Immunol.* **28,** 528–541.

Mahajan, S. D., Roy, I., Xu, G., Yong, K. T., Ding, H., Aalinkeel, R., Reynolds, J., Sykes, D., Nair, B. B., Lin, E. Y., Prasad, P. N., and Schwartz, S. A. (2010). Enhancing the delivery of anti retroviral drug "Saquinavir" across the blood brain barrier using nanoparticles. *Curr. HIV Res.* **8,** 396–404.

Manna, L., Scher, E. C., Li, L. S., and Alivisatos, A. P. (2002). Epitaxial growth and photochemical annealing of graded CdS/ZnS shells on colloidal CdSe nanorods. *J. Am. Chem. Soc.* **124,** 7136–7145.

Meyer, T. P., Zehnter, I., Hofmann, B., Zaisserer, J., Burkhart, J., Rapp, S., Weinauer, F., Schmitz, J., and Illert, W. E. (2005). Filter Buffy Coats (FBC): A source of peripheral blood leukocytes recovered from leukocyte depletion filters. *J. Immunol. Methods* **307,** 150–166.

Michalet, X., Pinaud, F. F., Bentolila, L. A., Tsay, J. M., Doose, S., Li, J. J., Sundaresan, G., Wu, A. M., Gambhir, S. S., and Weiss, S. (2005). Quantum rods for live cells, in vivo imaging, and diagnostics. *Science* **307,** 538–544.

Michalet, X., Weiss, S., and Jäger, M. (2006). Single-molecule fluorescence studies of protein folding and conformational dynamics. *Chem. Rev.* **106**(5), 1785–1813.

Patel, M. M., Goyal, B. R., Bhadada, S. V., Bhatt, J. S., and Amin, A. F. (2009). Getting into the brain: Approaches to enhance brain drug delivery. *CNS Drugs* **23,** 35–58.

Pathak, S., Cao, E., Davidson, M. C., Jin, S., and Silva, G. A. (2006). Quantum dot applications to neuroscience: New tools for probing neurons and glia. *J. Neurosci.* **26,** 1893–1895.

Persidsky, Y., and Gendelman, H. E. (1997). Development of laboratory and animal model systems for HIV-1 encephalitis and its associated dementia. *J. Leukoc. Biol.* **62,** 100–106.

Persidsky, Y., Stins, M., Way, D., Witte, M. H., Weinand, M., Kim, K. S., Bock, P., Gendelman, H. E., and Fiala, M. (1997). A model for monocyte migration through the blood-brain barrier during HIV-1 encephalitis. *J. Immunol.* **158,** 3499–3510.

Prasad, P. N. (2003). Introduction in Biophotonics. Wiley, New York.

Prasad, P. N. (2004). Nanophotonics. Wiley, New York.

Qian, J., Yong, K. T., Roy, I., Ohulchansky, T. Y., Bergey, E. J., Lee, H. H., Tramposch, K. M., He, S., Maitra, A., and Prasad, P. N. (2007). Imaging pancreatic cancer using surface-functionalized quantum rods. *J. Phys. Chem. B* **111,** 6969–6972.

Samia, A. C., Chen, X., and Burda, C. (2003). Semiconductor quantum dots for photodynamic therapy. *J. Am. Chem. Soc.* **125,** 15736–15737.

Schefe, J. H., Lehmann, K. E., Buschmann, I. R., Unger, T., and Funke-Kaiser, H. (2006). Quantitative real-time RT-PCR data analysis: Current concepts and the novel "gene expression's CT difference" formula. *J. Mol. Med. (Berl.)* **84,** 901–910.

Schroeder, U., Sommerfeld, P., Ulrich, S., and Sabel, B. A. (1998). Nanoparticle technology for delivery of drugs across the blood-brain barrier. *J. Pharm. Sci.* **87,** 1305–1307.

Shapshak, P., Kangueane, P., Fujimura, R. K., Commins, D., Chiappelli, F., Singer, E., Levine, A. J., Minagar, A., Novembre, F. J., Somboonwit, C., Nath, A., and Sinnott, J. T. (2011). Editorial neuroAIDS review. *AIDS* **25**(2), 123–141.

Silva, G. A. (2007). Nanotechnology approaches for drug and small molecule delivery across the blood brain barrier. *Surg. Neurol.* **67,** 113–116.

Singer, E. J., Valdes-Sueiras, M., Commins, D., and Levine, A. J. (2010). Neurologic presentations of AIDS. *Neurol. Clin.* **28,** 253–275.

Sparidans, R. W., Hoetelmans, R. M., and Beijnen, J. H. (2000). Sensitive liquid chromatographic assay for amprenavir, a human immunodeficiency virus protease inhibitor, in human plasma, cerebrospinal fluid and semen. *J. Chromatogr. B Biomed. Sci. Appl.* **742,** 185–192.

Tu, C., Ma, X., House, A., Kauzlarich, S. M., and Louie, A. Y. (2011). PET imaging and biodistribution of silicon quantum dots in mice. *ACS Med. Chem. Lett.* **2,** 285–288.

Wang, Y. Y., Lui, P. C., and Li, J. Y. (2009). Receptor-mediated therapeutic transport across the blood-brain barrier. *Immunotherapy* **1,** 983–993.

Yong, K. T., Qian, J., Roy, I., Lee, H. H., Bergey, E. J., Tramposch, K. M., He, S., Swihart, M. T., Maitra, A., and Prasad, P. N. (2007). Quantum rod bioconjugates as targeted probes for confocal and two-photon fluorescence imaging of cancer cells. *Nano Lett.* **7,** 761–765.

Yong, K. T., Ding, H., Roy, I., Law, W. C., Bergey, E. J., Maitra, A., and Prasad, P. N. (2009a). Imaging pancreatic cancer using bioconjugated InP quantum dots. *ACS Nano* **3,** 502–510.

Yong, K. T., Roy, I., Ding, H., Bergey, E. J., and Prasad, P. N. (2009b). Biocompatible near-infrared quantum dots as ultrasensitive probes for long-term in vivo imaging applications. *Small* **5,** 1997–2004.

Yong, K. T., Roy, I., Law, W. C., and Hu, R. (2010). Synthesis of cRGD-peptide conjugated near-infrared CdTe/ZnSe core-shell quantum dots for in vivo cancer targeting and imaging. *Chem. Commun. (Camb.)* **46,** 7136–7138.

CHAPTER FOUR

Antibacterial Activity of Doxycycline-Loaded Nanoparticles

Ranjita Misra *and* Sanjeeb K. Sahoo

Contents

1. Introduction 62
2. Polymeric Nanoparticles 65
3. Chitosan Nanoparticles 65
4. Polyalkylcyanoacrylate Nanoparticles 67
5. poly(d,l-lactide-co-glycolide) Nanoparticles 68
6. poly(ε-caprolactone) Nanoparticles 70
7. Doxycycline-Loaded PLGA:PCL Nanoparticles 70
   7.1. Preparation of PLGA:PCL nanoparticles 70
   7.2. Antibacterial activity studies 76
   7.3. Determination of minimum inhibitory concentration and minimum bactericidal concentration of doxycycline 76
   7.4. Comparison of the stability of native doxycycline and doxycycline in nanoparticles 77
   7.5. Effect of native doxycycline and doxycycline-loaded nanoparticles on bacterial growth kinetics 78
8. Conclusion 81
References 81

Abstract

Doxycycline is a tetracycline antibiotic with a potent antibacterial activity against a wide variety of bacteria. However, poor cellular penetration limits its use for the treatment of infectious disease caused by intracellular pathogens. One potential strategy to overcome this problem is the use of nanotechnology that can help to easily target the intracellular sites of infection. The antibacterial activity of these antibiotics is enhanced by encapsulating it in polymeric nanoparticles. In this study, we describe the improvement of the entrapment efficiency of doxycycline hydrochloride (doxycycline)-loaded PLGA:PCL nanoparticles up to 70% with variation of different formulation parameters such as polymer ratio, amount of drug loading (w/w), solvent selection, electrolyte addition, and pH alteration in the formulation. We have evaluated the efficacy of these

Institute of Life Sciences, Chandrasekharpur, Bhubaneswar, Orissa, India

Methods in Enzymology, Volume 509

ISSN 0076-6879, DOI: 10.1016/B978-0-12-391858-1.00004-6

nanoparticles over native doxycycline against a strain of *Escherichia coli* (DH5α) through growth inhibition and colony counting. The results indicate that doxycycline-loaded nanoparticles have superior effectiveness compared to native doxycycline against the above bacterial strain, resulting from the sustained release of doxycycline from nanoparticles. These results are encouraging for the use of these doxycycline-loaded nanoparticles for the treatment of infections caused by doxycycline-sensitive bacteria.

1. Introduction

Bacterial infection is one of the most frequent complications in the human society. There are some pathogenic bacteria that harbor some sets of genes that help in their prolonged persistence in the host cells, establishing lifelong, chronic infection (Chaturvedi *et al.*, 2011; Young *et al.*, 2002). The outcome of the infection depends solely on the balance between intrinsic virulence features displayed by the invading pathogen and the responses of the host that try to neutralize this attack (Casadevall and Pirofski, 1999, 2001, 2002). To circumvent these bacterial infections, a variety of novel methodologies have been used for the proper diagnosis and characterization of virulence determinants of these pathogenic microbes. Generally, bacteria are classified as Gram-negative or Gram-positive based on their membrane and peptidoglycan structure. A thin peptidoglycan layer (~2–3 nm) is present between the cytoplasmic membrane and the outer membrane in Gram-negative bacteria, whereas Gram-positive bacteria have a 30-nm thick layer of peptidoglycan outside the cell membrane (Murray *et al.*, 1965; Shockman and Barrett, 1983).

For the treatment of bacterial infections antibiotics are used, which may be classified into bactericidal (that kill the bacteria) or bacteriostatic (that inhibit the growth of bacteria). There exist different classes of antibiotics that inhibit vital cellular process in the pathogen without affecting the host. For example, some of the antibiotics like chloramphenicol and tetracycline exhibit selective toxicity by inhibiting the bacterial ribosome, but not the structurally different eukaryotic ribosome of the host (Waksman, 1947).

Thus, use of conventional drugs is one of the effective antibiotic therapies for bacterial infections. However, this approach still has some limitations, like the development of antibiotic resistance and an enduring need for new solutions (Martin and Ernst, 2003). The emergence of antibiotic resistant strains of bacteria has reduced the efficacy of these antibiotics; therefore, there is a strong incentive to develop new bactericides.

Tetracyclines are one of the broad spectrum of antibiotics used for the treatment of a number of infections, like the respiratory tract, sinuses, middle ear, urinary tract, and intestines, and is used in the treatment of

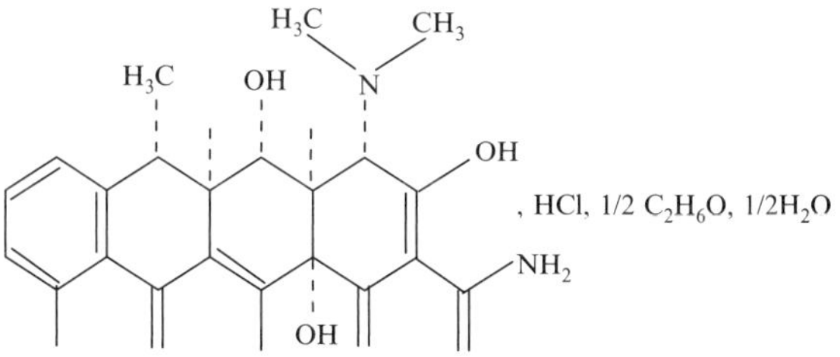

Figure 4.1 Molecular structure of doxycycline hydrochloride.

gonorrhea (Griffin *et al.*, 2011). These are defined as a subclass of polyketides containing an octahydrotetra-2-carboxamide skeleton. Tetracyclines remain the treatment of choice for various infections. Doxycycline is one of the most widely used tetracyclines because of its broad-spectrum antibiotic efficacy (Fig. 4.1; Mundargi *et al.*, 2007). Doxycycline, derived from the oxytetracycline, is used against a number of infections including anthrax, Lyme disease, cholera, syphilis, *Yersinia pestis*, malaria, filaria, and others (Griffin *et al.*, 2011). Doxycycline exerts its antibiotic effect primarily by halting protein synthesis (Hash *et al.*, 1964). During its course of action, doxycycline binds to the bacterial ribosome, thereby allosterically inhibiting the connection of aminoacyl tRNA with 30S ribosome subunits, leading to the inhibition of protein synthesis (Tritton, 1977). However, in spite of being a good antibiotic, it has a lot of drawbacks. It has several limitations, including instability in the biological environment and premature loss through rapid clearance and metabolism. Moreover, high concentrations of these agents may be toxic to healthy tissues, and resistance to tetracyclines is a worldwide problem. The increasing development of bacterial resistance to these traditional antibiotics has spurred a strong need to develop new antimicrobial agents (Kurek *et al.*, 2011). Attempts to overcome the above-mentioned problems associated with doxycycline have primarily focused on using slow release formulations of antibiotics (Yacoby *et al.*, 2007).

All the above problems associated with doxycycline have led to investigations of alternative strategies for delivering these antibiotics to the site of bacterial infections. In the past few decades, application of nanotechnology in the area of drug delivery (including antibiotics) has been explored widely. With the help of nanotechnology not only can the antibiotic be efficiently delivered to the infection site but also can the amount and frequency of dosage can be controlled thereby preventing toxicity related to conventional therapy (Kurek *et al.*, 2011). Nanotechnology plays a pivotal role especially for the treatment of disease like chronic bacterial infections that need a long duration of antibiotic therapy. In this regards, nanocarriers are being actively investigated for drug delivery to target site for various disease like intracellular bacterial and viral infections in the body (Singh *et al.*, 2008).

Due to the small size, these nanocarriers are able to localize the drug at the disease site for a longer period of time (Morones *et al.*, 2005). Moreover, nanocarriers deliver the drug in a sustained manner, thus avoiding the toxic systemic levels achieved during classical antibiotic therapy with the free drug (Ranjan *et al.*, 2009). The advent of bioadhesive drug delivery systems enhances drug bioavailability by prolonged residence at the site of adsorption owing to increased epithelial contact. Most of the bioadhesives are synthetic macromolecules available in the form of micro/nanoparticles that interact with the mucosal surface thus known as mucoadhesives. Nanotechnology provides various types of mucoadhesive nanoscale drug delivery systems such as liposomes, micelles, dendrimers, nanotubes, and nanoparticles (Fig. 4.2; Misra *et al.*, 2010). These nanocarriers are so engineered with the hope that they can act as likely candidates to carry therapeutic agents safely into the target site (Das *et al.*, 2009).

Among various nanocarriers, nanoparticles are found to be very useful for drug delivery. Nanoparticles are colloidal particles having a size of 1–1000 nm (Sahoo *et al.*, 2007). Therapeutic agents of interest are encapsulated within the polymeric matrix or adsorbed or conjugated onto the surface of the nanoparticles (Parveen and Sahoo, 2008). By using different polymers, nanoparticles and microparticles are formulated and are being intensively investigated as drug/gene delivery system as they protect the encapsulated drug from the degradation in endolysosomes besides being biodegradable and biocompatible (Parveen *et al.*, 2012). These nanoparticles are targeted to specific sites by conjugating targeting moiety on their surface that provide specific biochemical interactions with the receptors expressed on target cells (Das *et al.*, 2009). Recently, it has been reported that nanoparticles can rapidly escape from the endolysosomal compartment followed by intracellular uptake via an endocytic process. In this way, nanoparticles are able to

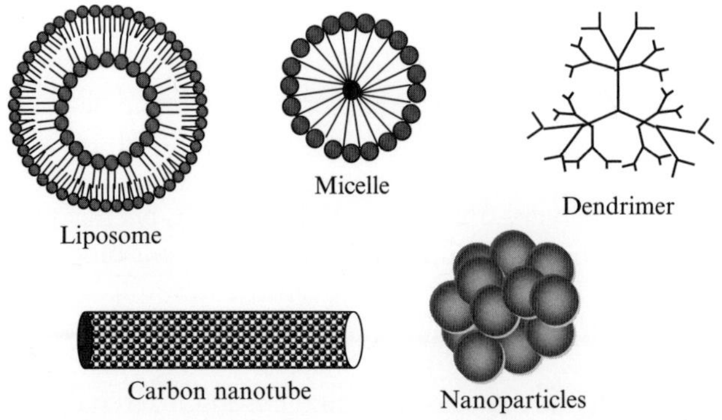

Figure 4.2 Schematic representation of different nanotechnology-based drug delivery vehicles. (For the color version of this figure, the reader is referred to the Web version of this chapter.)

protect the encapsulated drug from the degradative environment of the endolysosomes (Acharya and Sahoo, 2010).

Therapeutic agents encapsulated in nanoparticles have several advantages over the native drug not encapsulated in nanoformulations. Such as: (1) these nanoparticles are able to provide larger mass of drug to the target cells that helps in increasing the therapeutic efficacy of the drug, (2) it also helps in overcoming the drug resistance by inhibiting the cell's ability to pump the drug molecule back out, (3) these are targeted to the disease site by localized delivery agents that can be accomplished either by passive targeting or by active targeting, and (4) immunogenicity and other side effects of conventional method can be overcome by the help of these nanoparticles.

2. Polymeric Nanoparticles

To increase the therapeutic benefit and minimize the side effects of conventional drugs, various polymers are used in drug delivery research that can effectively deliver the drug to a target site. Biodegradable polymers are increasingly investigated as carriers in the design of drug delivery systems. These biodegradable polymers degrade into nontoxic small molecular weight compounds that are either metabolized or excreted from the body (Sahoo and Labhasetwar, 2003). The selection of polymers for the preparation of nanoparticles must meet some requirements like biocompatibility, drug compatibility, suitable degradation kinetics, and mechanical properties and ease of processing. Nanoparticles are fabricated from both synthetic and natural polymers that are biocompatible and biodegradable in nature (Fig. 4.3). Synthetic polymers include polyesters, polylactides, poly (ε-caprolactone) (PCL), poly(phosphoesters), poly (orthoesters), polyanhydrides, poly(d,l-lactide-co-glycolide) (PLGA), etc. Chitosan, dextran, gelatin, human serum albumin, etc., as natural polymers for the preparation of nanoparticles (Parveen and Sahoo, 2008). As compared to the natural polymers, the synthetic polymers have a sustained release of the therapeutic agent for a prolonged period of time.

3. Chitosan Nanoparticles

Chitosan is produced by deproteinization, demineralization, and deacetylation of chitinous material mostly found in arthropods, fungi, algae, or mollusca. It is a polysaccharide of β-1,4-linked 2-amino-2-deoxy-D-glucopyranose D-glucosamine; (Arya *et al.*, 2011). Chitosan has some ideal properties which makes it an attractive candidate for formulation of nanoparticles: it is biocompatible and does not cause allergic reactions or

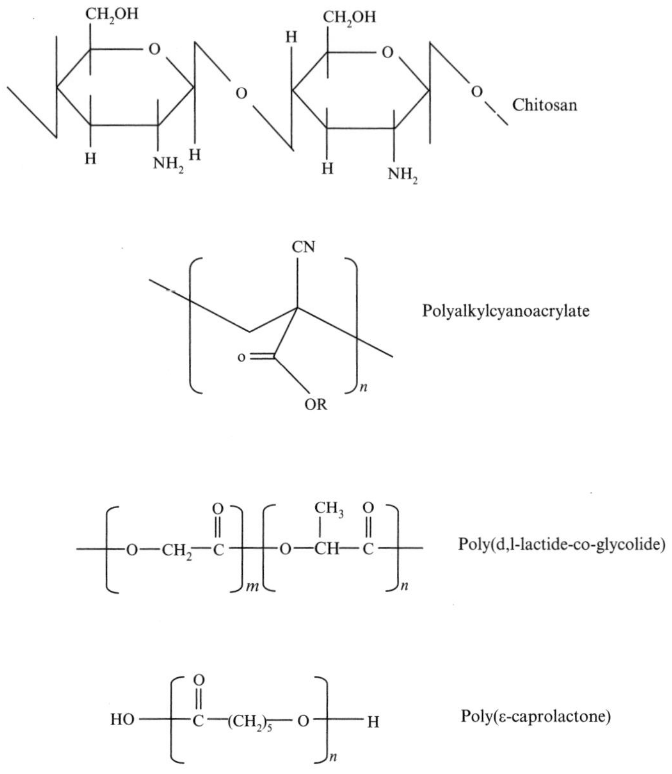

Figure 4.3 Representative structure of some of the synthetic and natural polymers used for the preparation of nanovehicles.

rejection; it is biodegradable, thus degrading slowly under the action of ferments *in vivo* to harmless products which are completely absorbed by the human body. In addition to this, chitosan is also known to possess antimicrobial properties and absorbs toxic metals such as mercury, cadmium, and lead. Another important feature of chitosan is that it has excellent mucoadhesive properties and also has the ability to penetrate large molecules across the mucosal surface (Parveen *et al.*, 2010). All these added features make chitosan an ideal polymer for the preparation of nanoparticles (Dai *et al.*, 2009; Yang *et al.*, 2005). Due to the antimicrobial and antifungal activity, chitosan has attracted considerable interest for the use against microbial growth (Rabea *et al.*, 2003). The antibacterial activity of chitosan is influenced by a number of factors including the species of bacteria, concentration, pH, solvent, and molecular mass (Tripathi *et al.*, 2011). The antimicrobial action of chitosan may be due to its efficient binding to the negatively charged bacterial cell wall with consequent destabilization of the cell envelope and altered permeability, followed by attachment to DNA with inhibition of its replication (Helander *et al.*, 2001). Chitosan exhibits higher antibacterial activity against Gram-positive bacteria than Gram-negative bacteria. Seo *et al.* have reported the antibacterial activity of chitosan against *Corynebacterium michiganense*, *Escherichia coli*, *Micrococcus luteus*, and *Staphylococcus aureus* (Seo *et al.*, 1992). Tsai and Su (1999) have found that *E. coli* cells were more sensitive to chitosan indicating the antimicrobial activity of the same. Moreover, at high temperature (25 and 37 °C) and acidic pH, chitosan shows its high antimicrobial activity (Tsai and Su, 1999). All these above properties of chitosan led to an intensive investigation on the preparation of chitosan nanoformulations for antibacterial therapy. The small size and quantum size effect of chitosan nanoparticles could make them unique also accelerating their effect. For gastric drug delivery, chitosan microspheres have also been used. Antimicrobial agents like amoxycillin and metronidazole have been released into the gastric cavity in a sustained manner with the use of reacetylated chitosan microspheres. Bharatwaj *et al.* have developed a pressurized metered-dose inhalers based platform polymeric nanocarriers consisting of a water soluble, hydrofluoroalkane-philic biodegradable copolymer of chitosan and poly(lactic acid), and a core of PLGA that is capable of efficiently delivering drugs to the lungs, and this formulation showed great potential of the formulation in the treatment of *Chlamydophila pneumoniae* related infections (Bharatwaj *et al.*, 2010). Another study done by Huang *et al.* reported the intranasal immunization of White Leghorn chickens using chitosan–DNA nanoparticles. They have demonstrated its efficacy in inducing specific immune responses and protection against *Campylobacter jejuni* infection (Huang *et al.*, 2010). Saboktakin *et al.* have examined the 5-aminosalicylic acid-loaded chitosan–carboxymethyl starch nanoparticles as drug delivery system to the colon. They have demonstrated that these nanoparticles developed based on the modulation of ratio which showed promise as a system for controlled delivery of drug to the colon (Saboktakin *et al.*, 2011).

4. Polyalkylcyanoacrylate Nanoparticles

The polyalkylcyanoacrylate (PACA) polymers are a family of biodegradable polymers, which have the ability to entrap a variety of biologically active compounds and thus have been extensively used in the preparation of nanoparticles for drug delivery. The length of the alkyl chain present in the polymer decides the rate of degradation of PACA by esterases (Andrieux and Couvreur, 2009). Moreover, the surface of these polymeric nanoparticles can easily be modified by coating of surfactants or by covalent linkage of poly(ethylene glycol) (PEG) chains in order to obtain "stealth" nanoparticles (Fontana *et al.*, 2001). To overcome the problems of administration of drugs like unstability in the gastrointestinal tract or inadequate absorption, PACA nanoparticles are widely investigated as oral drug carriers (Fontana *et al.*, 1998; Kreuter, 1991; Maincent *et al.*, 1986). For instance, a triple therapy is required for the treatment of *Helicobacter pylori* infection in peptic ulcer that includes antibiotic, antibacterial, and proton pump inhibitors. However, after *in vivo* administration no single antibiotic is effective in eradication of *H. pylori* due to poor permeability of the antibiotics across the mucus layer or due to the availability of subtherapeutic antibiotic concentrations at the infection site after oral administration in a conventional capsule or tablet dosage form (Talley *et al.*, 1999; Vakil and Cutler, 1999). To overcome these problems, the mucoadhesive PACA nanoparticles were developed to localize the drug at the infection site to achieve bactericidal concentrations (Patel and Amiji, 1996). Some of the commonly used antibiotic has a low lysosomal uptake leading to decreased efficacy in the treatment of many diseases caused by intracellular microorganisms. An interesting application of PACA nanoparticles is that these are used in order to develop endocytozable antimicrobial formulations for intracellular chemotherapy (Fontana *et al.*, 1998). Recently, Youssef *et al.* have reported that ampicillin-loaded polyisohexylcyano acrylate nanoparticles have a better therapeutic effect than free ampicillin in the treatment of infection caused by intracellular microorganisms such as *Listeria monocytogenes* (Youssef *et al.*, 1988).

5. Poly(d,l-lactide-co-glycolide) Nanoparticles

Poly(lactide-*co*-glycolides) (PLGA) are widely investigated biodegradable and biocompatible polymers and copolymers, which have been approved by food and drug administration (FDA) for human use. PLGA is synthesized by means of random ring opening copolymerization of two different monomers, the cyclic dimers (1,4-dioxane-2,5 dione) of glycolic and lactic acids and depending on the ratio of lactide to glycolide used for the polymerization, different forms of PLGA can be obtained. Alteration in the monomer ratio of PLGA influences the degradation time of the polymer. Moreover, various therapeutic agents can be encapsulated in PLGA nanoparticles. It is also soluble in a wide range of solvents thus it is widely used and preferred polymer for the preparation of nanoparticles. The drug entrapped inside the PLGA nanoparticles released at a sustained rate through diffusion followed by the degradation of the polymer matrix. In the field of nanotechnology, recent studies have considerably increased the research on drug delivery systems for the pharmaceutical industry. PLGA nanoparticles have shown several technological advantages such as biocompatibility, biodegradability, sustained release, and safety (Acharya and Sahoo, 2010).

These nanoparticles are specifically engineered to entrap different antibiotics inside the core of the matrix and at the same time it has shown different release kinetics (i.e., burst, intermediate, and slow; Sharma *et al.*, 2004). Malathi *et al.* have prepared a series of PLGA polymers with different molar feed ratios, that is, 90/10, 75/25, 50/50 by direct melt polycondensation method and encapsulated rifampicin in PLGA nanoparticles. They have reported that the rifampicin-loaded nanoparticles were remarkably advantageous in terms of high drug encapsulation efficiency, low polymer consumption, and better sustained release profile. Recently, the antibiotics ciprofloxacin and rifampicin and the antifungals fluconazole and voriconazole were incorporated in PLGA nanoformulations and their antibacterial and antifungal properties were extensively analyzed. Dalencon *et al.* have evaluated the efficacy of rifabutin-loaded PLGA nanocapsules against toxoplasmosis (Esmaeili *et al.*, 2007). In another study, Pandey *et al.* have evaluated the efficacy of three front-line antituberculosis drugs (rifampicin, isoniazid, and pyrazinamide) after encapsulating them in PLGA nanoparticles. They have reported that these nanoparticle-based antitubercular drugs form a sound basis that improved the drug bioavailability and reduced the dosing frequency for better management of pulmonary tuberculosis (Pandey *et al.*, 2003). Cheow *et al.* have prepared inhalable antibiotic encapsulating PLGA nanoparticles. They have examined the antibacterial efficacy and physical characteristics of different antibiotic-loaded polymeric nanoparticle formulations and have reported that ciprofloxacin-loaded PLGA nanoparticles was the most ideal formulation due to its high drug encapsulation efficiency and high antibacterial efficacy at a low dose against biofilm cells and biofilm-derived planktonic cells of *E. coli*. Moreover, the nanoparticulate suspension can be transformed into microscale dry-powder aerosols having aerodynamic characteristics ideal for inhaled delivery (Cheow *et al.*, 2010). Jeong *et al.* have prepared ciprofloxacin encapsulated nanoparticles using PLGA to treat urinary tract infection. They have demonstrated that ciprofloxacin encapsulated PLGA nanoparticles effectively inhibited the growth of bacteria due to the sustained release characteristics of nanoparticles, while free ciprofloxacin was less effective on the inhibition of bacterial growth (Jeong *et al.*, 2008). In another study, Gad *et al.* have formulated doxycycline hydrochloride and/or secnidazole-loaded *in situ* implants made up of biodegradable polymers [poly(lactide) (PLA) and poly(d,l-lactide-co-glycolide) (PLGA)]. They have compared the *in vitro* drug release and antimicrobial activity results with that of Atridox® (a commercial controlled release gel of doxycycline hyclate for the treatment of periodontal disease) and revealed that the pharmaceutical formulation based on PLGA/PLA containing secnidazole and doxycycline hydrochloride has promising activity in treating periodontitis in comparison with Atridox® (Gad *et al.*, 2008).

6. POLY(ε-CAPROLACTONE) NANOPARTICLES

poly(ε-caprolactone) is an important member of the aliphatic polyester family which is a semicrystalline, bioerodible, and biodegradable polymer. This is another class of FDA-approved polymer that can be used for the formulation of nanoparticles. The rate of degradation of homopolymer of PCL is slower than polyglycolic acid and polyglycolic acid-*co*-lactic acid, thus PCL is more suitable for formulation of long-term delivery systems. Moreover, PCL is compatible with a number of other polymers. It has been seen that when biodegradable PCL is used in combination with PLGA in different ratios we can vary the size, encapsulation efficiency and also the drug release of different formulations (Chang *et al.*, 1986). Studies have investigated the drug release behavior of PCL/PEG, PCL/poly(ethylene oxide) (PEO), and PCL/poly(butylenes succinate) microcapsules (Jin and Park, 2007). Park *et al.* have prepared erythromycin-loaded PCL and PCL/PEO microcapsules by emulsion solvent evaporation technique, and have evaluated the effects of PEO segments added to the PCL microcapsules on the degradation, size distribution, drug content in microcapsules, and drug release profiles. They have reported that the particle size of the microcapsules was decreased with increasing the emulsifier concentration and stirring rate. The surface free energy of the microcapsules increased as the PEO content increased, which is due to the higher hydrophilic features of PEO (Park *et al.*, 2005). Liu *et al.* have developed PCL electrospun fibers containing ampicillin sodium salt twisted into nanofiber yarns. They have tested the efficacy of these nanofiber yarns by doing zone inhibition study against both Gram-positive *S. aureus* and Gram-negative *Klebsiella pneumoniae* and have reported that the electrospun nanofibers yarns may have a great potential to be used as biomaterials, such as in surgical sutures, to decrease surgical site infection rate (Liu *et al.*, 2010).

In this chapter, we describe the improvement in encapsulation efficiency of doxycycline in PLGA:PCL blend formulation by varying different parameters during the preparation of nanoparticles and test the efficacy of these nanoformulations by doing several microbiological studies using DH5α strain of *E. coli*.

7. DOXYCYCLINE-LOADED PLGA:PCL NANOPARTICLES

7.1. Preparation of PLGA:PCL nanoparticles

Synthetic polymers such as PLGA and PCL have been intensively evaluated for the controlled release of pharmacologically active substances. However, studies have shown that the entrapment of hydrophilic drug substances inside the hydrophobic polymer capsules is a very difficult task. This may be due to the lower affinity of the hydrophilic drug for the hydrophobic polymer. In our laboratory, we have increased the encapsulation efficiency of a hydrophilic drug such as doxycycline hydrochloride in PLGA:PCL nanoparticles by varying different parameters. Doxycycline-loaded PLGA: PCL nanoparticles are prepared by the W/O/W double emulsion solvent evaporation method with few modifications (Sahoo *et al.*, 2002). In this method, doxycycline (Vetcare R&D Centre, Bangalore, India) equivalent to 20% (w/w) dry weight of polymer is dissolved in 300 μl of phosphate buffered saline (0.01 *M*, pH 7.4) to form doxycycline aqueous solution. The above prepared doxycycline aqueous solution is emulsified in an organic phase consisting of 90 mg of polymer (PLGA:PCL in different ratio) from Birmingham Polymers, Inc. (Birmingham, AL) dissolved in 3 ml of organic solvent (chloroform/dichloromethane) to form a primary water-in-oil emulsion. The emulsion is further emulsified in an aqueous PVA solution (12 ml, 2%, w/v) to form a water-in-oil-in-water emulsion. The emulsification is carried out using a microtip probe sonicator (VC 505, Vibracell Sonics, Newton, USA) set at 55 W of energy output for 2 min over an ice bath. The emulsion is stirred for 2 h on a magnetic stir plate at room temperature to allow the evaporation of organic solvent. Further, 1 h vacuum drying is also done to remove any residual organic solvent. Excess amount of PVA is removed by ultra centrifugation at 40,000 rpm, 4 °C for 20 min (Sorvall Ultraspeed Centrifuge, Kendro, USA) followed by washing with double distilled water for three times. The supernatant is collected and kept for estimation of doxycycline which is not encapsulated. The recovered nanoparticulate suspension is lyophilized for 2 days (−80 °C and <10 μm mercury pressure, LYPHLOCK 12, Labconco, Kansas City, MO) to get lyophilized powder for further use. The schematic representation of the preparation of doxycycline-loaded nanoparticles by the double emulsion solvent evaporation method is given in Fig. 4.4.

Different nanoformulations are prepared to achieve the high entrapment efficiency by varying different parameters such as: (i) the polymer ratio (PLGA:PCL) is varied from 100:0, 80:20, 70:30, and 60:40 while keeping other parameters constant (for further experiments only that polymer ratio will be chosen which exhibited lower size combined with higher encapsulation efficiency), (ii) the solvent is changed from chloroform ($CHCl_3$) to dichloromethane (DCM) with a constant polymer ratio (PLGA:PCL, 80:20, the ideal polymer ratio) without varying other parameters, (iii) pH of inner aqueous phase is changed from 7.4 to 4 and 4% NaCl was added to outer aqueous phase keeping other parameters like polymer ratio (PLGA: PCL, 80:20), solvent (DCM) and drug loading (20%, w/w) constant, and (iii) finally, the amount of drug loading was varied from 20%, 30%, and 60% (w/w) dry weight of polymer keeping polymer ratio (PLGA:PCL, 80:20), solvent (DCM), pH 4, and 4% NaCl concentration constant. The size

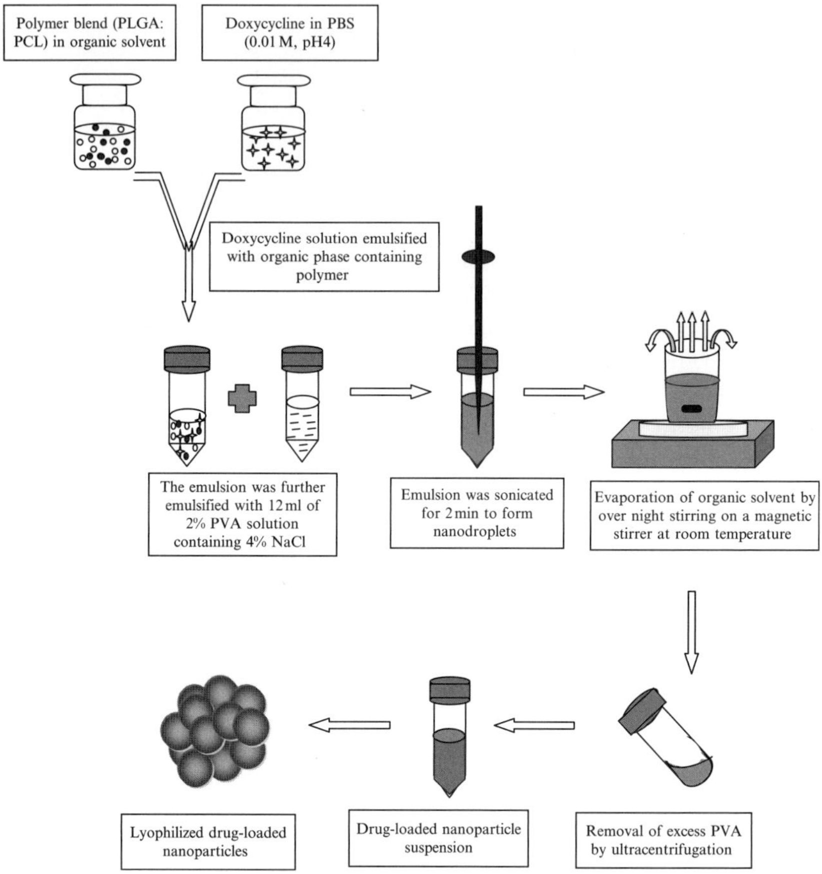

Figure 4.4 Schematic representation of the preparation of doxycycline-loaded PLGA: PCL nanoparticles (NPs) by the W/O/W solvent evaporation method. (For the color version of this figure, the reader is referred to the Web version of this chapter.)

distribution of the nanoparticles was determined in a dynamic light-scattering instrument (Nano-ZS) (Zeta sizer nano, Nano-ZS, ZEN3600, Malvern Instrument, UK). The mean particle size was found to range from 230 to 360 nm (Fig. 4.5).

Doxycycline content in nanoparticles is estimated by the reverse phase isocratic mode of high performance liquid chromatography using Agilent 1100 (Agilent Technologies, Waldbronn Analytical Division, Germany) which consists of a column (Zorbax Eclipse XDB-C18, 150 × 4.6 mm, i.d.) with internal standard of dimethylphthalate (Ruz *et al.*, 2004). Twenty microliters of sample is injected manually in the injection port and analyzed using a mobile phase of 5% acetic acid–acetonitrile–methanol (55:25:20)

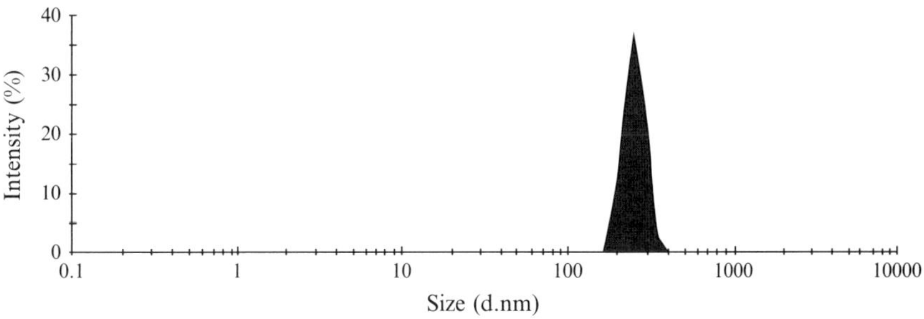

Figure 4.5 Representative figure showing the typical size distribution of nanoparticles measured by dynamic light scattering. To determine the particle size, ~1 mg/ml of nanoparticle solution was prepared in double distilled water. One hundred microliters of the sample was diluted to 1 ml, sonicated in an ice bath for 30 s and subjected to particle size measurement using the Zeta sizer (Nano-ZS) (Zeta sizer nano, Nano-ZS, ZEN3600, Malvern Instrument, UK).

(v/v/v). Separation is achieved by isocratic solvent elution at a flow rate of 0.5 ml/min with a quaternary pump (Model No. G1311A) at 10 °C with thermo start (Model No. G1316A). Doxycycline level is quantified by UV detection at 347 nm (with DAD, Model No. G1315A). The amount of doxycycline in the sample is determined from the peak area correlated with the standard curve. The standard curve of doxycycline is prepared under identical conditions. For measuring the drug concentration in the supernatant (as collected at the time of centrifugation mentioned during particle preparation), 20 μl of this supernatant is injected manually for analysis. The amount of drug present is calculated by using the standard plot of doxycycline prepared in HPLC. The amount of drug present in nanoparticles is calculated by subtracting the amount of drug in the supernatant from the total amount of the drug used in the formulation.

Different formulations of doxycycline-loaded nanoparticles are prepared by taking PLGA and PCL in different ratio (100:0, 80:20, 70:30, and 60:40) and then the size and encapsulation efficiency of all the formulations are compared. Results reveal that with increasing concentration of PCL in the formulation (80:20, 70:30, and 60:40), the size of the nanoparticles decreases along with a considerable decrease in encapsulation efficiency (Misra *et al.*, 2009; Table 4.1). The size of the particle decreases because of higher flexibility of PCL chains. Moreover, the carbonyl groups present in PCL may form hydrogen bonds with the PLGA chain, contributing to reducing the size of the particles. The low entrapment of doxycycline in the formulation (as we increase the concentration of PCL) may be due to high diffusion rate of doxycycline in PCL during solvent evaporation. Smaller size of particles may also be a reason for lower encapsulation efficiency of doxycycline in nanoparticles (Mundargi *et al.*, 2007). Thus

Table 4.1 Results of encapsulation efficiency and particle size of different formulations by varying different parameters

Formulation	Ratio of PLGA:PCL	Solvent	PVA% (w/v)	4% (w/v) NaCl	pH of inner aqueous phase	Drug loading (%, w/w)	Size (nm)[a]	Encapsulation efficiency (%)[b]
1	100:0	$CHCl_3$	2	−	7.4	20	360 ± 1.1	40
2	80:20	$CHCl_3$	2	−	7.4	20	290 ± 2.9	32
3	70:30	$CHCl_3$	2	−	7.4	20	256 ± 1.7	28
4	60:40	$CHCl_3$	2	−	7.4	20	237 ± 1.2	25
5	80:20	DCM	2	−	7.4	20	276 ± 2.7	47
6	80:20	DCM	2	+	4	20	285 ± 1.3	70
7	80:20	DCM	2	+	4	30	298 ± 1.0	60
8	80:20	DCM	2	+	4	60	310 ± 1.1	49

Reprinted with permission from Misra *et al.* (2009).
[a] Mean hydrodynamic diameter measured by dynamic light scattering.
[b] Encapsulation efficiency of doxycycline was measured by HPLC.

from above, the formulation with PLGA:PCL in 80:20 ratio having a smaller size and comparatively higher encapsulation efficiency was selected for further studies.

The polarity of the organic solvent used in the emulsion formation during the nanoparticle formulation might affect the entrapment efficiency. Therefore, nanoparticles with two different organic solvents $CHCl_3$ and DCM are formulated under identical conditions. With $CHCl_3$ as a solvent, the particles are round in shape but the encapsulation efficiency is slightly lower than with DCM (Misra *et al.*, 2009; Table 4.1). In the solvent $CHCl_3$, the encapsulation efficiency of PLGA:PCL (80:20) is found to be 32%, while when the solvent is changed to DCM, the encapsulation efficiency increases to 47%. The main reason for this difference is probably due to the less diffusion of $CHCl_3$ to the aqueous phase as compared to DCM. It has been reported that DCM is slightly more hydrophilic than $CHCl_3$ as DCM is soluble in ~50 parts of water while $CHCl_3$ is soluble in ~200 parts of water, and the polymer precipitates faster in DCM in comparison to $CHCl_3$ (Sahoo *et al.*, 2002). Thus, DCM is more appropriate solvent to encapsulate hydrophilic drugs in hydrophobic polymer (Bodmeier and McGinity, 1988; Nihant *et al.*, 1994).

The role of polymer ratio and choice of solvent helped to increase the encapsulation efficiency of doxycycline from 32% to 47% in a particular formulation (PLGA:PCL, 80:20 and organic solvent DCM; Misra *et al.*, 2009; Table 4.1). Thus, to improve the encapsulation efficiency of doxycycline in nanoparticles, three more additional parameters are taken into consideration. In our formulations (PLGA:PCL, 80:20 and organic solvent DCM) by changing the pH of inner aqueous phase from 7.4 to 4 and by adding 4% (w/v) NaCl to outer aqueous phase the encapsulation efficiency increased from 47% to 70% (Misra *et al.*, 2009; Table 4.1). The pH of the water phase affects the ionization of the drug substance and hence, the solubility. An ionic drug substance is prone to stay in the water phase, while the molecular form is more likely to be attached to the hydrophobic polymer phase thus, the drug substance is more efficiently encapsulated (Peltonen *et al.*, 2004). Based on this finding, by simply adjusting and controlling the pH value, the entrapment efficiency of doxycycline inside nanoparticles can be increased. By changing pH of inner aqueous phase from 7.4 to 4, the drug gets dissolved in this acidic aqueous solution and does not diffuse to the outer aqueous phase. As a result, the drug gets entrapped in the polymeric matrix leading to higher encapsulation efficiency in nanoparticles (Govender *et al.*, 1999; Niwa *et al.*, 1993). Addition of electrolyte affects the osmotic gradient between inner and outer aqueous phases; this may have an impact on drug entrapment. With the addition of salt, the concentration of the outer aqueous phase (PVA solution) increases and becomes hypertonic. So, the drug does not diffuse into the outer aqueous phase and remains in the polymeric matrix (Freytag *et al.*, 2000; Uchida *et al.*, 1997).

To further increase the encapsulation efficiency of doxycycline, varying concentrations of doxycycline are incorporated into nanoparticles. The percentage of drug loading is varied from 20% to 60% (w/w) corresponding to the amount of polymer dry weight (Misra *et al.*, 2009; Table 4.1). The encapsulation efficiency of the nanoparticles was found to be maximum with 20% doxycycline loading. With an increase in drug loading up to 60% the encapsulation efficiency of nanoparticles slightly decreases. This may be due to the saturation level of doxycycline inside the nanoparticles after 20% drug loading. With an increase in the amount of drug, a more porous polymeric matrix structure may be formed with large number of channels and hollow spaces, through which the drug could easily escape to the outer phase thereby decreasing the content of drug inside the polymeric matrix (Witschi and Doelker, 1998). Also, because of the increase in concentration of drug inside the polymer, a difference in osmotic pressure between the outer and inner aqueous phase results, which may cause the drug to escape from the inner aqueous phase (Lamprecht *et al.*, 2000). Thus by changing the pH and upon addition of electrolyte, the entrapment efficiency of the formulation consisting of 20% drug loading, PLGA:PCL in the ratio 80:20 with DCM as organic solvent can be enhanced to maximum value (70%).

7.2. Antibacterial activity studies

The present study includes preparation and characterization of doxycycline-loaded PLGA:PCL nanoparticles followed by confirmation of its antibacterial activity against doxycycline-sensitive bacteria such as *E. coli* (DH5α). The stability and sustained activity of the drug in nanoparticles was seen in the above bacterial culture and the results were compared with the native form of the drug.

The antibacterial activity of doxycycline is tested by the well diffusion method on a strain of *E. coil* (DH5α). Bacterial suspensions with a cell density equivalent to 0.5 McFarland (1.5×10^8 colony forming units/ml) are transferred individually on to the surface of Mueller-Hinton agar plates using sterile cotton swabs. Wells are prepared by punching a sterile cork borer on to the agar plates. Fifty microliters of doxycycline (10 μg/ml) in Mueller-Hinton broth (MHB) is added to these wells and incubated for overnight at 37 °C. Next day, the inhibition zones around the wells are seen by naked eye. Thus, DH5a strain is found to be susceptible to doxycycline because of a distinct inhibition zone formation around the wells.

7.3. Determination of minimum inhibitory concentration and minimum bactericidal concentration of doxycycline

Minimum inhibitory concentration (MIC) and minimum bactericidal concentration (MBC) are determined by the microdilution method (Wagenlehner *et al.*, 2006). Briefly, DH5α bacterial culture containing 0.5 McFarland

(1.5×10^8 colony forming units/ml) of organisms in Luria broth is added to various concentrations of doxycycline and doxycycline-loaded nanoparticles ranging from 0.1 to 8 μg/ml. The MIC concentration of doxycycline is defined as the lowest concentration inhibiting visible growth of bacteria after overnight incubation of above cultures at 37 °C. The MBC is measured by subculturing the broths used for MIC determination onto fresh agar plates. MBC is the lowest concentration of a drug that results in killing 99.9% of the bacteria being tested. The MIC or the lowest concentration inhibiting visible growth of bacteria is found to be 6 μg/ml in case of native doxycycline and 4 μg/ml in case of doxycycline-loaded nanoparticles in our study. MBC is found to be 8 μg/ml for native doxycycline and 6 μg/ml for doxycycline loaded nanoparticles for DH5α bacterial culture. This suggests that a higher concentration of drug is required to kill bacteria completely. However, it is noteworthy that even though 8 μg/ml concentrations of native doxycycline killed 99.9% of bacteria in our study, the microbes multiplied when transferred to fresh medium, indicating that even at a higher dose the antibiotics are not able to inhibit growth of bacteria completely. The native drug gradually loses its effect after 24 h, and bacteria which escaped drug action can multiply at a faster rate when given suitable condition. Therefore, a sustained release of a formulation is required which can control the growth of bacteria for a longer period of time. However, it is worth mentioning that MIC and MBC values are less in doxycycline-loaded nanoparticles than that of native doxycycline in our study. The reason may be the better penetration of smaller nanoparticles into the bacterial cells and better delivery of doxycycline to its site of action.

7.4. Comparison of the stability of native doxycycline and doxycycline in nanoparticles

The stability of doxycycline and doxycycline-loaded nanoparticles in the medium is studied for 10 days. To test this, 1 ml of MHB is taken per tube in two sets (10 tubes each). To one set, 10 μg of native doxycycline and to another set 10 μg of doxycycline loaded in nanoparticles are added. On day one, 20 μl of bacterial culture is added to first tubes of both sets containing doxycycline and doxycycline loaded in nanoparticles and incubated overnight at 37 °C in shaker. The next day, both the tubes are plated on Luria agar plates after a 10^4-fold serial dilution and incubated overnight at 37 °C to see bacterial growth in agar plates. This experiment is repeated for all the tubes up to 10 days and each day fresh culture medium is added. The medium is plated the next day to see bacterial growth. There is no bacterial growth in both the native and nanoparticle cases on first day. Third day results have shown growth in case of native doxycycline but no growth in case of doxycycline-loaded nanoparticles. On the fifth day, numerous colonies are found with native doxycycline whereas minimal growth is there in the case of doxycycline-loaded nanoparticles. On the tenth day, very few colonies are present in doxycycline–loaded nanoparticles, while profuse growth of bacteria colonies is seen with native doxycycline. This experiment confirms that native doxycycline is effective for 2 days only whereas nanoparticle formulation is effective up to 10 days. This shows that drug in native form loses its stability and hence its antibacterial property reduces after 2 days which is confirmed by a higher number of bacterial colonies. As nanoparticles are giving a control release of the encapsulated drug for a longer period of time, these are more effective than native doxycycline up to 10 days when used against the bacterial strain (Fig. 4.6). Similarly, to see the growth of bacteria visually, we have done similar type of experiment in another set and have taken the plate photos for different time points: for example, 1, 5, and 10 days and we found that these results corroborated with the above experiment results (Fig. 4.7).

7.5. Effect of native doxycycline and doxycycline-loaded nanoparticles on bacterial growth kinetics

To compare the effectiveness of native doxycycline and doxycycline-loaded nanoparticles on growth kinetics of bacteria, we have taken two different concentrations (0.1 and 0.2 μg/ml) of native doxycycline and doxycycline-loaded nanoparticles in different tubes. MHB and void nanoparticles are taken as control for this experiment. On the first day, 20 μl of DH5α bacterial culture containing 0.5 McFarland (1.5×10^8 colony forming units/ml) of organisms is added to all the tubes. The tubes are incubated at 37 °C in a shaker at 150 rpm (Wadegati Labequip, India) for 24 h.

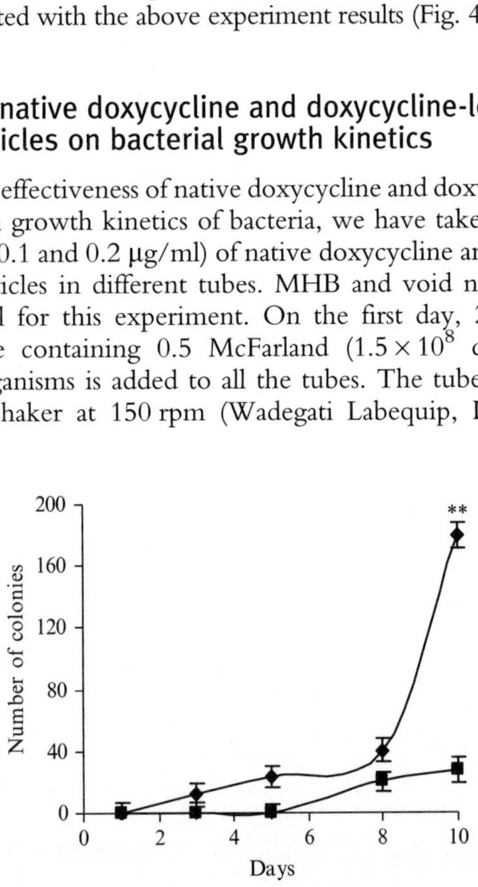

Figure 4.6 Comparison of stability of native doxycycline and doxycycline-loaded nanoparticles. Data as mean ± SEM ($n=3$). **$p<0.005$ Doxycycline-loaded NPs versus native doxycycline (—◆—, native doxycycline; —■—, doxycycline-loaded NP).

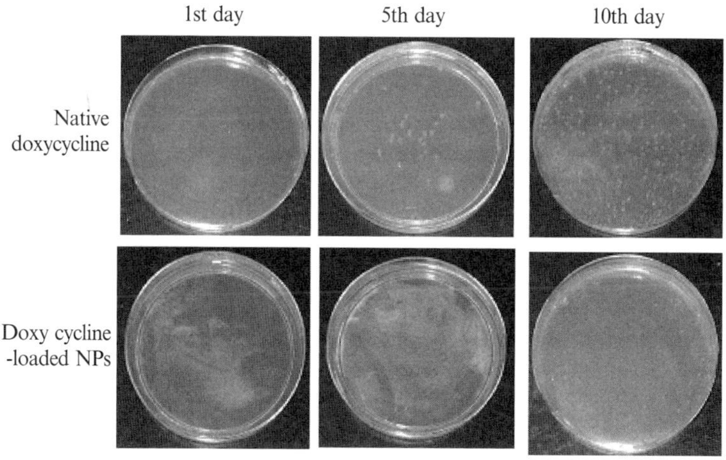

Figure 4.7 Visible growth of bacteria after treatment with 10 μg/ml of native doxycycline and doxycycline-loaded nanoparticles for 1, 5, and 10 days. (For the color version of this figure, the reader is referred to the Web version of this chapter.)

The next day, serial dilution followed by spread plating is done on Luria agar plates and incubated at 37 °C overnight. After 24 h, the colonies are counted by the visual method. On second day, the tubes are again plated to see the bacterial growth kinetics. In this similar way, the experiment is continued for 5 days. Growth graphs are plotted by taking number of colonies versus number of days for comparing the action of native doxycycline with doxycycline-loaded nanoparticles on growth of *E. coli*. To further confirm the fact that a sustained release formulation of doxycycline has more profound bactericidal effect than native doxycycline on growth kinetics of *E. coli*, viable colony count method is followed. Figure 4.8A shows that on first day native doxycycline of 0.1 μg/ml concentration is more effective than the same concentration of drug in the nanoparticulate formulation. On the second day, in case of native doxycycline the colony number increases by 2.4-fold whereas in case of doxycycline-loaded nanoparticles, a decrease in colony number (by 1.9-fold) is seen to the first day plate count results. A similar plate count of third day plated sample show a steep increase in number of bacterial colonies with native doxycycline while there is a decrease in number of colonies in case of nanoparticles. On fifth day, a decrease in number of colonies is observed in all the tubes including both the controls. This may be due to completion of bacterial life cycle. A difference in viable bacterial number was observed because native doxycycline of 0.1 μg concentration was able to kill maximum number of bacteria in the medium on first day. However, doxycycline-loaded

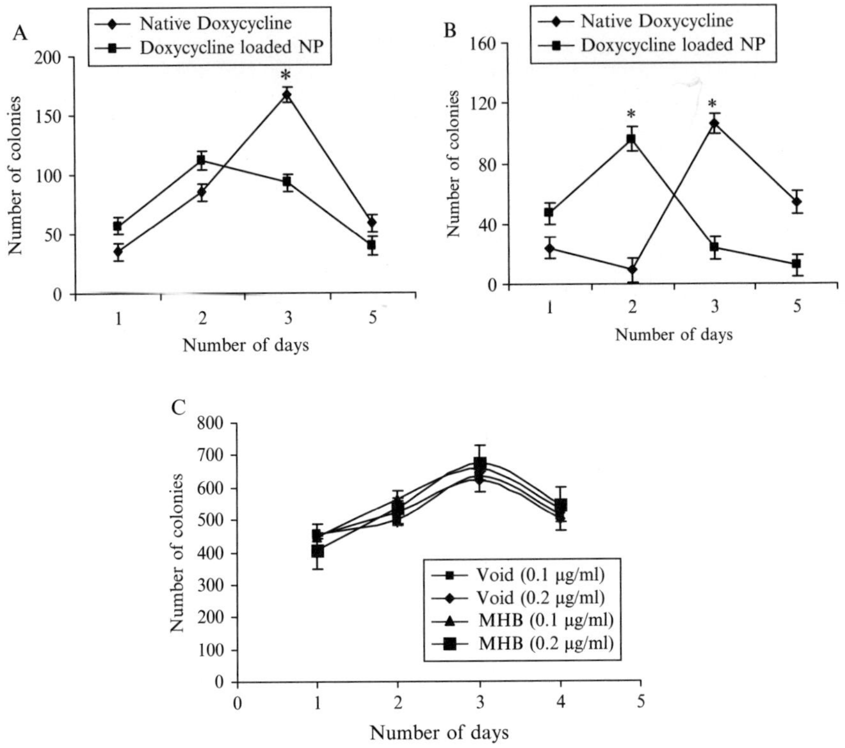

Figure 4.8 Comparative effect of native doxycycline and doxycycline-loaded NPs in bacterial culture. Data as mean ± SEM ($n = 3$). $^{*}p < 0.05$ Doxycycline-loaded NPs versus native doxycycline. (Reprinted with permission from Misra *et al.*, 2009.) (A) Colony comparison of native doxycycline with nanoparticles (0.1 μg/ml) (——◆——, native doxycycline; ——■——, doxycycline-loaded NP). (B) Colony comparison of native doxycycline with nanoparticles (0.2 μg/ml) (——◆——, native doxycycline; ——■——, doxycycline-loaded NP). (C) Colony comparison of void nanoparticles (0.1 and 0.2 μg/ml) and only MHB (0.1 and 0.2 μg/ml) (——◆——, void (0.1 μg/ml); ——■——, void (0.2 μg/ml); ——▲——, MHB (0.1 μg/ml); ——●——, MHB (0.2 μg/ml)).

nanoparticles is not able to release the entire encapsulated drug on the first day, as a result more number of colonies are present on agar plates. Following sustain release pattern, the drug which gets released from the nanoparticle (<0.1 μg) has effect but not as much as the native drug. By the second day, the native drug starts losing its activity, which is indicated by an increase in number of bacteria but the nanoparticles are able to control the multiplication of the bacterial strain as the amount of drug which gets released from it is enough to prevent further growth of bacteria. On the third day, the native drug completely loses its antibacterial activity but the nanoparticle efficiently prevents the bacteria from further multiplication.

Another experiment using a higher concentration of doxycycline and doxycycline-loaded nanoparticles (0.2 mg/ml) has been done that shows similar results, Fig. 4.8B. The noncytotoxic nature of void nanoparticles serving as control is confirmed in Fig. 4.8C. Observation reveals that the colony numbers are approximately same in both the controls (only MHB and void nanoparticles). Thus, the antibacterial studies conducted on *E. coli* proves that activity of doxycycline-loaded nanoparticle are much more than the native doxycycline. Drug in native form may become less effective after sometime however nanoparticles remain effective for a longer period of time following sustain release phenomena.

8. Conclusion

The method and studies illustrated in this chapter indicate that the encapsulation of a hydrophilic drug such as doxycycline hydrochloride inside a hydrophobic polymer can be increased up to 70% by changing various formulation parameters during preparation of the nanoparticles. The success of antibacterial activity of doxycycline-loaded nanoparticles was confirmed on DH5α strain of *E. coli*, and it suggested that doxycycline in nanoformulations is more effective as compared to its native form. It is also investigated that doxycycline-loaded nanoparticles were stable for 10 days, while native doxycycline loses its stability and its antibacterial property after 2 days; hence doxycycline-loaded nanoparticles were more efficient in controlling the growth kinetics of bacteria following sustain release phenomena. Thus, doxycycline-loaded nanoparticles can be used efficiently for antibacterial therapy, possibly against a large number of bacterial species.

REFERENCES

Acharya, S., and Sahoo, S. K. (2010). PLGA nanoparticles containing various anticancer agents and tumour delivery by EPR effect. *Adv. Drug Deliv. Rev.* **63,** 170–183.

Andrieux, K., and Couvreur, P. (2009). Polyalkylcyanoacrylate nanoparticles for delivery of drugs across the blood-brain barrier. *Wiley Interdiscip. Rev. Nanomed. Nanobiotechnol.* **1,** 463–474.

Arya, G., Vandana, M., Acharya, S., and Sahoo, S. K. (2011). Enhanced antiproliferative activity of Herceptin (HER2)-conjugated gemcitabine-loaded chitosan nanoparticle in pancreatic cancer therapy. *Nanomedicine* **7,** 859–870.

Bharatwaj, B., Wu, L., Whittum-Hudson, J. A., and da Rocha, S. R. (2010). The potential for the noninvasive delivery of polymeric nanocarriers using propellant-based inhalers in the treatment of Chlamydial respiratory infections. *Biomaterials* **31,** 7376–7385.

Bodmeier, R., and McGinity, J. W. (1988). Polylactic acid microspheres containing quinidine base and quinidine sulphate prepared by the solvent evaporation method. III. Morphology of the microspheres during dissolution studies. *J. Microencapsul.* **5,** 325–330.

Casadevall, A., and Pirofski, L. A. (1999). Host-pathogen interactions: Redefining the basic concepts of virulence and pathogenicity. *Infect. Immun.* **67,** 3703–3713.

Casadevall, A., and Pirofski, L. (2001). Host-pathogen interactions: The attributes of virulence. *J. Infect. Dis.* **184,** 337–344.

Casadevall, A., and Pirofski, L. A. (2002). What is a pathogen? *Ann. Med.* **34,** 2–4.

Chang, R. K., Price, J. C., and Whitworth, C. W. (1986). Dissolution characteristics of poly caprolactone-polylactide microspheres of chlorpromazine. *Drug Dev. Ind. Pharm.* **12,** 2355–2380.

Chaturvedi, A. K., Lazzell, A. L., Saville, S. P., Wormley, F. L., Jr., Monteagudo, C., and Lopez-Ribot, J. L. (2011). Validation of the tetracycline regulatable gene expression system for the study of the pathogenesis of infectious disease. *PLoS One* **6,** e20449.

Cheow, W. S., Chang, M. W., and Hadinoto, K. (2010). Antibacterial efficacy of inhalable antibiotic-encapsulated biodegradable polymeric nanoparticles against E. coli biofilm cells. *J. Biomed. Nanotechnol.* **6,** 391–403.

Dai, T., Tegos, G. P., Burkatovskaya, M., Castano, A. P., and Hamblin, M. R. (2009). Chitosan acetate bandage as a topical antimicrobial dressing for infected burns. *Antimicrob. Agents Chemother.* **53,** 393–400.

Das, M., Mohanty, C., and Sahoo, S. K. (2009). Ligand-based targeted therapy for cancer tissue. *Expert Opin. Drug Deliv.* **6,** 285–304.

Esmaeili, F., Hosseini-Nasr, M., Rad-Malekshahi, M., Samadi, N., Atyabi, F., and Dinarvand, R. (2007). Preparation and antibacterial activity evaluation of rifampicin-loaded poly lactide-co-glycolide nanoparticles. *Nanomedicine* **3,** 161–167.

Fontana, G., Pitarresi, G., Tomarchio, V., Carlisi, B., and San Biagio, P. L. (1998). Preparation, characterization and in vitro antimicrobial activity of ampicillin-loaded polyethylcyanoacrylate nanoparticles. *Biomaterials* **19,** 1009–1017.

Fontana, G., Licciardi, M., Mansueto, S., Schillaci, D., and Giammona, G. (2001). Amoxicillin-loaded polyethylcyanoacrylate nanoparticles: Influence of PEG coating on the particle size, drug release rate and phagocytic uptake. *Biomaterials* **22,** 2857–2865.

Freytag, T., Dashevsky, A., Tillman, L., Hardee, G. E., and Bodmeier, R. (2000). Improvement of the encapsulation efficiency of oligonucleotide-containing biodegradable microspheres. *J. Control. Release* **69,** 197–207.

Gad, H. A., El-Nabarawi, M. A., and Abd El-Hady, S. S. (2008). Formulation and evaluation of PLA and PLGA in situ implants containing secnidazole and/or doxycycline for treatment of periodontitis. *AAPS Pharm. Sci. Tech.* **9,** 878–884.

Govender, T., Stolnik, S., Garnett, M. C., Illum, L., and Davis, S. S. (1999). PLGA nanoparticles prepared by nanoprecipitation: Drug loading and release studies of a water soluble drug. *J. Control. Release* **57,** 171–185.

Griffin, M. O., Ceballos, G., and Villarreal, F. J. (2011). Tetracycline compounds with non-antimicrobial organ protective properties: Possible mechanisms of action. *Pharmacol. Res.* **63,** 102–107.

Hash, J. H., Wishnick, M., and Miller, P. A. (1964). On the mode of action of the tetracycline antibiotics in Staphylococcus aureus. *J. Biol. Chem.* **239,** 2070–2078.

Helander, I. M., Nurmiaho-Lassila, E. L., Ahvenainen, R., Rhoades, J., and Roller, S. (2001). Chitosan disrupts the barrier properties of the outer membrane of gram-negative bacteria. *Int. J. Food Microbiol.* **71,** 235–244.

Huang, J. L., Yin, Y. X., Pan, Z. M., Zhang, G., Zhu, A. P., Liu, X. F., and Jiao, X. A. (2010). Intranasal immunization with chitosan/pCAGGS-flaA nanoparticles inhibits Campylobacter jejuni in a White Leghorn model. *J. Biomed. Biotechnol.* doi:10.1155/2010/589476.20.

Jeong, Y. I., Na, H. S., Seo, D. H., Kim, D. G., Lee, H. C., Jang, M. K., Na, S. K., Roh, S. H., Kim, S. I., and Nah, J. W. (2008). Ciprofloxacin-encapsulated poly(DL-lactide-co-glycolide) nanoparticles and its antibacterial activity. *Int. J. Pharm.* **352,** 317–323.

Jin, F. L., and Park, S. J. (2007). Preparation and characterization of biodegradable antibiotic-containing poly(ε-caprolactone) microcapsules. *J. Ind. Eng. Chem.* **13,** 608–613.

Kreuter, J. (1991). Liposomes and nanoparticles as vehicles for antibiotics. *Infection* **19** (Suppl. 4), S224–S228.

Kurek, A., Grudniak, A. M., Kraczkiewicz-Dowjat, A., and Wolska, K. I. (2011). New antibacterial therapeutics and strategies. *Pol. J. Microbiol.* **60,** 3–12.

Lamprecht, A., Ubrich, N., Hombreiro Perez, M., Lehr, C., Hoffman, M., and Maincent, P. (2000). Influences of process parameters on nanoparticle preparation performed by a double emulsion pressure homogenization technique. *Int. J. Pharm.* **196,** 177–182.

Liu, H., Leonas, K. K., and Zhao, Y. (2010). Antimicrobial properties and release profile of ampicillin from electrospun poly(ε-caprolactone) nanofiber yarns. *J. Eng. Fiber. Fabr.* **5,** 10–19.

Maincent, P., Le Verge, R., Sado, P., Couvreur, P., and Devissaguet, J. P. (1986). Disposition kinetics and oral bioavailability of vincamine-loaded polyalkyl cyanoacrylate nanoparticles. *J. Pharm. Sci.* **75,** 955–958.

Martin, K. W., and Ernst, E. (2003). Herbal medicines for treatment of bacterial infections: A review of controlled clinical trials. *J. Antimicrob. Chemother.* **51,** 241–246.

Misra, R., Acharya, S., Dilnawaz, F., and Sahoo, S. K. (2009). Sustained antibacterial activity of doxycycline-loaded poly(D, L-lactide-co-glycolide) and poly(epsilon-caprolactone) nanoparticles. *Nanomedicine (Lond.)* **4,** 519–530.

Misra, R., Acharya, S., and Sahoo, S. K. (2010). Cancer nanotechnology: Application of nanotechnology in cancer therapy. *Drug Discov. Today* **15,** 842–850.

Morones, J. R., Elechiguerra, J. L., Camacho, A., Holt, K., Kouri, J. B., Ramirez, J. T., and Yacaman, M. J. (2005). The bactericidal effect of silver nanoparticles. *Nanotechnology* **16,** 2346–2353.

Mundargi, R. C., Srirangarajan, S., Agnihotri, S. A., Patil, S. A., Ravindra, S., Setty, S. B., and Aminabhavi, T. M. (2007). Development and evaluation of novel biodegradable microspheres based on poly(d, l-lactide-co-glycolide) and poly(epsilon-caprolactone) for controlled delivery of doxycycline in the treatment of human periodontal pocket: In vitro and in vivo studies. *J. Control. Release* **119,** 59–68.

Murray, R. G., Steed, P., and Elson, H. E. (1965). The location of the mucopeptide in sections of the cell wall of Escherichia coli and other Gram-negative bacteria. *Can. J. Microbiol.* **11,** 547–560.

Nihant, N., Schugens, C., Grandfils, C., Jerome, R., and Teyssie, P. (1994). Polylactide microparticles prepared by double emulsion/evaporation technique. I. Effect of primary emulsion stability. *Pharm. Res.* **11,** 1479–1484.

Niwa, T., Takeuchi, H., Hino, T., Kunou, N., and Kawashima, Y. (1993). Preparations of biodegradable nanospheres of water soluble and insoluble drugs with D, L-lactide/glycolide copolymer by a novel spontaneous emulsification solvent diffusion method, and the drug release behavior. *J. Control. Release* **25,** 89–98.

Pandey, R., Sharma, A., Zahoor, A., Sharma, S., Khuller, G. K., and Prasad, B. (2003). Poly (DL-lactide-co-glycolide) nanoparticle-based inhalable sustained drug delivery system for experimental tuberculosis. *J. Antimicrob. Chemother.* **52,** 981–986.

Park, S. J., Kim, K. S., and Kim, S. H. (2005). Effect of poly(ethylene oxide) on the release behaviors of poly(caprolactone) microcapsules containing erythromycin. *Colloids Surf. B Biointerfaces* **43,** 238–244.

Parveen, S., and Sahoo, S. K. (2008). Polymeric nanoparticles for cancer therapy. *J. Drug Target.* **16,** 108–123.

Parveen, S., Mitra, M., Krishnakumar, S., and Sahoo, S. K. (2010). Enhanced antiproliferative activity of carboplatin-loaded chitosan-alginate nanoparticles in a retinoblastoma cell line. *Acta Biomater.* **6,** 3120–3131.

Parveen, S., Misra, R., and Sahoo, S. K. (2012). Nanoparticles: A boon to drug delivery, therapeutics, diagnostics and imaging. *Nanomedicine* **8,** 147–166.

Patel, V. R., and Amiji, M. M. (1996). Preparation and characterization of freeze-dried chitosan-poly(ethylene oxide) hydrogels for site-specific antibiotic delivery in the stomach. *Pharm. Res.* **13,** 588–593.

Peltonen, L., Aitta, J., Hyvonen, S., Karjalainen, M., and Hirvonen, J. (2004). Improved entrapment efficiency of hydrophilic drug substance during nanoprecipitation of poly(l) lactide nanoparticles. *AAPS Pharm. Sci. Tech.* **5,** E16.

Rabea, E. I., Badawy, M. E., Stevens, C. V., Smagghe, G., and Steurbaut, W. (2003). Chitosan as antimicrobial agent: Applications and mode of action. *Biomacromolecules* **4,** 1457–1465.

Ranjan, A., Pothayee, N., Seleem, M. N., Tyler, R. D., Jr., Brenseke, B., Sriranganathan, N., Riffle, J. S., and Kasimanickam, R. (2009). Antibacterial efficacy of core-shell nanostructures encapsulating gentamicin against an in vivo intracellular Salmonella model. *Int. J. Nanomedicine* **4,** 289–297.

Ruz, N., Zabala, M., Kramer, M. G., Campanero, M. A., Dios-Vieitez, M. C., and Blanco-Prieto, M. J. (2004). Rapid and simple determination of doxycycline in serum by high-performance liquid chromatography. Application to particulate drug delivery systems. *J. Chromatogr. A* **1031,** 295–301.

Saboktakin, M. R., Tabatabaie, R. M., Maharramov, A., and Ramazanov, M. A. (2011). Synthesis and in vitro evaluation of carboxymethyl starch-chitosan nanoparticles as drug delivery system to the colon. *Int. J. Biol. Macromol.* **48,** 381–385.

Sahoo, S. K., and Labhasetwar, V. (2003). Nanotech approaches to drug delivery and imaging. *Drug Discov. Today* **8,** 1112–1120.

Sahoo, S. K., Panyam, J., Prabha, S., and Labhasetwar, V. (2002). Residual polyvinyl alcohol associated with poly (D, L-lactide-co-glycolide) nanoparticles affects their physical properties and cellular uptake. *J. Control. Release* **82,** 105–114.

Sahoo, S. K., Parveen, S., and Panda, J. J. (2007). The present and future of nanotechnology in human health care. *Nanomedicine* **3,** 20–31.

Seo, H., Mitsuhashi, K., and Tanibe, H. (1992). Antibacterial and antifungal fiber blended by chitosan. *In* "Advances in chitin and chitosan," (C. J. Brine, P. Sandford, and J. P. Zikakis, eds.). Elsevier Applied Science, New York.

Sharma, A., Sharma, S., and Khuller, G. K. (2004). Lectin-functionalized poly (lactide-co-glycolide) nanoparticles as oral/aerosolized antitubercular drug carriers for treatment of tuberculosis. *J. Antimicrob. Chemother.* **54,** 761–766.

Shockman, G. D., and Barrett, J. F. (1983). Structure, function, and assembly of cell walls of gram-positive bacteria. *Annu. Rev. Microbiol.* **37,** 501–527.

Singh, M., Singh, S., Prasad, S., and Gambhir, I. S. (2008). Nanotechnology in medicine and antibacterial effect of silver nanoparticles. *Dig. J. Nanomater. Biostruct.* **3,** 115–122.

Talley, N. J., Janssens, J., Lauritsen, K., Racz, I., and Bolling-Sternevald, E. (1999). Eradication of Helicobacter pylori in functional dyspepsia: Randomised double blind placebo controlled trial with 12 months' follow up. The Optimal Regimen Cures Helicobacter Induced Dyspepsia (ORCHID) Study Group. *BMJ* **318,** 833–837.

Tripathi, S., Mehrotra, G. K., and Dutta, P. K. (2011). Chitosan–silver oxide nanocomposite film: Preparation and antimicrobial activity. *Bull. Mater. Sci.* **34,** 29–35.

Tritton, T. R. (1977). Ribosome-tetracycline interactions. *Biochemistry* **16,** 4133–4138.

Tsai, G. J., and Su, W. H. (1999). Antibacterial activity of shrimp chitosan against Escherichia coli. *J. Food Prot.* **62,** 239–243.

Uchida, T., Nagareya, N., Sakakibara, S., Konishi, Y., Nakai, A., Nishikata, M., Matsuyama, K., and Yoshida, K. (1997). Preparation and characterization of polylactic acid microspheres containing bovine insulin by a w/o/w emulsion solvent evaporation method. *Chem. Pharm. Bull. (Tokyo)* **45,** 1539–1543.

Vakil, N., and Cutler, A. (1999). Ten-day triple therapy with ranitidine bismuth citrate, amoxicillin, and clarithromycin in eradicating *Helicobacter pylori*. *Am. J. Gastroenterol.* **94,** 1197–1199.

Wagenlehner, F. M., Kinzig-Schippers, M., Sorgel, F., Weidner, W., and Naber, K. G. (2006). Concentrations in plasma, urinary excretion and bactericidal activity of levofloxacin (500 mg) versus ciprofloxacin (500 mg) in healthy volunteers receiving a single oral dose. *Int. J. Antimicrob. Agents* **28,** 551–559.

Waksman, S. A. (1947). What is an antibiotic or an antibiotic substance? *Mycologia* **39,** 565–569.

Witschi, C., and Doelker, E. (1998). Influence of the microencapsulation method and peptide loading on poly(lactic acid) and poly(lactic-co-glycolic acid) degradation during in vitro testing. *J. Control. Release* **51,** 327–341.

Yacoby, I., Bar, H., and Benhar, I. (2007). Targeted drug-carrying bacteriophages as antibacterial nanomedicines. *Antimicrob. Agents Chemother.* **51,** 2156–2163.

Yang, T. C., Chou, C. C., and Li, C. F. (2005). Antibacterial activity of N-alkylated disaccharide chitosan derivatives. *Int. J. Food Microbiol.* **97,** 237–245.

Young, D., Hussell, T., and Dougan, G. (2002). Chronic bacterial infections: Living with unwanted guests. *Nat. Immunol.* **3,** 1026–1032.

Youssef, M., Fattal, E., Alonso, M. J., Roblot-Treupel, L., Sauzieres, J., Tancrede, C., Omnes, A., Couvreur, P., and Andremont, A. (1988). Effectiveness of nanoparticle-bound ampicillin in the treatment of *Listeria monocytogenes* infection in athymic nude mice. *Antimicrob. Agents Chemother.* **32,** 1204–1207.

CHAPTER FIVE

Antimicrobial Properties of Electrically Formed Elastomeric Polyurethane–Copper Oxide Nanocomposites for Medical and Dental Applications

Z. Ahmad,* M. A. Vargas-Reus,† R. Bakhshi,‡ F. Ryan,‡ G. G. Ren,§ F. Oktar,¶ *and* R. P. Allaker†

Contents

1. Introduction 88
2. Materials and Methods 90
   2.1. Materials 90
   2.2. Methods 90
3. Results and Discussion 91
   3.1. Antimicrobial fiber and film preparation 91
   3.2. Structural analysis 94
   3.3. Antimicrobial testing using MRSA 94
4. Concluding Remarks and Future Work 97
Acknowledgments 98
References 98

Abstract

With the rapidly advancing field of nanotechnology having an impact in several areas interfacing life and physical sciences, the potential applications of nanoparticles as antimicrobial agents have been realized and offer great opportunities in addressing several viral and bacterial outbreak issues. Polyurethanes (PUs) are a diverse class of polymeric materials which also have applications in several areas of biomedical science ranging from blood contact devices to

* School of Pharmacy and Biomedical Sciences, University of Portsmouth, Portsmouth, United Kingdom
† Queen Mary University of London, Barts and The London School of Medicine and Dentistry, Institute of Dentistry, London, United Kingdom
‡ Department of Mechanical Engineering, University College London, London, United Kingdom
§ School of Engineering and Technology, University of Hertfordshire, Hatfield, United Kingdom
¶ Nanotechnology and Biomaterials Application & Research Centre, Marmara University, Istanbul, Turkey

Methods in Enzymology, Volume 509
ISSN 0076-6879, DOI: 10.1016/B978-0-12-391858-1.00005-8

implantable dental technologies. In this report, copper oxide (CuO) nanoparticles (mean size ~50 nm) are embedded into a PU matrix via two electrical fabrication processes. To elucidate the antimicrobial activity, a range of different loading compositions of CuO within the PU matrix (0%, 1%, 5%, and 10% w/w) are electrospun to form thin porous films (thickness < 10 μm). After washing, the films are tested for their antimicrobial properties against methicillin-resistant *Staphylococcus aureus* (MRSA). Significant reduction of populations was demonstrated with 10% w/w CuO over a 4-h period. This approach demonstrates the potential of generating tailored antimicrobial structures for a host of applications, such as designer filters, patterned coatings, breathable fabrics, adhesive films (as opposed to sutures), and mechanically supporting structures.

1. Introduction

Nanoscaled particles have diverse applications ranging from medical device coatings to drug delivery carriers (De Jong and Borm, 2008; Dickinson *et al.*, 2011). More recently, and in a timely manner, their applications as antimicrobial agents have also been realized especially with metal and metal oxide particle systems (Borkow *et al.*, 2010; Ren *et al.*, 2009). The antimicrobial properties of copper (Cu) and its oxide (CuO) have been known for centuries (on the macroscale) and with current advances in technology; selected textile and material composite applications have demonstrated their potential as antimicrobial agents on the micro- and nanometer scales (Ruparelia *et al.*, 2008; Zhang *et al.*, 2006). Other types of metallic nanoparticles demonstrating such properties include gold and silver (Perni *et al.*, 2009; Ruparelia *et al.*, 2008); however, cost remains an important factor when considering the scale-up potential and the broad range of potential applications.

Polyurethanes (PUs) have several applications as biomaterials including coatings, blood bags, catheters, heart valves, dental fillers, protective clothing, and even tissue-engineering constructs (Bertoldi *et al.*, 2010; Luo *et al.*, 2010; Sui *et al.*, 2010). The mechanical, degradation, and biostability properties of these materials can be controlled by carefully selecting the various segments which constitute toward the polymeric backbone (Zdrahala and Zdrahala, 1999). They have also found several applications as antimicrobial materials and in many instances utilize an active agent, which is capable of killing bacteria or viruses (Ghosh, 2005). Numerous devices serving as filters, adhesive films, fillers, eluting structures, and coatings require a polymeric matrix to be impregnated with such components or secondary materials which possess the functional characteristics, for example, antimicrobial properties. Depending on the intended function, the polymeric matrix can contribute toward a mechanism for the release of active agents, the mechanical properties of the device, and the localization of the active agent at a specific point. In such cases, the loading volume of the active agent (e.g., possessing antimicrobial properties) must be sufficient to demonstrate the desired effect. In addition to this, polymers with elastomeric properties have potential uses in applications requiring considerable material or device flexibility and mechanical interaction. These properties are found in PUs with high molecular weights ($>$50 kD).

Various antimicrobial applications require a simple thin-film coating of polymer–active agent, which is sufficient in delivering the desired functionality; however, the ability to control the deposition rate of films (and subsequently, coating thickness) provides a method to estimate the optimal thickness for such structures. There is also a need to control the porosity of films, which is highly desirable in specific biomaterial applications, for example, in the case of biosensors or in the development of breathable fabrics, that is, filters for masks. Also, it is well established that rough and patterned surfaces provide an enhanced topography for cell adhesion (Hoffman-Kim *et al.*, 2010), which can be achieved using advanced material fabrication and processing techniques. All of these properties can be achieved using a fibrous coating methodology. Hence, utilizing a fiber-forming process or fabricating method will combine all these benefits, which can be readily coupled to intrinsic (e.g., antimicrobial) properties of the selected materials.

There are numerous ways to generate fibrous structures, which are rapidly finding increasing applications in several areas of biomedical science and engineering. One such method is electrospinning (Ahmad *et al.*, 2009; Pham *et al.*, 2006), which has been used to prepare biomaterial technologies in drug delivery, tissue engineering, and implantable devices (Almodovar and Kipper, 2011). The process has several advantages, one of which is the ability to utilize coarse processing needles to generate a high volume of ultrafine spun fibers. As the process takes place at ambient temperature, there is no heat-induced change to the chemical structure of the active agent or polymer, and the fibrous structures can be deposited directly onto a device or ready-to-deploy substrate. The method requires a flowing medium, which is perfused into an electrically conducting needle. A controllable electric field is generated, and at the optimized voltage window, fibers can be spun ranging from a few micrometers down to the nanometer scale (Xie *et al.*, 2010). Alternatively, patterns comprising the same material can be deposited in an ordered fashion using a direct write process, which also makes use of the electric field to generate a liquid writing-tip at the exit of the nozzle (Ahmad *et al.*, 2010). In both cases, the key controlling parameters are the applied voltage and flow rate, both of which have a direct impact on the size and structure of individual fiber morphologies. The deposition time is also a controlling parameter, but reflects the dimensions of the bulk structures, that is, film thickness and porosity.

This method combines biocompatible PU polymer and CuO nanoparticles to prepare a series of active composite solution blends, although a whole host of other nanoparticles can be selected and used in a similar fashion. Selected solutions are spun and result in dry fibrous mats. Here, PU–CuO composite mats (porous and elastomeric) are assessed for their antibacterial properties, using the epidemic methicillin-resistant strain of *Staphylococcus aureus* (EMRSA) 16. The activated porous elastomeric mats have a host of potential applications, which can reduce, control, or prevent various problems associated with microbial outbreaks such as MRSA. The same materials can also be plotted into patterned structures that are identical to those currently being used in several branches of biomedical science.

2. Materials and Methods

2.1. Materials

PU (poly [4,4′-methylenebis (phenyl-isocyanate)-*alt*-1,4-butanediol/poly-tetrahydrofuran]) elastomer is purchased from Sigma-Aldrich Company (Poole, UK). CuO nanoparticles are supplied by Intrinsiq Materials (Ren *et al.*, 2009); with a mean size of ~50 nm. The solvents dimethyl formamide (DMF, 99%) and tetrahydrofuran (THF, 99%) are purchased from Sigma-Aldrich Company. Ethanol (99%) and microscopic glass slides are purchased from VWR (Poole, UK). For the studies reported here, the MRSA epidemic strain is kindly provided by the Department of Infectious Disease Epidemiology, Imperial College.

2.2. Methods

2.2.1. Solution preparation

An initial stock polymer solution is prepared by suspending PU (10% w/w) in a cosolvent solution comprised of DMF and THF (30:70, respectively). The stock solution is sealed and allowed to stir mechanically at ambient temperature for 5 h. From this, four separate solutions are prepared by dissolving known quantities of CuO (0%, 1%, 5%, and 10% w/w) into individual containers (sealed), which are allowed to stir mechanically for 30 min and then utilized in the electrospinning process.

2.2.2. Fiber, film, and pattern generation

For fiber and film generation, individual solutions are used directly after being mechanically stirred. Solutions are infused into the processing needle via silicon tubing. Individual solutions are then introduced into a conducting needle (inner diameter of 330 μm) using a precision Harvard pump. An electrical field is generated using a Glassman high-voltage power supply. After preliminary testing of variables (applied voltage 0–30 kV, flow rate 10–30 μl/min, and collecting distance 20–100 mm), the correct processing parameters are determined. Samples are collected on microscopic glass slides directly below the processing orifice and the collection distance and deposition time are recorded. Patterned structures are generated using a printing method developed earlier (Ahmad *et al.*, 2009). Samples to be assessed for surface analysis and antimicrobial testing are washed with ethanol and then water. They are then allowed to stand alone to dry (24 h). The PU polymer used in this study is stable to degradation from ethanol and water.

2.2.3. Microscopy and elemental analysis (EDX)

CuO nanoparticles are characterized using transmission electron microscopy (JEOL 100-CX microscope) to determine the particle size. Electrospun fibers and directly deposited (writing) composite tracks are initially screened using a Nikon Eclipse optical microscope (ME600). Selected samples are then analyzed using a JEOL JSM-6301F scanning electron microscope (SEM) and are also characterized using an inbuilt elemental analysis (EDX) method. For this, samples are coated using a carbon and gold coating-sputtering device. Microscopy is carried out at an accelerating voltage of 5 kV. Other selected samples are also analyzed using an atomic force microscope (AFM) to observe the surface of the various fibers impregnated with different CuO concentrations.

2.2.4. Bacterial testing

EMRSA 16 is grown overnight in Tryptone Soya Broth (TSB, Oxoid). Optical density of the culture is adjusted to 0.1 ($\lambda = 540$ nm) with phosphate buffered saline (PBS). For this particular strain, it is found that this provides an approximate bacterial concentration of 10^8 colony-forming units per ml (cfu/ml). Each test slide is placed in a 50-ml Corning tube and filled with 45 ml of PBS + 550 μl of the OD adjusted bacterial culture. These are then placed into a shaker incubator at 200-rpm (37 °C). Twenty microliters of each sample is taken at 0, 1, 2, 3, and 4 h, and serial dilutions are made with PBS. The dilutions are plated on Tryptone Soya Agar (TSA, Oxoid) and incubated overnight at 37 °C for 24 h. After this period, colonies are counted. The experiment is performed in triplicate with a different set of slides on each occasion.

3. Results and Discussion

3.1. Antimicrobial fiber and film preparation

The electrospinning process is a versatile method, which can generate porous and dense films depending on the processing conditions. As it is a time-dependent deposition process, the thickness of films or membranes can also be controlled. In the process deployed to generate elastomeric films, the electric field is capable of spinning fibers on the micrometer scale from relatively coarse processing needles, in this instance 330 μm. The process, as shown in Fig. 5.1A, proceeds from conventional dripping (low voltage) to jet formation (progressive higher voltages), and at the optimal voltage, fibers are spun resulting from the jet apex (Kim and Dunn, 2010). In such processes, the solvent is lost rapidly due to various bending and whipping motions (Kim and Dunn, 2010). The CuO particles in the composite solutions (Pu–CuO) have a mean size of ~50 nm (Fig. 5.1B) which can

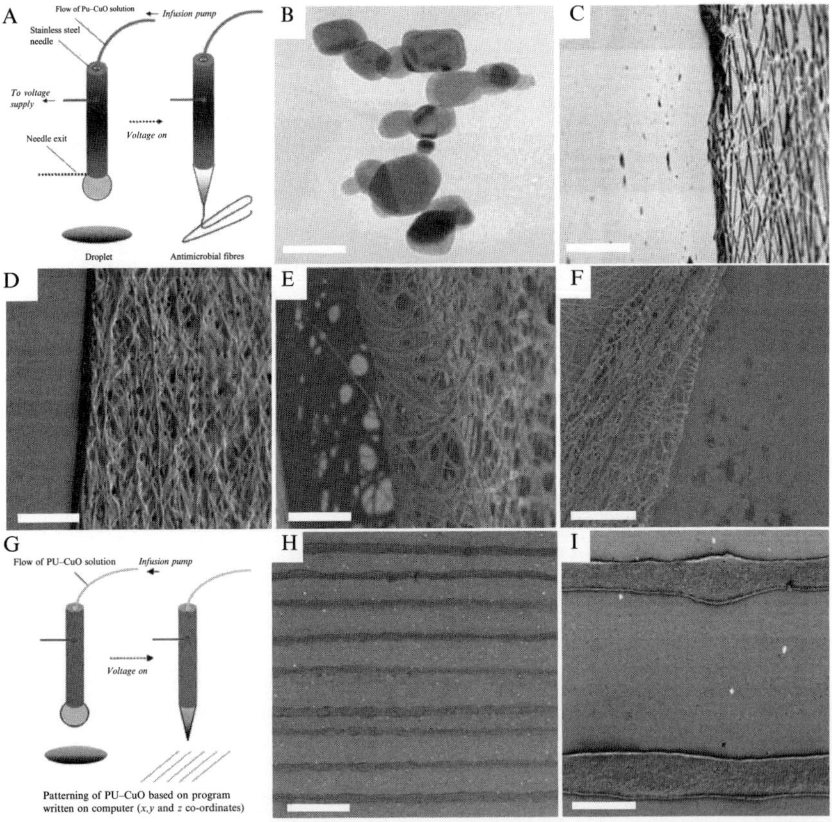

Figure 5.1 Fiber and thin-film (CuO) preparation. (A) Process method for spinning, (B) transmission electron micrograph of CuO nanoparticles alone, (C) micrograph of highly porous thin fibrous film on glass slide, (D) micrograph of dense film on glass slide, (E) micrograph of elastomeric fibrous coating on glass slide edge—complete cover, (F) micrograph of mechanically peeled elastomeric film, (G) process method for writing and an array or antimicrobial tracks at (H) low magnification, and (I) high magnification. (Scale bars: (B) = 50 nm, (C) = 20 μm, (D) = 20 μm, (E) = 10 μm, (F) = 20 μm, (H) = 200 μm, and (I) = 30 μm.). (For the color version of this figure, the reader is referred to the Web version of this chapter.)

be easily accommodated in micrometer-scaled fibers. The optimal electrospinning conditions are set at 20-μl/min (flow rate of media being introduced into the processing needle) and the applied voltage is variable between 10 and 13 kV. The deposition distance is another important factor as it permits greater time for any remaining residual solvent to evaporate. For example, collecting the spun fibers (CuO 10% w/w, in PU) under these optimal conditions and a working deposition distance of 150 mm (for 30 s) results in fiber morphologies that are well defined (Fig. 5.1C). This morphology is capable of providing antimicrobial properties for biomaterials requiring "breathable" and "interaction" characteristics, where gaseous exchange, moisture flow, and rough surface topography are made possible by fibers and interconnected fibers resulting in a porous meshwork. Generally at reduced deposition distances, for example, 20 mm, the effects of residual solvent become apparent and there is trivial merging of fibers at adjoining points (Fig. 5.1D). Here, the structures maintain their fibrous morphologies albeit not as distinctly as those afforded with the greater deposition distance. The antimicrobial fiber size obtained under these generating parameters is in the range of 1–3 μm and the various micrographs also suggest that the fibers have a smooth surface. Setting the sample collecting processing parameters (collecting distance at 150 mm, deposition time of 60 s), followed by washing with ethanol (to sterilize and remove any impurities or microscopic solvent residue), a microscopic glass slide (Fig. 5.1E), or any other medical device, can be coated on the surface, around edges and curves. Using these conditions, the deposited fibers provide intrinsic properties to the active base material (CuO). The polymeric material is elastomeric (Fig. 5.1F) which transpires into spun films that have been removed mechanically (peeled). The elastic deformation of these films is clearly evident once detached from the glass surface. These films can therefore potentially serve as filters, masks, and fillers. They may also have potential as controlled release or eluting devices, whereby antimicrobial agents are secreted (by diffusion or degradation) over a designated time period based on polymeric properties.

The electrical writing method (Fig. 5.1G) can generate ordered structures and can be used to deposit any pattern or parametric image onto a surface. It offers greater control on deposition and spatial arrangement. Hence, designer filters, grids, topographies, coatings, and antimicrobial spacers can be deposited at specific locations on or to fit inside an object. To demonstrate this, the 10% w/w (CuO) polymer solution was used for fabrication and for this a reduced flow rate of 15 μl/min was deployed at an applied voltage range of 10–13 kV. While the composite structures are ordered (Fig. 5.1H), the distance in between each thread can be varied by simplistic data entry into the uploading program for the plotter head. This can be further multifaceted by patterned structures using arches, angles, and also in three-dimensional (3D) formats, which would enable the movement of the plotter over a 3D object. A closer inspection of these structures reveals that these patterns are ~30 μm in width.

3.2. Structural analysis

Conventional electron micrographs (SEM) show that the fibrous polymer surface is smooth but the location of CuO in the PU fibers, after processing and washing, is not clear. Using various analysis techniques in addition to imaging indicates how well the CuO particles are dispersed in the polymeric matrix. Figure 5.2A1–A4 and B1–B4 show contrast-based images that indicate the location of CuO particles. At lower magnifications these appear as white dots scattered throughout fibers and there is an increase in these dots as the CuO loading is increased. High-magnification contrast imaging also reveals that the increase in the CuO concentration leads to appreciable particle agglomeration within the fibers, which appear as clusters rather than specs. The trend in loading volume is also supported by elemental analysis (Fig. 5.2C1–C4). However, from these graphs, it is clear that the presence of Cu is relatively low when compared to various other elements contributing toward the overall composite composition. Elemental mapping (Fig. 5.2D1–D4) confirms the presence of CuO-agglomerated clusters, some of which can also be found in the 1% w/w fibers, although at a reduced frequency. Figure 5.2D4 shows an inset of how the distribution of CuO particles appear at the fiber edge and on the glass surface interface, which confirms CuO particle entrapment in the PU matrix. Figure 5.2E1–E4 shows AFM images of all types of fibers prepared. The surfaces appear smooth and there is little difference between the fibers prepared from varying compositions. There is little evidence to suggest any fracture. The morphology at overlapping points demonstrate an increase in height, also supporting how the fibers were formed and collected at an increased deposition distance (150 mm) during the electrospinning process.

3.3. Antimicrobial testing using MRSA

Figure 5.3 shows the antimicrobial properties of the various coated slides against EMRSA 16. Although bacterial survival with the control slides (polymer free) was not 100%, the reduction was minimal over the 4-h test period. There was a clear relationship between the CuO concentration used in the matrix and the survival rates of MRSA; the highest concentration (10% w/w) reduced survival rates to below 50% after 1 h and close to 10% after 2 h. The 1% CuO concentration coatings reached this level of survival (~10%) after 4 h. It was also found that the PU fibers without any CuO demonstrated antimicrobial properties, although, after 4 h the survival rates were only down to ~35% which is appreciably higher than with the 1% CuO (~10%)-containing structures. A clear reduction in bacterial survival

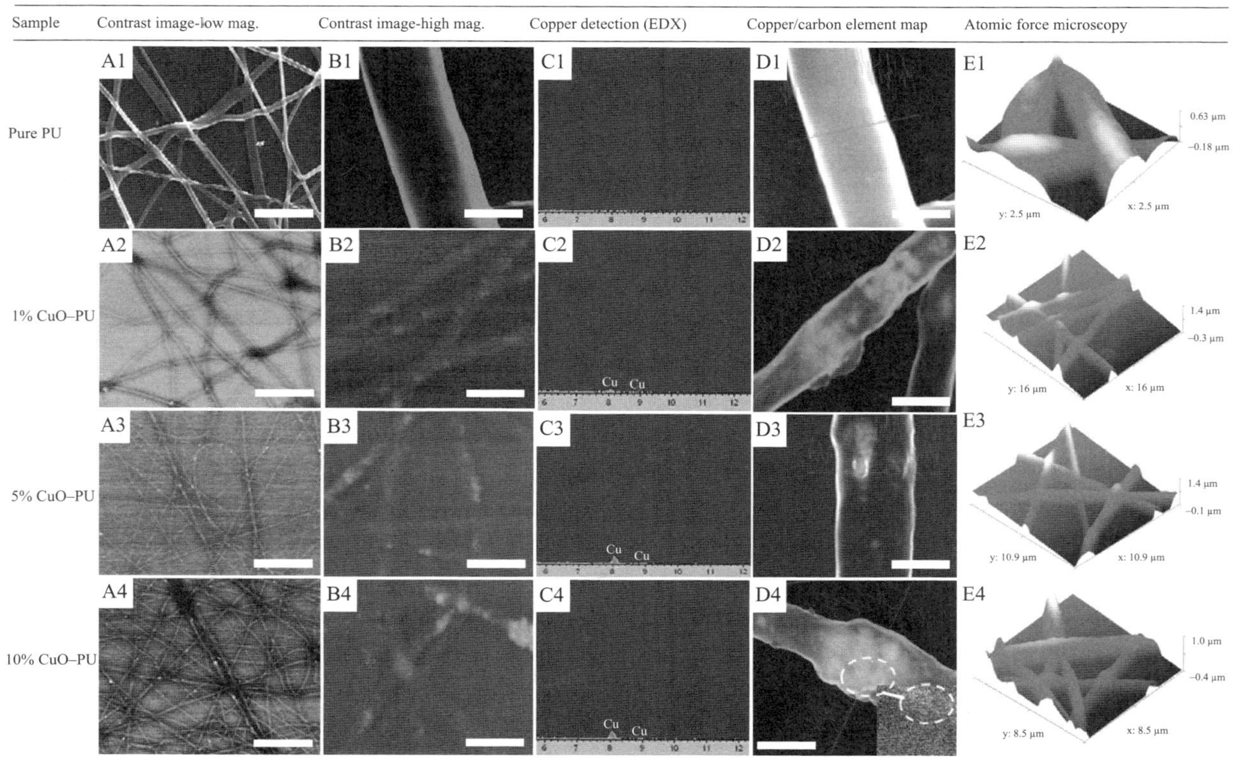

Figure 5.2 Fiber analysis on various electrospun compositions. Showing (A1–A4) low magnification contrast imaging—electron microscopy, (B1–B4) high magnification contrast imaging—electron microscopy, (C1–C4) element analysis, (D1–D4) element mapping, and (E1–E4) atomic force microscopy. (Scale bars: A1 = 10 μm, B1 = 1 μm, D1 = 1 μm, A2 = 10 μm, B2 = 5 μm, D2 = 1 μm, A3 = 10 μm, B3 = 5 μm, D3 = 1 μm, A4 = 10 μm, B4 = 5 μm, D4 = 1 μm.) (See Color Insert.)

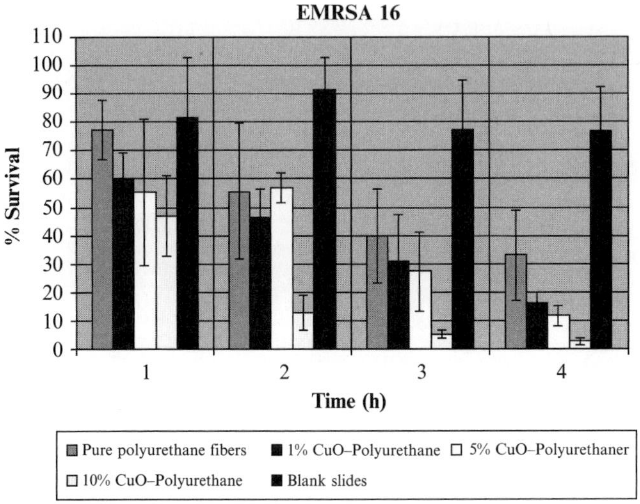

Figure 5.3 Survival of EMRSA 16 with respect to time when exposed to microscope slides coated with polyurethane blended with CuO nanoparticles at different concentrations (0%, 1%, 5%, and 10%). The survival of the bacterium in PBS (microscope slide free) was considered to be 100%. Standard deviation bars are shown ($n=3$). (For the color version of this figure, the reader is referred to the Web version of this chapter.)

with time was observed and a dose–response was noted. ANOVA analysis did not show any significant differences in the antimicrobial potential between PU alone and when CuO nanoparticles were incorporated at 1% and 5%. However, when this percentage was increased to 10%, significant differences ($p < 0.05$) were found with respect to PU alone in every case, and also when compared to the 1% and 5% slides at 2, 3, and 4 h. With 10% CuO slides, at the 2-h time point the bacterial survival had decreased to 10%, virtually the entire bacterial population was killed by 4 h. The antimicrobial properties shown by the PU polymer alone need to be investigated further as such characteristics have been reported with other polymeric systems (Kawahara *et al.*, 2009). These polymers are segmented and various constituents can be altered to observe any changes to their antimicrobial properties. As they are used extensively in biomaterial applications in medicine, this would provide excellent opportunities in developing stand-alone antimicrobial polymer devices. Although the fibrous film coatings are washed with ethanol and water to eliminate any remaining residual solvent, the bulk of which is evaporated during the spinning process, this may still be a contributing factor toward the antimicrobial effects observed, and also needs to be investigated further.

The antimicrobial activity is thought to be contact-based inhibition, as the polymer is stable from hydrolytic degradation and this would be potentially

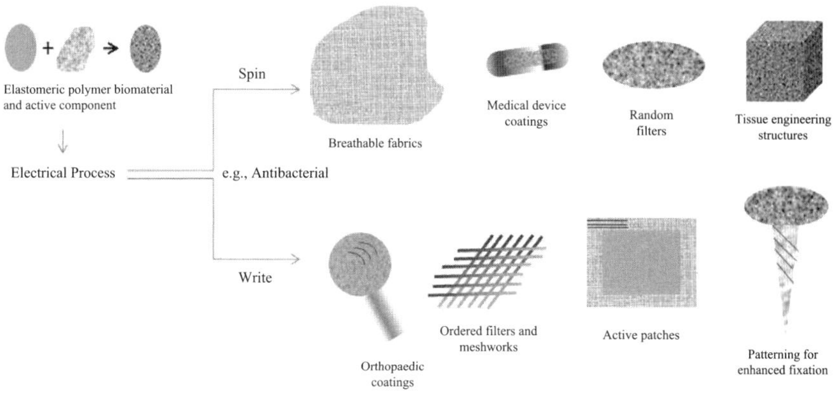

Figure 5.4 Potential applications of randomly and directly deposited elastomeric structures which combine a polymeric matrix and an active agent (i.e., CuO nanoparticles) system. Structures are formed using variations in electric field deposition. (For color version of this figure, the reader is referred to the Web version of this chapter.)

useful in material contact devices (fillers, coatings, filters, masks, etc.) with the ability to also directly spray onto surfaces. If the mechanism is release based, then this also suggests a sustained release profile over 4 h and applications in eluting devices (timely purifiers) are possible. For both of these mechanisms, altering the active agent or the polymer-degradation properties can further highlight the potential. There is also the prospect to encapsulate (Ahmad *et al.*, 2008, 2009) structures (core shell) with such materials using a two-tier inhibition mechanism that includes release and contact. These potential applications are highlighted in Fig. 5.4, where some of these areas are currently under detailed investigation for their potential application.

4. Concluding Remarks and Future Work

At this stage, there is evidence to suggest that the content of CuO particles in the PU fibrous matrix has an effect on the level of MRSA inhibition. With the ability to control the pore size and film thickness, the direct deposition onto devices, the type of active particle in use, the type of polymer in use, there is huge potential of such materials and agents in several biomedical applications as afore mentioned. With these preliminary findings in mind we hope to focus our attention on the mechanism of action, the properties of electrospun fibers, and their ability in inhibiting pathogens such as MRSA. We will also aim to develop a series of novel antimicrobial structures and develop a greater understanding into their inhibitory action, which in the current climate poses a huge burden on the healthcare remit.

ACKNOWLEDGMENTS

The authors would like to thank the Archaeology Department at University College London (UCL) for the use of their electron microscope facilities. They would also like to thank Kings College London (KCL) for the use of their atomic force microscope (AFM). The Royal Society is also gratefully acknowledged for their help in providing equipment. Finally they would like to thank The Department of Infectious Disease Epidemiology (DIDE), Imperial College, who provided the MRSA epidemic strain.

REFERENCES

Ahmad, Z., Zhang, H. R., Farook, U., Edirisinghe, M., Stride, E., and Columbo, P. (2008). Generation of multi-layered structures for biomedical applications using a novel tri-needle co-axial device and electrohydrodynamic flow. *J. R. Soc. Interface* **5,** 1255–1261.

Ahmad, Z., Nangrejo, M., Edirisinghe, M., Stride, E., Columbo, P., and Zhang, H. (2009). Engineering a material for biomedical applications with electric field assisted processing. *Appl. Phys. A Mater. Sci. Process* **97,** 31–37.

Ahmad, Z., Rasekh, M., and Edirisinghe, M. (2010). Electrohydrodynamic direct writing of biomedical polymers and composites. *Macromol. Mater. Eng.* **295,** 315–319.

Almodovar, J. And, and Kipper, M. J. (2011). Coating electrospun chitosan nanofibers with polyelectrolyte multilayers using the polysaccharides heparin and N, N, N-trimethyl chitosan. *Macromol. Biosci.* **11,** 72–76.

Bertoldi, S., Farè, S., Denegri, M., Rossi, M., Haugen, H. J., Parolini, O. And, and Tanzi, M. C. (2010). Ability of polyurethane foams to support placenta-derived cell adhesion and osteogenic differentiation: Preliminary results. *J. Mater. Sci. Mater. Med.* **21,** 1005–1011.

Borkow, G., Zhou, S. S., Page, T., and Gabbay, J. (2010). A novel anti-influenza copper oxide containing respiratory face mask. *PLoS One* **5,** e11295.

De Jong, W. H., and Borm, P. J. (2008). Drug delivery and nanoparticles: Applications and hazards. *Int. J. Nanomedicine* **3,** 133–149.

Dickinson, L. E., Kusuma, S., and Gerecht, S. (2011). Reconstructing the differentiation niche of embryonic stem cells using biomaterials. *Macromol. Biosci.* **11,** 36–49.

Ghosh, S. (2005). Antibacterial effects of silver doped polyethyleneglycol based polyamidoamine side chain dendritic polyurethane. *J. Macromol. Sci. Pure Appl. Chem.* **42,** 765–770.

Hoffman-Kim, D., Mitchel, J. A., and Bellamkonda, R. V. (2010). Topography, cell response and nerve regeneration. *Annu. Rev. Biomed. Eng.* **12,** 203–231.

Kawahara, T., Takeuchi, Y., Wei, G., Shirai, K., Yamauchi, T., and Tsubokawa, N. (2009). Preparation of antibacterial polymer-grafted silica nanoparticle and surface properties of composites filled with the silica. *Polym. J.* **41,** 744–751.

Kim, O. V., and Dunn, P. F. (2010). Controlled production of droplets by in-flight electrospinning. *Langmuir* **26,** 15807–15813.

Luo, J., Deng, Y., and Sun, Y. (2010). Antimicrobial activity and biocompatibility of polyurethane-iodine complexes. *J. Bioact. Compat. Polym.* **25,** 185–206.

Perni, S., Piccirillo, C., Pratten, J., Prokopovich, P., Chrzanowski, W., Parkin, I. P., and Wilson, M. (2009). The antimicrobial properties of light-activated polymers containing methylene blue and gold nanoparticles. *Biomaterials* **30,** 89–93.

Pham, Q. P., Sharma, U., and Mikos, A. G. (2006). Electrospinning of polymeric nanofibers for tissue engineering applications: A review. *Tissue Eng.* **5,** 1197–1211.

Ren, G., Hu, D., Cheng, E., Vargas-Reus, M. A., Reip, P., and Allaker, R. P. (2009). Characterisation of copper oxide nanoparticles for antimicrobial applications. *Int. J. Antimicrob. Agents* **33,** 587–590.

Ruparelia, J. P., Chatterjee, A. K., Duttagupta, S. P., and Mukerji, S. (2008). Strain specificity in antimicrobial activity of silver and copper nanoparticles. *Acta Biomater.* **4,** 707–716.

Sui, R., Han, J., Zhou, J., Hu, S., Zhou, X., and Feng, Z. (2010). In vitro evaluation of biodegradable cardiac tissue engineering polyurethane scaffold. *J. Clin. Rehab. Tiss. Eng. Res.* **8,** 1345–1348.

Xie, J., MacEwan, M. R., Schwartz, A. G., and Xia, Y. (2010). Electrospun nanofibers for neural tissue engineering. *Nanoscale* **2,** 35–44.

Zdrahala, R. J., and Zdrahala, I. J. (1999). Biomedical applications of polyurethanes: A review of past promises, present realities, and a vibrant future. *J. Biomater. Appl.* **14,** 67–90.

Zhang, W., Zhang, Y., Ji, J., Zhao, J., Yan, Q., and Chu, P. K. (2006). Antimicrobial properties of copper plasma-modified polyethylene. *Polymer* **47,** 7441–7445.

CHAPTER SIX

Gastrointestinal Delivery of Anti-inflammatory Nanoparticles

Hamed Laroui,* Shanthi V. Sitaraman,† *and* Didier Merlin*,‡

Contents

1. Introduction 102
2. Material According to Drug Application 103
3. Anti-inflammatory Compounds as Encapsulated Drug 104
   3.1. Anti-inflammatory peptides 106
   3.2. Anti-inflammatory proteins 107
   3.3. Small interfering RNA 107
   3.4. Plasmids 109
4. NSAIDs-Loaded NPs 110
   4.1. PVA-covered NPs 110
   4.2. Ab-covered NPs 110
5. Biomaterial Choice 111
6. Targeting NPs to the Colon: Hydrogel-Encapsulated NPs 112
   6.1. Biomaterials and hydrogel 114
   6.2. Preparation of the hydrogel 114
   6.3. Study of the hydrogel 117
7. Conclusion 118

Acknowledgments 121

References 121

Abstract

The concept of nanomedicine has risen to be the future of medicine. Advantages of using nanoobjects as vectors for drug delivery systems are numerous, such as fewer side effects due to a low drug dose, and high specificity between drug and target. Unlike systemic therapy, targeting a specific target is more efficient and less costly. In inflammatory bowel disease, including ulcerative colitis and Crohn disease, the colon represents the targeted organ. A large number of drugs are candidates for loading into nanoparticles (NPs). Small molecules, such as tripeptides and siRNA, or larger molecules, such as proteins (hormones,

* Department of Biology, Center for Diagnostics and Therapeutics, Georgia State University, Atlanta, Georgia, USA

† Department of Medicine, Division of Digestive Diseases, Emory University, Atlanta, Georgia, USA

‡ Veterans Affairs Medical Center, Decatur, Georgia, USA

Methods in Enzymology, Volume 509

ISSN 0076-6879, DOI: 10.1016/B978-0-12-391858-1.00006-X

antibodies (Ab), etc.), can be encapsulated alone or in a complex form inside the NPs. In our studies, once NPs are synthesized and loaded with anti-inflammatory compounds, they are delivered to the colon. An efficient technique has been developed for specific NP targeting to digestive tract regions, including the colon, using a hydrogel based on electrostatic interactions between positive ions and negative polysaccharides. An *in situ* double cross-linking process, mediated by Ca^{2+} and SO_4^{2-}, of chitosan and alginate administered to the mouse gastrointestinal (GI) tract by double gavage, is used for gel formation. When the drug is given in NPs, NPs are targeted to the colon, and NP degradation by aggressive environmental conditions in the GI tract is significantly reduced. Using a biomaterial (hydrogel) associated with nanotechnology, lower doses of drug can be loaded efficiently and delivered to the colon to reduce colonic inflammation.

1. Introduction

Techniques to deliver drugs into the gastrointestinal (GI) tract can include the provision of drugs in solution. However, such drugs will be directly affected by the pH of the stomach and are likely to be degraded under acidic pH conditions. To circumvent degradation by stomach acidic pH or small intestine digestive enzymes, high drug doses or frequent administration are commonly used, and side effects may be problematic. Recently, complicated techniques, such as sprays of micro- or miniemulsions with high drug doses, have been used. These approaches used nasal and esophageal drug delivery systems, as opposed to lower GI tract-directed systems (Eslamian and Shekarriz, 2009; Li *et al.*, 2002; Sintov *et al.*, 2010). Enemas are often used to target drugs to the colon, but the procedure is cumbersome and is associated with high risk of local complications, including bleeding or perforation, especially in small animals. Thus, there is an unmet need for targeted drug delivery to specific areas in the GI tract, particularly the colon. The colon is important to target, as it is one location for colon cancer or inflammatory bowel disease (IBD), including ulcerative colitis (UC) and Crohn disease (CD). The enema strategy is limited to the distal part of the colon and is not able to reach the ascending (proximal) part of the colon. As the last step of the digestive tract, the colon is challenging to target with intact and quantitative amounts of drug. From oral uptake (saliva enzymes) to colon (pH 7, higher pressure), through the stomach (pH 1–3) and the small intestine (enzymatic release and pH 3–6), drugs face a deleterious environment.

Pharmaceutical nanotechnologies are a promising future for medicine. They can be defined as nanoscale technological innovations or "nanomedicine." This high-interest field provides the opportunity to design and develop several modified or complexed molecules or vectors that can target, treat, and diagnose several diseases, including diseases of the colon. Several drug delivery nanosystems have been developed to target the GI tract since the early 1960s, including micelles (Bromberg, 2005; Gou *et al.*, 2011), liposomes (Garg and Kokkoli, 2011; Riviere *et al.*, 2011), dendrimers (Navarro and Tros de Ilarduya, 2009; Wiwattanapatapee *et al.*, 2003), and nanoparticles (NPs; Laroui *et al.*, 2010a, 2011a,b; Lasic, 1992). Drug delivery systems have been engineered for: (1) specificity: to target only tissues or cells related to the disorder detected; (2) efficiency: to protect a drug against early biological environmental degradation (pH, enzymes, and oxidative agents); and (3) modulation: to modulate and control drug pharmacokinetics.

NPs loaded with drug require a specific vector to efficiently deliver them to a strategic area like the colon. Two strategies can be performed to locally deliver nonsteroidal anti-inflammatory drugs (NSAIDs) using NPs. First, NP matrices can be engineered to favorably collapse in the targeted area, using specific characteristics of the inflamed colon, such as formation of a prodrug, time, pH (Camma *et al.*, 1997; Ewe *et al.*, 1999; Lamprecht *et al.*, 2005), release of oxidative species (Metz *et al.*, 2009; Wilson *et al.*, 2010), aberrant secretion of bacterial enzymes (Kabbaj and Phillips, 2001; Kinget *et al.*, 1998; Levitt *et al.*, 1987), local increase of heat by an external magnetic field (Sato *et al.*, 2009), osmotic systems (Chaudhary *et al.*, 2011), and pressure-controlled drug delivery systems (Jeong *et al.*, 2001; Takaya *et al.*, 1995). In addition, biomaterials such as hydrogels have biological, physical, and chemical characteristics that set them as major candidates for NP vectorization. Hydrogels are a network of polymer chains that are hydrophilic and can contain 99% of water. Hydrogels can protect the drug and/or the drug vector until the targeted organ is reached. They are then dissolved under particular conditions of pH, time, temperature, or enzyme activity. Again, the hydrogel has to be selected and/or engineered to be degraded at a specific GI location such as the colon.

In this review, we present a protocol to specially target the colon with NPs. The schematic shown in Fig. 6.1 illustrates our targeting strategy. As the GI tract is very aggressive, we load the NPs into a hydrogel with the specific ability to collapse only in the colon.

2. Material According to Drug Application

NPs are made based on the double emulsion/evaporation of solvent technique (Fig. 6.2). The hydrophilic drug is first loaded with bovine serum albumin (BSA; 5% in water) in D,L polylactic acid (PLA; 25 g/L in dichloromethane). Each hydrophilic drug that is loaded has to be studied to optimize the loading rate, the kinetic release profile, and to ensure the molecular integrity. Sonication at 50% P_{max} for 2 min of 50% active cycle is performed. Ultrasound provides energy to the system and allows emulsion formation. A water in oil (W/O) emulsion is obtained and dropped in a larger aqueous

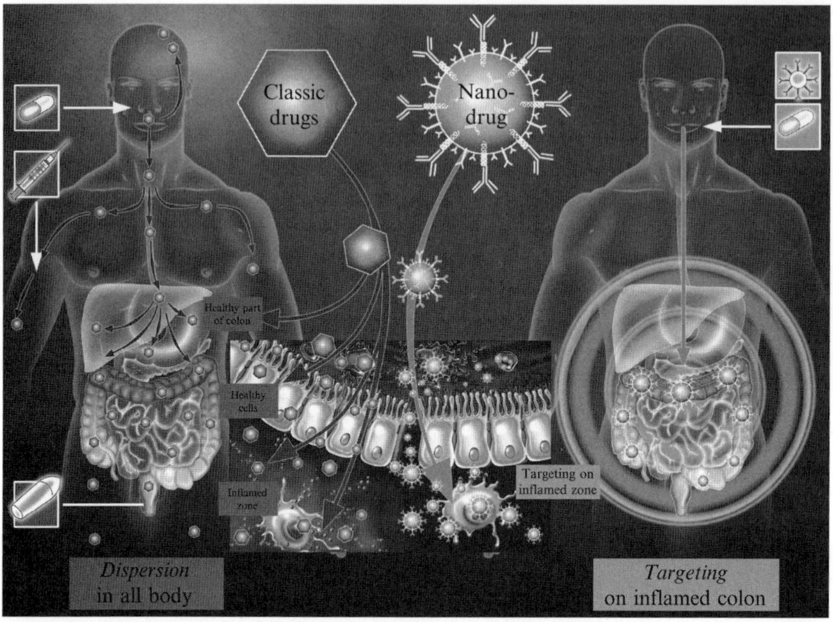

Figure 6.1 Illustration representing the different localization of drug in targeted strategy compared with systemic treatment. For oral intake or intravenous injection of the classical drug, the bioactive component is distributed throughout the body without any distinctions between healthy and inflamed tissue. Enema strategy can only target the distal part of colon. In targeting strategy, nanoparticles (NPs) are covered with an antibody whose ligands are overexpressed in inflamed areas. The NPs accumulate and the drug is released in the specific area (from Laroui *et al.*, 2010b). (See Color Insert.)

phase containing polyvinyl alcohol (PVA; 3 g/L in distilled water). The same protocol is performed to generate a water in oil in water (W/O/W) emulsion. Then, a rotating evaporator removes dichloromethane from the solution entrapping the drug into the NPs. Centrifugation removes the PVA that is not adsorbed on the surface of the NPs (Laroui *et al.*, 2007).

NPs allow for dramatically lowering the amount of the drug used by specifically and cellularly delivering the bioactive drug.

3. Anti-inflammatory Compounds as Encapsulated Drug

NSAIDs are important molecules to deliver to sites of inflammation. In this class of drugs, NSAIDs are widely used as a major treatment for IBD and colon cancer prevention and treatment. Studies from the 1980

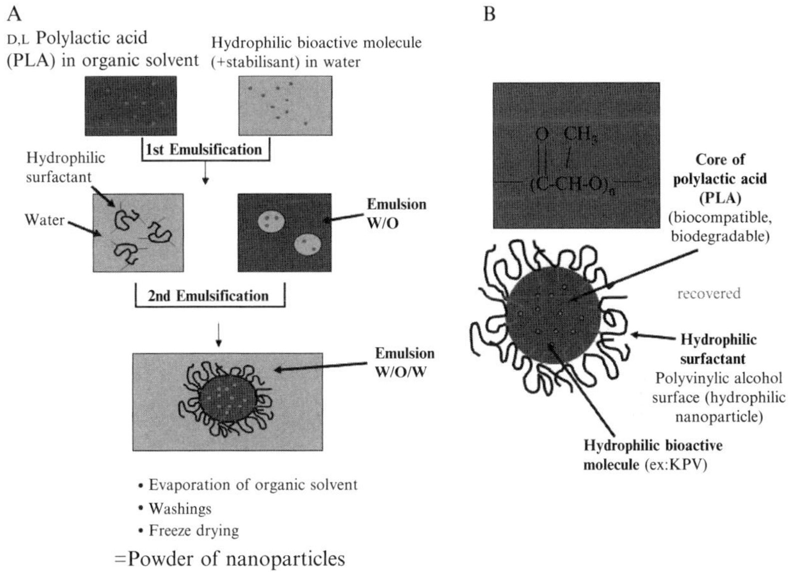

Figure 6.2 (A) Schematic process of NP synthesis. A hydrophilic drug is encapsulated by double emulsion water in oil in water (W/O/W). (B) Schematic representation of PLA NPs loaded with KPV, a hydrophilic drug, and coated with PVA (from Laroui *et al.*, 2010a). (See Color Insert.)

and 1990s showed beneficial effects of NSAIDs on inflammation and tumor regression in animal models (mouse and rat), including indomethacin (Kudo *et al.*, 1980; Narisawa *et al.*, 1983; Pollard and Luckert, 1981) and piroxicam (Rao *et al.*, 1991; Reddy *et al.*, 1987, 1990). Those studies have since been continued in humans, and NSAIDs have shown positive effects on inflammation and colon cancer. Clinical trials tested sulindac (Koornstra *et al.*, 2005; Waddell and Loughry, 1983), indomethacin (Itoh *et al.*, 1988), and acetyl salicylic acid (Paganinihill *et al.*, 1989; Rosenberg *et al.*, 1991a,b). However, in the GI field, classic oral prescriptions of NSAIDs are limited. NSAIDs are never to be used in individuals with IBD, including CD or UC, due to their tendency to cause gastric bleeding and form ulceration in the gastric lining. Alternative molecules used as pain relievers, such as paracetamol (also known as acetaminophen) or drugs containing codeine (which slows down bowel activity), are safer medications for pain relief in IBD.

To prevent any deleterious effect of the NSAID, the active molecules can be loaded into NPs and be delivered efficiently to the inflamed location. NP encapsulation of the bioactive compound allows for lower doses and side effects related to NSAIDs. Depending on the context, the size, and the charge of the drug loaded into NPs, an adjuvant can be used as a molecule of complexation. Complexation of the drugs inside NPs has multiple advantages, such as the protection of the drug from degradation (pH, enzyme, oxidation, pressure, etc.). Also, the complexation improves the kinetics of drug release from the NPs (mainly if the drug size is small and the release could be potentially fast). Finally, it can increase the drug efficiency as some positively charged adjuvants (polyethyleneimine (PEI), chitosan, or lysine) can boost the cytosolic uptake and release of the drug from the lysosomes.

Other classes of molecules have been used as anti-inflammatory compounds with different associated beneficial effects. The main applications for drug delivery systems use peptides, proteins, small interfering RNA (siRNA), or plasmids.

3.1. Anti-inflammatory peptides

Peptides are short polymers of amino acids linked by peptide bonds. They have the same peptide bonds as those in proteins, but are commonly shorter in length.

Peptides with anti-inflammatory properties are of great interest as drug delivery molecules. Several forms and origins of peptide make this drug class attractive for drug delivery. For example, depsipeptides and biooligomers found in microorganisms and marine invertebrates have been shown to have anti-inflammatory and anticancer potentialities (Ballard *et al.*, 2002). Several key advantages are known such as (1) high specificity, (2) high activity, (3) little unspecific binding with untargeted molecular structure, (4) less accumulation in tissues, (5) lower toxicity, (6) minimization of peptide intermolecular binding, and (7) unlimited potential for synthetic peptides. Low oral bioavailability requires intravenous or local injection, as peptides (used as a simple molecule alone) are difficult to deliver across biological membranes, are rapidly cleared from the body and nonstable. Anti-inflammatory peptides such as the somatostatin analogs octreotide, lanreotide, and vapreotide are now clinically available to treat GI tumors (Froidevaux and Eberle, 2002).

Once coupled to drug delivery nanocarriers (NPs, liposomes, etc.), anti-inflammatory peptides have tremendous results as the peptide has a high activity and specificity. In one recent study (Laroui *et al.*, 2010a), an anti-inflammatory peptide (KPV, proline–lysine–valine) was successfully delivered to the colon using PLA NPs transported to the colon in a hydrogel made of alginate and chitosan. The authors successfully coupled KPV to BSA in the inner phase of the NP to prevent the burst effect that leads to an early and inefficient drug delivery from the NPs. Time and bacterial enzyme effects specifically degraded the hydrogel in the colon and delivered KPV-loaded NPs. Once delivered to the colon, NPs interact with cells like epithelial cells and macrophages, and deliver KPV intracellularly (Laroui *et al.*, 2010a).

3.2. Anti-inflammatory proteins

Proteins are biochemical compounds consisting of one or more polypeptides typically folded into a globular or fibrous form, facilitating a biological function. A polypeptide is a single linear polymer chain of amino acids bonded together by peptide bonds between the carboxyl and amino groups of adjacent amino acid residues.

Many proteins have an anti-inflammatory effect, and two main classes of protein intensively used in drug delivery are enzymes and antibodies. Challenges of protein drug delivery include fast elimination from the systemic circulation due to renal clearance and enzymatic degradation, danger of developing an immune response from the use of Ab, nonspecific uptake, and nonefficient translocation into the cell cytosol. Despite these limitations, several enzymes and antibodies have been used and approved by the FDA.

Many protein drugs, such as antibodies, exert their action extracellularly through receptor interactions. Recently, Theiss *et al.* (2010) have shown that encapsulation of prohibitin 1 (PHB) efficiently reduced DSS-induced colitis in mice. PHB is an evolutionarily conserved protein that has pleiotropic functions including mitochondrial protein folding, inhibition of cell-cycle progression, and regulation of transcription. Theiss *et al.* showed that levels of PHB are decreased in colonic biopsies from CD patients and in experimental models of UC (Theiss *et al.*, 2007, 2009). Recently, the authors showed that villin-PHB transgenic mice, which exhibit intestinal epithelial cell-specific PHB overexpression, were protected from experimental colitis (Theiss *et al.*, 2009).

PHB-loaded NPs produce significant anti-inflammatory effects as assessed by clinical and endoscopic scores, and significantly reduced myeloperoxidase (MPO) activity and proinflammatory cytokine levels.

Clinically available Ab molecules are applied to IBD. Infliximab, Adalimumab, and Certolizumab pegol are FDA-approved and TNFα antibodies are commercially available. TNFα Ab treatment showed reduction of the severity of IBD (Dignass *et al.*, 2010; Kornbluth and Sachar, 2010; Lichtenstein *et al.*, 2009; Travis *et al.*, 2008). Despite many limitations as a problem with treating the immune response, TNFα Ab treatment remains an extensively used drug and the most efficient on the market.

3.3. Small interfering RNA

siRNA (also known as small interfering RNA or silencing RNA) is a class of double-stranded RNA molecules, 20–25 nucleotides in length, which play a variety of roles in biology. The most notable role of siRNA is its involvement in the RNA interference pathway, where it interferes with the expression of a specific gene (Ambros *et al.*, 2003; Elbashir *et al.*, 2001; Hammond *et al.*, 2000; Zamore *et al.*, 2000). siRNA duplexes are produced by processing of these longer double-stranded RNAs. Dicer ribonuclease

(Bernstein et al., 2001; Knight and Bass, 2001; Lau et al., 2001), and one strand of the duplex, is then incorporated into a ribonucleoprotein complex, the RNA-induced silencing complex (RISC) (Ambros et al., 2003; Martinez et al., 2002; Schwarz et al., 2002). The siRNA component guides RISC to mRNA molecules bearing a homologous antisense sequence, resulting in cleavage and degradation of that mRNA (Martinez et al., 2002; Schwarz et al., 2002). Figure 6.3 is an illustration of the mechanism required for the siRNA to be (1) chemically intact once interacting with the targeted cells and (2) to enter the cytosol of the cell and interact with the proinflammatory mRNA targeted. Due to these requirements, siRNA is a perfect candidate for therapeutic nanotechnology.

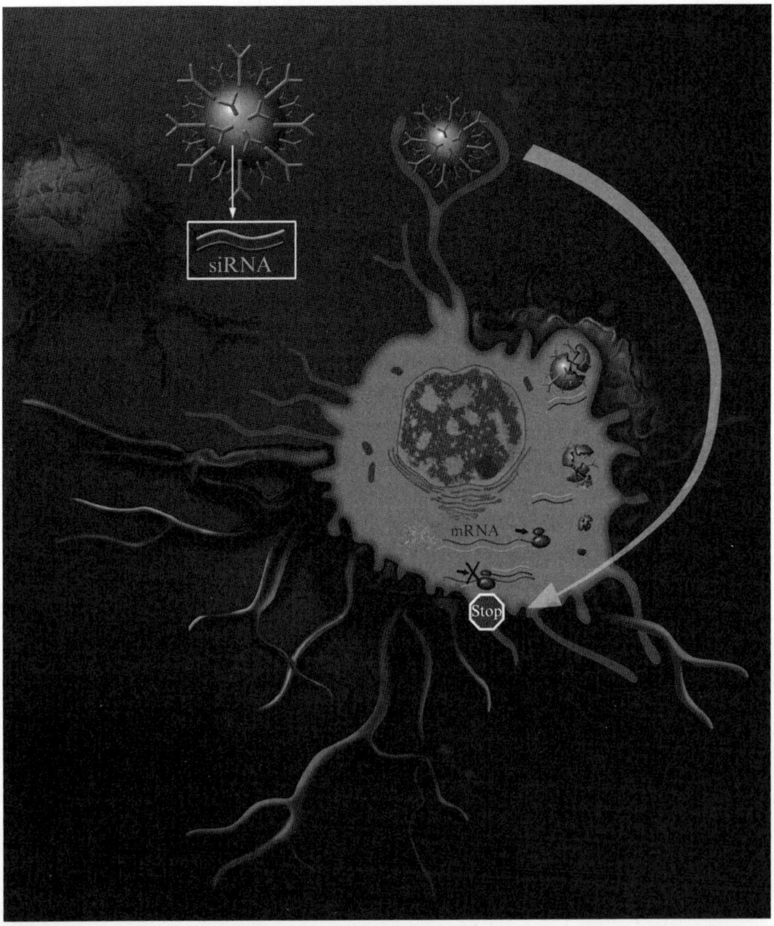

Figure 6.3 Schematic of the delivery of siRNA-loaded nanoparticles to a macrophage to stop the translation of mRNA and decrease the level of the corresponding protein. (See Color Insert.)

Since the discovery of siRNA (Fire *et al.*, 1998), knock-down of specific genes has been extensively used in drugs delivery (Leung and Whittaker, 2005; Novina and Sharp, 2004). The limitations of siRNA as a drug delivery molecule are size and enzymatic degradation. As siRNA is composed of 19–22 bases pairs, drug delivery systems usually associate it with a positive complex molecule (polylysine (Becker *et al.*, 2011), PEI (Laroui *et al.*, 2011a), or cationic lipid (Tseng *et al.*, 2009)). The formation of complexes allows for slower drug release (Laroui *et al.*, 2011a) and the prevention of siRNA degradation for RNAses. As Laroui *et al.* (2011a) demonstrated, PEI can also be used as a "proton sponge." Once the NP is taken up by the cell in an intracellular vesicle, lysosomes fuse to it. The pH of the lysosomes, known to be around 3, can quickly degrade the drug and lead to an inefficient treatment. In this case, as siRNA binds to excess PEI, the basic function of PEI quickly neutralizes the protons and disrupts the lysosomal membrane. siRNA is released into the cell cytosol and specifically knocks down proinflammatory gene expression by destroying mRNA. Recently, complexation agents have been associated with oligonucleotides (siRNA, miRNA, or plasmids used as anti-inflammatory agents) to increase the cell transfection rate. Positive molecules, such as lysine (Harada-Shiba *et al.*, 2002; Wagner, 1999; Wu and Wu, 1987), chitosan (Salva and Akbuga, 2010; Yu and Pishko, 2011), or PEI (Neu *et al.*, 2006; Thankappan *et al.*, 2011) are extensively used. PEI, alkaline and positively charged, can alone induce endocytosis of oligonucleotides in cells. An important notion is the N/P ratio that modifies transfection rate (N, nitrogen functional groups and P, phosphorous functional groups). Depending on the applications and the efficiency required, the N/P ratio can be calculated to fit expectations. The N/P ratio has to be set regarding the balance between positive charge, promoting the transfection efficiency, and negative charge, characterizing the importance of the DNA plasmid or siRNA in the complex. The N/P ratio is highly correlated to transfection results. Limitations to positive charges associated with molecules have to be observed. Cytotoxicity related to high molecular PEI and high N/P ratio is a concern. MTT and LDH tests are important as a proof of cytocompatibility of new biomaterials made of positive complexes known to be deleterious for cells in some cases.

In short, siRNAs are derived from long, double-stranded RNAs that are transcribed endogenously or introduced into cells by viral infection or transfection.

3.4. Plasmids

A plasmid is a DNA molecule that is separate from, and can replicate independently of, the chromosomal DNA. They are double-stranded and, in many cases, circular. Plasmids usually occur naturally in bacteria, but eukaryotic organisms can be transfected by plasmids and overexpress a molecule of interest.

Starting from the same approach used for siRNA and miRNA, plasmids are usually used in gene therapy for anti-inflammatory. They are also associated with positive molecules such as chitosan (CHI), PEI, or polylysine. Plasmid targeting is a "hot topic" of research for nanotechnology. Theoretically, plasmids could be delivered in a specific cell in a specific state. NPs could be covered with a specific Ab or overexpressed during inflammation in a specific cell type. Unlike siRNA or miRNA transfection, plasmids have the advantage of being permanently expressed during a cell lifetime and expression can be modulated by promoters sensitive to cellular inflammation state. On the other hand, designing effective plasmids can be challenging for molecular biologists.

4. NSAIDs-Loaded NPs

According to the solubility and hydrophilic characteristic of the drug, several protocols of NP synthesis can be proposed. Most of the NSAIDs are hydrophilic; therefore the double emulsion/evaporation of solvent represents an important and suitable technique.

Several types of surface engineering are possible based on the expectation for the targeting, such as nonspecific targeting (biologically neutral surface like polyvinylic alcohol, PVA) or specific targeting like antibodies, ligands as peptides or proteins adsorption on NPs.

4.1. PVA-covered NPs

Once the inner aqueous phase (1 mL of the aqueous drug) is sonicated with 25 g/L of PLA in dichloromethane (4 mL), the W/O emulsion" is dropped into an aqueous phase containing 3 g/L of PVA (8 mL) and sonicated (for 2 min, 50% active cycle). The second emulsion is called W/O/W emulsion. The final emulsion contains the hydrophilic active compound in an organic emulsion covered with PVA.

4.2. Ab-covered NPs

As targeting is a major challenge for any drug delivery system, NPs covered with ligands have been extensively used. One Ab has a high affinity for an antigen (even exclusive affinity to one antigen) and that characteristic makes it a major tool for NPs surfaces. Strategies using organic chemistry have led to a huge panel of potentialities to anchor an Ab on the NPs surface. Skewis and Reinhard (2010) used anti-EGFR Ab to cover gold NP and tested it in A431 cells. Using a specific bifunctional polyethylene glycol (PEG) (short thiolated alkyl-PEG-acetates

($HSC_{11}H_{22}(OC_2H_4)_6OCH_2COOH$)), Skewis and Reinhard were able to bind the Ab to the gold NP by activation of the carboxylic functional groups of the PEG. They used the carboxylic acid as the surface group for cross-linking to primary amines of the desired protein (e.g., antibody) by activating it with 1-ethyl-3-[3-dimethylaminopropyl] carbodiimide hydrochloride (EDC) and *N*-hydroxysulfosuccinimide (sulfo-NHS). Interestingly, this study leads to applications in advanced biosensing and biophotonics.

5. Biomaterial Choice

In our study, hydrogel formation is based on the association of two polysaccharides, alginate and chitosan (Fig. 6.4). Both are biocompatible and biodegradable, so they are widely used in the food industry, biology,

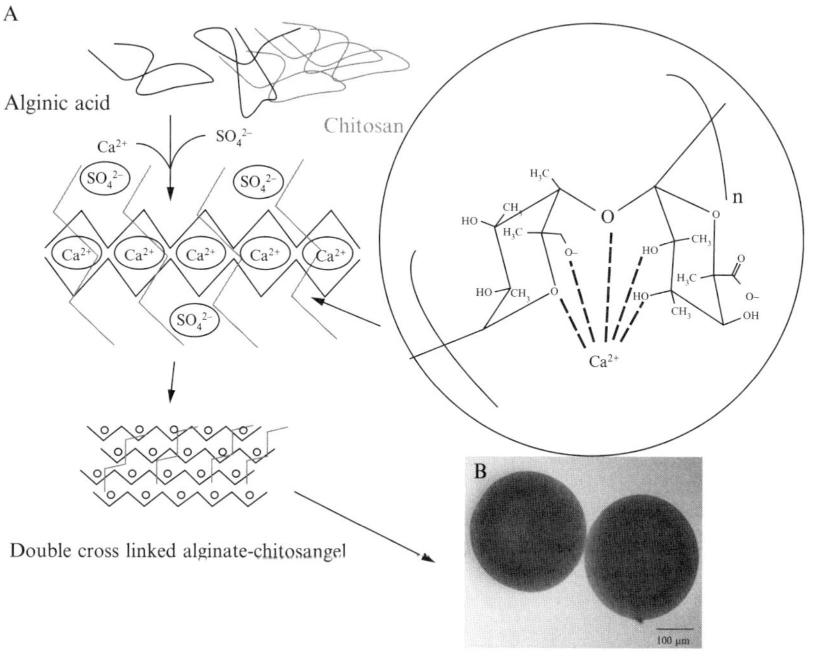

Figure 6.4 Schematic representation of biomaterial encapsulation of KPV-loaded NPs. (A) An alginate and chitosan hydrogel was formed by double linking of Ca^{2+} and SO_4^{2-} ions. Suspension of NPs in the polysaccharide solution loads the final formed hydrogel with NPs. (B) Optical microscopy image of KPV-loaded NPs encapsulated into a hydrogel bead of alginate-chitosan linked via Ca^{2+} and SO_4^{2-} ions (from Laroui *et al.*, 2010a). (For the color version of this figure, the reader is referred to the Web version of this chapter.).

and medicine (Hiorth *et al.*, 2010; Ho *et al.*, 2009; Laroui *et al.*, 2010b; Venkatesan *et al.*, 2011).

Alginate, extracted from brown algae, is a linear copolymer with homopolymeric blocks of (1-4)-linked β-D-mannuronate (M) and the C-5 epimer, α-L-guluronate (G). By gelation using divalent ions (such as calcium), alginate can be easily used as a biomaterial and is biocompatible. Alginate has been used to form beads carrying divalent ions for cell encapsulation (Mazumder *et al.*, 2009), as a sponge for drug delivery, or as a wound-healing gel (Thomas *et al.*, 2000).

Chitosan is a linear polysaccharide composed of randomly distributed β-(1-4)-linked D-glucosamine (deacetylated units) and *N*-acetyl-D-glucosamine (acetylated units). Chitin is susceptible to gelation when SO_4^{2-} ions are added. Chitosan is used in our work because: (1) it is specifically degraded in the colon by colonic bacteria, thus collapsing the whole hydrogel (Yamamoto, 2007) and (2) chitin has a well-known positive effect on inflammation because the material has antibacterial action (Qi *et al.*, 2004).

6. Targeting NPs to the Colon: Hydrogel-Encapsulated NPs

Biomaterials must show biocompatibility, biodegradability, and bioactivity. The latter is the main criterion for a biomaterial used as a drug delivery system. Such systems are varied, ranging from metallic implants to polymers. The latter is widely used because polymers can be associated with copolymers and can be grafted, degraded, and acquire hybrid characteristics when associated with cells. Polymers such as poly(lactic-*co*-glycolic) acid, PLA, and PEG are usually modified and used to form films (Gerhardt *et al.*, 2007) or for nano- or microparticle-mediated (Raynaud *et al.*, 2008) drug delivery. The range of possible polysaccharide uses is as wide as polysaccharides are diverse (Table 6.1). Polysaccharides are biodegradable and biocompatible natural polymers (and are approved by the U.S. Food and Drug Administration), so such materials are widely used in applications such as the cosmetic industry (Bais *et al.*, 2005; Gautier *et al.*, 2008), tissue engineering, or other biological projects, with attention to polysaccharide charge and mass characteristics. Thus, NPs (Boddohi *et al.*, 2009; Laroui *et al.*, 2007; Lemarchand *et al.*, 2006) can be used as drug delivery systems (de Guzman *et al.*, 2008; Ladet *et al.*, 2008) and, more specifically, as electrically charged molecules in electronic applications such as those requiring electroactive nanocomposites (Zampa *et al.*, 2007). When polysaccharides are used as biomaterials, size and charge regulate the kinetics of drug delivery (Chuah *et al.*, 2009; Sezer and Akbuga, 2006). Our technique is based on the formation of a hydrogel by ions (Ca^{2+} and SO_4^{2-}) that mediate cross-linking between alginate and chitosan. Other polysaccharides can be used in this technique because polysaccharides generally share similar electronic properties (Crouzier and Picart, 2009).

The principal point of the double-gavage method for drug delivery is that it allows a "macrohydrogel" to form in the stomach (Fig. 6.5). The first gavage contains the polysaccharide material that contains the drug. As the polysaccharide biomaterial is still liquid at the time of gavage, this technique overcomes the limitations of the size of the animal's oral cavity and allows an easier way to administer in addition to the ability to administer higher concentration of drug. A second gavage is performed with an ionic solution of calcium and sulfate. As soon as the ions and the polysaccharide solution are mixed, a hydrogel is formed within the stomach. The final volume of the biomaterial formed will be 150 μL.

An option technique is to perform the above protocol twice. The method can be called the "double-gavage" method. The purpose of this is to prevent early drug degradation during the digestion process. The first double gavage is made with a concentrated drug solution. The second double gavage is done with a drug = free polysaccharide solution that will recover the biomaterial (Fig. 6.5). This is an "onion-like" structure (Ladet *et al.*, 2008) and has two advantages (Fig. 6.5). First, this original structure can prevent a quick release or "burst effect" of the drug from the hydrogel because the external layer (containing no drug) is the first to be degraded. Secondly, the kinetics of drug release (mainly in GI tract with pH gradient from acid (pH 2) to neutral pH, and digestive enzymes) is surface dependent. This structure allows minimal surface contact between the hydrogel and the external medium and allows the drug to be completely separated from the degradation interface. After oral gavage of the polysaccharide solution into a small animal, the ions in the solution form a hydrogel within the stomach of the maximum possible size. Use of a "macrosize" biomaterial allows prevention of a large contact surface between the loaded drug and the aggressive digestive medium (Fig. 6.5). This technique allows all types of encapsulation, including that of NPs, liposomes, or drug molecules alone.

Using the unique "double-gavage" method, we have shown that a combination of alginate and chitosan at a weight ratio of 7:3 is appropriate for delivery of an encapsulated product to the colon. As shown in Fig. 6.6, most gel-loaded NPs labeled with dextran-FITC were released in the colon after complete collapse of the hydrogel. The alginate and chitosan can be made at different concentrations, ratios, or types of polysaccharides to make the hydrogel collapse in a specific region of the digestive tract.

6.1. Biomaterials and hydrogel

6.1.1. Polysaccharide hydrogels based on electrostatic interactions

Table 6.1 Common polysaccharides used in biology, medicine, and biotechnology with their structural units and specific applications

Polymer	Structural unit	Example of application
Dextran	Glucose unit (glc)	Nanoparticle core (Reis *et al.*, 2008)
Beta-glucan	Fructose unit (Fru)	Microparticles (Aouadi *et al.*, 2009)
Hyaluronan	D-glucuronic acid (GlcA) linked to D-*N*-acetylglucosamine (GlcNAc)	Wound healing (Pardue *et al.*, 2008)
Heparan sulfate	Glucuronic acid (GlcA) linked to *N*-acetylglucosamine (GlcNAc)	Regenerative tissue (DeCarlo and Whitelock, 2006)
Keratin sulfate	D-galactose (Gal) linked to *N*-acetylglucosamine (GlcNAc)	Synovial fluid supplementation
Heparin	2-*O*-sulfated iduronic acid [IdoA (2*S*)] and 6-*O*-sulfated, *N*-sulfated glucosamine GlcNS(6*S*) (variable)	Antithrombic agent
Chondroitin sulfate	D-glucuronic acid (GlcA) and *N*-acetyl-D-galactosamine (GalNAc)	Pain relief in osteoarthritis (Vangsness *et al.*, 2009)

6.2. Preparation of the hydrogel

6.2.1. Protocol of gavages of the hydrogel

Preparation of the sodium chloride solution

1. A 0.15-*M* solution of sodium chloride is prepared.
2. The solution is filtered through a 5-μm filter placed on a syringe to sterilize the solution.
3. The solution is kept under a cell culture hood in a 50-mL tube.

Preparation of the chelation solution

4. Sodium sulfate (60 m*M*) and calcium chloride (140 m*M*) solutions are prepared.

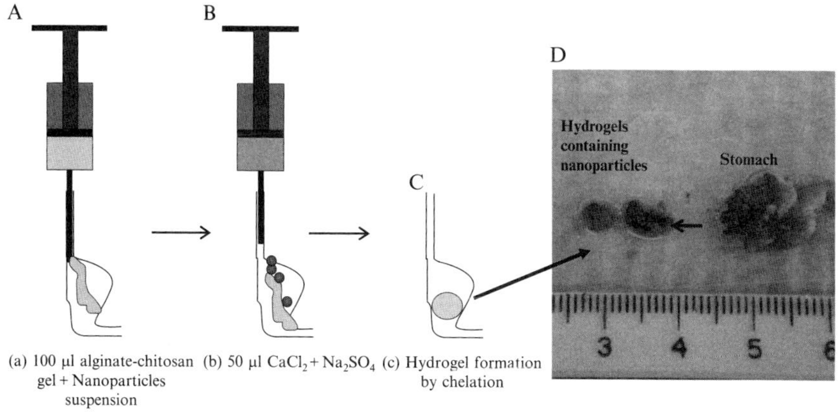

Figure 6.5 Double-gavage procedure of encapsulated KPV-loaded NPs. (A) The first gavage delivered 100 μL of the polymer mix solution (alginate, 7 g/L; chitosan, 3 g/L) containing a homogenous suspension of NPs (2 mg/mL). Red Ponceau has been added for visualization. (B) The second gavage delivered 50 μL of a solution containing 70 mmol/L calcium chloride and 30 mmol/L sodium sulfate. (C) Visualization of the mixed hydrogel formed by chelation of the polymers in the stomach. (D) The hydrogel after extraction from a mouse stomach 5 min after the double-gavage method (from Laroui *et al.*, 2010a). (For interpretation of the references to color in this figure legend, the reader is referred to the Web version of this chapter.).

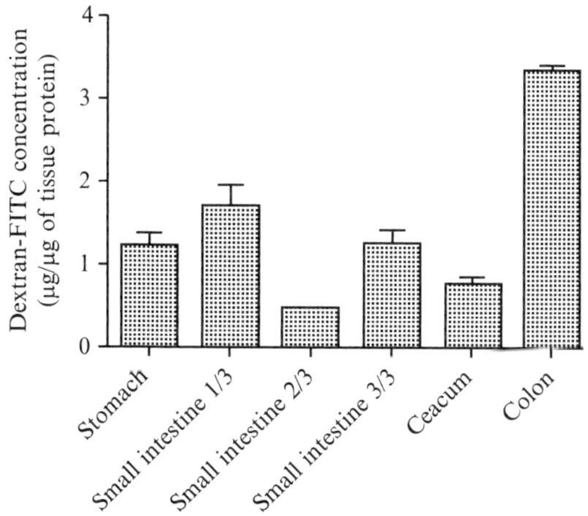

Figure 6.6 Localization of encapsulated dextran-FITC NPs (μg dextran-FITC per μg tissue protein) in the digestive tract after 3 days of gavage. The digestive tract was divided into six sections as follows: stomach, proximal (1/3) small intestine, medial (2/3) small intestine, distal (3/3) small intestine, cecum, and colon ($n = 12$) (from Laroui *et al.*, 2010a).

5. The sodium chloride solution is mixed with the same volume of each solution in (4) to produce a 30-m*M* solution of sodium sulfate and a 70-m*M* solution of calcium chloride.
6. The solutions are filtered through 5-μm filters placed on syringes to sterilize the solutions.
7. The solutions are kept under a cell culture hood in 50-mL tubes.

Preparation of the hydrogel solution

8. 140 mg of alginic acid sodium salt and 60 mg of chitosan are weighed for 7 and 3 g/L solutions, respectively.
9. Both powders are placed in a glass tube of 50-mL capacity and a magnetic bar is added to the tube.
10. The top of the tube is taped with autoclave tape and surrounded with aluminum foil.
11. The tube containing the polysaccharides is sterilized in an autoclave (120 °C under pressure) for 40 min.
12. The aluminum foil and the tape are removed under the cell culture hood and 20 mL of the sterile sodium chloride solution is added. At this stage, the bioactive compound (drug powder, particles) can be added. The tube is covered with Parafilm and placed on a stirring plate until total solubilization of the polysaccharides is evident.
13. The solution is stored at 4 °C overnight to remove all potential air bubbles formed during stirring or centrifuged at 3440 × *g* at room temperature for 15 min.

Synthesis of NPs

Preparation of the matrix of NPs (Solution 1)

14. 4 mL of dichloromethane containing 25 g/L of D,L PLA is prepared.

    PLA crystals are dissolved in dichloromethane kept at 4 °C in a closed beaker to prevent evaporation.

Preparation of the inside of the NPs (Solution 2)

15. Solution 2 is 800 μL of aqueous solution containing the water soluble molecule of interest (protein, peptide, or oligonucleotides, etc) to load inside the NPs.

    In a previous study (Laroui *et al.*, 2011a), we determined N/P = 30 (N/P is the ratio of the number of positive charges of PEI (N as the ammonium charge) and the negative charges of siRNA (TNFα siRNA or FITC-TNFα siRNA) (P as the phosphorous charge)). We complex 29 μL of siRNA (5 μ*M*) with 18 μL of PEI (5 m*M*) for 10 min at room temperature. To obtain the final solution 2, we add 750 μL of 5% BSA.

Preparation of the outside part of the NPs

16. Polyvinyl alcohol is dissolved in water to get 10 mL 0.3 g/L solution.

Synthesis of the NP

17. Solution 1 + solution 2: solution 1, containing the drug, is mixed with solution 2 to generate a W/O emulsion after 2 min of vortexing (Maxi Mix II, Thermodyne, Dubuque, Iowa, USA) and 1 min of sonication with 50% active cycle at 70% power (P^{max} ¼ 400 W) (Digital Sonifier 450, Branson, Danbury, CT, USA).

    This first emulsion is dropped in solution 3 containing 0.3 g/L of PVA to generate a W/O/W emulsion (same process as above).

    The W/O/W emulsion is dropped in a dispersing phase of 0.1 g/L PVA and stirred at 45 °C under a vacuum to remove dichloromethane. As each synthesis makes around 50 mg of dry NPs, each group of NPs is the accumulation of three independent syntheses.

    NPs are then centrifuged at 9953 × *g* and freeze-dried overnight at 50 °C under 0.1 mbar pressure.

Delivery of the biomaterial encapsulated with NPs into the mouse stomach

18. The chelation and the polysaccharide solution tubes are placed into a 37 °C water bath. The chelation solution is collected in a 1-mL syringe without creating any bubbles. This step is repeated using a second syringe to collect the polysaccharide solution.
19. The mouse is left for 3 h without food but with water to allow the animal to receive a biomaterial of a final volume of a 150-μL.
20. The gavage of the mouse starts with 100 μL polysaccharide solution containing the NPs and is followed by a second gavage of 50 μL chelation solution.
21. The hydrogel is formed as soon as the chelation solution reaches the polysaccharide solution and a hard gel forms in the stomach of the mouse.
22. A study is necessary to measure whether the biomaterial will reach the target, which depends on which part of the GI tract is targeted and what polysaccharide proportions are chosen.

    For colon targeting, Laroui *et al.* have shown that the biomaterial must be made of 3 g/L of chitosan and 7 g/L of alginate (Laroui *et al.*, 2010a).

6.3. Study of the hydrogel

6.3.1. Rate of release in colon versus other organs of NPs encapsulated in the hydrogel made of alginate and chitosan

Successful delivery of NPs to the colon is calculated using dextran-tagged FITC-loaded NPs (Laroui *et al.*, 2010a). Once the experiments are performed, all the organs of the digestive tract (from the stomach to the colon) are collected. Then, tissues are homogenized with a tissue homogenizer (Power Gen 125, Fisher Scientific, Pittsburgh, PA, USA).

Using dextran-FITC-loaded NPs, we measure the NP distribution throughout the GI tract. Five days after gavage of the NPs loaded with dextran-FITC encapsulated into the hydrogel, the stomach, jejunum, duodenum, ileum, and the colon are collected and sonicated. After centrifugation, supernatants are collected and the fluorescent signal is measured. We found that the biomaterial composition (mix of alginate and chitosan, respectively 7 and 3 g/L) was optimal for colonic release of the NPs. As shown in Fig. 6.6, dextran-FITC-loaded NPs were mainly released in the colon versus all other areas of the digestive tract.

6.3.2. Effect of NP release in the colon on local inflammation

Laroui *et al.* (2010a) encapsulated a tripeptide made of lysine–proline–valine (KPV, derived from alpha-melanocyte-stimulating hormone, α-MSH) in NPs made of PLA and covered with PVA. Once the biomaterial (mix of alginate and chitosan) specifically collapsed in the colon, NPs directly interact with colonic cells. Interestingly, they showed that a low dose of KPV loaded in NPs stimulates anti-inflammatory activity in the colon, reducing the intake amount of KPV and thus potential side effects of overdosing (Fig. 6.7). They showed that the dose contained in NPs was 12,000-fold lower than the efficient concentration of KPV used in a free drinkable solution. These results showed the principal advantages of nanotechnology applied as a drug delivery system.

6.3.3. Analysis of cells involved in NP uptake

In a recent paper, Laroui *et al.* (2011a) delivered TNFα siRNA-loaded NPs to the colon. The authors showed that TNFα protein expression was dramatically downregulated in the colon, which was the targeted organ, but also the liver by an indirect effect (Fig. 6.8). We recently performed an analysis by flow cytometry (FACS) to determine which cells in the colon were preferentially took up the NPs. We found that mainly the CD11b+F4/80+CD11c− cells (macrophages) and epithelial cells phagocytosed the NPs (data not shown). CD11b+F4/80−CD11+ positive cells (dendritic cells) were also involved but in a lower range.

Our FACS experiment is coherent with the literature as TNFα is mainly secreted in the mucosa by macrophages. The significant decrease of this proinflammatory cytokine was mainly due to the siRNA-loaded NPs taken up by the macrophages.

7. Conclusion

Our technique using an association of polymer NPs and biomaterials (hydrogel) allowed us to efficiently target the colon in a mouse model. Using the characteristic of the hydrogel to specifically collapse in the colon,

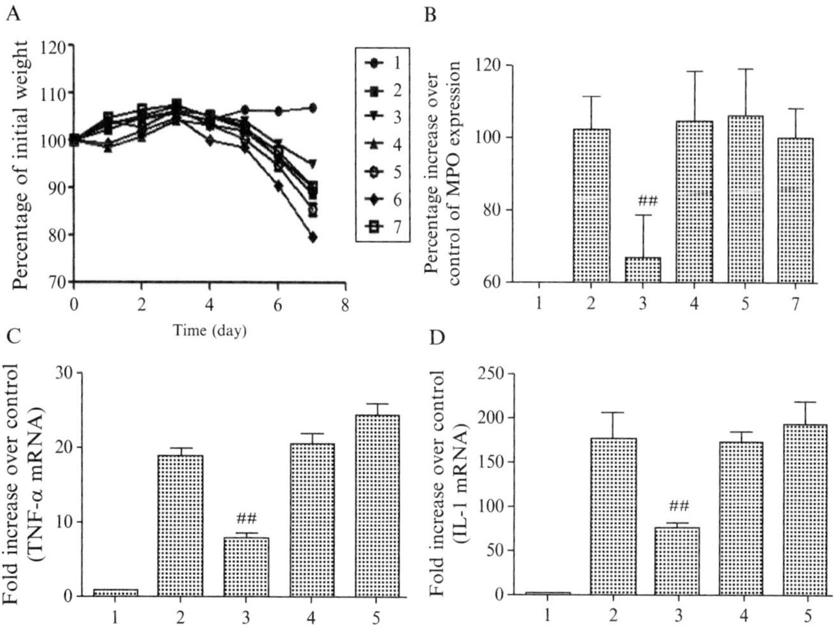

Figure 6.7 *In vivo* parameters of DSS-induced colitis in seven different biomaterial treatments on mice. During *in vivo* experiments, mice received (or not) a daily gavage of 150 μL of biomaterial with or without NPs. All groups had DSS as drinkable water except group 1 (water). Group numbers were defined as followed: number + biomaterial + daily drink; hydrogel, alginate 7 g/L, and chitosan 3 g/L; DSS, 30 g/L; Ø-NP, empty NP; 1, hydrogel + water; 2, hydrogel + DSS; 3, hydrogel + KPV–NP (encapsulated KPV-loaded NPs) + DSS; 4, hydrogel + Ø-NP + DSS; 5, hydrogel with 41 μg/L free KPV (free KPV in hydrogel) + DSS; 6, hydrogel + dextran-FITC NP (encapsulated FITC-loaded NPs); 7, DSS only. $^{\#\#}P = 0.001$. (A) Percentage of initial body weight (%) after 7 days of a daily gavage in groups 1–7. (B) Percentage increase over control of MPO activity after 7 days in groups 1, 2, 3 4, 5, and 7. (C) mRNA expression over control of TNF-α after 7 days in groups 1, 2, 3, 4, and 5. (D) mRNA expression over control of IL-1β after 7 days in groups 1, 2, 3, 4, and 5 (from Laroui *et al.*, 2010a).

we were able to treat colonic cells with bioactive compounds such as peptide, protein, or siRNA.

Polysaccharides have many advantages that make them essential for colon targeting strategies, including pH and bacterial enzyme sensitivity and biocompatibility. One idea will be to develop and optimize NPs made of polysaccharides that are resistant to the digestive tract and only degradable in the colon. Chemical engineering can modify the polymer without modifying the main physical and chemical characteristics and allow the polymer to form NPs.

Emerging applications for NP therapy in the GI tract are being discovered. In this review, we have described an example of a nanotechnology application for IBD. We specifically designed the NPs, the biomaterial and the technique of gavage to be suitable and efficient against IBD. However,

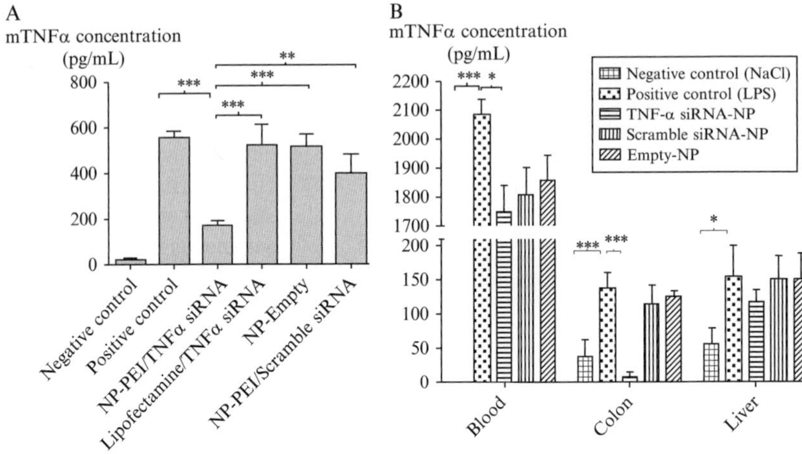

Figure 6.8 TNFα siRNA/polyethylenimine (PEI)-loaded nanoparticles (NPs) inhibit LPS-induced TNFα secretion in macrophages and in mice: (A) RAW 264-7 macrophages were pretreated with TNFα siRNA-loaded NPs suspension (800 μg/mL) for 24 h. As a control, RAW 264-7 macrophages were pretreated with empty NPs, scrambled siRNA-loaded NPs, and lipofectamine with the same amount of TNFα siRNA compared to TNFα siRNA-loaded NPs suspension. Cells were then treated with LPS (10 mg/mL) for 1 h (positive control cells were not pretreated with NPs). Cells were lysed and TNFα protein measured using ELISA test. Data is SEM ± S.E, $n=6$ (significantly different from ANOVA test $^{**}P<0.005$, $^{***}P<0.001$). (B) C57BL/6 female wild-type mice were gavaged with TNFα siRNA-loaded NPs (5 mg/mL), scrambled siRNA NPs (5 mg/mL), and empty NPs (5 mg/mL) contained in a hydrogel (alginate and chitosan) daily for four consecutive days. Positive control mice were gavaged only with the hydrogel without NPs. Mice were then administered LPS (100 μg Kg^{-1} i.p.) and euthanized after 1 h. The colon, spleen, and blood were collected and TNFα expression was measured using ELISA test. Data is represented as SEM ± S.E, $n=8$ (significantly different from ANOVA test $^{*}P<0.05$, $^{***}P<0.001$, ANOVA test) (from Laroui *et al.*, 2011a).

NPs are not limited to drug delivery and can also be an efficient tool to be used in diverse applications, such as bioimaging, diagnostics that can compete favorably with conventional molecular approaches, such as colon cancer therapy or diagnosis. NPs are not restricted to the digestive tract and applications in brain, liver, blood, bone, or cartilage disorders are being investigated. Recent evidence suggests that live oral Salmonella-HIV vaccine vectors have the potential to elicit HIV-specific T cell-mediated immunity in both the mucosal and systemic compartments. Oral vaccines using microfold cells are under investigation, and results could make several determinant advances in sciences as HIV vaccine. M cells, cells found in the follicle-associated epithelium of the Peyer's patch. They transport organisms and particles from the gut lumen to immune cells across the epithelial barrier, and thus are important in stimulating mucosal immunity.

ACKNOWLEDGMENTS

We dedicate this article to the memory of Dr. Shanthi V. Sitaraman, a brilliant scientist, dedicated physician, passionate humanitarian, and dearest friend.
This work was supported by grants from the Department of Veterans Affairs and the National Institutes of Health of Diabetes and Digestive and Kidney by RO1-DK-071594, RO1-DK55850 (D.M.). D.M. is a recipient of a Senior Research Award from the Crohn's and Colitis Foundation of America.

REFERENCES

Ambros, V., Bartel, B., Bartel, D. P., Burge, C. B., Carrington, J. C., Chen, X., Dreyfuss, G., Eddy, S. R., Griffiths-Jones, S., Marshall, M., Matzke, M., Ruvkun, G., *et al.* (2003). A uniform system for microRNA annotation. *RNA* **9,** 277–279.

Aouadi, M., Tesz, G. J., Nicoloro, S. M., *et al.* (2009). Orally delivered siRNA targeting macrophage Map4k4 suppresses systemic inflammation. *Nature* **458,** 1180–1184.

Bais, D., Trevisan, A., Lapasin, R., Partal, P., and Gallegos, C. (2005). Rheological characterization of polysaccharide-surfactant matrices for cosmetic O/W emulsions. *J. Colloid Interface Sci.* **290,** 546–556.

Ballard, C. E., Yu, H., and Wang, B. (2002). Recent developments in depsipeptide research. *Curr. Med. Chem.* **9,** 471–498.

Becker, A. L., Orlotti, N. I., Folini, M., Cavalieri, F., Zelikin, A. N., Johnston, A. P., Zaffaroni, N., and Caruso, F. (2011). Redox-active polymer microcapsules for the delivery of a survivin-specific siRNA in prostate cancer cells. *ACS Nano* **5,** 1335–1344.

Bernstein, E., Caudy, A. A., Hammond, S. M., and Hannon, G. J. (2001). Role for a bidentate ribonuclease in the initiation step of RNA interference. *Nature* **409,** 363–366.

Boddohi, S., Moore, N., Johnson, P. A., and Kipper, M. J. (2009). Polysaccharide-based polyelectrolyte complex nanoparticles from chitosan, heparin, and hyaluronan. *Biomacromolecules* **10,** 1402–1409.

Bromberg, L. (2005). Intelligent hydrogels for the oral delivery of chemotherapeutics. *Expert Opin. Drug Deliv.* **2,** 1003–1013.

Camma, C., Giunta, M., Rosselli, M., and Cottone, M. (1997). Mesalamine in the maintenance treatment of Crohn's disease: A meta-analysis adjusted for confounding variables. *Gastroenterology* **113,** 1465–1473.

Chaudhary, A., Tiwari, N., Jain, V., and Singh, R. (2011). Microporous bilayer osmotic tablet for colon-specific delivery. *Eur. J. Pharm. Biopharm.* **78,** 134–140.

Chuah, A. M., Kuroiwa, T., Ichikawa, S., Kobayashi, I., and Nakajima, M. (2009). Formation of biocompatible nanoparticles via the self-assembly of chitosan and modified lecithin. *J. Food Sci.* **74,** N1–N8.

Crouzier, T., and Picart, C. (2009). Ion pairing and hydration in polyelectrolyte multilayer films containing polysaccharides. *Biomacromolecules* **10,** 433–442.

DeCarlo, A. A., and Whitelock, J. M. (2006). The role of heparan sulfate and perlecan in bone-regenerative procedures. *J. Dent. Res.* **85,** 122–132.

De Guzman, R. C., Ereifej, E. S., Broadrick, K. M., Rogers, R. A., and Vandevord, P. J. (2008). Alginate-matrigel microencapsulated schwann cells for inducible secretion of glial cell line derived neurotrophic factor. *J. Microencapsul.* **25,** 487–498.

Dignass, A., Van Assche, G., Lindsay, J. O., Lemann, M., Soderholm, J., Colombel, J. F., Danese, S., D'hoore, A., Gassull, M., Gomollon, F., Hommes, D. W., Michetti, P., *et al.* (2010). The second European evidence-based Consensus on the diagnosis and management of Crohn's disease: Current management. *J. Crohns Colitis* **4,** 28–62.

Elbashir, S. M., Lendeckel, W., and Tuschl, T. (2001). RNA interference is mediated by 21- and 22-nucleotide RNAs. *Genes Dev.* **15,** 188–200.

Eslamian, M., and Shekarriz, M. (2009). Recent advances in nanoparticle preparation by spray and micro-emulsion methods. *Recent Pat. Nanotechnol.* **3,** 99–115.

Ewe, K., Schwartz, S., Petersen, S., and Press, A. G. (1999). Inflammation does not decrease intraluminal pH in chronic inflammatory bowel disease. *Dig. Dis. Sci.* **44,** 1434–1439.

Fire, A., Xu, S., Montgomery, M. K., Kostas, S. A., Driver, S. E., and Mello, C. C. (1998). Potent and specific genetic interference by double-stranded RNA in Caenorhabditis elegans. *Nature* **391,** 806–811.

Froidevaux, S., and Eberle, A. N. (2002). Somatostatin analogs and radiopeptides in cancer therapy. *Biopolymers* **66,** 161–183.

Garg, A., and Kokkoli, E. (2011). pH-sensitive PEGylated liposomes functionalized with a fibronectin-mimetic peptide show enhanced intracellular delivery to colon cancer cell. *Curr. Pharm. Biotechnol.* **12,** 1135–1143.

Gautier, S., Xhauflaire-Uhoda, E., Gonry, P., and Pierard, G. E. (2008). Chitin-glucan, a natural cell scaffold for skin moisturization and rejuvenation. *Int. J. Cosmet. Sci.* **30,** 459–469.

Gerhardt, L. C., Jell, G. M., and Boccaccini, A. R. (2007). Titanium dioxide (TiO(2)) nanoparticles filled poly(D,L lactid acid) (PDLLA) matrix composites for bone tissue engineering. *J. Mater. Sci. Mater. Med.* **18,** 1287–1298.

Gou, M., Men, K., Shi, H., Xiang, M., Zhang, J., Song, J., Long, J., Wan, Y., Luo, F., Zhao, X., and Qian, Z. (2011). Curcumin-loaded biodegradable polymeric micelles for colon cancer therapy in vitro and in vivo. *Nanoscale* **3,** 1558–1567.

Hammond, S. M., Bernstein, E., Beach, D., and Hannon, G. J. (2000). An RNA-directed nuclease mediates post-transcriptional gene silencing in Drosophila cells. *Nature* **404,** 293–296.

Harada-Shiba, M., Yamauchi, K., Harada, A., Takamisawa, I., Shimokado, K., and Kataoka, K. (2002). Polyion complex micelles as vectors in gene therapy—Pharmacokinetics and in vivo gene transfer. *Gene Ther.* **9,** 407–414.

Hiorth, M., Skoien, T., and Sande, S. A. (2010). Immersion coating of pellet cores consisting of chitosan and calcium intended for colon drug delivery. *Eur. J. Pharm. Biopharm.* **75,** 245–253.

Ho, Y. C., Mi, F. L., Sung, H. W., and Kuo, P. L. (2009). Heparin-functionalized chitosan-alginate scaffolds for controlled release of growth factor. *Int. J. Pharm.* **376,** 69–75.

Itoh, H., Ikeda, S., Oohata, Y., Iida, M., Inoue, T., and Onitsuka, H. (1988). Treatment of desmoid tumors in Gardner's syndrome. Report of a case. *Dis. Colon Rectum* **31,** 459–461.

Jeong, Y. I., Ohno, T., Hu, Z., Yoshikawa, Y., Shibata, N., Nagata, S., and Takada, K. (2001). Evaluation of an intestinal pressure-controlled colon delivery capsules prepared by a dipping method. *J. Control. Release* **71,** 175–182.

Kabbaj, M., and Phillips, N. C. (2001). Anticancer activity of mycobacterial DNA: Effect of formulation as chitosan nanoparticles. *J. Drug Target.* **9,** 317–328.

Kinget, R., Kalala, W., Vervoort, L., and Van Den Mooter, G. (1998). Colonic drug targeting. *J. Drug Target.* **6,** 129–149.

Knight, S. W., and Bass, B. L. (2001). A role for the RNase III enzyme DCR-1 in RNA interference and germ line development in *Caenorhabditis elegans*. *Science* **293,** 2269–2271.

Koornstra, J. J., Rijcken, F. E., Oldenhuis, C. N., Zwart, N., Van Der Sluis, T., Hollema, H., Devries, E. G., Keller, J. J., Offerhaus, J. A., Giardiello, F. M., and Kleibeuker, J. H. (2005). Sulindac inhibits beta-catenin expression in normal-appearing colon of hereditary nonpolyposis colorectal cancer and familial adenomatous polyposis patients. *Cancer Epidemiol. Biomarkers Prev.* **14,** 1608–1612.

Kornbluth, A., and Sachar, D. B. (2010). Ulcerative colitis practice guidelines in adults: American College Of Gastroenterology, Practice Parameters Committee. *Am. J. Gastroenterol.* **105,** 501–523quiz 524.

Kudo, T., Narisawa, T., and Abo, S. (1980). Antitumor activity of indomethacin on methylazoxymethanol-induced large bowel tumors in rats. *Gann* **71,** 260–264.
Ladet, S., David, L., and Domard, A. (2008). Multi-membrane hydrogels. *Nature* **452,** 76–79.
Lamprecht, A., Yamamoto, H., Takeuchi, H., and Kawashima, Y. (2005). A pH-sensitive microsphere system for the colon delivery of tacrolimus containing nanoparticles. *J. Control. Release* **104,** 337–346.
Laroui, H., Grossin, L., Leonard, M., Stoltz, J. F., Gillet, P., Netter, P., and Dellacherie, E. (2007). Hyaluronate-covered nanoparticles for the therapeutic targeting of cartilage. *Biomacromolecules* **8,** 3879–3885.
Laroui, H., Dalmasso, G., Nguyen, H. T., Yan, Y., Sitaraman, S. V., and Merlin, D. (2010a). Drug-loaded nanoparticles targeted to the colon with polysaccharide hydrogel reduce colitis in a mouse model. *Gastroenterology* **138**(3), 843–853e841-842.
Laroui, H., Dalmasso, G., Nguyen, H. T. T., Yan, Y. T., Sitaraman, S. V., and Merlin, D. (2010b). Drug-loaded nanoparticles targeted to the colon with polysaccharide hydrogel reduce colitis in a mouse model. *Gastroenterology* **138,** 843–877.
Laroui, H., Theiss, A. L., Yan, Y., Dalmasso, G., Nguyen, H. T., Sitaraman, S. V., and Merlin, D. (2011a). Functional TNFalpha gene silencing mediated by polyethyleneimine/TNFalpha siRNA nanocomplexes in inflamed colon. *Biomaterials* **32**(4), 1218–1228.
Laroui, H., Wilson, D. S., Dalmasso, G., Salaita, K., Murthy, N., Sitaraman, S. V., and Merlin, D. (2011b). Nanomedicine in GI. *Am. J. Physiol. Gastrointest. Liver Physiol.* **300** (3), G371–G383.
Lasic, D. D. (1992). Mixed micelles in drug delivery. *Nature* **355,** 279–280.
Lau, N. C., Lim, L. P., Weinstein, E. G., and Bartel, D. P. (2001). An abundant class of tiny RNAs with probable regulatory roles in *Caenorhabditis elegans*. *Science* **294,** 858–862.
Lemarchand, C., Gref, R., Passirani, C., Garcion, E., Petri, B., Muller, R., Costantini, D., and Couvreur, P. (2006). Influence of polysaccharide coating on the interactions of nanoparticles with biological systems. *Biomaterials* **27,** 108–118.
Leung, R. K., and Whittaker, P. A. (2005). RNA interference: From gene silencing to gene-specific therapeutics. *Pharmacol. Ther.* **107,** 222–239.
Levitt, M. D., Hirsh, P., Fetzer, C. A., Sheahan, M., and Levine, A. S. (1987). H2 excretion after ingestion of complex carbohydrates. *Gastroenterology* **92,** 383–389.
Li, L., Nandi, I., and Kim, K. H. (2002). Development of an ethyl laurate-based microemulsion for rapid-onset intranasal delivery of diazepam. *Int. J. Pharm.* **237,** 77–85.
Lichtenstein, G. R., Hanauer, S. B., and Sandborn, W. J.Practice Parameters Committee of the American College of Gastroenterology (2009). Management of Crohn's disease in adults. *Am. J. Gastroenterol.* **104,** 465–483.
Martinez, J., Patkaniowska, A., Urlaub, H., Luhrmann, R., and Tuschl, T. (2002). Single-stranded antisense siRNAs guide target RNA cleavage in RNAi. *Cell* **110,** 563–574.
Mazumder, M. A., Burke, N. A., Shen, F., Potter, M. A., and Stover, H. D. (2009). Core-cross-linked alginate microcapsules for cell encapsulation. *Biomacromolecules* **10,** 1365–1373.
Metz, K. M., Mangham, A. N., Bierman, M. J., Jin, S., Hamers, R. J., and Pedersen, J. A. (2009). Engineered nanomaterial transformation under oxidative environmental conditions: Development of an in vitro biomimetic assay. *Environ. Sci. Technol.* **43,** 1598–1604.
Narisawa, T., Satoh, M., Sano, M., and Takahashi, T. (1983). Inhibition of initiation and promotion by N-methylnitrosourea-induced colon carcinogenesis in rats by non-steroid anti-inflammatory agent indomethacin. *Carcinogenesis* **4,** 1225–1227.
Navarro, G., and Tros De Ilarduya, C. (2009). Activated and non-activated PAMAM dendrimers for gene delivery in vitro and in vivo. *Nanomedicine* **5,** 287–297.
Neu, M., Sitterberg, J., Bakowsky, U., and Kissel, T. (2006). Stabilized nanocarriers for plasmids based upon cross-linked poly(ethylene imine). *Biomacromolecules* **7,** 3428–3438.

Novina, C. D., and Sharp, P. A. (2004). The RNAi revolution. *Nature* **430,** 161–164.
Paganinihill, A., Chao, A., Ross, R. K., and Henderson, B. E. (1989). Aspirin use and chronic diseases—A cohort study of the elderly. *Br. Med. J.* **299,** 1247–1250.
Pardue, E. L., Ibrahim, S., and Ramamurthi, A. (2008). Role of hyaluronan in angiogenesis and its utility to angiogenic tissue engineering. *Organogenesis* **4,** 203–214.
Pollard, M., and Luckert, P. H. (1981). Treatment of chemically-induced intestinal cancers with indomethacin. *Proc. Soc. Exp. Biol. Med.* **167,** 161–164.
Qi, L., Xu, Z., Jiang, X., Hu, C., and Zou, X. (2004). Preparation and antibacterial activity of chitosan nanoparticles. *Carbohydr. Res.* **339,** 2693–2700.
Rao, C. V., Tokumo, K., Rigotty, J., Zang, E., Kelloff, G., and Reddy, B. S. (1991). Chemoprevention of colon carcinogenesis by dietary administration of piroxicam, alpha-difluoromethylornithine, 16 alpha-fluoro-5-androsten-17-one, and ellagic acid individually and in combination. *Cancer Res.* **51,** 4528–4534.
Raynaud, J., Choquenet, B., Marie, E., Dellacherie, E., Nouvel, C., Six, J. L., and Durand, A. (2008). Emulsifying properties of biodegradable polylactide-grafted dextran copolymers. *Biomacromolecules* **9,** 1014–1021.
Reddy, B. S., Maruyama, H., and Kelloff, G. (1987). Dose-related inhibition of colon carcinogenesis by dietary piroxicam, a nonsteroidal antiinflammatory drug, during different stages of rat colon tumor development. *Cancer Res.* **47,** 5340–5346.
Reddy, B. S., Nayini, J., Tokumo, K., Rigotty, J., Zang, E., and Kelloff, G. (1990). Chemoprevention of colon carcinogenesis by concurrent administration of piroxicam, a nonsteroidal antiinflammatory drug with D, L-alpha-difluoromethylornithine, an ornithine decarboxylase inhibitor, in diet. *Cancer Res.* **50,** 2562–2568.
Reis, C. P., Ribeiro, A. J., Veiga, F., *et al.* (2008). Polyelectrolyte biomaterial interactions provide nanoparticulate carrier for oral insulin delivery. *Drug Deliv.* **15,** 127–139.
Riviere, K., Kieler-Ferguson, H. M., Jerger, K., and Szoka, F. C., Jr. (2011). Anti-tumor activity of liposome encapsulated fluoroorotic acid as a single agent and in combination with liposome irinotecan. *J. Control. Release* **153,** 288–296.
Rosenberg, L., Palmer, J. R., and Shapiro, S. (1991a). Aspirin use and incidence of large-bowel cancer in a California retirement community—Response. *J. Natl. Cancer Inst.* **83,** 1182–1183.
Rosenberg, L., Palmer, J. R., Zauber, A. G., Warshauer, M. E., Stolley, P. D., and Shapiro, S. (1991b). A hypothesis: Nonsteroidal anti-inflammatory drugs reduce the incidence of large-bowel cancer. *J. Natl. Cancer Inst.* **83,** 355–358.
Salva, E., and Akbuga, J. (2010). In vitro silencing effect of chitosan nanoplexes containing siRNA expressing vector targeting VEGF in breast cancer cell lines. *Pharmazie* **65,** 896–902.
Sato, M., Yamashita, T., Ohkura, M., Osai, Y., Sato, A., Takada, T., Matsusaka, H., Ono, I., Tamura, Y., Sato, N., Sasaki, Y., Ito, A., *et al.* (2009). N-propionyl-cysteaminylphenol-magnetite conjugate (NPrCAP/M) is a nanoparticle for the targeted growth suppression of melanoma cells. *J. Invest. Dermatol.* **129,** 2233–2241.
Schwarz, D. S., Hutvagner, G., Haley, B., and Zamore, P. D. (2002). Evidence that siRNAs function as guides, not primers, in the Drosophila and human RNAi pathways. *Mol. Cell* **10,** 537–548.
Sezer, A. D., and Akbuga, J. (2006). Fucosphere—New microsphere carriers for peptide and protein delivery: Preparation and in vitro characterization. *J. Microencapsul.* **23,** 513–522.
Sintov, A. C., Levy, H. V., and Botner, S. (2010). Systemic delivery of insulin via the nasal route using a new microemulsion system: In vitro and in vivo studies. *J. Control. Release* **148,** 168–176.
Skewis, L. R., and Reinhard, B. M. (2010). Control of colloid surface chemistry through matrix confinement: Facile preparation of stable antibody functionalized silver nanoparticles. *ACS Appl. Mater. Interfaces* **2,** 35–40.

Takaya, T., Ikeda, C., Imagawa, N., Niwa, K., and Takada, K. (1995). Development of a colon delivery capsule and the pharmacological activity of recombinant human granulocyte colony-stimulating factor (rhG-CSF) in beagle dogs. *J. Pharm. Pharmacol.* **47,** 474–478.

Thankappan, U. P., Madhusudana, S. N., Desai, A., Jayamurugan, G., Rajesh, Y. B., and Jayaraman, N. (2011). Dendritic poly(ether imine) based gene delivery vector. *Bioconjug. Chem.* **22,** 115–119.

Theiss, A. L., Idell, R. D., Srinivasan, S., Klapproth, J. M., Jones, D. P., Merlin, D., and Sitaraman, S. V. (2007). Prohibitin protects against oxidative stress in intestinal epithelial cells. *FASEB J.* **21,** 197–206.

Theiss, A. L., Vijay-Kumar, M., Obertone, T. S., Jones, D., Hansen, J., Gewirtz, A. T., Merlin, D., and Sitaraman, S. V. (2009). Prohibitin (PHB) is a novel antioxidant that attenuates colonic inflammation in mice. *Gastroenterology* **136**A41-A41.

Thomas, A., Harding, K. G., and Moore, K. (2000). Alginates from wound dressings activate human macrophages to secrete tumour necrosis factor-alpha. *Biomaterials* **21,** 1797–1802.

Travis, S. P., Stange, E. F., Lemann, M., Oresland, T., Bemelman, W. A., Chowers, Y., Colombel, J. F., D'haens, G., Ghosh, S., Marteau, P., Kruis, W., Mortensen, N. J., *et al.* (2008). European evidence-based Consensus on the management of ulcerative colitis: Current management. *J. Crohns Colitis* **2,** 24–62.

Tseng, Y. C., Mozumdar, S., and Huang, L. (2009). Lipid-based systemic delivery of siRNA. *Adv. Drug Deliv. Rev.* **61,** 721–731.

Vangsness, C. T., Jr., Spiker, W., and Erickson, J. (2009). A review of evidence-based medicine for glucosamine and chondroitin sulfate use in knee osteoarthritis. *Arthroscopy* **25,** 86–94.

Venkatesan, P., Puvvada, N., Dash, R., Kumar, B. N. P., Sarkar, D., Azab, B., Pathak, A., Kundu, S. C., Fisher, P. B., and Mandal, M. (2011). The potential of celecoxib-loaded hydroxyapatite-chitosan nanocomposite for the treatment of colon cancer. *Biomaterials* **32,** 3794–3806.

Waddell, W. R., and Loughry, R. W. (1983). Sulindac for polyposis of the colon. *J. Surg. Oncol.* **24,** 83–87.

Wagner, E. (1999). Application of membrane-active peptides for nonviral gene delivery. *Adv. Drug Deliv. Rev.* **38,** 279–289.

Wilson, D. S., Dalmasso, G., Wang, L., Sitaraman, S. V., Merlin, D., and Murthy, N. (2010). Orally delivered thioketal nanoparticles loaded with TNF-alpha-siRNA target inflammation and inhibit gene expression in the intestines. *Nat. Mater.* **9,** 923–928.

Wiwattanapatapee, R., Lomlim, L., and Saramunee, K. (2003). Dendrimers conjugates for colonic delivery of 5-aminosalicylic acid. *J. Control. Release* **88,** 1–9.

Wu, G. Y., and Wu, C. H. (1987). Receptor-mediated in vitro gene transformation by a soluble DNA carrier system. *J. Biol. Chem.* **262,** 4429–4432.

Yamamoto, A. (2007). Study on the colon specific delivery of prednisolone using chitosan capsules. *Yakugaku Zasshi* **127,** 621 630.

Yu, X., and Pishko, M. V. (2011). Nanoparticle based biocompatible and targeted drug delivery: Characterization and in vitro studies. *Biomacromolecules* **12,** 3205–3212.

Zamore, P. D., Tuschl, T., Sharp, P. A., and Bartel, D. P. (2000). RNAi: Double-stranded RNA directs the ATP-dependent cleavage of mRNA at 21 to 23 nucleotide intervals. *Cell* **101,** 25–33.

Zampa, M. F., De Brito, A. C., Kitagawa, I. L., Constantino, C. J., Oliveira, O. N., Jr., Da Cunha, H. N., Zucolotto, V., Dos Santos, J. R., Jr., and Eiras, C. (2007). Natural gum-assisted phthalocyanine immobilization in electroactive nanocomposites: Physicochemical characterization and sensing applications. *Biomacromolecules* **8,** 3408–3413.

CHAPTER SEVEN

Chitosan-Based Nanoparticles as a Hepatitis B Antigen Delivery System

Filipa Lebre,*,† Dulce Bento,*,† Sandra Jesus,*,† *and* Olga Borges*,†

Contents

1. Introduction	128
2. Chitosan-Based Particle Preparation	129
2.1. Alginate-coated chitosan particles	129
2.2. Chitosan nanoparticles	130
2.3. Chitosan/alginate particles	130
2.4. Chitosan/poly-ε-caprolactone particles	131
3. Physicochemical Characterization of the Particles	132
3.1. Size measurement	132
3.2. Zeta potential titration	133
3.3. Scanning electron microscopy	133
4. Antigen Adsorption Studies	134
5. *In vitro* Release Studies	135
6. Evaluation of the Bioactivity of the Antigen	137
7. Cell Viability Studies with Spleen Cells	137
7.1. Preparation of spleen cell suspensions	138
7.2. MTT viability assay	138
8. Studies on Uptake into Peyer's Patches	139
9. Concluding Remarks	140
References	140

Abstract

The design of antigen delivery systems, particularly for mucosal surfaces, has been a focus of interest in recent years. In this chapter, we describe the preparation of chitosan-based particles as promising antigen delivery systems for mucosal surfaces already tested by our group with hepatitis B surface antigen. The final proof of the concept is always carried out with immunization studies performed in an appropriate animal model. However, before these

* Center for Neuroscience and Cell Biology, University of Coimbra, Coimbra, Portugal
† Faculty of Pharmacy, University of Coimbra, Pólo das Ciências da Saúde Azinhaga de Santa Comba, Coimbra, Portugal

Methods in Enzymology, Volume 509
ISSN 0076-6879, DOI: 10.1016/B978-0-12-391858-1.00007-1

important studies, it is advisable that the delivery system should be submitted to a variety of *in vitro* tests. Among several tests, the characterization of the particles (size, morphology, and zeta potential), the studies of antigen adsorption onto particles, the evaluation of toxicity of the particles, and the studies of particle uptake into lymphoid organs are the most important and will be described in this chapter.

1. Introduction

At present, the term "vaccination" is generally considered to be identical to "injection." This conception is due to the fact that vaccines are typically given by intramuscular injection. In the new era of vaccine development, with the emergence of subunit vaccines, the formulation of needle-free vaccines is undoubtedly more challenging. Novel vaccines obtained by recombinant technology are, in principle, safer with regard to toxicity; however, they are also less immunogenic, making it mandatory to include adjuvants in the formulation of such vaccines. Numerous efforts made by the scientific community to develop needle-free vaccine formulations are justifiable by several distinct advantages. An obvious one is the possibility of painless self-administration of the vaccine. Moreover, vaccine delivery via mucosal surfaces elicits mucosal immune responses at the site of pathogen entry and enhances cellular immunity through stimulation of Toll-like receptors (Bessa and Bachmann, 2010), thus improving overall effectiveness. Taking these facts into account, needle-free vaccination could have a big impact on the efficacy of immunization against mucosal transmitted diseases such as hepatitis B. In 1981, FDA approved the first hepatitis B vaccine which consisted of the surface antigen of the hepatitis B (HBsAg) virus present in the blood of human carriers of the infection, replaced in 1986 by the currently available vaccine which represents the world's first subunit vaccine and the world's first recombinant expressed vaccine. Since the hepatitis B virus can be transmitted perinatally or by exchange of body fluids (e.g., blood, semen, and vaginal fluid), the design of new hepatitis B vaccines with the additional possibility to induce mucosal antibodies (e.g., secretory IgA) is particularly attractive. The only available hepatitis B vaccines to date are injectable formulations, adjuvanted with aluminum salts, which are evidently not appropriate for oral or intranasal administration owing to two main reasons. One, mucosally administered antigens will be exposed to enzymatic degradation, and second, the adjuvant is not adequate for application at mucosal surfaces. Therefore, formulations with enhanced adjuvant properties are needed for the application at mucosal surfaces to reduce the high antigen doses normally required to increase the low immune response and decrease the variability of the individual immune responses frequently observed. Preclinical investigation of new needle-free hepatitis B vaccines relies on the development of adjuvants/ new formulations with additional capability to increase the immunogenicity of the antigen. To achieve this goal, several strategies are currently being discussed (Lebre *et al.*, 2011; Thanavala *et al.*, 2009). A good example is the development of nanosized carrier systems that adsorb or encapsulate antigens, protect them from proteolytic enzymes, allow the increase of antigen retention time at the nasal mucosa, and finally target antigens to M-cells present on the mucosa (Jabbal-Gill, 2010). Lastly, the loading of particles not only with antigens but also with immunopotentiators such as combinations of Toll-like receptor ligands (Kasturi *et al.*, 2011) may modulate the quantity and the quality of the immune response.

Chitosan is a cationic polymer consisting of β-(1-4)-linked D-glucosamine (deacetylated unit) and *N*-acetyl-D-glucosamine (acetylated unit) monomers that can be obtained by deacetylation of chitin (Illum, 1998). It has been considered a nontoxic, biodegradable, and biocompatible polymer (Baldrick, 2010), thus extensive research has been directed toward its use in medical applications such as drug and vaccine delivery (Lebre *et al.*, 2011; Panos *et al.*, 2008; van der Lubben *et al.*, 2001b). One major advantage of this polymer is its ability to easily produce nanoparticles under mild conditions without the application of harmful organic solvents. This has been one of the main reasons for its wide applicability to the encapsulation of different molecules such as therapeutic proteins, DNA, and antigens. Chitosan is also known to be mucoadhesive, and its ability to stimulate cells of the immune system has been shown in many studies (Borges *et al.*, 2007a). These unique features make chitosan an attractive polymer to act as an adjuvant.

2. Chitosan-Based Particle Preparation

2.1. Alginate-coated chitosan particles

Chitosan nanoparticle preparation can be achieved by several techniques. One of the most common is the precipitation/coacervation method, which is a process of spontaneous phase separation that occurs when two oppositely charged polyelectrolytes are mixed in an aqueous solution.

The protocol used in our laboratory results from an adaptation and optimization of a previously described method (Berthold *et al.*, 1996). The preparation of this delivery system contains three main steps: manufacturing of the chitosan particles, their loading by adsorption, and finally coating with sodium alginate (Borges *et al.*, 2005). Low molecular weight chitosan is dissolved at a concentration of 0.25% (w/v) in a solution with 2% (v/v) of acetic acid and 1% (w/v) of Tween™ 80. The formation of the particles is achieved after the addition of 3.5 ml of sodium sulfate solution (10%, w/v) to 200 ml of the chitosan solution. The addition is made at a rate of 1 ml/min under mild agitation ($<$50 rpm) and continuous sonication (VibraCell sonicator, 600-watt model; Sonics & Materials, Inc., Newtown, CT, USA). Sonication is maintained for an additional 15 min and the agitation for 60 min at room temperature (RT). The suspension is centrifuged for 30 min at $2800 \times g$ and the supernatant is discarded. The particles are resuspended twice in Milli-Q water, centrifuged again for 30 min, and the supernatants are discarded. The particles are frozen in liquid nitrogen and freeze–dried overnight using a freeze-dryer. The dry powder is kept frozen until further use.

At this point, particles can be loaded with proteins of biological interest. In our case, we loaded them with HBsAg so they can act as a delivery system for the antigen. Loading studies will be described later in the chapter. In order to ensure the stability and protection of the antigen, loaded particles are coated with sodium alginate. For this purpose, equal volumes of the antigen-loaded nanoparticle suspension and a buffer phosphate solution of sodium alginate (1%, w/v) are mixed under magnetic stirring. The agitation is maintained during a 20-min period. The suspension is then centrifuged for 10 min at $460 \times g$ and the supernatant is discarded. The particles are resuspended in 0.524 m*M* $CaCl_2$ in 50 m*M* HEPES buffer solution and kept under agitation for another 10 min. A laboratory temperature below 20 °C is crucial for these experiments.

2.2. Chitosan nanoparticles

A second protocol used in our laboratory results from an adaptation of a method previously described (Roy *et al.*, 1999). Briefly, equal volumes of a solution of chitosan (0.1% in sodium acetate buffer, 25 m*M*, pH 5.0) and a sodium sulfate solution (0.625%) are mixed under high-speed vortexing for 20 s. The resultant nanoparticles are left to rest at RT for approximately 1 h. In order to remove compounds that did not react, the resulting nanoparticle suspension is centrifuged for 30 min at $4500 \times g$. The supernatant is discarded and the obtained pellet resuspended in sodium acetate buffer, 25 m*M*, pH 5.5. Nanoparticles should be used immediately after resuspension to avoid particle aggregation.

2.3. Chitosan/alginate particles

Alginate is a biodegradable and a biocompatible natural polyanionic polysaccharide with a good safety profile. Its molecular structure consists of linear copolymers of L-guluronic and D-mannuronic acid residues joined linearly by 1,4-glycosidic linkages. Divalent cations such Ca^{2+}, Ba^{2+}, and Sr^{2+} work as alginate cross-link agents inducing gel formation via a sol–gel transformation (Wee and Gombotz, 1998). Calcium ions have higher affinity to guluronic acid residues, so the relative composition of the alginate can have impact in characteristics of the delivery system. Therefore, alginates with higher guluronic acid content tend to form more rigid structure and higher porosity than alginates rich in mannuronic acids (De and Robinson, 2003).

Chitosan/alginate (Chi/Alg) particles are prepared using a two-step method modified from Rajaonarivony *et al.* (1993). In order to prepare the pregel, 3 ml of a calcium chloride solution (2 mg/ml) is added dropwise to 47 ml of sodium alginate solution 0.063% (pH 5.1) in an ultrasound bath while stirring for 15 min at 25,000 rpm with a homogenizer (Ystral GmbH, Dottingen). Ca^{2+}/alginate pregel is stirred for another 20 min with a magnetic stirrer. Finally, particles are formed upon mixing 1.5 ml of pregel with an equal volume of chitosan 0.1% (acetic acid solution; pH 5.4) under high-speed vortexing following additional 30 min of magnetic stirring allowing nanoparticle maturation. Nanoparticles are isolated by centrifugation at $5000 \times g$ for 40 min at 20 °C. The supernatant is discarded and the pellet resuspended in the intended buffer (e.g., phosphate buffer (PB), pH 7.4 for protein adsorption studies).

2.4. Chitosan/poly-ε-caprolactone particles

As we have discussed above, chitosan nanoparticles offer some advantages as drug delivery systems. Nevertheless, the inclusion of a hydrophobic polymer like poly-ε-caprolactone (PCL) might confer additional useful properties to the delivery system, like the possibility to establish hydrophobic interactions between delivery system and loaded proteins.

The procedure for the production of chitosan/PCL particles in our laboratory resulted from the adaptation of and experimentation of different techniques described in the literature, in particular, one described by Bilensoy *et al.* (2009) that is based on the nanoprecipitation technique patented by Fessi *et al.* (1992). An aqueous phase of 0.1% acetic acid containing 0.1% chitosan and 5% Tween™ 80 is placed under high-speed homogenization. Then, the organic phase, consisting of 0.2% PCL diluted on acetone, is added dropwise to the first solution at a proportion of 1:3 (v/v) to a final volume of 18 ml. The resultant particle suspension is placed under magnetic stirring for additional 45 min for the maturation process. Finally, the organic phase is removed by evaporating acetone with a nitrogen flux in a warm bath (40 °C maximum). The nanoparticles suspended in the original medium can be isolated, concentrated, and resuspended in other diluents by centrifugation at $16{,}000 \times g$, for 75 min at 4 °C. To guarantee minimal aggregation of the particles during the centrifugation, a 200 μl glycerol bed for each 18 ml batch is recommended.

3. Physicochemical Characterization of the Particles

3.1. Size measurement

It is generally accepted that the size and size distribution of the particles are important for their adjuvant activity. Therefore, size characterization is an important step in vaccine formulation development even if the attempts to correlate particle size and the resultant immune responses lead to conflicting findings (Oyewumi *et al.*, 2010).

The size of particles can be measured by Dynamic Light Scattering techniques. Among those techniques, Photon Correlation Spectroscopy (PCS) has been widely used as routine standard technique in biophysics, colloid and polymer laboratories. PCS is based on the fact that the intensity of light scattered from a dispersion of particles into a given scattering angle is the result of interference on the surface of a square-law detector between light scattered from different particles in the medium (Pecora, 2000). PCS gives the translational self-diffusion coefficient of the nanoparticle (Pecora, 2000), which, for particles in a dilute dispersion, can be related with hydrodynamic diameter of the particle (nonspherical or flexible particle) through the Stokes–Einstein equation (Eq. 7.1):

$$d(H) = \frac{kT}{3\pi\eta D} \tag{7.1}$$

d(*H*), hydrodynamic diameter; *D*, translational diffusion coefficient; *k*, Boltzmann's constant; *T*, absolute temperature; η, viscosity.

Additionally, particles in suspension undergo Brownian motion due to random bombardment by the solvent molecules that surround them. When particles are illuminated with a laser, the intensity of the scattered light fluctuates at a rate dependent of the size of the particles (Malvern, 2004). The smaller particles move quickly and induce the intensity to fluctuate more rapidly than the larger ones. Thus, the analysis of the rate of intensity fluctuations using the autocorrelation function allows the determination of the particle size distribution of the sample.

It is important to be aware that PCS measures the hydrodynamic diameter, which refers to how a particle diffuses within a fluid and corresponds to the diameter of a sphere that has the same translational diffusion coefficient as the particle that is being measured. The size of the particle "core" is not the only determinant of the translational diffusion coefficient. Thus, factors like ionic strength of the medium and surface structures that can affect the particles' diffusion speed will possibly change the apparent size of the particle. Samples for PCS analysis should consist of a well-dispersed phase in a suspending medium, and both the refractive index of the solvent and the viscosity at the selected measurement temperature must be known.

3.2. Zeta potential titration

In a colloidal system, when a particle is dispersed in a fluid, a range of processes causes the interface to become electrically charged. The liquid layer surrounding the particle can be divided into two parts: the inner region where ions are strongly bound and the outer region where they are less firmly associated to particle. On this outer region named the diffuse layer, there is a notional boundary inside which the ions and particles form a stable entity. The potential at this boundary is the zeta potential. Although zeta potential occurs at a distance from the particle, it is related to it and therefore influences a wide range of properties of colloidal systems, such as stability, interaction with electrolytes, and suspension rheology.

Theoretically, nanoparticles with a zeta potential above (+/−) 30 mV have been shown to be stable in suspension, as the surface charge prevents aggregation of the particles. Several researchers have been shown that the stability of the nanoparticles is highly dependent on the pH values, and optimal pH value can result in the highest stability of the nanoparticles. The other important parameter is the ionic strength of the medium, when it is high, zeta potential becomes closer to zero. Electrostatic repulsion disappears due to an increase of salt concentration which compresses the electrical field double layer of the particles.

The pH of the medium is one of the most important factors that affect zeta potential. When a particle is in suspension and alkali is added to the suspension, the particles will acquire negative charges, decreasing its zeta potential and on its turn, if acid is added, particles will acquire positive charges, increasing its zeta potential. Considering this phenomenon, we can perform a zeta potential titration to characterize nanoparticles, once this variable is highly dependent on the conditions of the suspending medium. The results can be expressed on a graph like the one represented in Fig. 7.1. Measuring zeta potential variation at different pHs, we can observe positive values at low pH and lower or negative values at high pH. The point where the plot passes through zero is the isoelectric point of the particles, normally the point where the colloidal system is less stable. In a very simplistic way, we can say that two pH values above or below the isoelectric point, the colloidal suspension starts to be stable (Malvern, 2004).

3.3. Scanning electron microscopy

Size and morphology of nanoparticles can be observed by scanning electron microscopy (SEM) using an electron microscope such as JSM-700 1 FA (JOEL, Japan). Prior to image acquisition, one drop of nanoparticle

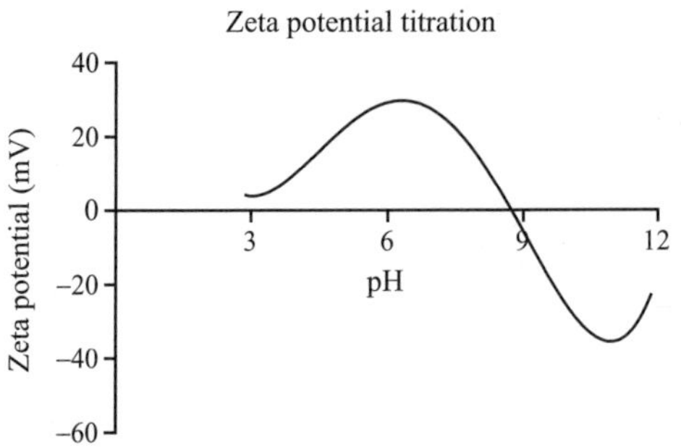

Figure 7.1 An example of zeta potential titration of chitosan/PCL nanoparticles. Measurements of the nanoparticle zeta potential were performed on the Delsa™ Nano C, by resuspending 100 µl of a nanoparticle suspension on 1.9 ml of an acidic or basic solution.

suspension is placed over a copper surface and let to dry overnight. Afterward, samples are mounted on microscope stub, coated with gold, and then observed on microscopy. Figure 7.2 represents SEM images of chitosan particles prepared in our laboratory. It is frequently observed that during the drying, the particles tend to stick together.

4. Antigen Adsorption Studies

Polymeric particles can be used as delivery systems for molecules with biological interest such as proteins or more specifically antigens. The adsorption of the antigens onto particles is a mild process since can be performed simply by the incubation of the particles with the solution of the antigens at RT. Furthermore, given that the antigens are located at the surface of the particles, it is expected that they are more available to be presented by antigen-presenting cells. However, protein adsorption is a complex process that is affected by a number of factors concerning protein (charge, size, and structure), polymer (size, composition, hydrophobicity, and zeta potential), and medium properties (pH, ionic strength, and viscosity) (Gonzalez Ferreiro *et al.*, 2002; Kim *et al.*, 2002). The adsorption of a protein onto the hydrophilic chitosan particles is mainly caused by electrostatic interaction of the protonated chitosan amino groups with the carboxyl groups of the protein substrate in a buffer. It is prudent to start the adsorption studies with model antigens which are less expensive than the real antigens to obtain preliminary information. According to the objective of the assay, these studies are performed suspending the particles into a buffer solution (PB) with the model antigen. In order to investigate the best conditions that generate the highest particle loading capacity (LC) and antigen loading efficacy (LE), different

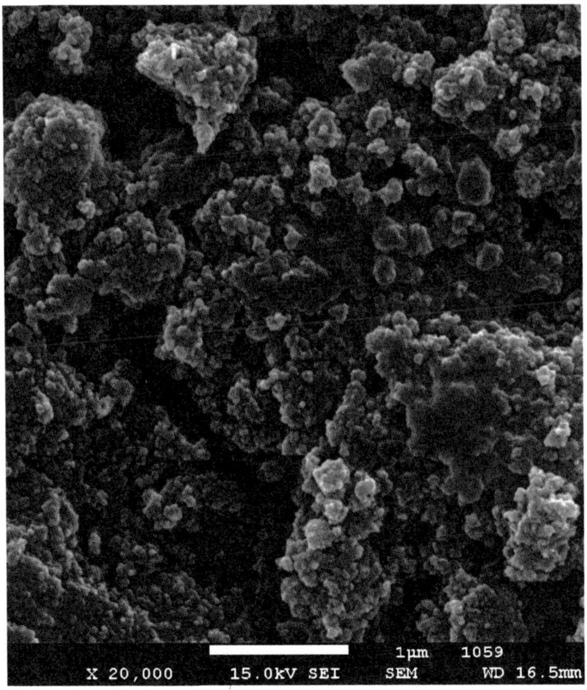

Figure 7.2 SEM image of chitosan nanoparticles (JEOL JSM 6400 scanning electron microscopy) revealed the presence of small rough rounded particles (~100 nm).

concentrations of the antigens should be experimented at a fixed time and fixed particle concentration (see results on Fig. 7.3). After incubation, aliquots of the particle suspension are centrifuged at 18,000 × *g* for 30 min and supernatant is collected to measure nonbound protein (indirect method) and determine the LE (Eq. 7.2) and the LC (Eq. 7.3):

$$\mathrm{LE}(\%) = \frac{(\text{total amount of protein} - \text{nonbound protein})}{\text{total amount of protein}} \times 100 \qquad (7.2)$$

$$\mathrm{LC}\ (\%) = \frac{(\text{total amount of protein} - \text{nonbound protein})}{\text{weight of the particles}} \times 100 \qquad (7.3)$$

5. *In vitro* Release Studies

To evaluate the suitability of nanoparticles as a delivery system for proteins with biological interest such as the HBsAg, an *in vitro* release study should be performed. These studies are performed with different buffers

Nanoparticle loading capacity — comparison of the delivery systems

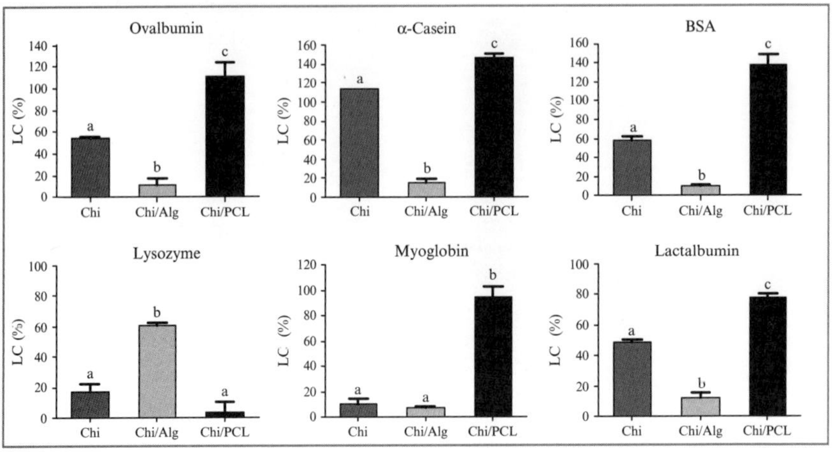

	Chi	Chi/Alg	Chi/PCL
Size (nm)	592.0 ± 25.6	3613.9 ± 1655.6	169.0 ± 7.4
Zeta potential (mV)	+ 5.50 ± 0.55	−15.74 ± 2.17	−8.18 ± 1.77

Figure 7.3 Comparison of the particle's loading capacity (LC) using different proteins. The results illustrated on this figure were obtained during adsorption studies performed with buffer phosphate, pH 7.1 ± 0.2 with six proteins with different isoelectric points. Chi/PCL nanoparticles have a significantly higher LC when compared with Chi and Chi/Alg particles for all the proteins except lysozyme. The inclusion of a hydrophobic polymer like PCL into chitosan particles allows the establishment of hydrophobic interactions between proteins and particles, which explains, at least in part, the result. In contrast, Chi/Alg particles have the lowest LC for almost all the proteins. The adsorption of proteins with low IEP (<7.0) is below 20% and about 60% for lysozyme (IEP ~ 11.0). This last result can be explained by the negative charge of particles in PB 7.4 which favors electrostatic interactions only with the positively charged proteins. The table below the chart illustrates the size and zeta potential of chitosan-based delivery systems suspended in phosphate buffer, pH 7.4 (mean ± SD, $n = 3$), during adsorption studies. (See Color Insert.)

which mimetize the physiological conditions. To mimic the digestive tract, simulated gastric fluid and simulated intestinal fluid, both described by United States Pharmacopeia, are normally used as release buffers, by our group to study oral antigen delivery systems. The antigen delivery systems developed to be administered by nasal or by one of the parenteral routes of the administration should be tested with PB or even better with phosphate-buffered saline (PBS), pH 7.4. For the release studies, aliquots of the antigen-loaded nanoparticle suspension are added to individual tubes containing release medium previously equilibrated at 37 °C and placed in a shaker bath adjusted to 50 rpm. At appropriate time intervals, samples from each tube are collected and filtered with a low protein-binding filter (MILLEX® GV—0.22 μm; durapore PVDF membrane; Millipore, Molsheim, France) followed by centrifugation for 20 min at 18,000 × *g* and the protein in supernatant assayed with an appropriated method. Suspensions of unloaded particles also have to be analyzed under the same conditions to evaluate possible interferences on the protein quantification method.

6. Evaluation of the Bioactivity of the Antigen

The preliminary evaluation of the bioactivity of the antigen, after their association with particles, is normally performed using the Western blotting technique. The integrity and bioactivity of hepatitis B antigen are done after antigen being released from nanoparticles. Therefore, the samples resulting from the release studies mentioned above need to be centrifuged at 14,000 rpm in order to separate the released antigen from the particles and then an aliquot is solubilized with the SDS-PAGE loading buffer and treated 5 min at 100 °C. The SDS-PAGE is performed in accordance with standard protocols (Gallagher and Smith, 2003) with 12% resolving gel, cast and run in Tris–glycine buffer at 25 mA. The antigenicity of the entrapped hepatitis B antigen is assessed by immunoblotting using a mouse antiserum raised against the native antigen. The hepatitis B antigen samples were transferred from the unstained gel onto a nitrocellulose membrane, using a semi-dry electroblotting system (115 mA; 1 h) and the membrane is blocked overnight at 4 °C with PBS-T (0.05% of Tween 20) containing 5% of milk. After washing with PBS-T, the membrane is incubated for 2 h at RT with the positive anti-HBsAg IgG mouse antiserum, diluted 1:500 in PBS-T with 5% of low-fat milk. After washing with PBS-T, the membrane is incubated with antimouse IgG conjugated to alkaline phosphatase, diluted 1:750. The ability of the mouse antiserum to recognize hepatitis B antigen released from the nanoparticles is demonstrated colorimetrically using 5 ml of phosphatase buffer with 33 μl NBT (50 μg/ml) and 16.7 μl BCIP (50 μg/ml). The reaction is stopped by washing the membrane with water (Borges *et al.*, 2007b). It is expected that the HBs-specific antibodies recognize the antigen epitopes from the sample in a similar way as for the original antigen. With this analysis, it is then possible to confirm if the antigenicity of the hepatitis B antigen is altered after antigen adsorption with nanoparticles which can compromise the efficacy of the formulation developed (Borges *et al.*, 2007b).

7. Cell Viability Studies with Spleen Cells

There are different methods for assessing the *in vitro* cytotoxicity. Alterations in plasma membrane permeability can be evaluated both by the release of cytoplasmic enzymes (e.g., lactate dehydrogenase) or by the uptake of dyes (e.g., trypan blue, propidium iodide). Alternatively, cytotoxicity can be evaluated by changes in cell metabolic activity. Tetrazolium salts are widely used in these metabolic assays (e.g., MTT, XTT, WST-1). In our laboratory, the MTT assay (Sigma-Aldrich, St. Louis, MO, USA) is normally used to evaluate the cytotoxicity of the particles. In this assay, the tetrazolium salt is reduced to purple formazan crystals by metabolically active cells. The principal reason for the choice of spleen cells to assess the cytotoxicity of nanoparticles intended for mucosal immunization is related to the fact that they are a very good and sensitive representative of the different immune cells and are obtained and cultured easier, compared to other lymphoid organs, like Peyer's patches (Borges *et al.*, 2006).

7.1. Preparation of spleen cell suspensions

Mice are euthanized by cervical dislocation and their spleens are aseptically removed. Individual spleen cell suspensions are prepared in a Petri dish using curved needles. One needle is used to hold the spleen and the other to detach cells from the capsule by moving the needle along the length of the spleen. The cell suspension is then transferred into a 15-ml sterile conical tube to allow large fragments to settle down for 5 min. The cell suspension is decanted into another sterile centrifuge tube and is centrifuged for 10 min at $259 \times g$. The resultant supernatant is discarded and cells are resuspended in 5 ml of RPMI 1640. This washing step is repeated two times, and finally, the cells are resuspended in complete RPMI 1640 medium (supplemented with 10% (v/v) fetal bovine serum, 1% (v/v) glutamine, 1% (v/v) gentamicin, and 2% (v/v) 1 *M* HEPES buffer). The final suspension is adjusted to a final concentration of 5×10^6 cells per ml.

7.2. MTT viability assay

One-hundred microliters of aseptically prepared nanoparticles are resuspended in complete RPMI and platted in a 96-well plate. One-hundred microliters of spleen cell suspension (5×10^5 cells/well) is then added to the wells. Cell and particles are incubated for 24 h (95% relative humidity and 5% CO_2.) at 37 °C. MTT solution (5 mg/ml in PBS, pH 7.4) is filtered to remove any precipitate (0.22-μm filter), preheated at 37 °C, and added to each well (20 μl/well). The plate is then incubated for additional 4 h. In the end, the plate is centrifuged for 25 min ($800 \times g$) and supernatant removed using a multichannel pipette. To dissolve the formazan crystals, 200 μl of preheated (37 °C) DMSO are added to each well and pipetted up and down (carefully, to avoid any bubble formation). The plate is mixed in a plate shaker for 10 min and incubated at 37 °C for 30 min. After the incubation time, optical density (OD) of plate solutions is read at 540 nm with 630 nm as wavelength reference. The relative cell viability (%) related to control wells containing spleen cells in culture medium without nanoparticles is calculated by Eq. (7.4):

$$\text{cell viability}(\%) = \frac{\text{OD sample }(540\text{nm}) - \text{OD sample }(630\text{nm})}{\text{OD control }(540\text{nm}) - \text{OD control }(630\text{nm})} \times 100 \quad (7.4)$$

It is important to notice that this protocol is optimized for spleen cells. For a different cell line, the linear relationship between metabolically active cell number and signal produced (color), as well as the incubation time, should be established, allowing an accurate quantification of cell viability.

8. Studies on Uptake into Peyer's Patches

Oral vaccination presents advantages over parenteral injection, nevertheless the degradation of the vaccine and the low uptake by the gut-associated lymphoid tissue are determinant factors that limit the success of this strategy (Jung *et al.*, 2000; Van Der Lubben *et al.*, 2001a). The uptake of inert particles across the GI tract is known to occur mainly transcellularly through normal enterocytes and Peyer's patches via M-cells (Hussain *et al.*, 2001). Considering nanoparticulate vaccines for oral administration, the antigen is only released in the lymphoid tissue to induce the immune response after Peyer's patch internalization of the particles (Van Der Lubben *et al.*, 2001a). Therefore, uptake studies into Peyer's patches have extreme importance on the evaluation of the potential of the particles as an oral antigen delivery system. The uptake studies can be performed with rats with a weight ranging between 250 and 350 g. On the day before the experiment, animals are starved overnight, only with free access to water.

The rats are anesthetized by IM administration of 0.5 ml/kg of Hypnorm® (fentanyl citrate 0.315 mg/ml and fluanisone 10 mg/ml) and 0.5 ml/kg of Dormicum® (midazolam 5 mg/lm). The animals need to remain anesthetized throughout the experiment and are placed on electrical heating mats. A small incision is made in the lower stomach and a Teflon tube (Ø: 0.5 mm I.D. × 1.0 mm O.D.) is introduced through the pylorus approximately 3–5 cm into the duodenum. Fluorescent particles are placed (~500 μl) into the duodenum through the Teflon tube, and the incision is closed after the removal of the tube from the stomach. The rats are sacrificed after 2 h by cervical dislocation. The whole intestine is removed and flushed with 20 ml of cold (~4 °C) PBS. Between four and five Peyer's patches can be excised from each intestine. They are fixed with 2% paraformaldehyde, and rinsed again with PBS (4 °C), and the tissue is then permeabilized by immersion in 0.1%

Triton X-100 (in PBS) for 20 min. The tissue is rinsed again and stained with a 0.0617% solution of BODIPY® 665/676 (Pierce) in methanol for 60 min. Finally, the Peyer's patches are mounted on glass slides and observed using a confocal laser scanning microscope.

9. Concluding Remarks

The development of novel vaccine adjuvants is becoming as important as the development of novel antigens itself. Presently, most of the vaccines are given by intramuscular injection, which requires the use of needles that are painful, are potentially dangerous, requires trained medical personnel, and are therefore unsuitable for mass vaccination campaigns, especially in developing countries. Recently, many researchers have focused their interest on needle-free technologies for immunization, including a variety of approaches for mucosal and topical immunization. Although several strategies have been proposed, some with very promising results, there is not an approved needle-free vaccine against HBV so far, most likely because regulatory entities tend to adopt a cautious approach toward novel adjuvants and administration routes in terms of safety in humans. The design of chitosan-based antigen delivery systems has been explored by a considerable number of researchers mainly with the purpose of finding a good mucosal adjuvant. Therefore, it is prudent that a considerable number of *in vitro* tests would be performed to prove the efficacy of the delivery system as an adjuvant, before starting the immunization studies.

REFERENCES

Baldrick, P. (2010). The safety of chitosan as a pharmaceutical excipient. *Regul. Toxicol. Pharmacol.* **56,** 290–299.

Berthold, A., Cremer, K., and Kreuter, J. (1996). Preparation and characterization of chitosan microspheres as drug carrier for prednisolone sodium phosphate as model for anti-inflammatory drugs. *J. Control. Release* **39,** 17–25.

Bessa, J., and Bachmann, M. F. (2010). T cell-dependent and -independent IgA responses: Role of TLR signalling. *Immunol. Invest.* **39,** 407–428.

Bilensoy, E., Sarisozen, C., Esendagli, G., Dogan, A. L., Aktas, Y., Sen, M., and Mungan, N. A. (2009). Intravesical cationic nanoparticles of chitosan and polycaprolactone for the delivery of Mitomycin C to bladder tumors. *Int. J. Pharm.* **371,** 170–176.

Borges, O., Borchard, G., de Sousa, A., Junginger, H. E., and Cordeiro-da-Silva, A. (2007a). Induction of lymphocytes activated marker CD69 following exposure to chitosan and alginate biopolymers. *Int. J. Pharm.* **337,** 254–264.

Borges, O., Borchard, G., Verhoef, J. C., de Sousa, A., and Junginger, H. E. (2005). Preparation of coated nanoparticles for a new mucosal vaccine delivery system. *Int. J. Pharm.* **299,** 155–166.

Borges, O., Cordeiro-da-Silva, A., Romeijn, S. G., Amidi, M., de Sousa, A., Borchard, G., and Junginger, H. E. (2006). Uptake studies in rat Peyer's patches, cytotoxicity and release studies of alginate coated chitosan nanoparticles for mucosal vaccination. *J. Control. Release* **114,** 348–358.

Borges, O., Tavares, J., de Sousa, A., Borchard, G., Junginger, H. E., and Cordeiro-da-Silva, A. (2007b). Evaluation of the immune response following a short oral vaccination schedule with hepatitis B antigen encapsulated into alginate-coated chitosan nanoparticles. *Eur. J. Pharm. Sci.* **32,** 278–290.

De, S., and Robinson, D. (2003). Polymer relationships during preparation of chitosan-alginate and poly-L-lysine-alginate nanospheres. *J. Control. Release* **89,** 101–112.

Fessi, H., Devissaguet, J. P., Puisieux, F., and Theis, C. (1992). Process for the preparation of dispersible colloidal systems of a substance in the form of nanoparticles. United States Patent 5118528.

Gallagher, S., and Smith, J. (2003). One-dimensional gel electrophoresis of proteins. *Curr. Protoc. Immunol.* **8.4,** 1–21.

Gonzalez Ferreiro, M., Tillman, L., Hardee, G., and Bodmeier, R. (2002). Characterization of alginate/poly-L-lysine particles as antisense oligonucleotide carriers. *Int. J. Pharm.* **239,** 47–59.

Hussain, N., Jaitley, V., and Florence, A. T. (2001). Recent advances in the understanding of uptake of microparticulates across the gastrointestinal lymphatics. *Adv. Drug Deliv. Rev.* **50,** 107–142.

Illum, L. (1998). Chitosan and its use as a pharmaceutical excipient. *Pharm. Res.* **15,** 1326–1331.

Jabbal-Gill, I. (2010). Nasal vaccine innovation. *J. Drug Target.* **18,** 771–786.

Jung, T., Kamm, W., Breitenbach, A., Kaiserling, E., Xiao, J. X., and Kissel, T. (2000). Biodegradable nanoparticles for oral delivery of peptides: Is there a role for polymers to affect mucosal uptake? *Eur. J. Pharm. Biopharm.* **50,** 147–160.

Kasturi, S. P., Skountzou, I., Albrecht, R. A., Koutsonanos, D., Hua, T., Nakaya, H. I., Ravindran, R., Stewart, S., Alam, M., Kwissa, M., Villinger, F., Murthy, N., *et al.* (2011). Programming the magnitude and persistence of antibody responses with innate immunity. *Nature* **470,** 543–547.

Kim, B., Bowersock, T., Griebel, P., Kidane, A., Babiuk, L. A., Sanchez, M., Attah-Poku, S., Kaushik, R. S., and Mutwiri, G. K. (2002). Mucosal immune responses following oral immunization with rotavirus antigens encapsulated in alginate microspheres. *J. Control. Release* **85,** 191–202.

Lebre, F., Borchard, G., de Lima, M. C., and Borges, O. (2011). Progress towards a needle-free hepatitis B vaccine. *Pharm. Res.* **28**(5), 986–1012.

Malvern. (2004). Zetasizer Nano Series. User Manual. MAN 031, Issue 2.1.

Oyewumi, M. O., Kumar, A., and Cui, Z. (2010). Nano-microparticles as immune adjuvants: Correlating particle sizes and the resultant immune responses. *Expert Rev. Vaccines* **9,** 1095–1107.

Panos, I., Acosta, N., and Heras, A. (2008). New drug delivery systems based on chitosan. *Curr. Drug Discov. Technol.* **5,** 333–341.

Pecora, R. (2000). Dynamic light scattering measurement of nanometer particles in liquids. *J. Nanopart. Res.* **2,** 123–131.

Rajaonarivony, M., Vauthier, C., Couarraze, G., Puisieux, F., and Couvreur, P. (1993). Development of a new drug carrier made from alginate. *J. Pharm. Sci.* **82,** 912–917.

Roy, K., Mao, H. Q., Huang, S. K., and Leong, K. W. (1999). Oral gene delivery with chitosan–DNA nanoparticles generates immunologic protection in a murine model of peanut allergy. *Nat. Med.* **5,** 387–391.

Thanavala, Y., Lavelle, E., and Ogra, P. (2009). All things mucosal. *Expert Rev. Vaccines* **8,** 139–142.

Van Der Lubben, I. M., Konings, F. A., Borchard, G., Verhoef, J. C., and Junginger, H. E. (2001a). *In vivo* uptake of chitosan microparticles by murine Peyer's patches: Visualization studies using confocal laser scanning microscopy and immunohistochemistry. *J. Drug Target.* **9,** 39–47.

van der Lubben, I. M., Verhoef, J. C., Borchard, G., and Junginger, H. E. (2001b). Chitosan for mucosal vaccination. *Adv. Drug Deliv. Rev.* **52,** 139–144.

Wee, S., and Gombotz, W. R. (1998). Protein release from alginate matrices. *Adv. Drug Deliv. Rev.* **31,** 267–285.

CHAPTER EIGHT

Targeting Nanoparticles to Dendritic Cells for Immunotherapy

Luis J. Cruz,*,[1] Paul J. Tacken,* Felix Rueda,† Joan Carles Domingo,† Fernando Albericio,‡ *and* Carl G. Figdor*

Contents

1. Introduction 144
2. Passive Targeting 144
3. Active Targeting 146
4. Particulate Vaccines 147
5. Targeting Gold NP Vaccines to DCs 148
6. Experimental Procedure for Targeted AuNP Preparation 150
   6.1. Preparation of AuNPs 150
   6.2. Conjugation of peptide and Fc fragment to AuNPs 150
   6.3. Preactivation of Fc fragment 151
7. Targeting Liposome-Based Vaccines to DCs 151
   7.1. Preparation of liposomes and peptide encapsulation 152
   7.2. Preparation of targeted liposomes with the Fc fragment of human IgG conjugated to PEG 153
8. Targeting Poly(Lactic-*co*-Glycolic Acid)-Based Vaccines to DCs 154
9. Experimental Procedure for the Preparation of Targeted PLGA NP 155
   9.1. PLGA NP preparation 155
   9.2. Quantification of antigen in NPs 156
   9.3. Quantification of TLR-L in NPs 156
   9.4. Conjugating antibodies to NPs 157
10. Conclusion 157
References 158

* Department of Tumor Immunology, Nijmegen Centre for Molecular Life Sciences, Radboud University Medical Centre, Nijmegen, The Netherlands
† Department of Biochemistry and Molecular Biology, University of Barcelona, Barcelona, Spain
‡ Institute for Research in Biomedicine, Barcelona Science Park, Barcelona, Spain
[1] Present address: Endocrinology Research Lab and Molecular Imaging, Leiden University Medical Center, Leiden, The Netherlands

Methods in Enzymology, Volume 509
ISSN 0076-6879, DOI: 10.1016/B978-0-12-391858-1.00008-3

Abstract

Dendritic cells (DCs) are key players in the initiation of adaptive immune responses and are currently exploited in immunotherapy for treatment of cancer and infectious diseases. Development of targeted nanodelivery systems carrying vaccine components, including antigens and adjuvants, to DCs *in vivo* represents a promising strategy to enhance immune responses. Delivering particulate vaccines specifically to DCs and preventing nonspecific uptake by other endocytotic cells are challenging. Size represents a critical parameter determining whether particulate vaccines can penetrate lymph nodes and reach resident DCs. Specific delivery is further enhanced by actively targeting DC-specific receptors. This chapter discusses the rationale for the use of particle-based vaccines and provides an overview of antigen-delivery vehicles currently under investigation. In addition, we discuss how vaccine delivery systems may be developed, focusing on liposomes, PLGA polymers, and gold nanoparticles, to obtain safe and efficacious vaccines.

1. Introduction

Dendritic cells (DCs) are key antigen-presenting cells (APCs) that initiate and control adaptive immune responses, thereby resulting in immunity or tolerance (Reis e Sousa, 2011). Apart from inducing humoral responses, DCs can cross-present exogenous antigens to $CD8^+$ T cells, thus leading to the induction of cellular immune responses responsible for the clearance of virus-infected or tumor cells, among others. Consequently, vaccines benefit from effective delivery of their components to DCs. Targeted delivery of antigens to DC surface receptors enhances presentation to T cells (Tacken *et al.*, 2007). In addition to antigen, DCs require activation signals that drive their maturation to induce immunity. These signals can be provided by coadministration of immunopotentiators, such as Toll-like receptor (TLR) ligands. Initially, these immunopotentiators were applied locally or systemically, but several studies suggest that codelivery of antigen and immunopotentiator to DCs enhances vaccine efficacy (Blander and Medzhitov, 2006; Schlosser *et al.*, 2008; Wille-Reece *et al.*, 2005). This finding has resulted in an increased interest in particle-based vaccine carriers to simultaneously target multiple vaccine components to DCs.

2. Passive Targeting

Particle-based vaccines are generally easily recognized and ingested by APCs. The efficiency of these vaccines to passively target DCs is determined by parameters such as their size, surface charge, hydrophobicity, hydrophilicity, and interactions with serum proteins and cell-surface receptors (Bachmann and Jennings, 2010).

Particulate vaccines are taken up less efficiently by DCs when their size exceeds 500 nm (Foged *et al.*, 2005). Such relatively large particles are ingested mainly by macrophages (Xiang *et al.*, 2006). Size is also a critical parameter in determining whether particles present in interstitial fluid will enter the lymphatic capillaries and be retained in lymph nodes (Oussoren *et al.*, 1997; Swartz, 2001). Particles smaller than 200 nm rapidly enter these capillaries and are captured by lymph node-resident DCs. By contrast, larger particles remain at the site of injection and require uptake by local APCs and active cellular transport for antigens to be presented to T cells in the draining lymph node (Manolova *et al.*, 2008; Reddy *et al.*, 2007). Consequently, small nanoparticles (NPs) can reach relatively high numbers of DCs, while larger ones depend on uptake by relatively scarce DC populations present at the injection site. Moreover, skin-derived and lymph node-resident DCs comprise phenotypically and functionally distinct DC subsets that differ in their capacity to induce T cell activation. Although lymph node-resident DC subsets have a less mature phenotype than skin-derived subsets, they show enhanced levels of inflammatory cytokine release and superior T cell priming capacity (van de Ven *et al.*, 2011). A preclinical study in mice confirms the notion that the optimal size for a particle-based vaccine for prophylactic and therapeutic cancer strategies is 40–50 nm (Fifis *et al.*, 2004), although particles up to 300 nm induce potent $CD4^+$ and $CD8^+$ T cell responses that protect mice from lethal influenza infections (Kasturi *et al.*, 2011).

In vitro studies show that particles with a positive surface charge are generally ingested more efficiently by DCs than those with a neutral or negative charge (Foged *et al.*, 2005; Wischke *et al.*, 2006). However, *in vivo* research has revealed that positively charged particles are not by definition more effective vaccine carriers, thereby suggesting that the effect of surface charge also depends on the model system used (Nakanishi *et al.*, 1999; Yotsumoto *et al.*, 2004). The lipid composition of liposomes, for example, not only affects the surface charge and efficiency of uptake by APCs but also determines whether the liposomal content is released in early or late endosomal compartments, which has a major impact on the presentation of distinct peptide epitopes (Belizaire and Unanue, 2009; Harding *et al.*, 1991). In addition, a positive surface charge may also immobilize the vaccine carrier through electrostatic interactions with negatively charged components present in the extracellular matrix (van den Berg *et al.*, 2010), thereby hampering vaccine efficacy by reducing tissue penetration. Charged liposomes are opsonized with complementary proteins when they come into contact with plasma, thus resulting in rapid clearance by the macrophages of the reticuloendothelial system (Chonn *et al.*, 1991; Hillaireau and Couvreur, 2009). For polymeric NPs, hydrophobicity appears to be the key factor for opsonization (Hillaireau and Couvreur, 2009). Although rapid clearance by this process is generally considered a negative property for drug delivery systems, vaccination strategies might benefit from complement activation as a signal to activate DCs (Reddy *et al.*, 2007).

In conclusion, given the many factors that determine the efficiency by which particles are taken up by DCs and induce immune responses, it is difficult to predict the efficacy of novel vaccine strategies. Nevertheless, vaccine carrier size is a crucial determinant for optimal DC delivery and should not exceed 500 nm.

3. Active Targeting

Ligands for DC-specific surface receptors can be grafted onto the surface of particle-based vaccines to increase DC-specific delivery. For this purpose, nonspecific uptake by other phagocytes can be reduced by applying a hydrophilic surface coat consisting of poly(ethyleneglycol) (PEG) to reduce interactions with cell surfaces and plasma constituents (Hillaireau and Couvreur, 2009). Like passive targeting, active targeting is effective only when the vaccine carrier is relatively small. Although the same size restrictions apply for both targeting strategies for efficient entry into lymphatics, uptake by DCs is strongly enhanced by surface grafting of DC-specific antibodies on PEGylated nano- but not microparticles (Cruz *et al.*, 2010a).

Numerous DC-specific surface receptors, including Fc receptors and a range of C-type lectin receptors (CLRs), have been harnessed for targeted delivery of vaccine components (Tacken *et al.*, 2007). NPs can be targeted to Fc receptors by full antibodies or Fc fragments (Cruz *et al.*, 2010a; Mi *et al.*, 2008), with most studies targeting Fcγ receptors. However, the immunological outcome of these strategies depends on the balance between activating and inhibitory signals induced by the triggering of the various Fcγ receptors expressed by DCs (Nimmerjahn and Ravetch, 2006). This can be circumvented by using receptor-specific antibodies to specifically target immune activating Fc receptors that signal through an immunoreceptor tyrosine-based activation motif (Heijnen *et al.*, 1996; Keler *et al.*, 2000).

CLRs represent a family of lectins that bind specific carbohydrate residues in a calcium-dependent manner via their carbohydrate-recognition domain. Several CLRs show a relatively DC-specific expression pattern. Initially, carbohydrates, such as mannose and mannan, were used to target protein antigen or vaccine carriers to DCs (Karanikas *et al.*, 1997). However, most of the carbohydrates used in these early studies lacked CLR specificity (Keler *et al.*, 2004). Identification of carbohydrate ligands with more selective binding to specific CLRs led to an enhancement of the specificity of these vaccines (Sanchez-Navarro and Rojo, 2010; Singh *et al.* 2011), although many studies currently use CLR-specific antibodies to target DCs. CLRs harnessed for DC targeting include mannose receptor (He *et al.*, 2007), DEC-205 (Bonifaz *et al.*, 2004; Hawiger *et al.*, 2001; Kwon *et al.*, 2005; van Broekhoven *et al.*, 2004), DC-SIGN (Cruz *et al.*, 2010a; Kretz-Rommel *et al.*, 2007; Tacken *et al.*, 2005), Langerin (Flacher *et al.*, 2010), DCIR2 (Dudziak *et al.*, 2007), and clec9A (Caminschi *et al.*, 2008; Idoyaga *et al.*, 2011; Sancho *et al.*, 2008). The choice of specific target receptor may have a significant impact on the immunological outcome of vaccination. A study by Dudziak and coworkers demonstrates that targeting antigens to DEC-205 results mainly in cross-presentation of antigen to $CD8^+$ T cells, whereas antigens targeted to DCIR2 are preferentially presented via MHC class II molecules to $CD4^+$ T cells (Dudziak *et al.*, 2007). This difference is explained by the fact that these CLRs are expressed on distinct DC subsets that differ in their functional properties and antigen-processing capacity. DEC-205 is expressed by $CD8^+$ DCs, which are specialized in cross-presentation and initiation of cellular immune responses, while DCIR2 is expressed on $CD8^-$ DCs, which show poor antigen cross-presentation capacity (Den Haan *et al.*, 2000). Recent studies have identified the $BDCA3^+$ DC subset as the human homologue of the mouse $CD8^+$ DC subset (Bachem *et al.*, 2010; Crozat *et al.*, 2010; Jongbloed *et al.*, 2010; Poulin *et al.*, 2010; Robbins *et al.*, 2008). Clec9A expression is relatively specific for $BDCA3^+$ DCs and may therefore represent a promising target to induce cellular immune responses in humans (Caminschi *et al.*, 2008; Joffre *et al.*, 2010; Sancho *et al.*, 2008).

4. Particulate Vaccines

Many types of polymer have been used to prepare NPs for antigen delivery. Particle-based delivery systems, such as ISCOMs (immunostimulatory complexes) (Sun *et al.*, 2009), metallic NPs (Bolhassani *et al.*, 2011), liposomes (Rosenkrands *et al.*, 2011), polymers (Mata *et al.*, 2011), exosomes (Rountree *et al.*, 2011), virosomes (Moser *et al.*, 2011), and bioconjugates (Patel and Swartz, 2011), are yielding promising results in the field of vaccine development, probably because they meet many of the above-mentioned demands. Liposomes and polymeric and metallic NPs are the most extensively studied particulate vaccine carriers showing favorable characteristics for co-delivery of antigens and immunopotentiators.

One of the greatest benefits of particle-based antigen delivery systems resides in their capacity to carry antigens and immunomodulators concomitantly to the same APC, which is crucial for efficient induction of immune responses. The selection of the immunomodulator (adjuvant) for co-delivery will determine the preferential induction of Th1 or Th2 responses. Furthermore, the physical association of immunomodulators with particles may prevent adverse effects accompanying systemic immune activation by only activating the cells that are targeted. In summary, targeted nanodelivery systems for antigens offer the following advantages: (i) they protect antigens against exacerbated degradation before reaching DCs, (ii) they facilitate uptake by and/or activation of APCs, (iii) they allow concomitant delivery of antigens and adjuvants to the same APC, and (iv) they facilitate cell-mediated immune responses.

5. Targeting Gold NP Vaccines to DCs

Gold nanoparticles (AuNPs) are inert, available in many sizes (Merchant, 1998), and can be efficiently coated with a wide range of biomolecules, such as DNA (Alivisatos *et al.*, 1996), proteins (Gole *et al.*, 2001), and peptides (Hosta-Rigau *et al.*, 2010) (Fig. 8.1). AuNPs have been applied in phase I immunization studies to carry genes for the tumor-associated antigen gp100 and the cytokine GM-CSF into the skin of melanoma patients (Cassaday *et al.*, 2007). Coating antigen onto AuNPs induces humoral responses against poor immunogens, provided an adjuvant is coadministered (Parween *et al.*, 2011). Conjugation of a targeting moiety

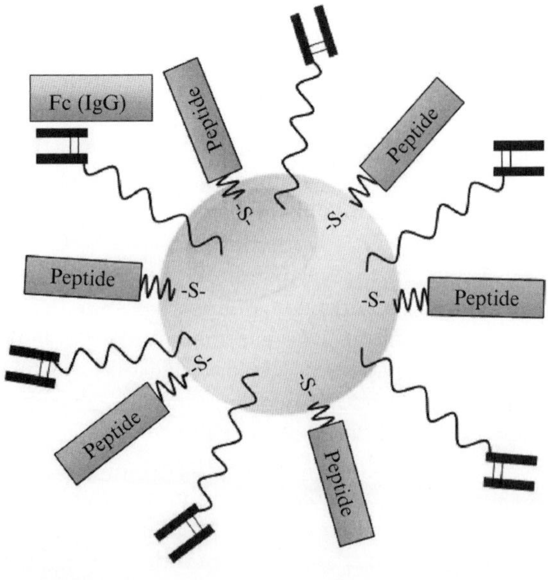

Figure 8.1 AuNPs carrying peptide antigen and targeted to FcR using the Fc fragment of IgG. (See Color Insert.)

to the gold surface allows for specific targeting of cell subsets (Cruz *et al.*, 2010b; Hosta-Rigau *et al.*, 2010). AuNPs are solid and, in contrast to liposomes or biodegradable polymer particles, they do not allow the encapsulation of antigen or immunopotentiator. However, the main advantage of AuNPs is that they can be manufactured in sizes between 5 and 100 nm, which should allow efficient delivery of conjugated antigens and adjuvants to DCs residing in the lymph nodes. In addition, AuNPs have the advantage that they show an unusually high affinity for sulfhydryl groups. This affinity facilitates efficient attachment of vaccine components that are functionalized by introducing extra cysteine residues, such as peptides or immunopotentiators (Chen *et al.*, 2010).

In mice, 25-nm-diameter AuNPs are transported efficiently into lymphatic capillaries and draining lymph nodes, whereas NPs with a diameter of 100 nm are 10 times less efficient. In another study, NPs measuring 8–17 nm in diameter were reported to be optimal in stimulating the highest antibody levels and accumulated at the highest numbers in the spleen. Particles in this size range are stable in solution and cause a visually identifiable color change upon aggregation (Chen *et al.*, 2010).

Upon reaching DCs, AuNPs can release the antigen intracellularly, so that it can be processed by MHC class I and class II pathways (Fig. 8.2).

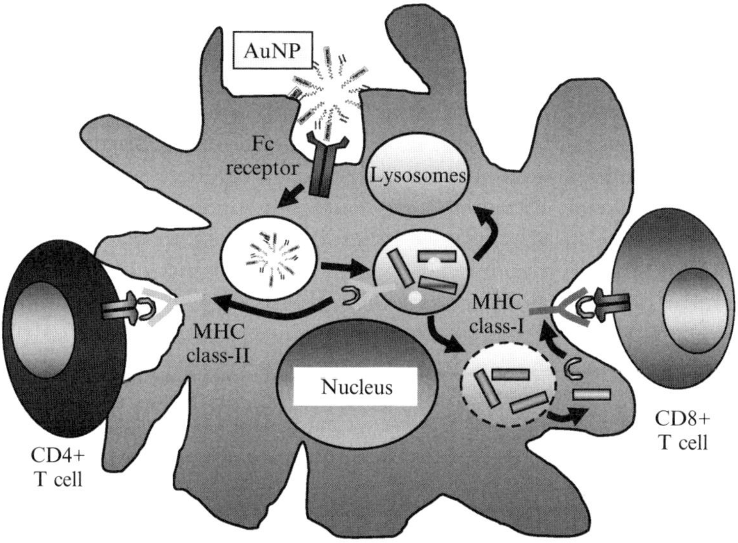

Figure 8.2 AuNPs targeted to Fc receptors on DCs are phagocytosed and processed by endosomal and lysosomal acidification and enzymes. Subsequently, they are presented in the context of MHC class II. Endosomal scape allows the processing and cross-presentation of these particles by MHC class I. (See Color Insert.)

In our experience, targeting AuNP-containing antigens to specific receptors on DCs, in our case Fcγ receptors, prevents antigen dispersion in the bloodstream. Moreover, we have shown that this strategy not only increases antigen uptake with respect to the naked peptides but also almost doubles the lymphoproliferative response induced under *in vitro* conditions (Cruz *et al.*, 2010b).

6. Experimental Procedure for Targeted AuNP Preparation

6.1. Preparation of AuNPs

AuNPs are produced by reduction of hydrogen tetrachloroaurate ($HAuCl_4 \times H_2O$) (Aldrich, Milwaukee, WI, USA). $HAuCl_4 \times H_2O$ (8.7 mg) is added to a sodium citrate solution (100 ml, 2.2 m*M*) in a round-bottom flask and reflux system. The temperature 150 °C is achieved using a magnetic stirrer hotplate. The reducing agent is added rapidly and the reaction is allowed to continue under uniform and vigorous stirring until the solution turns red. This method allows the synthesis of NPs of about 10–25 nm in diameter (Bauer *et al.*, 2003). The AuNPs can be characterized using UV–Vis spectroscopy and TEM. UV–Vis spectra of Au colloids display a single absorption peak in the visible range at about 525 nm.

6.2. Conjugation of peptide and Fc fragment to AuNPs

Conjugation is performed in the presence of an excess of peptide and Fc fragment (Kogan *et al.*, 2006). Peptides (1 mg solubilized in 1 ml of water) and the activated Fc fragment (see below for method of activation) are added dropwise to a 10-ml solution of AuNPs at room temperature with magnetic stirring, and agitation is maintained for 1 h. The AuNP complexes are purified by centrifugation using an Amicon 100-kDa filter device (Centriplus, Millipore Corporation, USA) to remove excess peptide (around 3.5 kDa) and Fc (around 50 kDa).

To determine the amount of peptide and Fc fragment per AuNP previous to purification, a 1-ml aliquot of the conjugate is used for the quantification. First, the peptide and Fc are separated by an Amicon 10-kDa filter. Thus, the Fc fraction is retained and concentrated by the filter, while the peptide fraction is collected in the flow through. The peptide fraction is lyophilized and measured by RP-HPLC (reverse-phase high performance liquid chromatography). Generally, 20–30% of peptide is conjugated to the AuNP. Second, the unbound Fc fragment is removed by an Amicon 100-kDa filter. Thus, Fc protein is collect in the flow through and lyophilized. The concentration of Fc protein is determined by the coomassie dye protein assay (Thermo Scientific) (Bradford, 1976). It is observed that 10–20% of Fc protein is conjugated to the AuNP solution.

6.3. Preactivation of Fc fragment

To bind the Fc fragment to AuNPs, the Fc (2 mg/ml in PBS, 0.5 ml) is conjugated to the heterobifunctional reagent SPDP (SPDP/Fc, 15/1 molar ratio) to introduce a thiol group. Excess SPDP is removed by a Sephadex G-50 spin column (Pharmacia Biotech, Sweden). As estimated spectrophotometrically, there are four PDP residues per Fc molecule. Thiolated Fc (Fc-SH) is obtained by reducing the Fc-PDP with 50 m*M* DTT at pH 4.5. Excess DTT is removed using a Sephadex G-50 spin column. The protein with reduced thiols is kept in a N_2 atmosphere and used immediately.

7. Targeting Liposome-Based Vaccines to DCs

The basic liposome structure used for drug and antigen delivery has not undergone a significant alteration since its introduction more than 40 years ago. Conventional unilamellar liposomes have one aqueous compartment, delimited by a single bilayer membrane, and they are classically composed of natural, biodegradable, nontoxic, and nonimmunogenic phospholipids (Düzgüneş, 2009). This bilayer defines the interior space and protects the liposomal content, while simultaneously regulating release rates. Moreover, liposomes, which range from 50 nm to several micrometers in size (Zhou and Neutra, 2002), show flexibility with regard to structure/size and lipid composition, parameters that determine the charge, and the fluidity of the liposomal membrane bilayer, as well as the capacity to incorporate large amounts of hydrophilic or hydrophobic compounds, including antigens and immunopotentiators. Owing to their particulate nature, liposomes are internalized efficiently by APCs. This feature is possibly the most important characteristic of these structures with respect to their use as vaccine delivery systems, because it allows for ingested antigen to be simultaneously processed and presented on MHC molecules (Ahsan *et al.*, 2002). In addition, the efficacy of liposome-based vaccines can be improved by delivering them specifically to APCs by exploiting various scavenger and other receptors as their targets. Co-encapsulation of antigen and TLR ligand within liposomes shows enhanced immune responses compared to vaccination with antigen and TLR ligand alone (Bal *et al.*, 2011). Targeting moieties can be engrafted on the liposomal surface for DC-specific delivery. Liposomes carrying antigen and harboring TLR5 ligand-related peptides on their surface are preferentially taken up by mouse DCs and induce potent humoral and cellular immune responses that inhibit the growth of established tumors (Faham and Altin,

2010). Alternatively, liposomes can be coated with carbohydrates as ligands for CLRs (Thomann *et al.*, 2011) or Fc fragments to target Fc receptors (Cruz *et al.*, 2010b). DC-specific targeting can be further enhanced using receptor-specific antibodies, such as those that recognize DEC-205 (van Broekhoven *et al.*, 2004) or DC-SIGN (Gieseler *et al.*, 2004). Liposomes engrafted with single-chain antibody fragments directed against DEC-205 and carrying antigen protect against tumor growth in mice, provided that DC-activating agents such as interferon-γ or the TLR4 ligand LPS are co-encapsulated in the liposome (van Broekhoven *et al.*, 2004).

Other *in vivo* studies reveal that the delivery of targeted liposome vaccines to DCs enhances immune responses. For example, subcutaneously injected targeted liposomes induce stronger humoral and CTL responses than non-targeted liposomes (Arigita *et al.*, 2003; Fukasawa *et al.*, 1998; Tacken *et al.*, 2006). Several ligands have been used for targeted delivery of antigen-loaded liposomes. For example, phosphatidylserine was incorporated into liposomes to facilitate interactions with surface receptors on monocytes (Foged *et al.*, 2004). Ligands with terminal mannose, fucose, or *N*-acetylglucosamine promote phagocytosis by binding to lectin-like receptors.

The advantage of liposome-based vaccines is that they allow simultaneous encapsulation of antigen and immunomodulators. In contrast to the AuNP strategy, conventional liposomes are generally between 50 and 250 nm in size and are not stable below 50 nm because of the high lipid curvature this requires. Although larger liposomes have the advantage that they allow entrapment of more molecules, they are probably less effective in penetrating the lymphatics.

7.1. Preparation of liposomes and peptide encapsulation

Unilamellar liposomes are prepared by freeze–thawing of lipid/peptide mixtures followed by sequential filter extrusion. Liposomes are prepared by dissolving egg phosphatidylcholine/phosphatidylglycerol/palmitoylpeptide (80/20/10 molar ratio) in methanol/methylene chloride (1:1, v/v) in a round-bottom flask. Antigens and targeting moieties can be incorporated by addition of various components to this lipid mixture. After removal of the organic solvents by rotary evaporation (40–45 °C, 30–60 min), the dry lipid mixture is suspended in PBS buffer by vigorous agitation containing peptide (2 mg/ml) in the case of aqueous soluble peptide. In addition to the peptide antigens, immunostimulatory compounds can be added to the lipid mixture or the aqueous phase. The multilamellar vesicles are subjected to 3–5 freeze–thaw cycles (liquid nitrogen–water 40 °C), followed by repetitive extrusion through Nuclepore filters (800, 400, 200, and 100 nm pore size) using a Lipex™ Extruder (Northern Lipids Inc., Vancouver, BC, Canada). Liposomes are filter-sterilized (using 0.45- or 0.2-mm sterile filters). Non-entrapped peptides and other material are removed by spin column chromatography on Sepharose 4B. Peptide encapsulation is estimated by phospholipid and FITC-peptide fluorescence determinations and is calculated to range between 70% and 85%. Vesicle size is characterized by dynamic laser light scattering using a PCS41 optic unit (Malvern Autosizer IIC, Malvern Instruments, Worcestershire, UK). The polydispersity index is an estimation of the particle size distribution width and is calculated from a simple two-parameter fit to the correlation data called a cumulants analysis of the DLS intensity autocorrelation function. The mean diameter of the liposomes is around 100–200 nm.

7.2. Preparation of targeted liposomes with the Fc fragment of human IgG conjugated to PEG

Targeted unilamellar liposomes are prepared as above, but now 1% molar ratio of DSPE-PEG_{3400}-Mal (Avanti Polar Lipids, Alabaster, AL, USA) is added to the lipid mixture (Fig. 8.3). Thiol groups are incorporated into Fc

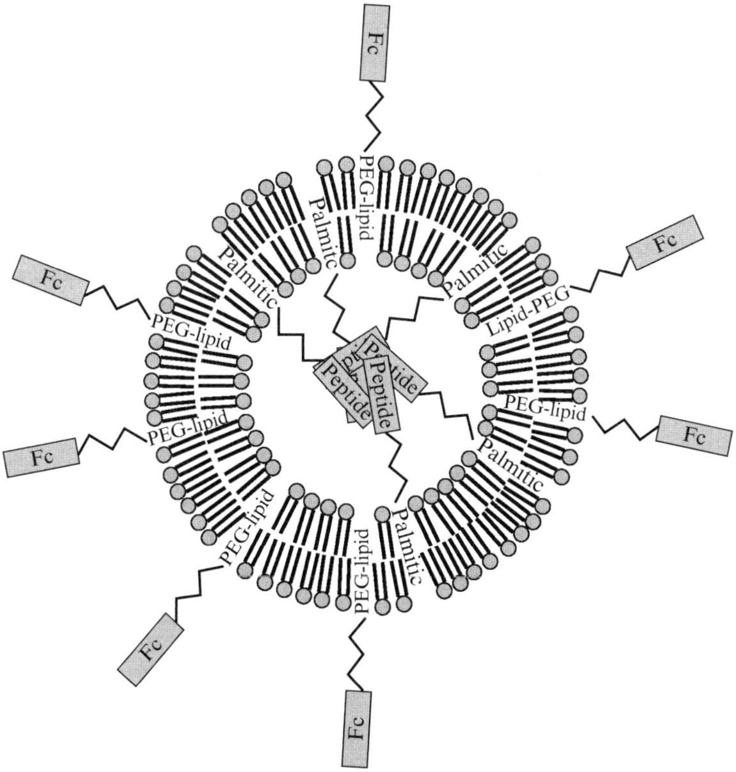

Figure 8.3 Schematic illustration of targeted liposomes. Palmitoyl-peptide-encapsulated liposomes for targeting to DC via FcR. (For the color version of this figure, the reader is referred to the Web version of this chapter.)

fragments as described above. Freshly prepared Fc-SH (2 mg/ml) is mixed with liposomes containing the coupling lipid, DSPE-PEG-Mal (a molar ratio of 20 Mal/Fc and a 1 mg/ml of lipid concentration), and incubated for 24 h at room temperature, while stirring to avoid precipitation. Free Fc-SH is removed by spin column chromatography on Sepharose 4B. The amount of Fc coupled to liposomes is calculated using FITC-Fc in comparison to a standard curve. These methods result in the incorporation of approximately 60 Fc fragments per vesicle.

8. Targeting Poly(Lactic-*co*-Glycolic Acid)-Based Vaccines to DCs

The biodegradable polymer poly(lactic-*co*-glycolic acid) (PLGA) has been used for various therapeutic applications for several decades. PLGA hydrolyzes into its monomeric components, lactic and glycolic acid, which are natural metabolites of the human body. PLGA has attracted much attention for drug delivery and vaccination purposes because of its slow release properties (Mundargi *et al.*, 2008). Many peptide and protein antigens have been successfully encapsulated within PLGA micro- and nanoparticles (Hamdy *et al.*, 2011). Co-encapsulation of antigen and TLR ligands within PLGA further improves vaccine efficacy (Elamanchili *et al.*, 2007; Schlosser *et al.*, 2008). These nontargeted PLGA vaccines induce potent humoral and cellular immune responses directed against viral infections and tumors (Kasturi *et al.*, 2011; Zhang *et al.*, 2011). Cell-specific delivery is established by conjugating targeting moieties directly to the PLGA particle core (Mo and Lim, 2005) or to ligands incorporated in the PLGA matrix (Fahmy *et al.*, 2005). However, to improve targeting efficiency, polymer NPs are often shielded with a layer of PEG or lipid-PEG, to which the targeting moiety is attached (Cruz *et al.*, 2011a; Duncanson *et al.*, 2007). We have shown that PEG chain length is a crucial factor determining the targeting efficiency of PLGA NPs carrying DC-specific antibodies. Long PEG chains inhibit antibody–receptor interactions and thereby hamper particle binding and uptake (Cruz *et al.* 2011a). So far, studies showing successful targeting of PLGA NPs to DCs are limited to *in vitro* studies. We have performed preliminary experiments targeting antigen and TLR ligands to DEC-205$^+$ DCs in mice. The results show that cellular responses are induced at much lower doses when TLR ligands are targeted together with the antigen than when administered in soluble form, thereby reducing serum cytokine levels and related toxicity (unpublished data). This observation suggests that targeted delivery of adjuvants and antigens improves the efficacy and safety of DC-targeted NP vaccines.

9. Experimental Procedure for the Preparation of Targeted PLGA NP

9.1. PLGA NP preparation

PLGA NPs with entrapped FITC-TT peptide (FITC-**KK**QYIKANSKFI-GITEL**KK**-COOH) and maturation stimulus, such as R848, are prepared using an o/w emulsion and solvent evaporation–extraction method (Fig. 8.4). Ninety milligrams of PLGA in 3 ml of dichloromethane

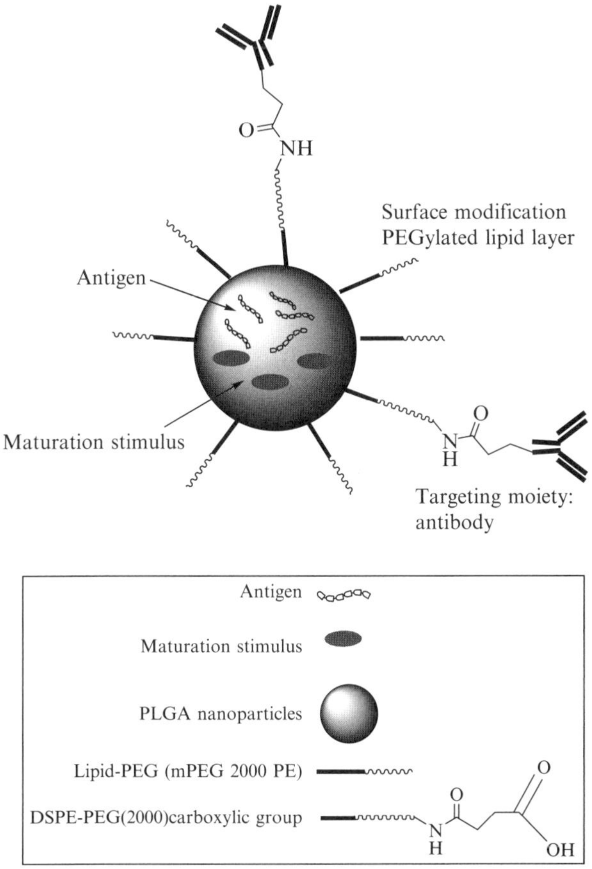

Figure 8.4 Schematic illustration of targeted PLGA NPs. NPs were generated to contain an antigen and maturation stimulus and encapsulated into the PLGA matrix by a single emulsion method. NPs were shielded by a combination of PEG-lipid and carboxyl functionalized PEG-lipid layers. The PEG-lipid prevents nonspecific interactions and the carboxyl functionalized PEG-lipid allows the introduction of antibodies onto the PLGA surface. (For the color version of this figure, the reader is referred to the Web version of this chapter.)

(DCM) containing the FITC-TT peptide (2 mg in 100 μl of water) and R848 (0.8 mg) are added dropwise to 25 ml of aqueous 2% (w/v) PVA in distilled water and emulsified for 90 s using a sonicator (Branson, Sonifier 250). A combination of lipids (DSPE-PEG(2000)carboxylic acid (6 mg) and mPEG 2000 PE (6 mg)) is dissolved in DCM and added to the first emulsion. Following overnight evaporation of the solvent at 4 °C to remove the organic solvent (DCM), the NPs are collected by ultracentrifugation at 60,000 × *g* for 30 min, washed four times with distilled water, and lyophilized. Particle size and polydispersity can be measured by DLS. Zeta potential of the particles is determined on a Malvern Zetasizer 2000. The NP diameter range is between 200 and 400 nm. The zeta potential of the particles generally ranges between −10 and −30 mV because of the negatively charged carboxylic acid groups of the PLGA polymer and the functional carboxyl group on the PEG-lipid.

9.2. Quantification of antigen in NPs

Entrapment efficiency of FITC-labeled antigen is determined by digesting 10 mg of particles in 2 ml 0.8 N NaOH overnight at 37 °C, then passing this solution through a 0.2-mm sterile syringe, and finally measuring fluorescence relative to a standard curve (492 nm excitation and 520 nm emission) using a CytoFluor II (Applied Biosystems, Foster City, CA). Antigen loading is determined by dividing the amount of encapsulated antigen by the theoretical amount assuming that the entire amount of antigen added was encapsulated. The entrapment efficiency of antigen is normally 60–90%, depending on the hydrophilicity or hydrophobicity of the antigen.

9.3. Quantification of TLR-L in NPs

The encapsulation efficiency of TLR-L (R848) is determined by RP-HPLC. After digesting 10 mg of particles in 2 ml 0.8 N NaOH overnight at 37 °C, neutralization with 1 mol/l HCl solution until neutral pH, and passage through a 0.2-mm sterile syringe filter, HPLC analysis is performed at room temperature using a Shimadzu system (Shimadzu, Kyoto, Japan) equipped with a reverse-phase Symmetric C18 column (250 × 4.6 mm). The flow rate is fixed at 1 ml/min and detection is obtained by UV detection at 220 nm. A linear gradient of 0–100% of acetonitrile (containing 0.036% trifluoroacetic acid) in water (containing 0.045% trifluoroacetic acid) is used for the separation of R848 and FITC-antigen. The peak of R848 is well separated from that of the FITC-antigen in the chromatographic conditions established. Regression analysis is performed by plotting the peak/area ratio of R848 versus concentration (mg/ml).

The calibration curves are linear within the 1–10 mg/ml range for R848, with a correlation coefficient (R2) that is greater than 0.99. The encapsulation efficiency of R848 is generally between 40% and 60%.

9.4. Conjugating antibodies to NPs

Antibody is conjugated to the DSPE-PEG(2000)carboxyl-containing PLGA NP preparation. NPs are activated with 1-ethyl-3-[3-dimethylaminopropyl] carbodiimide (EDAC) and *N*-hydroxysuccinimide (NHS) in MES buffer (pH 6.0) for 2 h at room temperature. The activated carboxyl-NP is washed three times with MES buffer by centrifugation to remove the excess of EDAC/NHS and the water-soluble isourea. Activated NPs are resuspended in 1 ml PBS buffer and reacted with antibody (200 μg per mg NP) by stirring for 3 h at room temperature. Unbound antibody is removed by centrifugation (25,000 × *g*, for 10 min) and washing of the NPs four times with PBS. The presence of antibody on the NP surface is confirmed by staining NPs with goat antihuman secondary antibodies, followed by analysis on a FACSCalibur flow cytometer using CellQuest software (BD Biosciences, USA). Quantification of antibody on the NP surface is performed by coomassie dye protein assay. In general, this conjugation procedure attaches 20–40 μg of antibody per mg of NP.

10. Conclusion

Here we described distinct methods to deliver vaccine components specifically to DCs, using targeted NP delivery systems consisting of gold, lipids, and PLGA polymer. Although all particulates have been successfully used to induce humoral and cellular responses, we favor the use of PLGA NP vaccines. PLGA NPs are relatively stable and have the advantage over AuNPs that biomolecules that should become available only after cell entry can be encapsulated. For example, AuNPs carrying a DC-targeting moiety and TLR ligands as an immunopotentiator will be recognized and ingested not only by DCs but also by cells recognizing the TLR ligands. Various TLR ligands, including the TLR2 ligand Pam3Cys and the TLR9 ligand CpG, are recognized and taken up by cells in a TLR-independent fashion by yet undefined receptors (Khan *et al.*, 2007). In addition, PLGA particles have been described to escape from the endosomal compartment (Cruz *et al.*, 2011b), which probably favors cross-presentation. Therefore, targeted PLGA NPs constitute an attractive delivery system for vaccines aiming to induce potent cellular immune responses, which are required to combat persistent viral infections and cancer.

REFERENCES

Ahsan, F., Rivas, I. P., Khan, M. A., and Torres Suarez, A. I. (2002). Targeting to macrophages: Role of physicochemical properties of particulate carriers-liposomes and microspheres-on the phagocytosis by macrophages. *J. Control. Release* **79,** 29–40.

Alivisatos, A. P., Johnsson, K. P., Peng, X., Wilson, T. E., Loweth, C. J., Bruchez, M. P., Jr., and Schultz, P. G. (1996). Organization of "nanocrystal molecules" using DNA. *Nature* **382,** 609–611.

Arigita, C., Bevaart, L., Everse, L. A., Koning, G. A., Hennink, W. E., Crommelin, D. J., van de Winkel, J. G., van Vugt, M. J., Kersten, G. F., and Jiskoot, W. (2003). Liposomal meningococcal B vaccination: Role of dendritic cell targeting in the development of a protective immune response. *Infect. Immun.* **71,** 5210–5218.

Bachem, A., Guttler, S., Hartung, E., Ebstein, F., Schaefer, M., Tannert, A., Salama, A., Movassaghi, K., Opitz, C., Mages, H. W., Henn, V., Kloetzel, P. M., *et al.* (2010). Superior antigen cross-presentation and XCR1 expression define human CD11c+CD141+ cells as homologues of mouse CD8+ dendritic cells. *J. Exp. Med.* **207,** 1273–1281.

Bachmann, M. F., and Jennings, G. T. (2010). Vaccine delivery: A matter of size, geometry, kinetics and molecular patterns. *Nat. Rev. Immunol.* **10,** 787–796.

Bal, S. M., Hortensius, S., Ding, Z., Jiskoot, W., and Bouwstram, J. A. (2011). Co-encapsulation of antigen and Toll-like receptor ligand in cationic liposomes affects the quality of the immune response in mice after intradermal vaccination. *Vaccine* **29,** 1045–1052.

Bauer, G., Hassmann, J., Walter, H., Haglmuller, J., Mayer, C., and Schalkhammer, T. (2003). Resonant nanocluster technology: From optical coding and high quality security features to biochips. *Nanotechnology* **14,** 1289–1311.

Belizaire, R., and Unanue, E. R. (2009). Targeting proteins to distinct subcellular compartments reveals unique requirements for MHC class I and II presentation. *Proc. Natl. Acad. Sci. USA* **106,** 17463–17468.

Blander, J. M., and Medzhitov, R. (2006). Toll-dependent selection of microbial antigens for presentation by dendritic cells. *Nature* **440,** 808–812.

Bolhassani, A., Safaiyan, S., and Rafati, S. (2011). Improvement of different vaccine delivery systems for cancer therapy. *Mol. Cancer* **10,** 3.

Bonifaz, L. C., Bonnyay, D. P., Charalambous, A., Darguste, D. I., Fujii, S., Soares, H., Brimnes, M. K., Moltedo, B., Moran, T. M., and Steinman, R. M. (2004). In vivo targeting of antigens to maturing dendritic cells via the DEC-205 receptor improves T cell vaccination. *J. Exp. Med.* **199,** 815–824.

Bradford, M. M. (1976). A rapid and sensitive method for the quantitation of microgram quantities of protein utilizing the principle of protein-dye binding. *Anal. Biochem.* **72,** 248–254.

Caminschi, I., Proietto, A. I., Ahmet, F., Kitsoulis, S., Shin Teh, J., Lo, J. C., Rizzitelli, A., Wu, L., Vremec, D., van Dommelen, S. L., Campbell, I. K., Maraskovsky, E., *et al.* (2008). The dendritic cell subtype-restricted C-type lectin Clec9A is a target for vaccine enhancement. *Blood* **112,** 3264–3273.

Cassaday, R. D., Sondel, P. M., King, D. M., Macklin, M. D., Gan, J., Warner, T. F., Zuleger, C. L., Bridges, A. J., Schalch, H. G., Kim, K. M., Hank, J. A., Mahvi, D. M., *et al.* (2007). A phase I study of immunization using particle-mediated epidermal delivery of genes for gp100 and GM-CSF into uninvolved skin of melanoma patients. *Clin. Cancer Res.* **13,** 540–549.

Chen, Y.-S., Hung, Y.-C., Lin, W.-H., and Huang, G. S. (2010). Assessment of gold nanoparticles as a size-dependent vaccine carrier for enhancing the antibody response against synthetic foot-and-mouth disease virus peptide. *Nanotechnology* **21,** 195101.

Chonn, A., Cullis, P. R., and Devine, D. V. (1991). The role of surface charge in the activation of the classical and alternative pathways of complement by liposomes. *J. Immunol.* **146,** 4234–4241.

Crozat, K., Guiton, R., Contreras, V., Feuillet, V., Dutertre, C. A., Ventre, E., Vu Manh, T. P., Baranek, T., Storset, A. K., Marvel, J., Boudinot, P., Hosmalin, A., *et al.* (2010). The XC chemokine receptor 1 is a conserved selective marker of mammalian cells homologous to mouse CD8alpha+ dendritic cells. *J. Exp. Med.* **207,** 1283–1292.

Cruz, L. J., Tacken, P. J., Fokkink, R., Joosten, B., Stuart, M. C., Albericio, F., Torensma, R., and Figdor, C. G. (2010a). Targeted PLGA nano—but not microparticles specifically deliver antigen to human dendritic cells via DC-SIGN in vitro. *J. Control. Release* **144,** 118–126.

Cruz, L. J., Rueda, F., Cordobilla, B., Simón, L., Hosta, L., Albericio, F., and Domingo, J. C. (2010b). Targeting nanosystems to human DCs via Fc receptor as an effective strategy to deliver antigen for immunotherapy. *Mol. Pharm.* **8,** 104–116.

Cruz, L. J., Tacken, P. J., Fokkink, R., and Figdor, C. G. (2011a). The influence of PEG chain length and targeting moiety on antibody-mediated delivery of nanoparticle vaccines to human dendritic cells. *Biomaterials* **32,** 6791–6803.

Cruz, L. J., Tacken, P. J., Bonetto, F., Buschow, S. I., Croes, H. J., Wijers, M., de Vries, I. J., and Figdor, C. G. (2011b). Multimodal imaging of nanovaccine carriers targeted to human dendritic cells. *Mol. Pharm.* **8,** 520–531.

Den Haan, J. M. M., Lehar, S. M., and Bevan, M. J. (2000). CD8+ but not CD8- dendritic cells cross-prime cytotoxic T cells in vivo. *J. Exp. Med.* **192,** 1685–1696.

Dudziak, D., Kamphorst, A. O., Heidkamp, G. F., Buchholz, V. R., Trumpfheller, C., Yamazaki, S., Cheong, C., Liu, K., Lee, H. W., Park, C. G., Steinman, R. M., and Nussenzweig, M. C. (2007). Differential antigen processing by dendritic cell subsets in vivo. *Science* **315,** 107–111.

Duncanson, W. J., Figa, M. A., Hallock, K., Zalipsky, S., Hamilton, J. A., and Wong, J. Y. (2007). Targeted binding of PLA microparticles with lipid-PEG-tethered ligands. *Biomaterials* **28,** 4991–4999.

Düzgüneş, N. (2009). Liposomes, Part G. Preface. *In* "Methods in Enzymology," (N. Düzgünes, ed.), **465,** p. xix. Elsevier/Academic Press, San Diego.

Elamanchili, P., Lutsiak, C. M., Hamdy, S., Diwan, M., and Samuel, J. (2007). "Pathogen-mimicking" nanoparticles for vaccine delivery to dendritic cells. *J. Immunother.* **30,** 378–395.

Faham, A., and Altin, J. G. (2010). Antigen-containing liposomes engrafted with flagellin-related peptides are effective vaccines that can induce potent antitumor immunity and immunotherapeutic effect. *J. Immunol.* **185,** 1744–1754.

Fahmy, T. M., Samstein, R. M., Harness, C. C., and Mark-Saltzman, W. (2005). Surface modification of biodegradable polyesters with fatty acid conjugates for improved drug targeting. *Biomaterials* **26,** 5727–5736.

Fifis, T., Gamvrellis, A., Crimeen-Irwin, B., Pietersz, G. A., Li, J., Mottram, P. L., McKenzie, I. F., and Plebanski, M. (2004). Size-dependent immunogenicity: Therapeutic and protective properties of nano-vaccines against tumors. *J. Immunol.* **173,** 3148–3154.

Flacher, V., Tripp, C. H., Stoitzner, P., Haid, B., Ebner, S., Del Frari, B., Koch, F., Park, C. G., Steinman, R. M., Idoyaga, J., and Romani, N. (2010). Epidermal Langerhans cells rapidly capture and present antigens from C-type lectin-targeting antibodies deposited in the dermis. *J. Invest. Dermatol.* **130,** 755–762.

Foged, C., Arigita, C., Sundblad, A., Jiskoot, W., Storm, G., and Frokjaer, S. (2004). Interaction of dendritic cells with antigen-containing liposomes: Effect of bilayer composition. *Vaccine* **22,** 1903–1913.

Foged, C., Brodin, B., Frokjaer, S., and Sundblad, A. (2005). Particle size and surface charge affect particle uptake by human dendritic cells in an in vitro model. *Int. J. Pharm.* **298,** 315–322.

Fukasawa, M., Shimizu, Y., Shikata, K., Nakata, M., Sakakibara, R., Yamamoto, N., Hatanaka, M., and Mizuochi, T. (1998). Liposome oligomannose-coated with neoglycolipid, a new candidate for a safe adjuvant for induction of CD8(+) cytotoxic T lymphocytes. *FEBS Lett.* **441,** 353–356.

Gieseler, R. K., Marquitan, G., Hahn, M. J., Perdon, L. A., Driessen, W. H., Sullivan, S. M., and Scolaro, M. J. (2004). DC-SIGN-specific liposomal targeting and selective intracellular compound delivery to human myeloid dendritic cells: Implications for HIV disease. *Scand. J. Immunol.* **59,** 415–424.

Gole, A., Dash, C., Soman, C., Sainkar, S. R., Rao, M., and Sastry, M. (2001). On the preparation, characterization, and enzymatic activity of fungal protease-gold colloid bioconjugates. *Bioconjug. Chem.* **12,** 684–690.

Hamdy, S., Haddadi, A., Hung, R. W., and Lavasanifar, A. (2011). Targeting dendritic cells with nano-particulate PLGA cancer vaccine formulations. *Adv. Drug Deliv. Rev.* **63,** 943–955.

Harding, C. V., Collins, D. S., Slot, J. W., Geuze, H. J., and Unanue, E. R. (1991). Liposome-encapsulated antigens are processed in lysosomes, recycled, and presented to T cells. *Cell* **64,** 393–401.

Hawiger, D., Inaba, K., Dorsett, Y., Guo, M., Mahnke, K., Rivera, M., Ravetch, J. V., Steinman, R. M., and Nussenzweig, M. C. (2001). Dendritic cells induce peripheral T cell unresponsiveness under steady state conditions in vivo. *J. Exp. Med.* **194,** 769–779.

He, L. Z., Crocker, A., Lee, J., Mendoza-Ramirez, J., Wang, X. T., Vitale, L. A., O'Neill, T., Petromilli, C., Zhang, H. F., Lopez, J., Rohrer, D., Keler, T., *et al.* (2007). Antigenic targeting of the human mannose receptor induces tumor immunity. *J. Immunol.* **178,** 6259–6267.

Heijnen, I. A., Vugt, M. J., Fanger, N. A., Graziano, R. F., de Wit, T. P., Hofhuis, F. M., Guyre, P. M., Capel, P. J., Verbeek, J. S., and van de Winkel, J. G. (1996). Antigen targeting to myeloid-specific human Fc gamma RI/CD64 triggers enhanced antibody responses in transgenic mice. *J. Clin. Invest.* **97,** 331–338.

Hillaireau, H., and Couvreur, P. (2009). Nanocarriers' entry into the cell: Relevance to drug delivery. *Cell. Mol. Life Sci.* **66,** 2873–2896.

Hosta-Rigau, L., Olmedo, I., Arbiol, J., Cruz, L. J., Kogan, M. J., and Albericio, F. (2010). Multifunctionalized gold nanoparticles with peptides targeted to gastrin-releasing peptide receptor of a tumor cell line. *Bioconjug. Chem.* **21,** 1070–1078.

Idoyaga, J., Lubkin, A., Fiorese, C., Lahoud, M. H., Caminschi, I., Huang, Y., Rodriguez, A., Clausen, B. E., Park, C. G., Trumpfheller, C., and Steinman, R. M. (2011). Comparable T helper 1 (Th1) and CD8 T-cell immunity by targeting HIV gag p24 to CD8 dendritic cells within antibodies to Langerin, DEC205, and Clec9A. *Proc. Natl. Acad. Sci. USA* **108,** 2384–2389.

Joffre, O. P., Sancho, D., Zelenay, S., Keller, A. M., and Reis e Sousa, C. (2010). Efficient and versatile manipulation of the peripheral CD4+ T-cell compartment by antigen targeting to DNGR-1/CLEC9A. *Eur. J. Immunol.* **40,** 1255–1265.

Jongbloed, S. L., Kassianos, A. J., McDonald, K. J., Clark, G. J., Ju, X., Angel, C. E., Chen, C. J., Dunbar, P. R., Wadley, R. B., Jeet, V., Vulink, A. J., Hart, D. N., *et al.* (2010). Human CD141+ (BDCA-3)+ dendritic cells (DCs) represent a unique myeloid DC subset that cross-presents necrotic cell antigens. *J. Exp. Med.* **207,** 1247–1260.

Karanikas, V., Hwang, L. A., Pearson, J., Ong, C. S., Apostolopoulos, V., Vaughan, H., Xing, P. X., Jamieson, G., Pietersz, G., Tait, B., Broadbent, R., Thynne, G., *et al.* (1997). Antibody and T cell responses of patients with adenocarcinoma immunized with mannan-MUC1 fusion protein. *J. Clin. Invest.* **100,** 2783–2792.

Kasturi, S. P., Skountzou, I., Albrecht, R. A., Koutsonanos, D., Hua, T., Nakaya, H. I., Ravindran, R., Stewart, S., Alam, M., Kwissa, M., Villinger, F., Murthy, N., *et al.* (2011). Programming the magnitude and persistence of antibody responses with innate immunity. *Nature* **470,** 543–547.

Keler, T., Guyre, P. M., Vitale, L. A., Sundarapandiyan, K., van De Winkel, J. G., Deo, Y. M., and Graziano, R. F. (2000). Targeting weak antigens to CD64 elicits potent humoral responses in human CD64 transgenic mice. *J. Immunol.* **165,** 6738–6742.

Keler, T., Ramakrishna, V., and Fanger, M. W. (2004). Mannose receptor-targeted vaccines. *Expert Opin. Biol. Ther.* **4,** 1953–1962.

Khan, S., Bijker, M. S., Weterings, J. J., Tanke, H. J., Adema, G. J., van Hall, T., Drijfhout, J. W., Melief, C. J., Overkleeft, H. S., van der Marel, G. A., Filippov, D. V., van der Burg, S. H., and Ossendorp, F. (2007). Distinct uptake mechanisms but similar intracellular processing of two different toll-like receptor ligand-peptide conjugates in dendritic cells. *J. Biol. Chem.* **282,** 21145–21159.

Kogan, M. J., Bastus, N. G., Amigo, R., Grillo-Bosch, D., Araya, E., Turiel, A., Labarta, A., Giralt, E., and Puntes, V. F. (2006). Nanoparticle-mediated local and remote manipulation of protein aggregation. *Nano Lett.* **6,** 110–115.

Kretz-Rommel, A., Qin, F., Dakappagari, N., Torensma, R., Faas, S., Wu, D., and Bowdish, K. S. (2007). In vivo targeting of antigens to human dendritic cells through DC-SIGN elicits stimulatory immune responses and inhibits tumor growth in grafted mouse models. *J. Immunother.* **30,** 715–726.

Kwon, Y. J., James, E., Shastri, N., and Frechet, J. M. (2005). In vivo targeting of dendritic cells for activation of cellular immunity using vaccine carriers based on pH-responsive microparticles. *Proc. Natl. Acad. Sci. USA* **102,** 18264–18268.

Manolova, V., Flace, A., Bauer, M., Schwarz, K., Saudan, P., and Bachmann, M. F. (2008). Nanoparticles target distinct dendritic cell populations according to their size. *Eur. J. Immunol.* **38,** 1404–1413.

Mata, E., Igartua, M., Patarroyo, M. E., Pedraz, J. L., and Hernández, R. M. (2011). Enhancing immunogenicity to PLGA microparticulate systems by incorporation of alginate and RGD-modified alginate. *Eur. J. Pharm. Sci.* **44,** 32–40.

Merchant, B. (1998). Gold, the noble metal and the paradoxes of its toxicology. *Biologicals* **26,** 49–59.

Mi, W., Wanjie, S., Lo, S. T., Gan, Z., Pickl-Herk, B., Ober, R. J., and Ward, E. S. (2008). Targeting the neonatal Fc receptor for antigen delivery using engineered Fc fragments. *J. Immunol.* **181,** 7550–7561.

Mo, Y., and Lim, L. Y. (2005). Preparation and in vitro anticancer activity of wheat germ agglutinin (WGA)-conjugated PLGA nanoparticles loaded with paclitaxel and isopropyl myristate. *J. Control. Release* **107,** 30–42.

Moser, C., Amacker, M., and Zurbriggen, R. (2011). Influenza virosomes as a vaccine adjuvant and carrier system. *Expert Rev. Vaccines* **10,** 437–446.

Mundargi, R. C., Babu, V. R., Rangaswamy, V., Patel, P., and Aminabhavi, T. M. (2008). Nano/micro technologies for delivering macromolecular therapeutics using poly(d, l-lactide-co-glycolide) and its derivatives. *J. Control. Release* **125,** 193–209.

Nakanishi, T., Kunisawa, J., Hayashi, A., Tsutsumi, Y., Kubo, K., Nakagawa, S., Nakanishi, M., Tanaka, K., and Mayumi, T. (1999). Positively charged liposome functions as an efficient immunoadjuvant in inducing cell-mediated immune response to soluble proteins. *J. Control. Release* **61,** 233–240.

Nimmerjahn, F., and Ravetch, J. V. (2006). Fc-gamma receptors: Old friends and new family members. *Immunity* **24,** 19–28.

Oussoren, C., Zuidema, J., Crommelin, D. J. A., and Storm, G. (1997). Lymphatic uptake and biodistribution of liposomes after subcutaneous injection: II. Influence of liposomal size, lipid composition and lipid dose. *Biochim. Biophys. Acta* **1328,** 261–272.

Parween, S., Gupta, P. K., and Chauhan, V. S. (2011). Induction of humoral immune response against PfMSP-119 and PvMSP-119 using gold nanoparticles along with alum. *Vaccine* **29,** 2451–2460.

Patel, K. G., and Swartz, J. R. (2011). Surface functionalization of virus-like particles by direct conjugation using azide-alkyne click chemistry. *Bioconjug. Chem.* **22,** 376–387.

Poulin, L. F., Salio, M., Griessinger, E., Anjos-Afonso, F., Craciun, L., Chen, J. L., Keller, A. M., Joffre, O., Zelenay, S., Nye, E., Le Moine, A., Faure, F., *et al.* (2010). Characterization of human DNGR-1^{+} BDCA3^{+} leukocytes as putative equivalents of mouse CD8α^{+} dendritic cells. *J. Exp. Med.* **207,** 1261–1271.

Reddy, S. T., van der Vlies, A. J., Simeoni, E., Angeli, V., Randolph, G. J., O'Neil, C. P., Lee, L. K., Swartz, M. A., and Hubbell, J. A. (2007). Exploiting lymphatic transport and complement activation in nanoparticle vaccines. *Nat. Biotechnol.* **25,** 1159–1164.

Reis e Sousa, C. (2011). Harnessing dendritic cells. *Semin. Immunol.* **23,** 1.

Robbins, S., Walzer, T., Dembele, D., Thibault, C., Defays, A., Bessou, G., Xu, H., Vivier, E., Sellars, M., Pierre, P., Sharp, F. R., Chan, S., *et al.* (2008). Novel insights into the relationships between dendritic cell subsets in human and mouse revealed by genome-wide expression profiling. *Genome Biol.* **9,** R17.

Rosenkrands, I., Vingsbo-Lundberg, C., Bundgaard, T. J., Lindenstrø, T., Enouf, V., van der Werf, S., Andersen, P., and Agger, E. M. (2011). Enhanced humoral and cell-mediated immune responses after immunization with trivalent influenza vaccine adjuvanted with cationic liposomes. *Vaccine* **29,** 6283–6291.

Rountree, R. B., Mandl, S. J., Nachtwey, J. M., Dalpozzo, K., Do, L., Lombardo, J. R., Schoonmaker, P. L., Brinkmann, K., Dirmeier, U., Laus, R., and Delcayre, A. (2011). Exosome targeting of tumor antigens expressed by cancer vaccines can improve antigen immunogenicity and therapeutic efficacy. *Cancer Res.* **71,** 5235–5244.

Sanchez-Navarro, M., and Rojo, J. (2010). Targeting DC-SIGN with carbohydrate multivalent systems. *Drug News Perspect.* **23,** 557–572.

Sancho, D., Mourao-Sa, D., Joffre, O. P., Schulz, O., Rogers, N. C., Pennington, D. J., Carlyle, J. R., and Reis e Sousa, C. (2008). Tumor therapy in mice via antigen targeting to a novel, DC-restricted C-type lectin. *J. Clin. Invest.* **118,** 2098–20110.

Schlosser, E., Mueller, M., Fischer, S., Basta, S., Busch, D. H., Gander, B., and Groettrup, M. (2008). TLR ligands and antigen need to be coencapsulated into the same biodegradable microsphere for the generation of potent cytotoxic T lymphocyte responses. *Vaccine* **26,** 1626–1637.

Singh, S. K., Streng-Ouwehand, I., Litjens, M., Kalay, H., Burgdorf, S., Saeland, E., Kurts, C., Unger, W. W., and van Kooyk, Y. (2011). Design of neo-glycoconjugates that target the mannose receptor and enhance TLR-independent cross-presentation and Th1 polarization. *Eur. J. Immunol.* **41,** 916–925.

Sun, H. X., Xie, Y., and Ye, Y. P. (2009). ISCOMs and ISCOMATRIX. *Vaccine* **27,** 4388–4401.

Swartz, M. A. (2001). The physiology of the lymphatic system. *Adv. Drug Deliv. Rev.* **50,** 3–20.

Tacken, P. J., de Vries, I. J., Gijzen, K., Joosten, B., Wu, D., Rother, R. P., Faas, S. J., Punt, C. J., Torensma, R., Adema, G. J., and Figdor, C. G. (2005). Effective induction of naive and recall T cell responses by targeting antigen to human dendritic cells via a humanized anti-DC-SIGN antibody. *Blood* **106,** 1278–1285.

Tacken, P. J., Torensma, R., and Figdor, C. G. (2006). Targeting antigens to dendritic cells in vivo. *Immunobiology* **211,** 599–608.

Tacken, P. J., de Vries, I. J., Torensma, R., and Figdor, C. G. (2007). Dendritic-cell immunotherapy: From ex vivo loading to in vivo targeting. *Nat. Rev. Immunol.* **7,** 790–802.

Thomann, J. S., Heurtault, B., Weidner, S., Brayé, M., Beyrath, J., Fournel, S., Schuber, F., and Frisch, B. (2011). Antitumor activity of liposomal ErbB2/HER2 epitope peptide-based vaccine constructs incorporating TLR agonists and mannose receptor targeting. *Biomaterials* **32,** 4574–4583.

van Broekhoven, C. L., Parish, C. R., Demangel, C., Britton, W. J., and Altin, J. G. (2004). Targeting dendritic cells with antigen-containing liposomes: A highly effective procedure for induction of antitumor immunity and for tumor immunotherapy. *Cancer Res.* **64,** 4357–4365.

van de Ven, R., van den Hout, M. F., Lindenberg, J. J., Sluijter, B. J., van Leeuwen, P. A., Lougheed, S. M., Meijer, S., van den Tol, M. P., Scheper, R. J., and de Gruijl, T. D. (2011). Characterization of four conventional dendritic cell subsets in human skin-draining lymph nodes in relation to T-cell activation. *Blood* **118,** 2502–2510.

Van den Berg, J. H., Oosterhuis, K., Hennink, W. E., Storm, G., van der Aa, L. J., Engbersen, J. F., Haanen, J. B., Beijnen, J. H., Schumacher, T. N., and Nuijen, B. (2010). Shielding the cationic charge of nanoparticle-formulated dermal DNA vaccines is essential for antigen expression and immunogenicity. *J. Control. Release* **141,** 234–240.

Wille-Reece, U., Flynn, B. J., Lore, K., Koup, R. A., Kedl, R. M., Mattapallil, J. J., Weiss, W. R., Roederer, M., and Seder, R. A. (2005). HIV Gag protein conjugated to a Toll-like receptor 7/8 agonist improves the magnitude and quality of Th1 and $CD8^+$ T cell responses in nonhuman primates. *Proc. Natl. Acad. Sci. USA* **102,** 15190–15194.

Wischke, C., Borchert, H. H., Zimmermann, J., Siebenbrodt, I., and Lorenzen, D. R. (2006). Stable cationic microparticles for enhanced model antigen delivery to dendritic cells. *J. Control. Release* **114,** 359–368.

Xiang, S. D., Scholzen, A., Minigo, G., David, C., Apostolopoulos, V., Mottram, P. L., and Plebanski, M. (2006). Pathogen recognition and development of particulate vaccines: Does size matter? *Methods* **40,** 1–9.

Yotsumoto, S., Aramaki, Y., Kakiuchi, T., and Tsuchiya, S. (2004). Induction of antigen-dependent interleukin-12 production by negatively charged liposomes encapsulating antigens. *Vaccine* **22,** 3503–3509.

Zhang, Z., Tongchusak, S., Mizukami, Y., Kang, Y. J., Ioji, T., Touma, M., Reinhold, B., Keskin, D. B., Reinherz, E. L., and Sasada, T. (2011). Induction of anti-tumor cytotoxic T cell responses through PLGA-nanoparticle mediated antigen delivery. *Biomaterials* **32,** 3666–3678.

Zhou, F., and Neutra, M. R. (2002). Antigen delivery to mucosa-associated lymphoid tissues using liposomes as a carrier. *Biosci. Rep.* **22,** 355–369.

CHAPTER NINE

Protein–Carbon Nanotube Sensors: Single Platform Integrated Micro Clinical Lab for Monitoring Blood Analytes[☆]

Sowmya Viswanathan,* Pingzuo Li,[†,‡] Wonbong Choi,[§] Slawomir Filipek,[¶] T. A. Balasubramaniam,[||] *and* V. Renugopalakrishnan[†]

[☆] Dedicated to Varun

Contents

1. Introduction 166
2. Prototype 167
3. Concept 168
4. Biosensors 169
5. Microfluidics 169
6. Amperometric Sensor for the Detection of Blood Analytes 169
7. Protein Probes for Detection and Monitoring Molecular Components of Serum 171
   7.1. Glucose oxidase 171
   7.2. Cholesterol oxidase 173
   7.3. Cholesterol esterase 173
   7.4. Lipase 175
   7.5. Fructosyl valine amino oxidase 176
8. Protein Engineering and Molecular Biology of Probe Proteins 180
9. Control of the Fermentation Process 181
10. Gene Structure and Purification of Overexpressed Protein 181
11. Immobilization of Proteins on SWCNT 183
12. Immobilization of Proteins on Self-Assembled Monolayers 185
13. Characterization of SWCNT Enzyme Adduct 186
14. Mediators 186

* Newton - Wellesley Hospital/Partners Healthcare System, Newton, Massachusetts, USA
[†] Children's Hospital, Harvard Medical School, Boston, Massachusetts, USA
[‡] Shanghai Research Center of Biotechnology, Chinese Academy of Sciences, Shanghai, PR China
[§] Nanomaterials and Device Lab, College of Engineering, Florida International University, Miami, Florida, USA
[¶] Faculty of Chemistry, University of Warsaw, Warsaw, Poland
[||] Biomedical Engineer, 184 Algonquin Trail, Ashland, Massachusetts, USA

Methods in Enzymology, Volume 509
ISSN 0076-6879, DOI: 10.1016/B978-0-12-391858-1.00010-1

Abstract

Design of a unique, single-platform, integrated, multichannel sensor based on carbon nanotube (CNT)–protein adducts specific to each one of the major analytes of blood, glucose, cholesterol, triglyceride, and Hb1AC is presented. The concept underlying the sensor, amperometric detection, is applicable to various disease-monitoring strategies. There is an urgent need to enhance the sensitivity of glucometers to $<5\%$ level instead of greater than the present 15% standard in these detectors. CNTs enhance the signals derived from the interaction of the enzymes with the different analytes in blood. Fabricated sensors using the new methodology is a point-of-care device that is targeted for home, clinical, and emergency use and can be redesigned for continuous monitoring for critical care patients.

1. Introduction

Health care is poised to revolutionize the present century with scalable technology and the collaborative power of the internet. The quest for greater efficiency in the delivery of health care services is eternal for a country that spends far more on health care that outstrips many other segments of the economy. In several countries, healthcare expenditure will grow at a faster pace than their economic growth (20% of GDP by 2014 in United States). So there is a need to establish a robust health information system for effective delivery of health services. This should begin with prevention and continue with an integrated approach to manage chronic illnesses, treat ongoing healthcare needs, and address life-threatening diseases. In order to achieve this, there will be a compelling need to integrate an ever-increasing body of scientific knowledge of both generalized and individualized practice. Nanotechnology affords cost-effective medical devices that can be used for self-monitoring. Automated monitoring devices that can effectively track serum lipids, serum glucose, and glycosylated hemoglobin (HbA1C) in a single platform will make home monitoring, outpatient and inpatient monitoring easy.

Serum glucose, cholesterol, triglyceride, and HbA1C monitoring are valuable tools in the management of diabetes and cardiovascular diseases. Point-of-care devices, continuous glucose monitoring systems, and noninvasive glucose monitoring systems have significantly improved. However, there continues to be several challenges related to the achievement of accurate and reliable glucose monitoring (Klonoff, 2011). Currently, glucometers are required under an international standard to produce results within a 20% margin of error. Food and Drug Administration is contemplating more stringent standards (http://www.aacc.org/members/nacb/NACBBlog/lists/posts/post.aspx?ID=25).

Protein–carbon nanotube (CNT) sensors typically are several orders of magnitude greater in their sensitivity (Roy *et al.*, 2006, 2008). Further technical improvements in glucose biosensors, standardization of the analytical goals for their performance, and continuously assessing and training lay users are required (Yoo and Lee, 2010). Protein tagged to CNT offers enhanced sensitivity in the measurement of serum glucose levels and other analytes in blood by enhancing the signals emanating from the electron charge transfer occurring between the chosen candidate protein and CNT which are excellent electrical conductors.

With this motivation, a unique, integrated, highly efficient, lab-on-a-chip device based on the technology of CNT is proposed that will measure cholesterol, triglycerides, glucose, and HbA1C from a single drop of blood. Recent experiments in our laboratory have shown graphene will be a more versatile platform than CNTs (see Pumera, 2011).

2. Prototype

Prototype device exhibits three functions: (i) a compartment to introduce and remove a disposable chip that accommodates microchambers with multichannels, (ii) a micropump, and (iii) a reader and it requires only a few microliters for detection. The chip contains filtration and electrochemical reaction microchambers which are interconnected through microchannels (Children's Hospital Boston, a teaching affiliate of Harvard, and FIU, Miami, Florida have been working on a multichannel SW CNT tagged to geneticallly engineered GOx, ChE, ChO to enhance thermal properties by protein engineering and the detection of HbA1C using fructosyl valine amino acid oxidase tagged to SW CNT, the first of its kind to detect HbA1C levels in serum (IP disclosures, Provisional Patents, pending). Carbon nanotube based MultiSensor Biochip for point-of-care clinical diagnostics (W.Choi and S. Roy, FIU provisional Patent). The sensors are several field effect transistors (FETs), the channel of which are prefunctionalized semiconducting single wall carbon nanotube (SWCNT) and will be operated in back−gated mode. Each covalently anchored protein on the SWCNTs would be sensitive only to a particular targeted species. When the targeted species interact with the proteins, electrochemical reactions happen and conductance of the FET changes resulting in a signal. The ease in functionalizing SWCNT to detect desired species compared to other nanowires or nanoelements makes the detection of species at the sub-ppm level. Conventional and state-of-the-art-directed assembly methods will be employed to fabricate the prototype devices. The electrohydrodynamic micropump would be integrated into the micro sample holder. The filtration chamber consisting of SWCNTs separates out the cholesterol and lipoproteins from their background cellular products in the blood.

Microscale prototype lab-on-a-chip consists of four parts, microchamber for blood and reagent loading, fluidic channel, nanotube-based filter, and nanotube sensor. This chip fabrication requires microchannels, chambers construction by lithography, micron scale CNT sensor installation followed by probe immobilization, surface treatment (hydrophilicity/hydrophobicity), quartz/silicon bonding, rendering the fabrication cost effective compared to conventional fabrication. The substrate should also be capable of withstanding nanotube growth temperatures since it is going to be used for *in situ* nanotube growth substrate. The candidate for the substrate is silicon and quartz glass. The housing for the chambers, channels, filter, and the sensor can be obtained by selectively etching the substrate. Nanotube growth takes place selectively on the predefined area which has deposition of selected catalysts. The growing temperature can be controlled as low as 400 °C using thermal chemical vapor deposition technique. Quartz and silicon channels are fabricated by using e-beam lithography. The chip is covered with glass plate to contain the samples and reagents. Electrodes will be placed in reservoirs that connect to the end of the various channels. Potential is applied to the various reservoirs to move the fluid in the desired direction.

3. Concept

Bio-inspired systems attempt to harness various functions of biological macromolecules and integrate them with engineering for technological applications (Renugopalakrishnan and Lewis, 2006). Biosensors combine the selectivity of biological macromolecules with the processing power of modern microelectronics and optoelectronics to offer new powerful analytical tools with major application in military as well as in medical and environmental diagnostics.

When proteins are covalently bonded to CNTs, electron charge transfer occurs which is facilitated by CNT providing an excellent platform for high-performance devices (Hu *et al.*, 2010; Roy *et al.*, 2005). The nano-dimensions of SWCNTs and their electronic properties make them an ideal candidate for anchoring the proteins for biochemical sensing. The monitoring of glucose, triglycerides, low-density lipoprotein (LDL), high-density lipoprotein (HDL), as well as HbA1c is very important due to their crucial role in diabetes and its associated complications like hyperlipidemia and coronary heart disease (Roy, *et al.*, 2006; Viswanathan *et al.*, to be submitted). Reengineering protein functionalities for technological applications draws upon garnered knowledge in protein structural biology and engineering. Process of self-assembly promotes spontaneous formation of ordered supramolecular aggregates or with other organic/inorganic nanomaterials for easy design of protein-based nanodevices (Renugopalakrishnan and Lewis, 2006). Bio–nano interface between the enzyme and CNT is the corridor through which electron charge transfer takes place in a biosensor. The hot spots in enzymes/proteins that are at the interface which make the physical contact between the enzymes and CNT are potential candidates for protein engineering. We have recently examined a number of protein hot spots in enzymes and other proteins (Audette *et al.*, 2011) in biosolar cell and bio fuel cells in addition to biosensors. A multidisciplinary approach to detect major components of blood by an innovative fusion of protein engineering and nanoscience is promising.

4. Biosensors

Biosensors typically consist of a candidate enzyme which is anchored on a CNT platform and more recently on graphene by covalent attachment. Interaction of a biosensor consists of three steps: molecular recognition (Roy, *et al.*, 2006) where the enzyme interacts with an analyte, signal generation produced by molecular recognition and the transfer of electronic charge, and finally there is a current variation which is proportional to the concentration of analyte (Arya *et al.*, 2008). Selection of an enzyme candidate is carefully done, which is expected to manifest optimal molecular recognition and the ease of molecular biological and protein engineering methods to clone, overexpress, purify, and perform site-directed mutagenesis to further enhance the chosen candidate enzyme's binding to a given analyte. When multiple enzymes are chosen to interact with different analytes, the design of a single integrated biosensor to detect and quantify the different analytes becomes technologically challenging. A schematic of a biosensor is shown in Fig. 9.1 reproduced from Pumera (2011).

5. Microfluidics

Also the flow of a viscous fluid like blood requires understanding and manipulating the microfluidics (Chakraborty, 2005) to ensure the efficient functioning of the designed biosensor. Microfluidics provides an understanding of the behavior, precise control, and manipulation of fluids that are geometrically constrained to a small, typically submillimeter scale.

6. Amperometric Sensor for the Detection of Blood Analytes

A generalized scheme of an integrated four-channel biosensor is shown in Fig. 9.2.

The designed platform can be generalized to any pathological/metabolic disorder that has enzyme detection or assay titration as the basis for diagnosing, monitoring, and treating diseases. This could include comprehensive

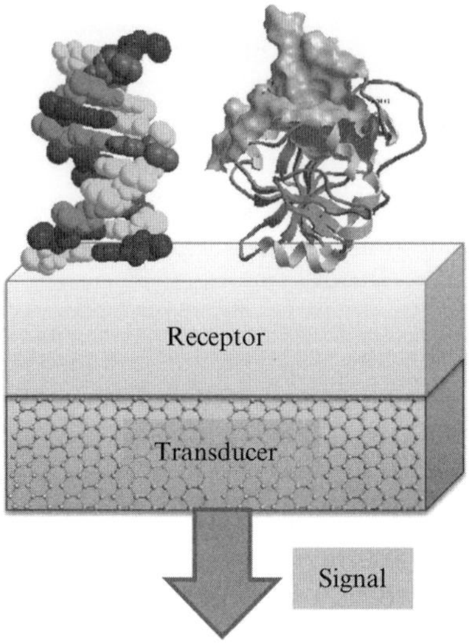

Figure 9.1 A schematic of a biosensor reproduced with copyright permission from Pumera (2011). (For the color version of this figure, the reader is referred to the Web version of this chapter.)

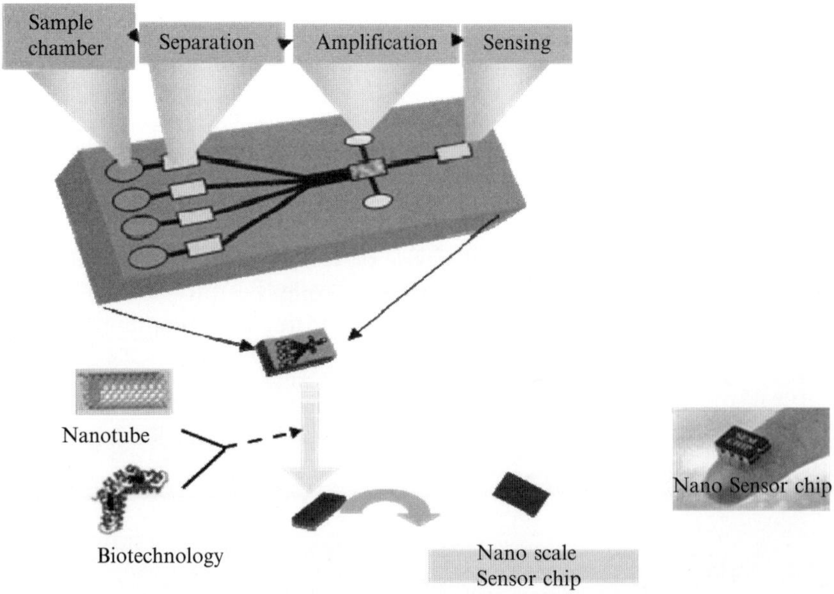

Figure 9.2 Schematic of a multichannel single platform CNT–protein sensor for quantitative monitoring of blood analytes. (For the color version of this figure, the reader is referred to the Web version of this chapter.)

metabolic panels such as liver enzymes such as AST/ALT; kidney tests such as BUN/creatinine, thyroid function tests and the prospects are endless. Oral fluid is a perfect medium to be explored for health and disease surveillance and the method developed here may be suitably modified for salivary diagnostics (Wong 2006).

Protein-based biosensors combine the selectivity of proteins with the processing power of modern nano and optoelectronics to offer new powerful analytical tools with major applications in medical diagnostics. The nano-dimensions of SWCNTs and their electronic properties make them an ideal substrate candidate for anchoring the proteins for biochemical sensing.

We have demonstrated vertically aligned CNTs for monitoring blood cholesterol by promoting heterogeneous electron transfer between the enzyme and the working electrode (Roy *et al.*, 2006). Surface modification of the CNT with a biocompatible polymer converted the hydrophobic nanotube surface into a highly hydrophilic one, which facilitates efficient attachment of biomolecules. The fabricated working electrodes showed a linear relationship between cholesterol concentration and the output signal. The efficacy of the CNT in promoting heterogeneous electron transfer was evident by distinct electrochemical peaks and higher signal-to-noise ratio as compared to the Au electrode with identical enzyme immobilization protocol.

Total blood cholesterol consists of molecules of LDL and HDL, both of which are very important because of their crucial role in two of the major diseases, coronary heart disease and diabetes. Our goal is to develop a low-cost, hand-held device capable of measuring the three molecular species and Hb1AC rapidly. An ultra-sensitive sensor has been fabricated using new innovative materials and concepts of nanotube lab-on-a-chip. CNT has been used for the development of the detector, exploiting its unique property to measure extremely weak electrical signals due to its intrinsic characteristics of fast electron transfer, small surface area, and capacity of carrying high current (Hwang *et al.*, 2010). The nanoscale arrangement of CNTs paves the way for high sensitivity and signal/noise ratio, leading to mediatorless biosensors. The CNT surfaces provide a biocompatible and highly conductive platform for anchoring proteins (Wang *et al.*, 2004). Use of amperometric electrical measurement methods, versus conventional fluorescent or luminescent methods, has made it possible to develop small-size, portable, hand-held, and cost-effective device.

7. Protein Probes for Detection and Monitoring Molecular Components of Serum

7.1. Glucose oxidase

Glucose oxidase (GOx) has been widely used in monitoring glucose levels in serum (see Bankar *et al.*, 2009; Yoo and Lee, 2010, for recent reviews). GOx catalyzes oxidation of β-D-glucose to β-D-glucono-1,5-lactone and hydrogen peroxide, using molecular oxygen as the electron acceptor as shown in Fig. 9.3. The conversion of β-D-glucose to gluconic acid involves the transfer of two protons and two electrons from the substrate to flavin adenine dinucleotide (FAD).

GOx is a dimeric protein, MW ~150 kDa, with 1FAD cofactor per monomer. Crystal structure of GOx has been resolved, see Fig. 9.4. GOx, shown here from PDB entry code: 1gpe, is a small, stable enzyme that oxidizes glucose into glucolactone, converting oxygen into hydrogen peroxide in the process. Its normal biological function appears to be centered on toxic

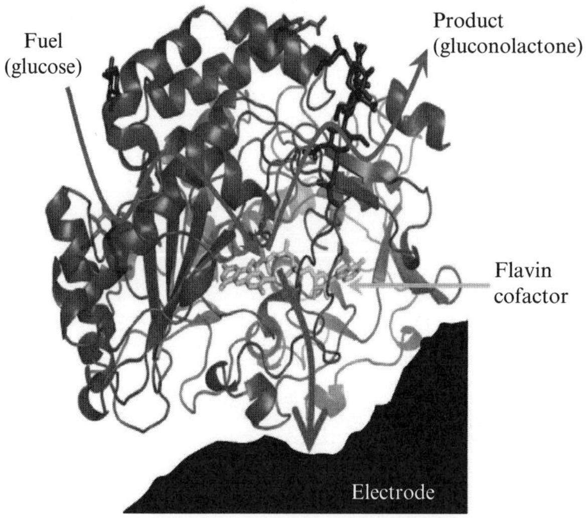

Figure 9.3 Glucose oxidase catalyzes glucose oxidation releasing electrons. (For the color version of this figure, the reader is referred to the Web version of this chapter.)

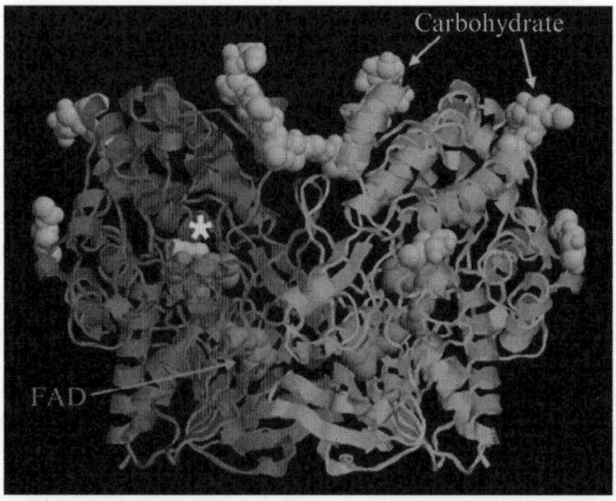

Figure 9.4 X-ray structure of glucose oxidase, PDB code 1gpe. (See Color Insert.)

hydrogen peroxide that is formed which is cytotoxic to bacteria. For instance, GOx is found on the surface of fungi, where it helps protect against bacterial infection, and it is also found in honey, where it acts as a natural preservative. Eremin *et al.*, 2001 have discussed thermal stability of GOx from Penicillum adametzii.

7.2. Cholesterol oxidase

Cholesterol oxidase, ChoA (EC 1,1,3,6; cholesterol:oxygen oxidoreductase), a 53 kDa, monomeric flavoenzyme that catalyzes oxidation of cholesterol by oxygen, see Fig. 9.5A and B. Cholesterol oxidase is an alcohol dehydrogenase/oxidase flavoprotein that catalyzes the dehydrogenation of C(3)—OH of cholesterol.

$$\text{cholesterol} + O_2 \rightleftharpoons \text{cholest} - 4 - \text{en} - 3 - \text{one} + H_2O_2$$

Cholesterol oxidases of type I are those containing the FAD cofactor tightly but not covalently bound to the protein moiety, whereas type II members contain covalently bound FAD. A number of X-ray structures have been reported, one of them from *Streptomyces* sp. CO at pH 9 is shown in Fig. 9.6.

Cloning of cholesterol oxidase is shown in Flowchart 9.1 in Section 8. The ChoA gene has been previously cloned and sequenced (Ishizaki *et al.*, 1989; Murooka *et al.*, 1986) and the secretory production in the Streptomyces host-vector has been demonstrated. Nomura *et al.* (1995) have reported the expression of ChoA in *Escherichia coli*. Crystal structure of ChoA from *Brevibacterium sterolicum* (ChoB) has been reported by Vrielink *et al.* (1991). Substrate binding features of possible importance to the catalytic mechanisms have been discussed by Vrielink *et al.* (1991) (Lario *et al.*, 2003). ChoA sequence is homologous to ChoB (59.2%). ChoA suffers from lower thermal stability which hampers its technological applications. From previous studies, we have successfully cloned, expressed, purified a number of mutants by mutating Ser 103, Val121, Arg135, and Val145 to create a repertoire of thermally stable mutants. We are focusing on many more mutations to increase, in general hydrophobicity, which, as a rule of thumb, enhances thermal stability (Renugopalakrishnan *et al.*, 2005).

Recently Ghasemian *et al.* (2008) have reported mutation of the Gln 145 by Glu which facilitates a hydrogen bond between Glu145 and Asp 134 and a salt bridge between Glu 145 and Arg 147 enhancing the thermal stability of cholesterol oxidase.

7.3. Cholesterol esterase

Cholesterol esterase (ChE: EC 3.1.1.13), obtained from the fungus, *Candida cylindracea*, and the closely related *Pseudomonas fluorescens*, is a glycoprotein that belongs to the lipase/esterase family (Schrag and Cygler, 1993). It is a

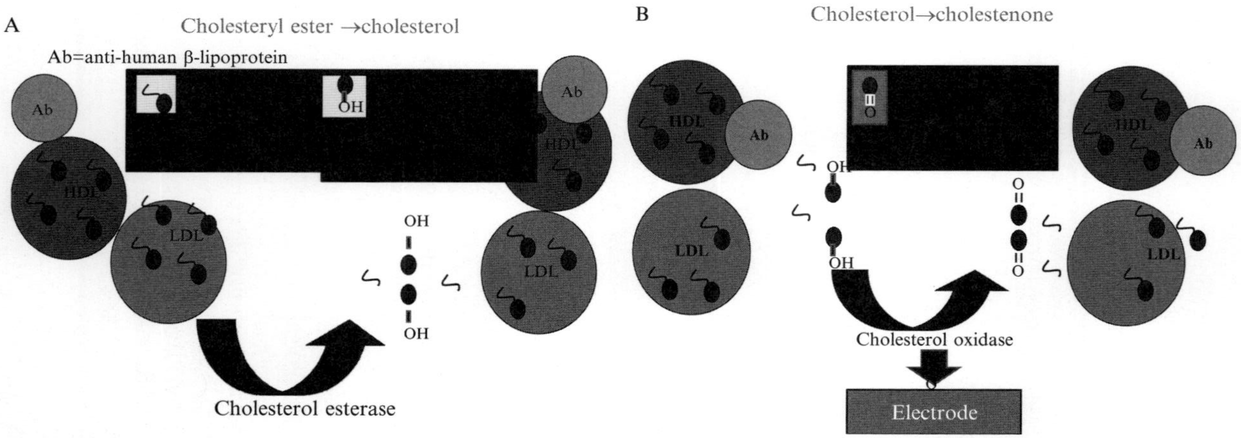

Figure 9.5 Catalysis of cholesterol. (For the color version of this figure, the reader is referred to the Web version of this chapter.)

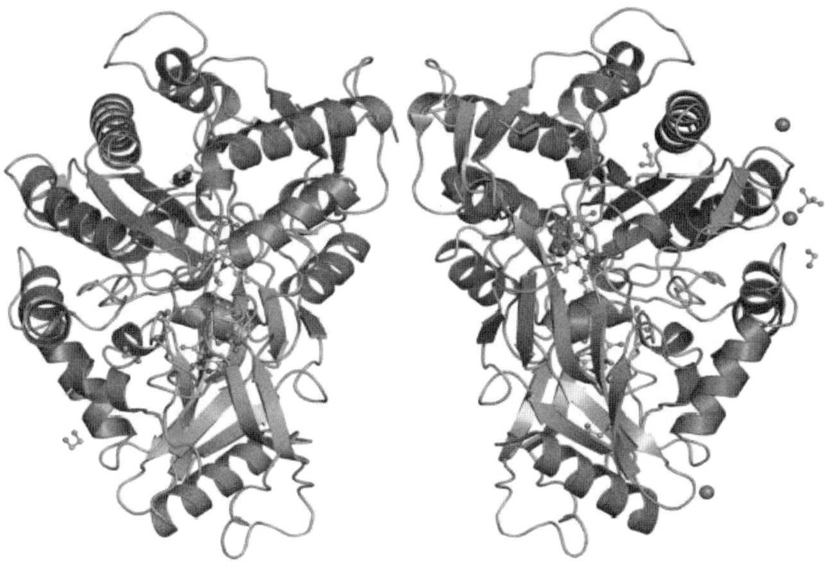

Figure 9.6 X-ray structure of cholesterol oxidase, PDB code 1B4V. (For the color version of this figure, the reader is referred to the Web version of this chapter.)

homodimer (534 × 2 amino acids). ChE hydrolyzes many fatty acid esters of cholesterol. The rate of hydrolysis is greater for long-chain than for short-chain fatty acid esters and is greater for unsaturated than for saturated fatty acid esters. The 3D structure of a *C. cylindracea* cholesterol esterase (ChE) homodimer (534 × 2 amino acids) in complex with a ligand has been determined at 1.4 Å resolution, space group *P*1, using synchrotron low-temperature data (Pletnev *et al.*, 2003), and is shown in Fig. 9.7.

7.4. Lipase

Lipases are water-soluble enzymes that catalyze the hydrolysis of esters and are a subclass of esterases. Triglyceride measurement uses bovine pancreatic lipase to convert triglycerides into glycerol. Further treatment of glycerol using glycerol kinase converts it into glycerol-3-phosphate which results in the formation of H_2O as shown in Fig. 9.8.

Dhand *et al.* (2010) have reported the fabrication of amperometric triglyceride biosensor based on electrophoretically deposited polyaniline–SWCNT composite film onto indium-tin-oxide (ITO) glass plate which differs significantly from our approach described here.

Pancreatic lipase RP2 from the guinea pig (Withers-Martinez *et al.*, 1996) is shown in Fig. 9.9.

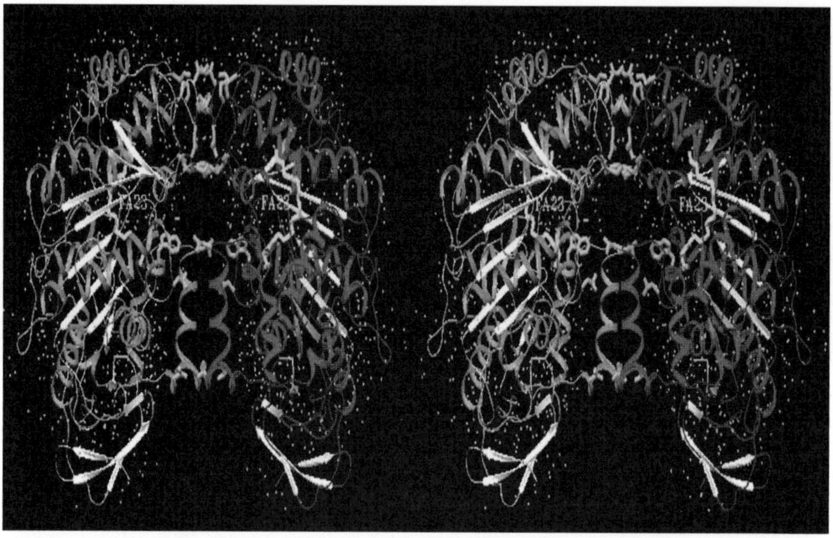

Figure 9.7 X-ray structure of cholesterol esterase, PDB code 1LLF. (For the color version of this figure, the reader is referred to the Web version of this chapter.)

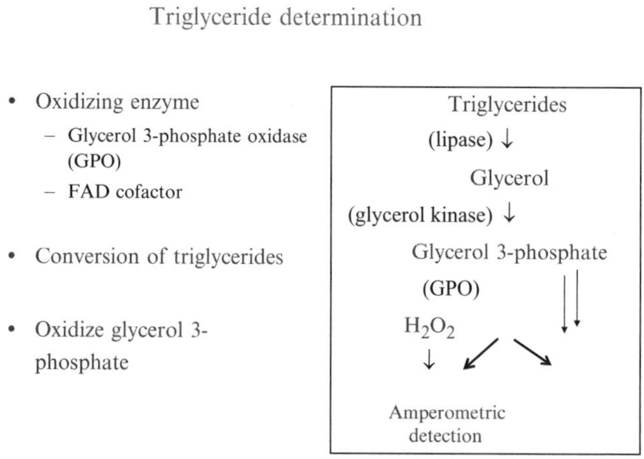
Triglyceride determination
• Oxidizing enzyme
– Glycerol 3-phosphate oxidase (GPO)
– FAD cofactor
• Conversion of triglycerides
• Oxidize glycerol 3-phosphate
Triglycerides
(lipase) ↓
Glycerol
(glycerol kinase) ↓
Glycerol 3-phosphate
(GPO)
H_2O_2
↓
Amperometric detection

Figure 9.8 Triglyceride determination using lipase. (For the color version of this figure, the reader is referred to the Web version of this chapter.)

7.5. Fructosyl valine amino oxidase

Selvin *et al.* (2010) have reported a community-based study of population of black or white nondiabetic adults, glycated hemoglobin was shown to be superior to fasting glucose for assessment of the long-term risk of subsequent cardiovascular disease, especially at values above 6.0%. Such prognostic data

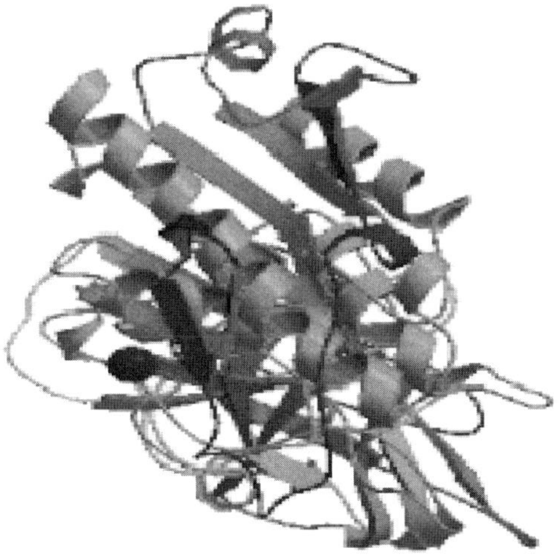

Figure 9.9 X-ray structure of guinea pig pancreatic lipase RP2, PDB code 1GPL. (For the color version of this figure, the reader is referred to the Web version of this chapter.)

may add to the evidence supporting the use of glycated hemoglobin as a diagnostic marker for diabetes.

HbA1C is a glycated hemoglobin in which the N-terminal Val residue of the β subunit has been modified by blood glucose. This modification, called glycation to distinguish it from the enzymatic glycosylation of proteins, is a nonenzymatic reaction of glucose with free amino groups, proceeding through a Schiff base intermediate to produce a relatively stable product. Fructosyl amino acid oxidase (FAO) catalyzes the oxidation of the C—N bond linking the C1 of the fructosyl moiety and the nitrogen of the amino group of fructosyl amino acids (see Fig. 9.10).

The reaction proceeds to an unstable Schiff base intermediate, which hydrolyzes to produce glucosone and an amino acid. The enzyme's reduced FAD cofactor is then reoxidized by molecular oxygen with the release of hydrogen peroxide, amenable for electrochemical detection, and quantification of HbA1C levels in serum. FAOs have been isolated from a number of different microorganisms, for example, production of fungal FAO useful for diabetic diagnosis in the peroxisome *Candida boidinii*, which was reported by Sakai *et al.* (1999). Cloning and expression of FAO gene from *Corynebacterium* sp. 2-4-1 in *E. coli* was reported by Sakaue *et al.* (2002). Thermostabilization of bacterial FAO by directed evolution was reported by Sakaue *et al.* (2002) and Sakaue and Kajiyama (2003).

Electrochemical measurement of HbA1c level is a convenient method for the point-of-care application and can be integrated with the current glucose, cholesterol, triglyceride multichannel lab-on-a-chip for comprehensive

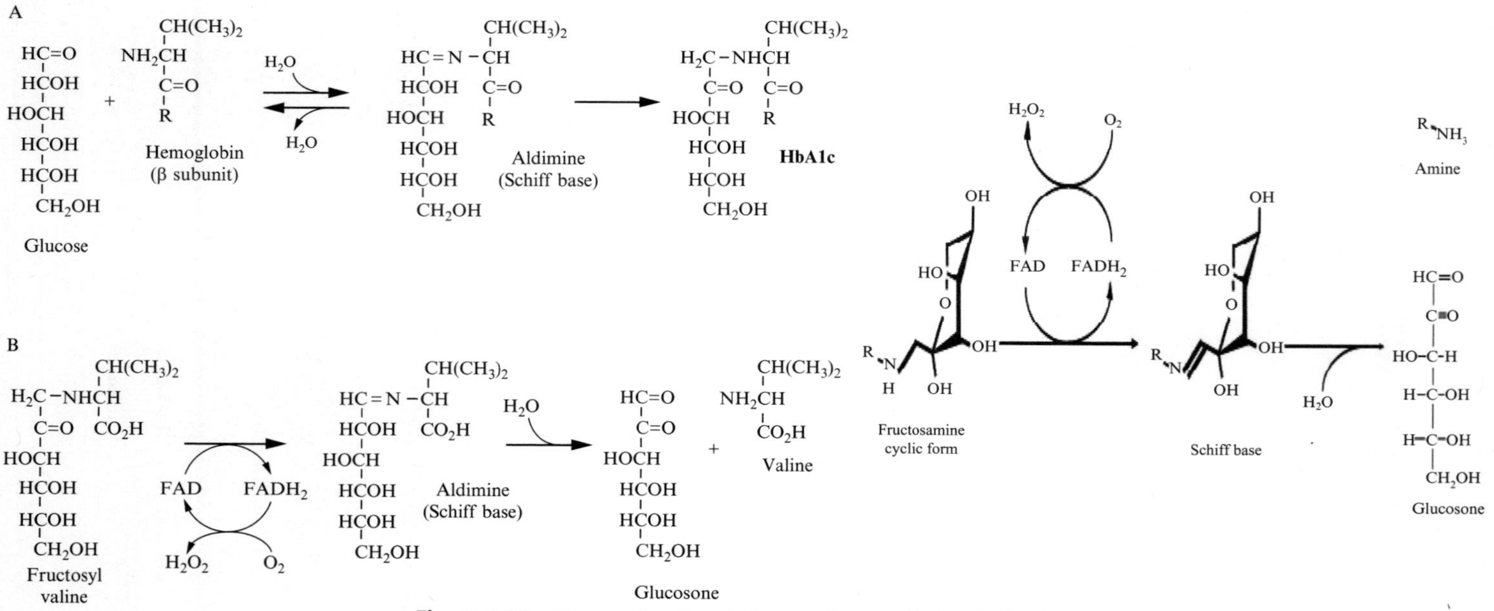

Figure 9.10 Fructosyl valine amino oxidase catalysis of HbA1C.

patient health care management. The electrochemical measurement of HbA1c level involves the total hemoglobin concentration measurement and the HbA1c molecular concentration measurement. Therefore, the development of electrochemical measurement of HbA1c is required for the electrochemical HbA1c level biosensing. Enzymatic-based detection method is relatively rapid and selective. As a result, there is a great interest to develop an electrochemical enzyme biosensor for HbA1c. Enzymatic assay of HbA1c is supposed to consist of the following steps: first, HbA1c proteolysis produces glycated hexapeptides; second, glycated hexapeptides is further decomposed to fructosyl amine, namely, FV; finally, FV is catalyzed by enzyme FAO (Ferri *et al.*, 2009) to produce hydrogen peroxide, which can be detected electrochemically.

The crystal structure of the deglycating enzyme fructosamine oxidase (Amadoriase II) has been reported by Collard *et al.* (2008). Based on this approach, the development of an FV biosensor is relatively simpler and will demonstrate the feasibility of the development of an HbA1c biosensor. X-ray structure of deglycating enzyme fructosamine oxidase Collard *et al.* (2008 is shown in Fig. 9.11.

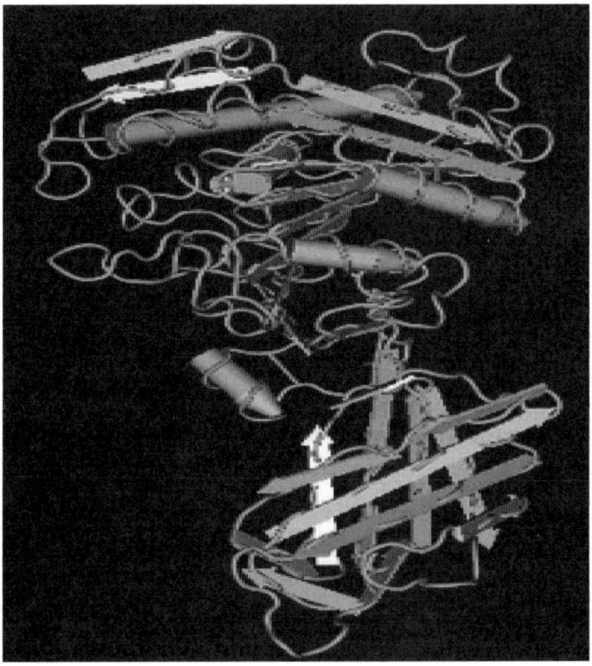

Figure 9.11 X-ray structure of fructosyl valine amino oxidase, PDB code 3DJD. (For the color version of this figure, the reader is referred to the Web version of this chapter.)

8. Protein Engineering and Molecular Biology of Probe Proteins

Protein engineering of GOx, cholesterol oxidase, cholesterol esterase, pancreatic lipase, and fructosyl valine amino oxidase have two important goals: enhanced thermal stability and electrochemical responsiveness. Amperometric sensors require thermally robust enzymes which are designed by rational site-directed mutations; general methodology is described by Renugopalakrishnan *et al.* (2005, 2006). Electrochemical responsiveness is more challenging to engineer in proteins.

ChoE suffers from lower thermal stability, which hampers its technological applications. From previous studies (to be submitted, patents pending), we have successfully cloned, expressed, purified a number of mutants by mutating Ser 103, Val121, Arg135, and Val145 to create a repertoire of thermally stable mutants (Viswanathan *et al.*, to be published). We are focusing on many more extensive mutations to increase, in general hydrophobicity, which, as a rule of thumb, enhances thermal stability (Renugopalakrishnan *et al.*, 2006).

Recently, Ghasemian *et al.* (2008) have reported the substitution of the glutamine (Q) 145 by glutamic acid (E) which confers a hydrogen bond between Glu145 and Asp134 and a salt bridge between Glu 145 and Arg 147 improving the thermal activity of cholesterol oxidase.

Kohno *et al.* (2001) optimized thermal properties of *Rhizopus niveus* lipase (RNL) activity by random mutagenesis (Kohno *et al.*, 2001). The lipase gene was mutated using the error-prone PCR technique. Three amino acids were mutated, Pro 18 His, Ala 36Thr, and Glu 218 Val. The wild-type and randomly mutated lipases were both purified and characterized. The specific activity of the mutant lipase was 80% that of the wild type. The optimum temperature of the mutant lipase was higher by 15 °C than that of the wild type.

In attempts to define the molecular basis of difference in catalytic activity and stability, mutagenesis studies have demonstrated the importance of optimized charge interactions on the surface of the protein, which can significantly increase the thermal stability. Therefore, the rational site-directed mutagenesis will be applied to the following amino acid residues substitution. Another basis for choosing the mutation site is by comparison between different COX and CES. The mutation will be conducted via overlap PCR method. These proteins will be expressed in *E. coli* and *Pichia pastoris*, respectively. The expression level, enzyme solubility, and activity would be compared, and then appropriate expression system will be selected. The construction of the expression vectors will follow standard molecular biology protocols or Novagen and Invitrogen manuals. To meet the requirement for coupling these candidates' proteins with CNT, site-directed mutagenesis will be applied to modify or enhance protein functionalities. The construction of the expression vectors are listed below. The column chromatography methods will be used to purify the recombinant or engineered lipase, cholesterol esterase, GOx, and cholesterol oxidase.

9. Control of the Fermentation Process

For *E. coli*, the DO-controlled fed-batch fermentation will be applied to achieve high cell density, especially the toxic metabolite formation, acetic acid will be controlled by adjusting the specific growth rate, and the concentration of acetic acid will be assayed by gas chromatography in *vivo*.

For *P. pastoris*, although the general fermentation protocols have been suggested by Invitrogen, it still needs to be optimized according to individual process. Methanol induction strategy is critical for successful expression of the heterologous proteins in *P. pastoris*, and the methanol feeding is closely linked with DO level. In this project, we will utilize these two parameters together to optimize the induction phase strategies and to increase the yield.

10. Gene Structure and Purification of Overexpressed Protein

To scale up the purification process, in order to decrease the costs, column chromatography process would be applied. For *P. pastoris*, the secreted COX and CES were first recovered by ammonium sulfate precipitation and then purified by DEAE-Sepharose and Sephacryl S-300; finally, the highly purified COX and CES were obtained by FPLC. For the intracellular COX and CES expressed in *E. coli*, after extraction from the mycelia, the following purification steps were the same as from *P. pastoris*. Structure of gene encoding lipase, cholesterol oxidase and esterase, and GOx is shown in Fig. 9.12 and Flowchart 9.1 describes various steps in the production of the enzyme.

Genes encoding lipase, cholesterol esterase, and cholesterol oxidase are cloned from *Streptomyces coelicolor* according to the literature reports (Bentley *et al*., 2002; Hatzinikolaou *et al*., 1996), and GOx is cloned from *Aspergillus niger* (Fig. 9.13). These proteins are expressed in *E. coli* and *P. pastoris*, respectively (Li *et al*., 2007). The expression level, enzyme solubility, and activity are compared, and the best expression system is selected. The construction of the expression vectors follows the common molecular biology protocols, or Novagen and Invitrogen manuals. The construction

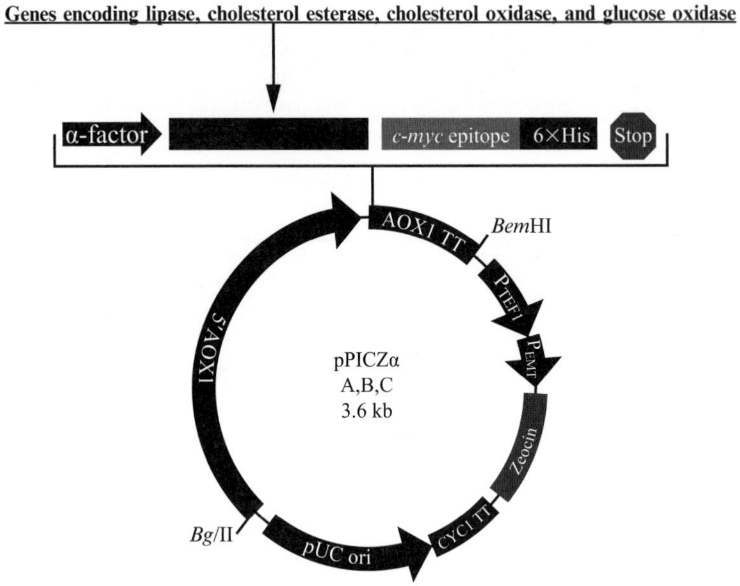

Figure 9.12 Genes encoding lipase, cholesterol esterase, cholesterol oxidase, and glucose oxidase.

Transform into *Pichia pastoris* strain

⇩

Plate transformants on medium containing Zeocin or Blasticidin

⇩

Select 10–20 colonies for small scale expression

⇩

Choose clones with highest expression levels for large-scale expression

⇩

Control methanol and glycerol feeding rate according to DO level in the fermentation

⇩

Purify CES/COX

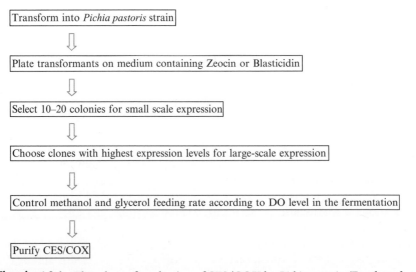

Flowchart 9.1 Flowchart of production of CES/COX by *Pichia pastoris*. (For the color version of this figure, the reader is referred to the Web version of this chapter.)

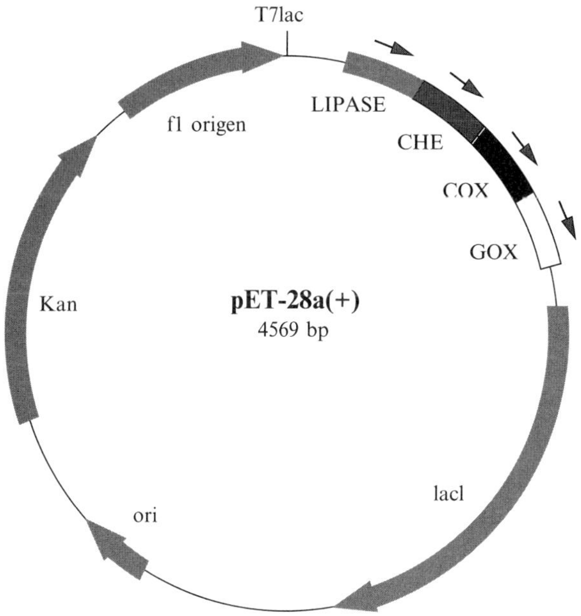

Figure 9.13 Expression vector pET 28a(+)—lipase, cholesterol easterase, cholesterol oxidase, and glucose oxidase.

of the expression vectors are listed below. The column chromatography methods are used to purify the recombinant or engineered lipase, cholesterol esterase, GOx, and cholesterol oxidase (Flowchart 9.2).

11. Immobilization of Proteins on SWCNT

In our laboratory we have developed a four-step process to covalently attach enzymes and proteins to SWCNT:

Step 1: Treatment with 1-ethyl-3-[3-dimethylamino propyl] carbodiimide, EDAC to create stable, active succinimidyl ester groups.
Step 2: Acid treatment (2:1 ratio of H_2SO_4 and HNO_3) of SWCNT creates carboxyl functionality.
Step 3: Active ester groups react with the side chains in the enzymes.
Step 4: Focused ion beam bombardment (Raghuveer *et al.*, 2006a) or microwave sonication of SWCNT (Raghuveer *et al.*, 2006b) to create reactive groups, for example, carboxyl functionality on the surface of SWCNT.

Prepare pET28 vector: Digest pET28 with restriction enzymes, purified by agarose gel

↓

Prepare inserted gene (CEX, COX, or GOX): Digest genes with restriction enzymes, purified by agarose gel

↓

Cloning the inserted genes to the prepared pET28 vector: Ligate the inserted genes with the prepared vector, transform the ligated vector to the nonexpression host, and identify positive clones

↓

Transform into expression host: Extract the pET28 vector with the inserted CEX, COX, or GOX gene, transform to host carrying T7 RNA polymerase gene or non-DES host compatible with CE6 infection.

↓

Optimization of the expression process: Determine time course and temperature for expression in cell; apply DO-controlled fed-batch fermentation with acetic acid *in vivo* assay by GC

↓

Purification of CES, COX, and GOX: Purified by Centrifugation, Ultrafiltration, Chromatography of DEAE Sepharose

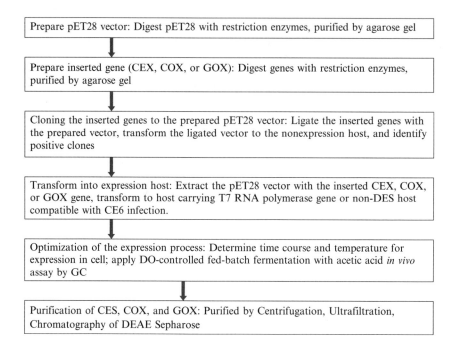

Flowchart 9.2 Overview of applying pET28 system to produce CES and COX.

Layer-by-layer fabrication and direct electrochemistry of GOx on SWCNT has been reported by Zhang *et al.* (2007). Liu *et al.* (2005) have described a procedure which differs in Step 4 used in our laboratory (http://www.chem.unsw.edu.au/research/groups/gooding/nanotubes.html).

The electronic coupling of the redox protein to the electrode is of fundamental importance in the design of nanodevices like bionanosensors. We have pursued several approaches to this problem. A controlled attachment of the redox enzyme GOx to the nanotube side wall is achieved through a linker molecule. We follow the noncovalent functionalization approach of Chen *et al.* (2003). The noncovalent functionalization involves a bifunctional molecule 1-pyrenebutonic acid, succinimidyl ester (Molecular Probes, Inc.) irreversibly adsorbed onto the inherently hydrophobic surfaces of CNT in an organic solvent, for example, dimethylformamide (DMF). The pyrenyl group being highly aromatic in nature, is known to interact strongly with the basal plane of graphite via π-stacking. The anchored molecules are highly stable against desorption in aqueous solutions. The succinimidyl ester groups are highly reactive to nucleophilic substitution by primary and secondary amines that exist in abundance on the surface of the proteins.

Two-nanosecond molecular dynamics simulation of GOx covalently attached to CNT was reported by Tatke *et al.* (2004). A snapshot of 20-ns MD simulation of GOx and CNT (Flipek, unpublished data) is shown in Fig. 9.14.

12. Immobilization of Proteins on Self-Assembled Monolayers

For utilization of highly sophisticated functions of biomaterials in nanoscale functional systems, immobilization of enzymes, for example, on artificial devices such as electrodes via thin-film technology is one of the most powerful strategies. There are at least four methods for immobilization of enzymes: organic ultrathin films (Forrest, 1997), self-assembled monolayers (SAM), Langmuir–Blodgett (LB) films, and layer-by-layer (LBL) assemblies. The SAM method allows facile contact between enzymes and inorganic interface like CNT/graphene used in biosensors. In addition, recent microfabrication techniques such as microcontact printing and dip-pen nanolithography have been successfully applied to preparation of enzyme patterning. A monolayer at the air–water interface, which is a unit structure of LB films, provides a unique environment for recognition

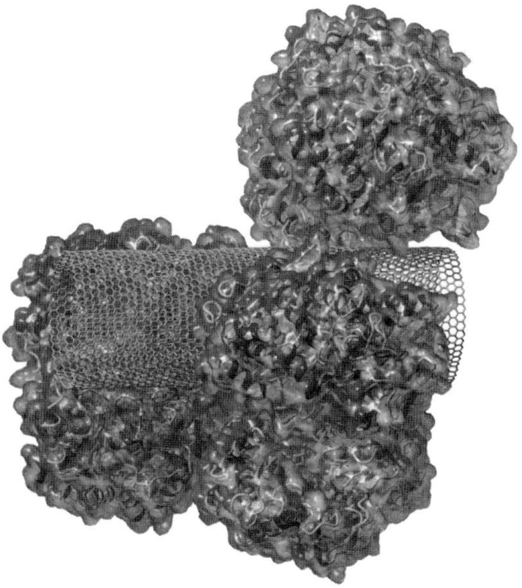

Figure 9.14 A snapshot from 20 ns molecular dynamics simulation of glucose oxidase. (See Color Insert.)

of aqueous biomaterials. The LB film can be also used for immobilization of enzymes in an ultrathin film on an electrode, resulting in sensor application. The LBL assembling method is available for wide range of biomaterials and provides great freedom in designs of layered structures. These advantages are reflected in preparation of thin-film bio-reactors where multiple kinds of enzymes sequentially operate. LBL assemblies were also utilized for sensors and drug delivery systems. This kind of assembling structures can be prepared on microsized particles and are very useful for the preparation of hollow capsules with biological functions. More recently, prefunctionalized SWCNT have become commercially available (Sigma Aldrich).

13. Characterization of SWCNT Enzyme Adduct

We have used X-ray photoelectron spectroscopy (XPS) to establish covalent linkage between the SWCNT and enzyme. Fourier self-deconvolution of C1s, N1s, O1s signals are used to corroborate covalent linkage between the side chains of the enzyme with CNT. Further corroboration is obtained from FT-IR where the carboxyl stretching frequency, 1700 cm^{-1}, of enzyme undergoes a shift when bound to SWCNT. Raman studies provide complimentary corroboration to FT-IR. Atomic force microscopy (AFM), scanning tunneling microscopy (STM), atomic resolution probe microscopy techniques such as AFM, STM, magnetic force microscopy (MFM), capacitance probe microcopy (CPM), and total internal reflection fluorescence are generally recognized to be the most critical and unique techniques utilized for obtaining the most detailed information regarding the surface texture, electron properties, magnetic properties, charge properties, and the orientation of the immobilized proteins on CNT.

14. Mediators

A mediator is an electron transfer agent, an usually reversible oxidizing agent that will transfer electrons between an enzyme and an electrode.

Suitable mediators should fulfill the following criteria:

1. Rapid heterogeneous kinetics with the electrode
2. Low one-electron redox potentials
3. pH-independent redox potentials
4. Nontoxic
5. High chemical stability in both redox states

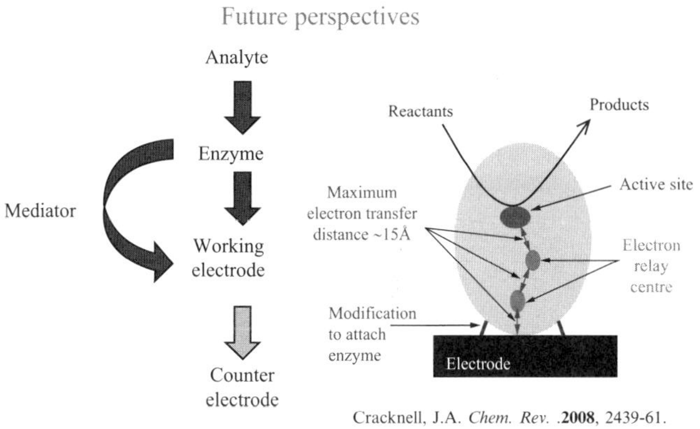

Figure 9.15 Directed and mediated transfer of electron charge transfer. (For the color version of this figure, the reader is referred to the Web version of this chapter.)

We are closely monitoring the progress that has been made for mediatorless biosensors (Tender *et al.*, 2002). Reagentless transduction of molecular recognition at electrode surfaces has been the subject of several research efforts with the goal of developing electrochemical biosensors not limited by manipulation of reagents (Tang *et al.*, 2003). Earlier experiments have proven that CNT itself is a good mediator (Balavoine *et al.*, 1999; Wang and Musameh, 2003). Even, the difficult electron transfer protein myoglobin was shown to function effectively with multiwalled CNT (Zhao *et al.*, 2003).

Figure 9.15 shows sensors utilizing direct electron and mediated electron transfer. Electron charge transfer is effective at ~15Å distance, depending on the 3D structure of the enzyme, and the spatial disposition of the FAD cofactor (Cracknell *et al.*, 2008).

15. Surface Characterization

We have used atomic resolution probe microscopy techniques such as AFM (Alessandrini and Facci, 2005). STM, MFM, CPM, and total internal reflection fluorescence are generally recognized to be the most critical and unique techniques utilized for obtaining the most detailed information regarding the surface texture, electron properties, charge properties, and the orientation of the immobilized proteins on the substrate surface. The ability of AFM to create 3D micrographs with resolution down to the nanometer and Angstrom scales has made it an essential tool for imaging surfaces in applications ranging from semiconductor processing to cell biology. AFM at single molecule resolution is to date one of the most powerful methods to observe single protein molecules on a substrate surface. MFM is an extension of AFM that images magnetization patterns with submicron resolution providing additional probe to understand the surface properties of the biosensor.

16. Fabrication

The patterning of biomolecules on semiconducting surfaces is of central importance in the fabrication of novel biodevices. In the process of patterning, the structural integrity of the enzyme should be retained as it is in the native state.

17. Signal Detection

There is close attachment between the biomaterial and the transducer. The transducer is very adaptable and reliable. There is also close attachment between the enzyme and CNT in the transducer (see Fig. 9.16).

A schematic of experimental detection for measuring electric current (resistance) through a nanoscale sensor is shown in Fig. 9.17.

18. Experimental Protocol

Detailed experimental protocol can be found in the following link: http://www.searo.who.int/en/Section10/Section17/Section53/Section481_1753.htm.

The patient's blood (30 μl) flows through capillaries into four compartments. In the first compartment, GOx covalently linked to functionalized CNT is the sensor to detect selectively serum glucose. Similarly, second, third, and fourth compartments consist of cholesterol oxidase, cholesterol esterase. Esterase, lipase, and fructosyl valine amino oxidase, which are the sensors to detect cholesterol, triglycerides, and HbA1C. GOx will be mediated by ferrocene (Yoo and Lee, 2010). The amperometric response is proportional to the serum glucose concentration (Patel *et al.*, 2003). GOx is genetically modified by adding a poly-Lys segment at the C-terminal with a peptide chain linker between the enzyme and poly-Lys. This chain anchors more electron transfer through ferrocene and improves the S/N ratio (Chen *et al.*, 2002). The second part goes through a compartment that consists of three serial cells. The first cell consists of ChE, antihuman lipoprotein antibody, and horse radish peroxidase. The above buffer, due to the bonding

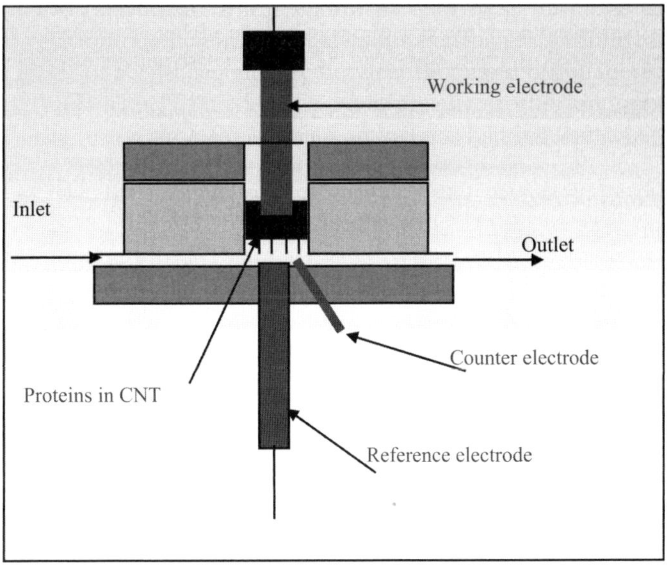

Figure 9.16 Close attachment of protein and transducer. (For the color version of this figure, the reader is referred to the Web version of this chapter.)

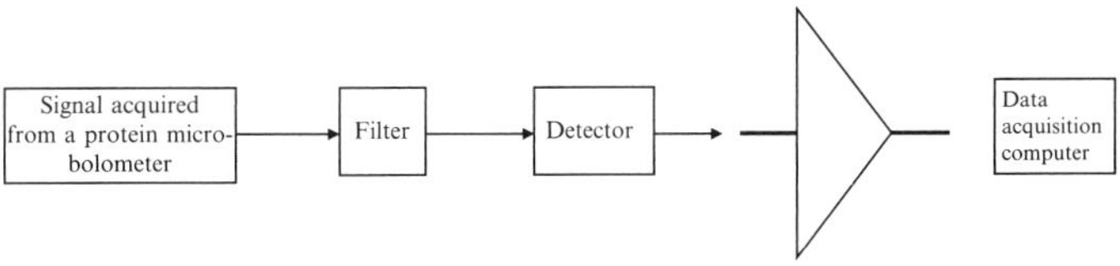

Figure 9.17 A schematic of experimental detection of generated current.

to the surface of the apo B containing lipoproteins (LDL, VLDL, and chylomicrons), blocks the enzymatic measurement of cholesterol on the non-HDL fraction. The HDL is measured amperometrically by cholesterol esterase. The mediator would be FDAOS (*N*-ethyl *N*(2-hydroxy-3-sulfopropyl)-3,5-dimethoxy-4-fluroaniline). In this compartment, deoxycholate in the buffer solubilizes cholesterol for the system, disrupting the antibody–apo B lipoprotein complex. The cholesterol assay consists of cholesterol oxidase which acts as the probe. Serum enters through the next compartment where lipase, glycerol 3-phosphate oxidase, and ATP are in MES buffer. Glycerol oxidase will serve as an amperometric detector. Triglyceride quantification in serum will utilize GPO-Trinder reaction. Buffers for each cell will be stored in separate pouches and injected into the serial cells directly. LDL will be calculated from total cholesterol and HDL using Friedewald's formula (Friedewald *et al.*, 1972). HbA1C measurement will utilize Fructosyl valine amino oxidase as the detector.

Pancreatic lipase will be the sensor for measuring triglyceride levels. From total cholesterol, HDL, triglyceride, LDL will be calculated using the formula of Friedewald *et al.* (1972).

Detection of HbA1C will consist of three steps: (1) In the first step, endoproteinase will cleave the glycosylated segment of HbA1C into glycated hexapeptides, fru-Val-His-Leu-Thr-Pro-Glu. (2) Proteinase treatment will result in fructosyl amine. (3) In the last step, fructosyl amine will react with fructosyl valine amino acid oxidase releasing hydrogen peroxide which will be detected amperometrically.

The sample molecules are detected by varying the bias voltage and capturing the current variations in the sensor. The thin film from the sample, which is biological in nature, forms the basis for the measurement. The thin film's interaction with the CNTs and hence the current formed reveal vital information on sample molecules. The measurement system would comprise of the computer or a portable device to evaluate and detect intricate details in the sample molecule.

Depending upon the detected molecules, the nature of the blood sample can be estimated and further course of treatment can be carried out. A software module can be written to analyze the measured data and reflect the necessary output. Actually, the portable device will be envisaged to be a stand-alone system to measure and analyze the probe information and the current information from the sample. An LCD display for the same is also planned.

One of the technical problems is how to move small amounts of solution through the tiny capillaries from station to station within the chip. Minute puffs of hot air can be used to separate and propel the samples. Other companies use an electrical charge, air pressure, or air-filled tubes that rhythmically constrict the small pipes like squeezing toothpaste from a tube. Here we are planning to use micropumps to move the sample from one place to the other. The micropump will render solution flow from a microvolume liquid sample chamber to filtration chamber to a detection chamber, and the reader will read electrical signal transmitted from a sensor in the detection chamber in the chip.

19. Conclusion

The structure, function, and dynamics of molecules have a triangular correlation that permeates all through chemistry and biochemistry. This triangular correlation poses daunting challenges due to their inherent structural complexity, stereospecific physiological function, and lability in the case of biological macromolecules. Since extremely small perturbations to molecular systems can cause unpredictable exponential spikes in their physiological function, capturing in functional states is challenging, unlike their organic and inorganic counterparts. The need to preserve structural integrity and their function requires delicate lithographic techniques to etch them on passive scaffolding. Nanoscience has ushered in fabrication procedures, by which biological macromolecules can be captured in their physiologically active state on substrates which provide passive scaffolding. Bionanoscience is a natural progression of nanoscience, which is a bottom up approach that is integrative, utilizing molecular systems as building blocks, which can be changed at will, by the methods of molecular biology and biotechnology in *a priori* predictable manner. Biological macromolecules manifest a wide spectrum of physiological function, providing myriads of windows in their technological application ranging from sensing, energy generation, smell transmission, imparting more human-like characteristics to robots, storing information at ultra high density, and many others.

The fusion of protein engineering, nano fabrication, and materials science and engineering holds much promise in auguring a new era in point-of-care devices for self-monitoring, hospital use especially in continuous monitoring systems, reducing errors in diagnosis, and reducing the cost of health care.

ACKNOWLEDGMENTS

V. R. and S. V. express their thanks to the Rothschild Foundation, NIH, and Harvard Medical School for generous financial support. This project was initiated at the Florida International University, Miami, FL when V. R. was a Coulter Eminent Scholar supported by the Coulter Foundation, Miami, FL, led by W.C. The authors also wish to acknowledge the Pittsburgh Supercomputing Center for generous allocation of Supercomputer time on TeraGrid through Project Serial Number: TG-CH090102 for performing molecular dynamics simulation of glucose oxidase-SWCNT. Authors thank Elsevier for copyright permission to reproduce Fig. 1 from Martin Pumera, Materials Today, 2011. Authors wish to thank Ms. Priya Kumaraguruparan, Elsevier, Chennai for her superb help in editing and type-setting.

REFERENCES

Alessandrini, A., and Facci, P. (2005). AFM: A versatile tool in biophysics. *Meas. Sci. Technol.* **16,** R65–R92.

Arya, S. K., Datta, M., and Malhotra, B. D. (2008). Recent advances in cholesterol biosensor. *Biosens. Bioelectron.* **23,** 1083–1100.

Audette, G. F., Lombardo, S., Dudzik, J., Arruda, T. M., Kolinski, M., Filipek, S., Mukerjee, S., Kannan, A. M., Thavasi, V., Ramakrishna, S., Chin, M., Somasundaran, P., *et al.* (2011). Protein hot spots at bio-nano interfaces. *Mater. Today* **14,** 360–365.

Balavoine, F., Schultz, P., Richard, C., Mallouh, V., Ebbesen, T. W., and Mioskowski, C. (1999). Helical crystallization of proteins on carbon nanotubes: A first step towards the development of new biosensors. *Angew. Chem. Int. Ed.* **38,** 1912–1915.

Bankar, S. B., Bule, M. V., Singhal, R. S., and Ananthanarayan, L. (2009). Glucose oxidase—An overview. *Biotechnol. Adv.* **27,** 489–501.

Bentley, S. D., Chater, K. F., Cerdeño-Tárraga, A.-M., Challis, G. L., Thomson, N. R., James, K. D., Harris, D. E., Quail, M. A., Kieser, H., Harper, D., Bateman, A., Brown, S., *et al.* (2002). Complete genome sequence of the model actinomycete *Streptomyces coelicolor* A3(2). *Nature* **417,** 141–147.

Chakraborty, S. (2005). Dynamics of capillary flow of blood into a microfluidic channel. *Lab Chip* **5,** 421–430.

Chen, J., Barell, A. K., Collis, G. E., Officer, D. L., Swiegers, G. F., Too, C. O., and Wallace, G. C. (2002). Preparation, characterisation and biosensor application of conducting polymers based on ferrocene, substituted thiophene and terthiophene. *Electrochim. Acta* **47,** 2715–2724.

Chen, R. J., Bangsaruntip, S., Drouvalakis, K. A., Kam, N. W. S., Shim, M., Li, Y., Kim, W., Utz, P. J., and Dai, H. (2003). Noncovalent functionalization of carbon nanotubes for highly specific electronic biosensors. *Proc. Natl. Acad. Sci. USA* **29,** 4984–4989.

Collard, F., Zhang, J., Nemet, I., Qanungo, K. R., Monnier, V. M., and Yee, V. C. (2008). Crystal structure of the deglycating enzyme fructosamine oxidase (Amadoriase II). *J. Biol. Chem.* **283,** 27007–27016.

Cracknell, J. A., Vincent, K. A., and Armstrong, F. A. (2008). Enzymes as working or inspirational electrocatalysts for fuel cells and electrolysis. *Chem Rev.* **108,** 2439–2461, http://www.ncbi.nlm.nih.gov/pubmed/18620369.

Dhand, C., Solanki, P. R., Datta, M., and Malhotra, B. D. (2010). Polyaniline/single-walled carbon nanotubes composite based triglyceride biosensor. *Electroanalysis* **22,** 2683–2693.

Eremin, A. N., Metelitsa, D. I., Shishko, Zh.F., Mikhailova, R. V., Yasenko, M. I., and Lobanok, A. G. (2001). Thermal stability of glucose oxidase from Penicillium adametzii. *Appl. Biochem. Microbiol.* **37,** 578–586.

Ferri, S., Kim, S., Eng, M., Tsugawa, W., and Sode, K. (2009). Review of fructosyl amino acid oxidase engineering research: A glimpse into the future of hemoglobin A1C biosensing. *J. Diabetes Sci. Technol.* **3,** 585–592.

Forrest, S. R. (1997). Ultrathin organic films grown by organic molecular beam deposition and related techniques. *Chem. Rev.* **97,** 1793–1896.

Friedewald, W. T., Levy, R. I., and Fredrickson, D. S. (1972). Estimation of the concentration of low-density lipoprotein cholesterol in plasma, without use of the preparative ultracentrifuge. *Clin. Chem.* **18,** 499–502.

Ghasemian, A., Yazdi, M. T., and Sepehrizadeh, Z. (2008). Construction of a thermally stable cholesterol oxidase mutant by site-directed mutagenesis. *Biotechnology* **7,** 826–829.

Hatzinikolaou, D. G., Hansen, O. C., Macris, B. J., Tingey, A., Kekos, D., Goodenough, P., and Stougaard, P. (1996). A new glucose oxidase from Aspergillus niger: Characterization and regulation studies of enzyme and gene. *Appl. Microbiol. Biotechnol.* **46,** 371–381.

Hu, P.-A., Zhang, J., Li, L., Wang, Z., O'Neill, W., and Pedro Estrela, P. (2010). Carbon nanostructure-based field-effect transistors for label-free chemical/biological sensors. *Sensors* **10,** 5133–5159.

Hwang, S., Vedala, H., Kim, T., Choi, H., Choi, W., and Jeon, M. (2010). Fabrication of nanoelectrodes using individual multi-walled carbon nanotubes and their cyclic voltammetric properties. *J. Electrochem. Soc.* **157**(4), J139–142.

Ishizaki, T., Hirayama, N., Shinkawa, H., Nimi, O., and Murooka, Y. (1989). Nucleotide sequence of the gene for cholesterol oxidase from a Streptomyces sp. *J. Bacteriol.* **171,** 596–601.

Klonoff, D. C. (2011). Improving the safety of blood glucose monitoring. *J. Diabetes Sci. Technol.* **5,** 1307–1311.

Kohno, M., Enatsu, M., Funatsu, J., Yoshiizumi, M., and Kugimiya, W. (2001). Improvement of the optimum temperature of lipase activity for *Rhizopus niveus* by random mutagenesis and its structural interpretation. *J. Biotechnol.* **87,** 203–210.

Lario, P. I., Sampson, N., and Vrielink, Alice (2003). Sub-atomic resolution crystal structure of cholesterol oxidase: What atomic resolution crystallography reveals about enzyme mechanism and the role of the FAD cofactor in redox activity. *J. Mol. Biol.* **326,** 1635–1650.

Li, P., Anumanthan, A., Gao, X.-G., Ilangovan, K., Suzara, V. V., Düzgünes, N., and Renugopalakrishnan, V. (2007). Expression of recombinant proteins in Pichia pastoris. *Appl. Biochem. Biotechnol.* **142,** 105–124.

Liu, J., Chou, A., Rahmat, W., Paddon-Row, M. N., and Gooding, J. J. (2005). Achieving direct electrical connection to glucose oxidase using aligned single walled carbon nanotube arrays. *Electroanalysis* **17,** 38–46.

Murooka, Y., Ishizaki, T., Nimi, O., and Maekawa, N. (1986). Cloning and expression of a Streptomyces cholesterol oxidase gene in Streptomyces lividans with, plasmid pIJ702. *Appl. Environ. Microbiol.* **52**(6), 1382–1385.

Nomura, N., Choi, K.–P, Yamashita, M., and Murooka, Y. (1995b). Genetic modification of Streptomyces cholesterol oxidase gene for expression in Escherichia coli and development of promoter-probe vectors in enteric bacteria. *J. Ferment. Bioeng.* **79,** 410–416.

Patel, H., Li, X., and Karan, H. (2003). Amperometric glucose sensors based on ferrocene containing polymeric electron transfer systems – a preliminary report Biosens. *Bioelectron* **18,** 1073–1076.

Pletnev, V., Addlagatta, A., Wawrzak, Z., and Duax, W. (2003). Three-dimensional structure of homodimeric cholesterol esterase–ligand complex at 1.4 A resolution. *Acta. Crystallogr. D Biol. Crystallogr.* **59,** 50–56.

Pumera, S. (2011). Graphene in biosensing. *Mater. Today* **14,** 308–315.

Raghuveer, M. S., Kumar, A., Frederick, M. J., Louie, G. P., Ganesan, P. G., and Ramanath, G. (2006a). Site-selective functionalization of carbon nanotubes. *Adv. Mater.* **18,** 547–552.

Raghuveer, M. S., Agrawal, S., Bishop, N., and Ramanath, G. (2006b). Microwave-assisted single-step functionalization and in situ derivatization of carbon nanotubes with gold nanoparticles. *Chem. Mater.* **18,** 1390–1393.

Renugopalakrishnan, V., and Lewis, R. V. (2006). Bionanotechnology: Proteins to nanodevices. Springer BV, Dordrecht, The Netherlands.

Renugopalakrishnan, V., Garduno-Juarez, R., Narasimhan, G., Verma, C. S., Wei, X., and Li, P. (2005). Rational design of thermally stable proteins: Relevance to bionanotechnology. *J. Nanosci. Nanotechnol.* **5,** 1759–1767.

Renugopalakrishnan, V., Wei, X., Narasimhan, G., Verma, C. S., Li, P., and Anumanthan, A. (2006). Enhancement of protein thermal stability: Towards the design of robust proteins for bionanotechnological applications. *In* "Bionanotechnology: Proteins to Devices," (V. Renugopalakrishnan and R. V. Lewis, eds.), pp. 117–139. Springer, Heidelberg, Germany.

Roy, S., Vedala, H., and Choi, W. (2006). Vertically aligned carbon nanotube probes for monitoring blood cholesterol. *Nanotechnology* **17,** S14–S18.

Roy, S., Vedala, H., Roy, A., Kim, D., Doud, M., Mathee, K., Shin, H., Shimamoto, N., Prasad, V., and Choi, W. (2008). Electronic detection of single-molecule DNA hybridization using single-walled carbon nanotubes. *Nano Lett.* **8,** 26–30.

Sakai, Y., Yoshida, H., Yurimoto, H., Yoshida, N., Fukuya, H., Takabe, K., and Kato, N. (1999). Production of fungal fructosyl amino acid oxidase useful for diabetic diagnosis in the peroxisome *Candida boidinii. FEBS Lett.* **459,** 233–237.

Sakaue, R., and Kajiyama, N. (2003). Thermostabilization of bacterial fructosyl-amino acid oxidase by directed evolution. *Appl. Environ. Microbiol.* **69,** 139–145.

Sakaue, R., Hiruma, M., Kajiyama, N., and Koyama, Y. (2002). Cloning and expression of fructosyl-amino acid oxidase gene from *Corynebacterium* sp. 2-4-1 in *Escherichia coli. Biosci. Biotechnol. Biochem.* **66,** 1256–1261.

Schrag, J. D., and Cygler, M. (1993). 1.8 A refined structure of the lipase from Geotrichum candidum. *J. Mol. Biol.* **230,** 575–591.

Selvin, E., Steffes, M. W., Zhu, H., Matshusita, K., Wagenknecht, L., Pankow, J., Coresh, J., and Brancati, F. L. (2010). Glycated hemoglobin, diabetes, and cardiovascular risk in nondiabetic adults. *N. Engl. J. Med.* **362,** 800–811.

Somenath, R., Harindra, V., and Won Bong, C. (2005). Selective detection of cholesterol using carbon nanotube based biochip. MRS Proceedings, 900, 0900-O09-02. doi:10.1557/PROC-0900-O09-02.

Tang, Y., Tehan, E. C., Tao, Z. Y., and Bright, F. V. (2003). Sol-Gel-Derived Sensor Materials that Yield Linear Calibration Plots, High Sensitivity, and Long Term Stability. *Anal. Chem.* **75,** 2407–2413.

Tatke, S. S., Renugopalakrishnan, V., and Prabhakaran, M. (2004). Interfacing biological macromolecules with carbon nanotubes and silicon surfaces: A computer modelling and dynamic simulation study. *Nanotechnology* **15,** S684.

Tender, L. M., Reimers, C. E., Stecher, H. A., III, Holmes, D. E., Bond, D. R., Lowy, D. A., Pilobello, K., Fertig, S. J., and Lovley, D. R. (2002). Harnessing microbially generated power on the seafloor. *Nat. Biotechnol.* **20,** 821–825.

Vrielink, A., Lloyd, L. F., and Blow, D. M. (1991). Crystal structure of cholesterol oxidase from Brevibacterium sterolicum refined at 1.8 A resolution. *J. Mol. Biol.* **219,** 533–554.

Wang, J., and Musameh, M. (2003). Carbon nanotube/teflon composite electrochemical sensors and biosensors. *Anal. Chem.* **75,** 2075–2079.

Wang, J., Liu, G., and Jan, M. (2004). Ultrasensitive electrical biosensing of proteins and DNA: Carbon-nanotube derived amplification of the recognition and transduction events. *J. Am. Chem. Soc.* **126,** 3010.

Withers-Martinez, C., Carriere, F., Verger, R., Bourgeois, D., and Cambillau, C. (1996). A pancreatic lipase with a phospholipase A1 activity: Crystal structure of a chimeric pancreatic lipase-related protein 2 from guinea pig. *Structure* **4,** 1363–1374.

Wong, D. T. (2006). Salivary diagnostics powered by nanotechnologies, proteomics and genomics. *J. Am. Dent. Assoc.* **137,** 313–321.

Yoo, E.-H., and Lee, S.-Y. (2010). Review glucose biosensors: An overview of use in clinical practice. *Sensors* **10,** 4558–4576.

Zhang, J., Feng, M., and Tachikawa, H. (2007). Layer-by-layer fabrication and direct electrochemistry of glucose oxidase on single wall carbon nanotubes. *Biosens. Bioelectron.* **22,** 3036–3041.

Zhao, G. C., Zhang, L., Wei, X. W., and Yang, Z. S. (2003). Myoglobin on multi-walled carbon nanotubes modified electrode: Direct electrochemistry and electrocatalysis. *Electrochem. Commun.* **5,** 825–829.

CHAPTER TEN

Investigating the Toxic Effects of Iron Oxide Nanoparticles

Stefaan J. Soenen,* Marcel De Cuyper,† Stefaan C. De Smedt,* *and* Kevin Braeckmans*

Contents

1. Introduction 196
2. Materials 197
   - 2.1. Spectrophotometric determination of cell-associated iron 197
   - 2.2. Induction of reactive oxygen species 198
   - 2.3. Cell morphology 198
   - 2.4. Rapid and quantitative cell functionality assay using PC12 (rat pheochromocytoma) cells 198
   - 2.5. Analysis of possible pH-dependent IONP degradation 198
3. Methods 199
   - 3.1. Cell culture 199
   - 3.2. Spectrophotometric determination of cell-associated iron 200
   - 3.3. Quantitative and qualitative assessment of cell viability 202
   - 3.4. Induction of ROS 207
   - 3.5. Assessment of cell proliferation 208
   - 3.6. Investigation of cell morphology 211
   - 3.7. Rapid and quantitative cell functionality assay using PC12 cells 214
   - 3.8. Investigating possible IONP degradation 217
4. Concluding Remarks 221

Acknowledgments 221

References 222

Abstract

The use of iron oxide nanoparticles (IONPs) in biomedical research is steadily increasing, leading to the rapid development of novel IONP types and an increased exposure of cultured cells to a wide variety of IONPs. Due to the large variation in incubation conditions, IONP characteristics, and cell types

* Laboratory of General Biochemistry and Physical Pharmacy, Department of Pharmaceutical Sciences, Ghent University, Ghent, Belgium

† Laboratory of BioNanoColloids, IRC, Kortrijk, Belgium

Methods in Enzymology, Volume 509
ISSN 0076-6879, DOI: 10.1016/B978-0-12-391858-1.00011-3

studied, it is still unclear whether IONPs are generally safe or should be used with caution. During the past years, several contradictory observations have been reported, which highlight the great need for a more thorough understanding of cell–IONP interactions. To improve our knowledge in this field, there is a great need for standardized protocols and toxicity assays, that would allow to directly compare the cytotoxic potential of any IONP type with previously screened particles. Here, several approaches are described that allow to rapidly but thoroughly address several parameters which are of great impact for IONP-induced toxicity. These assays focus on acute cytotoxicity, induction of reactive oxygen species, measuring the amount of cell-associated iron, assessing cell morphology, cell proliferation, cell functionality, and possible pH-induced or intracellular IONP degradation. Together, these assays may form the basis for any detailed study on IONP cytotoxicity.

1. Introduction

The past two decades, the application of inorganic nanoparticles (NPs) such as quantum dots, carbon nanotubes, gold, or iron oxide nanoparticles (IONPs) for biomedical purposes has increased enormously, owing to the great variation in NP types, each with their own specific features (Chan *et al.*, 2002; Han *et al.*, 2007; Soenen *et al.*, 2009b; Sun *et al.*, 2008). Furthermore, the great versatility in biomedical applications made possible by these NPs, such as non-invasive imaging, targeted and triggered drug release, or cancer therapy, has boosted the exposure of NPs to biological environments (Hussain *et al.*, 2009; Lewinski *et al.*, 2008; Maurer-Jones *et al.*, 2009). For IONPs, one important application is the non-invasive monitoring of stem or immune cell migration after transplantation *in vivo*. For this purpose, the cells are isolated and then grown *in vitro* to reach sufficient numbers, after which they are labeled with IONPs to make them detectable by magnetic resonance imaging (MRI) (Himmelreich and Dresselaers, 2009). As MRI is inherently a rather insensitive technique, high contrast generation and the ability to monitor the cells for longer times require that the cells are loaded with high amounts of IONPs (Shapiro *et al.*, 2005). Recent findings have, however, shown that the *in vitro* labeling of cultured cells with high doses of IONPs can be detrimental to the cells (Soenen *et al.*, 2009c, 2010b). This has stimulated the development of novel IONP types to reduce IONP-induced toxicity while maintaining or even augmenting the magnetic contrast. The great versatility in IONPs of different sizes and surface coatings, along with the lack of any standardized procedure for cell labeling, resulting in a great variation in incubation times, IONP concentrations added, types of cells labeled, and types of assays used to assess cell viability, has made it nearly impossible to compare the labeling efficacy and potential toxicity of different IONPs (Soenen and De Cuyper, 2011). Prior to performing any cytotoxic evaluations, it is important to characterize the IONPs thoroughly to be able to interpret any obtained cytotoxicity data. More specifically, at least the following points should be carefully checked: core size of the particles (commonly by electron microscopy), hydrodynamic radius and zeta-potential (by means of electrophoretic mobility measurements), the type of stabilizing coating and its characteristics, and also the purity of the particles and the presence of any contaminating compounds. In terms of colloidal stability, the potential aggregation of the IONPs in physiological saline and in serum-containing cell medium should also be evaluated.

To allow the assessment of IONP labeling efficiency and safety in comparison to already available IONP types, there is an urgent need for a standardized procedure which enables to quickly, but thoroughly, evaluate the cytotoxic profile of novel IONPs (Rivera Gil *et al.*, 2010; Soenen and De Cuyper, 2010; Soenen *et al.*, 2011b). As NPs can interact with cellular components on many levels, it is essential to check a great number of cellular parameters to get a more complete overview. In the present work, a few basic and more advanced procedures are described which can be used to assess IONP toxicity. This panel should, however, not be considered as a complete analysis. It is important for anyone working on nanotoxicology to note that even the most elaborate studies only indicate the potential safety or toxicity of the NP type studied within a certain time-frame, a limited concentration range, and with respect to the few selected parameters which have been evaluated. Using the panel of procedures described here, the combination of all these methods provide a sound basis to evaluate general IONP toxicity and can then be used easily to compare the toxicological profile of the IONP type tested to that of other IONP types tested using the same procedure (Soenen *et al.*, 2011a). Depending on the envisaged application of the IONP or any toxic effects which are found, additional assays can be included, as also indicated in the following sections.

2. Materials

2.1. Spectrophotometric determination of cell-associated iron

The quantitative spectrophotometric determination of cellular iron uses Tiron [4,5-dihydroxy-1,3-benzenedisulfonic acid, disodium salt] (Acros Organics, Geel, Belgium). For the preparation of a standard curve, a stock solution of 1 mg Fe/ml is used (solution of $Fe(NO_3)_3 \cdot 4$ aq. in HNO_3 0.5 N) (Panreac Quimica, Barcelona, Spain). The assay further requires KOH 4 N, HCl_{conc} (37%), HNO_{3conc} (65%), and a 0.2 *M* phosphate buffer, pH 9.5 (71.6 g $Na_2HPO_4 \cdot 12$ aq. and 4 ml KOH (1 N) diluted to 1 l).

2.2. Induction of reactive oxygen species

The level of reactive oxygen species (ROS) is evaluated using dichlorodihydrofluorescein diacetate acetyl ester (H_2DCFDA; Molecular Probes, Merelbeke, Belgium). For a positive control, cells are stimulated using hydrogen peroxide (1%). As for a negative control, ROS inhibitors such as *N*-*t*-butyl-α-phenylnitrone (PBN; Sigma-Aldrich, Bornem, Belgium) or iron chelators such as desferrioxamine (Sigma-Aldrich, Bornem, Belgium) are used.

2.3. Cell morphology

Cell morphology is evaluated directly with non-stained cells, but to be able to detect more subtle changes and to assess the organization of the cell cytoskeleton, staining for F-actin and α-tubulin is highly recommended. For F-actin staining, fluorescently conjugated phalloidin is applied. A great variety of fluorophore–phalloidin conjugates is commercially available. At the same time, primary anti-α-tubulin antibodies are used for immunostaining of α-tubulin, where it is important to check whether the antibody selected is reactive toward the species that the target cells are from. For fluorescent detection, a fluorophore-conjugated secondary antibody that can recognize the primary antibody is used. For simultaneous 2-color detection, it is important to select the correct fluorophores for both stainings (e.g., Alexa Fluor 488-conjugated secondary antibody and Alexa Fluor 633–phalloidin) to reduce any possible fluorescent bleed-through or spectral overlap.

2.4. Rapid and quantitative cell functionality assay using PC12 (rat pheochromocytoma) cells

The functionality of cells can be evaluated readily using rat PC12 cells, which can be obtained from the European Collection of Cell Cultures (ETACC; http://www.hpacultures.org.uk/collections/ecacc.jsp) or the American Type Culture Collection (ATCC; http://www.atcc.org/CulturesandProducts/CellBiology/CellLinesandHybridomas/tabid/169/Default.aspx). For induction of neurite outgrowth, cells are stimulated with nerve growth factor (NGF, 7S-subunit; Sigma-Aldrich, Bornem, Belgium) of which a stock solution of 1 mg/ml in sterile dimethyl sulfoxide can be prepared. For typical cell exposure experiments, NGF is used at 100 ng/ml for 3 days.

2.5. Analysis of possible pH-dependent IONP degradation

To investigate the effect of pH on the stability of IONPs, the particles can be exposed to incubation buffers of various pH. These buffers consist of phosphate-buffered saline (PBS), supplemented with 40 m*M* sodium citrate. [When the particles would undergo acid hydrolysis, Fe^{2+} will be converted to Fe^{3+}, which is not well soluble. Therefore, sodium citrate can be added, because the citrate can help as a low molecular weight scaffold to solubilize the Fe^{3+}.] This buffer is then mixed with cell medium containing 10% fetal bovine serum in a 1:1 ratio (v:v). Next, the buffer is split into three equal parts and the pH is adjusted to 7.0, 5.5, or 4.5 using HCl (2 *M*). Finally, the different buffers are filter-sterilized using 0.20-μm filters and are then ready to be used for further experiments.

3. Methods

3.1. Cell culture

The majority of the methods described in this work can be applied to nearly all cell types, which all have their own specific culture conditions. In the present work, three different types of cells are used as testing multiple cell types gives a more complete overview of the potential cytotoxic profile of the IONPs, and every cell type has specific features that make them better suited for a certain protocol. Please note that, for toxicity studies, the use of cancer cells is not recommended, but rather highly sensitive cell lines (e.g., neural cells) or primary human cells will give a better idea on the cytotoxic profile of the studied NPs. The culture conditions for every specific cell type are described below.

a. *C17.2 murine neural progenitor cells:* These cells are cultured in high glucose containing Dulbecco's modified Eagle's medium (DMEM), supplemented with 10% fetal bovine serum, 5% horse serum, 1 m*M* sodium pyruvate, 2 m*M* L-glutamine, and 1% penicillin/streptomycin (Gibco/Invitrogen, Merelbeke Belgium). Cells are kept in 75-cm^2 tissue culture flasks and need to be passaged with 0.05% trypsin (Gibco) every 48 h and split 1/5, reaching maximal confluency levels of 80–90%.
b. *PC12 rat pheochromocytoma cells:* The cells are cultured in the same medium as the C17.2 cells, but as the cells are semi-adherent, the tissue culture flasks (25 cm^2) can be coated with, for instance, rat tail collagen type I (Invitrogen) to increase cell adhesion. [Collagen coating of tissue culture plates is achieved by preparing an acidic stock solution (e.g., sterile, cell culture-grade H_2O (200 ml) to which 0.23 ml of 17.4 N acetic acid is added). Then, an appropriate amount of collagen stock solution (stock concentration is usually 2–5 mg/ml, but this depends on the distributor) is added to reach a final collagen concentration of 50 μg/ml. For 25-cm^2 tissue culture plates, 3 ml of this solution is sufficient, where 75-cm^2 tissue culture plates use 9 ml. After a 1-h incubation (minimal, if possible, it is recommended to incubate these flasks overnight) in a humidified atmosphere at 37 °C and 5% CO_2, the plates can be washed

once with sterile PBS and then kept in PBS until the cells can be applied.] The cells are passaged with 0.05% trypsin when reaching near 70% confluency and split 1/5. [PC12 cells tend to grow in small clumps, so they will never reach high levels of confluency. Normally, when clumps of about 5 cells are formed, the cells should be passaged for further growth.] The medium has to be changed every 2 days.

For induction with NGF, PC12 cells were maintained in NGF induction medium, consisting of high glucose DMEM, supplemented with 1% fetal calf serum, 1 m*M* sodium pyruvate, 2 m*M* L-glutamine, 1% penicillin/streptomycin, and 100 ng/ml NGF (Sigma-Aldrich, Bornem, Belgium).

c. Primary human (or bovine) endothelial cells such as blood outgrowth endothelial cells (BOECs) or umbilical vein endothelial cells (HUVECs) can also be obtained through the ETACC or ATCC or can be freshly isolated, as described elsewhere (De Meyer *et al.*, 2006). These cells are cultured on collagen-coated tissue culture flasks (75 cm^2) in endothelial basal/growth culture medium (e.g., complete endothelial cell growth medium; Cell Applications, Inc., San Diego, USA) with medium changes every 48 h. Cells are passaged when reaching near 80% confluency by lifting the cells with 0.05% trypsin and are reseeded (1/5) on collagen-coated culture plates.

These cells can generally be used for over 18 passages, but it is best to perform the toxicity experiments on cells with low passage numbers (preferably between 3 and 10). The use of primary human cells further enhances the significance of the obtained results compared to long-lived cell lines.

3.2. Spectrophotometric determination of cell-associated iron

As the IONPs are used to magnetically label the cells, it is essential to determine the amount of iron (the number of particles) that is taken up by the cells and/or adheres to the cell surface, to verify whether this is sufficient to enable any further biomedical use of these labeled cells. Transmission electron microscopy can aid in determining the precise cellular localization of the particles, but is less suited for a quantitative analysis. To this end, researchers can use inductively coupled plasma as a tool to determine the cellular iron levels (Raynal *et al.*, 2004). An alternative method, which is less sensitive, but more practical and faster, is a spectrophotometric determination using Tiron, which will bind Fe^{3+} and result in the formation of a red complex (De Cuyper and Soenen, 2010).

a. Cells (e.g., C17.2 cells) are seeded in 96-well plates at 5×10^4 cells/well and allowed to settle overnight. [For toxicological cell labeling studies, a wide IONP concentration range is recommended, where IONP concentrations between 10 and 200 μg Fe/ml are most commonly used

(Soenen and De Cuyper, 2010).] Then, fresh media is applied to the cells containing the IONPs at the desired concentration and the cells are allowed to incubate in a humidified atmosphere at 37 °C and 5% CO_2 for the appropriate time. [Incubation times can vary quite a lot, but most IONPs appear to have an optimal incubation time between 4 and 24 h, after which the cells appear to be saturated (Soenen *et al.*, 2007). Employing multiple incubation times (2, 4, 8, and 24 h) can help to indicate the optimal labeling procedure.] It is important to seed enough cells to allow for controls, negative controls, and take 5–10 wells for every specific IONP concentration. To verify reproducibility, this procedure should be repeated at least three times.

b. After cell labeling, cells should be washed three times with sterile PBS (150 μl/well) after which any remaining liquid is discarded and 25 μl/well of a mixture of H_2O, HCl, and HNO_3 (5/3/1) is added, after which the cells are placed on a shaker (200 rpm) at 37 °C for 1 h.

c. Next, 50 μl of sterile H_2O is added, followed by 96 μl/well of an alkalic Tiron solution (0.25 *M* Tiron in H_2O/KOH 4 N; 1/5) [The formation of Tiron–Fe^{3+} complexes is highly dependent on the pH, where the number of Fe^{3+} ions which can complex with 1 molecule of Tiron can vary between 1 (for lower pH) and 3 (for higher pH), which also affects the color of the complexes (varying from green to red). Therefore, it is important to use a KOH 4 N solution to neutralize the acids which were used to lyse the cells and degrade the IONPs and to apply the phosphate buffer to reach a stable pH], and subsequently 160 μl of a 0.2 *M* phosphate buffer (pH 9.5). [After addition of the alkaline Tiron solution, the red color develops immediately. The phosphate buffer is added to stabilize the complex, after which the color remains stable for several hours.] The mixtures are then allowed to incubate for 15 min at ambient temperature, after which the absorbance is measured at 480 nm.

d. To obtain quantitative data, a standard curve made from the 1 mg Fe/ml stock solution is prepared, as described elsewhere (De Cuyper and Soenen, 2010) with iron concentrations ranging from 0 to 150 μg Fe/ml. [In short, the iron stock solution is diluted with HCl (37%) and HNO_3 (65%) and distilled water, following the steps as indicated in Table 10.1.] From these standard solutions, 75 μl can directly be transferred into the appropriate wells, followed by the addition of 96 μl of Tiron solution and 160 μl phosphate buffer.

e. With regard to controls, cells which were not incubated with the IONPs but otherwise treated identically can be used as negative controls. To get an estimate of the amount of cell-adhering particles and actually cell-internalized IONPs, IONP incubation can occur at temperatures between 4 °C and 15 °C, where all active endocytosis mechanisms are blocked. The value obtained here then refers to the amount of cell surface-attached IONPs, which can then be subtracted from the total

Table 10.1 Preparation of the iron stock solutions for quantitative iron determination

Tube	Fe stock (μl)	HCl (37%) (ml)	HNO_3 (65%) (ml)	H_2O dist (ml)	[Fe] (μg/ml)
0	0	0.6	0.2	4.20	0
1	50	0.6	0.2	4.15	10
2	100	0.6	0.2	4.10	20
3	150	0.6	0.2	4.05	30
4	250	0.6	0.2	3.95	50
5	350	0.6	0.2	3.85	70
6	500	0.6	0.2	3.70	100
7	750	0.6	0.2	3.45	150

amount of cell-associated iron to obtain the amount of intracellular iron. [The "amount" of intracellular iron will be expressed in μg/ml, which can then be calculated to pg/cell by determining the number of cells in every condition. Generally, this can easily be done using a protein determination test, such as the bicinchoninic acid assay (Pierce, Rockford, USA) which can be conducted according to the manufacturer's protocol for a 96-well plate. By first measuring the protein levels of a known number of cells (ranging from 10,000 to 100,000/well), a standard curve can be produced, which can be used to determine the number of cells per well after incubation with the IONPs. Also note that for MRI, the mass of iron per cell is indeed an indicative value, but for toxicological purposes, the number of IONPs per cell is more meaningful. By multiplying the mass of iron obtained by 1.38, the mass of Fe_3O_4 can be calculated. By taking into account the density of iron oxide (5.1 g/cm^3) (Razzaq *et al.*, 2007) and the size of the iron oxide core (e.g., 2 nm diameter, leading to a total volume of 4.19 nm^3), the weight of a single IONP can then easily be calculated. By dividing the cellular iron level over the weight of a single IONP, the number of IONPs per cell can be determined.] To check for any possible aggregation, the same procedure can also be applied for IONP-containing media without any cells. If larger aggregates are formed and sediment on the bottom of the well, they can then be detected.

3.3. Quantitative and qualitative assessment of cell viability

To assess cell viability, many commonly used assays are readily available. Most of these assays determine cell viability by means of a single parameter, which can sometimes give seemingly contradicting results. For instance, cells with an impaired metabolism or slowed proliferation will show a lower viability when using an MTT or similar assay (measures mitochondrial metabolism) compared to when a Trypan blue exclusion assay or lactate dehydrogenase (LDH) assay (measures plasma membrane permeability) is performed. Also, depending on the specific features of the IONP used (fluorescently labeled, size, presence of amine groups), the particles may interfere with the assay readout, so appropriate controls are essential (Fig. 10.1A). To get a clear and reliable idea on cell viability, it is therefore recommended to apply multiple assays and validate the data obtained by one assay by means of the second assay (Fig. 10.1). Here, the LDH assay is selected as an excellent assay for a rapid, sensitive and quantitative determination of cell death, where cell staining using calcein AM and ethidium homodimer-1 provide an attractive means of a semi-quantitative approach, also allowing visual confirmation (Soenen and De Cuyper, 2009).

3.3.1. LDH assay

The present protocol makes use of the LDH assay as available from Pierce (Rockford, USA) (Fig. 10.1A). Depending on the manufacturer, small changes in incubation times and reagent concentrations may be possible, but the overall setup of the protocol can be kept as is.

a. Cells (e.g., C17.2 cells) are seeded in 96-well plates at 25,000 cells/well in 150 μl of full cell medium/well and allowed to settle overnight in a humidified atmosphere at 37 °C and 5% CO_2. Next, medium is removed

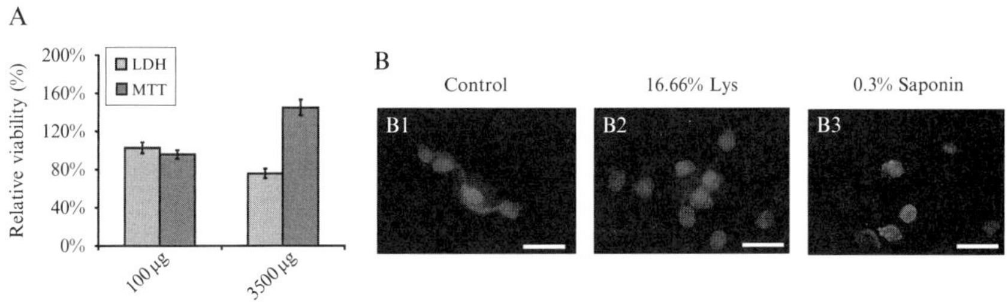

Figure 10.1 Assessing cell viability of magnetoliposome (ML)-treated 3T3 fibroblasts. (A) Results of an MTT and an LDH assay for 3T3 fibroblasts incubated with 3% cationic lipid-containing MLs at 100 and 3500 μg Fe/ml for 24 h. Values are given relative to those of untreated control cells. The error bars indicated are mean ± SEM ($n = 10$). (B) Representative images depicting the viability of untreated NIH 3T3 cells (B1) or cells incubated for 24 h with 16% (B2) cationic lipid-containing MLs at 100 μg Fe/ml, as determined by calcein AM (green; live cells) and ethidium homodimer (red; damaged cells) treatment; blue color indicates DAPI nuclear staining. B3 shows control cells treated with 0.3% saponin; scale bars: 75 μm. See Soenen and De Cuyper (2009) for more details. Reproduced with permission from Soenen and De Cuyper (2009), © Wiley VCH. (See Color Insert.)

and fresh medium is added containing the IONPs at the appropriate concentration (the same ones as those used for the iron determination) and allowed to incubate for the desired time (the same durations as for the iron determination). For all assays, using multiple incubation times is beneficial, but for the LDH assay, the time factor is even more critical. As the assay is based on the enzymatic activity of LDH, it requires an active enzyme to give reliable results. The enzyme itself is normally located solely in the cell cytoplasm, but upon plasma membrane rupture, the enzyme can be released into the surrounding cell medium, after which it will slowly be degraded. [The half-life of LDH is in the order of 9 h, rendering a single LDH assay after 24 h quite useless to determine any acute cytotoxic effect as nearly all of the early released enzymes will have been degraded. By employing multiple time points, especially after short incubation times, more reliable results can be obtained and the exact timing of maximal LDH release can also be determined more accurately.]

b. Twenty minutes prior to the assay, 25 μl of a 2% Triton X-100 or any other detergent is added to some of the untreated control cells (e.g., 5 wells = negative control), where the other control cells and the IONP-treated cells get 25 μl of full medium, after which the cells can further incubate for 20 min.

c. After incubation, 50 μl/well of the cell medium is collected and transferred to a new 96-well plate. [When collecting the medium, care must be taken not to collect any cells as only LDH present in the extracellular medium must be measured. Therefore, it is best to gently pipette off some of the medium at the top of the wells.] Then, 50 μl of LDH substrate is added to the new plate containing 50 μl of transferred cell medium and the plate is incubated at ambient temperature for 30 min.

d. After incubation, 50 μl of STOP solution (from the Pierce kit) is added, after which the absorbance can be measured at 490 nm. The amount of dead cells can then be calculated.[1] The absorbance at 690 nm can also be measured as a background control, after which this value can be subtracted from the value measured at 490 nm.

e. An advantage of using the LDH assay is the fact that it allows to normalize all the obtained data in terms of cell numbers and to correct for any variability that is generally inherent to cell seeding procedures. Therefore, when collecting the first 50 μl of cell medium for the LDH assay, 25 μl of a 2% Triton X-100 solution is added to all wells after

[1] Calculation of the percentage of dead cells can occur easily using the following simple formula:

$$\text{cytotoxicity}(\%) = \frac{A_{\text{sample}} - A_{\text{negcontrol}}}{A_{\text{poscontrol}} - A_{\text{negcontrol}}} \times 100$$

With A_{sample}: average A480 measured for the test sample.
$A_{\text{negcontrol}}$: average A480 measured for untreated control cells.
$A_{\text{poscontrol}}$: average A480 measured for lysed control cells.

which the plate is incubated for 20 min at 37 °C and 5% CO_2. Then, 50 μl/well can be taken off again and transferred to a new plate, where the LDH substrate and STOP solution can be added as described above. By determining the total number of dead cells for every well, the data obtained for the IONP-only treated cells can then be normalized using these data. [As NPs have been described to bind serum proteins, the degree of which is determined for a large part by the particle's surface properties, the possibility exists that cell-internalized or cell surface-attached IONPs bind LDH and impede its release into the extracellular medium. Furthermore, it is possible that fluorescently tagged IONPs may interfere with the absorbance readout. This can be checked by directly comparing the amount of LDH released for Triton-lysed cells which were not exposed to IONPs (as described in Section 3.3.1.b) and Triton-lysed cells which were previously exposed to IONPs (as described in Section 3.3.1.b). When the IONPs do not interact with the LDH assay, these values should be similar, whereas reduced LDH release for the IONP-treated cells hints at a direct interaction of LDH and the IONPs, making the assay useless for determining the viability of IONP-treated cells.]

3.3.2. Cell staining with fluorescent viability agents

As a semi-quantitative assay to assess cell viability while allowing optical confirmation, an easy tool consists of cellular co-staining with fluorescent agents such as calcein AM and ethidium homodimer-1 (Fig. 10.1B).

a. Sterile glass coverslips are placed into every well of a 12-well plate and are coated with rat collagen type 1 as described in Section 3.1 to increase cell adhesion. Then, the cells (e.g., C17.2 cells) are seeded at 5×10^4 cells/well and allowed to settle overnight in a humidified atmosphere at 37 °C and 5% CO_2. Next, the medium is removed, fresh medium containing the IONPs at the appropriate concentrations is added, and the cells are incubated for the desired times. [Controls are easily obtained by staining cells which are not incubated with IONPs but otherwise treated identically (should normally be all green cells). Negative controls can be obtained by exposing some control cells to 1% Triton X-100 for 15 min at 37 °C prior to removal of the medium and further staining (should normally lead to only cells with red nuclei). If at the end, the control cells also have red-colored nuclei, this may indicate that the level of ethidium homodimer-1 was too high and should be lower or, alternatively, that more precautions should be taken while handling the cells to avoid any damage.]
b. After removal of the medium and washing the cells twice with sterile PBS (1 ml/well), cells can be incubated for 45 min at ambient temperature with calcein AM (recommended concentration: 2–5 μ*M*) and ethidium homodimer-1 (recommended concentration: 1.5–2 μ*M*) in sterile, tissue-grade PBS.

c. Next, the incubation medium is removed, and cells are washed three times with sterile PBS, fixed for 15 min in 2% paraformaldehyde (PFA) at ambient temperature and subsequently washed again with PBS (three times).

d. The coverslips can then be removed from the wells by using sterile tweezers and mounted onto glass microscopy slides. [For mounting, the glass microscopy slides should be cleaned by ethanol to remove any dust or smears. Then a drop of mounting medium (e.g., Vectashield® or ProLong Gold Antifade reagent®) can be added onto the microscopy slide, after which the coverslip (with the side where the cells are adhering faced toward the glass microscopy slide) can be placed on top of the mounting medium. The mounting medium must first dry out, where clear nail polish can then be used to cover the sides of the coverslips to prevent the cells from drying out (the latter is only required if the cells cannot be visualized in the first few days).] The slides can be incubated in the dark at room temperature for approximately 24 h, after which the mounting medium has dried and the slides can be kept at 4 °C for several days or at −20 °C for several weeks prior to viewing by fluorescence microscopy.

e. The samples can then be viewed using the fluorescence microscope (calcein AM: ex/em: 495/515; ethidium homodimer-1: ex/em: 495/635), preferably at a somewhat lower magnification (e.g., 20 ×) to get a nice overview of the average condition of every cell population. [As this procedure relies on fluorescent analysis, the use of fluorescently tagged IONPs is not recommended as this could harden the interpretation of any results obtained. Furthermore, when high intracellular or cell surface-attached levels of IONPs are obtained, this may result in a loss of fluorescence. Especially in the case of high levels of cell surface attachment, this assay is therefore not recommended. Furthermore, Fe^{3+} can lead to quenching of calcein fluorescence, so rapid IONP degradation could also compromise the outcome of the assay, although the rate of IONP degradation is generally not that fast to result in such high levels of Fe^{3+} to quench the complete cellular signal.] Dead cells will have a red nuclear compartment, and live cells will have a green-colored cytoplasm. Cells with a bright green cytoplasm and red nucleus indicate cells which are still alive but with a comprised plasma membrane. For semi-quantitative analysis, the number of green cells can be counted and divided by the total number of cells, giving the percentage of live cells (preferably at least 300 cells are to be counted for every condition, but this can occur semi-automatically by software analysis tools such as ImageJ, the Java-based image processing software developed at NIH, Bethesda, MD, USA). [If the LDH assay has also been performed, the percentages of cell viability for both assays can be compared. Theoretically, these should be similar, or else this would indicate interactions of the IONPs with either of the two assays. If different results are obtained, a third assay (e.g., Trypan blue staining or an MTT assay) can be performed to confirm one of both results.]

3.4. Induction of ROS

One of the most common toxic effects for NPs in general and IONPs in particular is the induction of ROS (Nel *et al.*, 2009). For IONPs, quite often a transient induction on ROS is observed, with maximal ROS levels after 4 or 24 h (Arbab *et al.*, 2003; Soenen *et al.*, 2009a). Although often only transient, high levels of ROS can lead to cellular damage by resulting in mitochondrial damage, lipid peroxidation, or protein oxidation, which can then induce a cascade of Ca^{2+}-dependent signaling mechanisms resulting in cell death (Soenen and De Cuyper, 2009; Stroh *et al.*, 2004). Therefore, for every IONP type, it is important to verify whether the induction of ROS is indeed only transient and what the maximal level of ROS is.

a. Cells can be seeded in sterile, tissue culture-grade, black 96-well plates at a density of 5×10^4 cells/well and allowed to settle overnight in a humidified atmosphere at 37 °C and 5% CO_2. Then, the medium is removed, and cells are given fresh medium containing the particles at the desired concentrations and incubated for the appropriate time points. [To verify whether any observed increase in ROS is transient or more long-lasting, it is imperative that multiple time points are used. To this end, it may be interesting to incubate the cells with the particles for the same time points as used for the other assays (e.g., 2, 4, 8, or 24 h), but then include multiple time points after NP removal where the reagent is added (e.g., after removal of the IONP-medium and washing of the cells, the cells can be treated with the reagent immediately or first allowed to further grow at 37 °C and 5% CO_2 where the reagent is added after 2, 4, 6, 8, 12, 24, 36, 48 h) for continuous culture.]
b. After incubation, the medium is removed, and cells are washed three times with sterile PBS (150 μl/well) and incubated with 150 μl/well of H_2DCFDA (10 μ*M* in sterile, tissue-grade PBS) for 45 min at 37 °C. [Although the assay generally works well using standard H_2DCFDA, several derivatives are also available, containing carboxyl moieties (which are better retained within the cell) or fluorinated derivatives, which are more photostable.]
c. Following incubation, the incubation mixture is removed, and cells are washed three times with sterile PBS (150 μl/well) and incubated in complete cell medium for 30 min at 37 °C.
d. Next, the medium is removed, and the cells are washed again with 150 μl PBS/well (2 ×) and kept in sterile PBS (150 μl/well) for readout using a fluorescence plate reader, equipped with emission and excitation filters

of approximately 495 and 525 nm, respectively. [Apart from using a fluorescence plate reader, the samples can also be prepared on glass coverslips as described in Section 3.3.2, where the concentration and incubation conditions are identical to those described for this procedure. However, using the fluorescence plate reader setup, the procedure is less laborious and fast (compared to flow cytometry or microscopy setups) and generally gives the most straightforward results. The DCF fluorophores produced after oxidation have only a very limited photostability, and the longer illumination times typically associated with fluorescence microscopy will induce photobleaching, which greatly impairs the use of this approach for quantification of fluorescence intensities.] The fluorescence intensity can then be measured directly and expressed in terms of control cells (which were not incubated with the IONPs) which equal 100% ROS levels. For positive controls, cells which were not incubated with IONPs but exposed to H_2O_2 (1%) equally as long as the IONP duration can be used. [When high levels of ROS (e.g., more than a twofold increase compared to control cells) are observed, or when the effects appear to be long-lasting rather than transient, it would be useful to check whether the ROS induction has any effect on cell viability. This can be done by using ROS scavengers such as *N*-*t*-butyl-α-phenylnitrone (1–5 m*M*) to pre-incubate the cells for 30 min and also apply the ROS scavengers at the same concentration during IONP incubation and then performing the viability assays as described above. If ROS directly induces cell death, the use of the ROS scavenger should be able to diminish this effect. Also, if toxicity is observed, additional parameters such as mitochondrial membrane potential or cellular calcium levels can be measured to get a better idea on the mechanism by which ROS results in cell death.]

3.5. Assessment of cell proliferation

Recently, it has been described that high intracellular levels of IONPs can reduce cell proliferation (Hu *et al.*, 2006) by affecting cell cytoskeleton and associated signaling pathways (Soenen *et al.*, 2010b). A clear concentration-dependent decrease in cell viability was observed, which was often only transient and where maximal effects were found after approximately 3 days (Soenen *et al.*, 2009c). As secondary or long-term effects are somewhat hard to study, the cellular proliferation of cells is a rather straightforward approach to monitor IONP effects on cell homeostasis and to elucidate any major secondary effects on intracellular signaling pathways. For a quantitative analysis, cells can be manually counted in, for instance, a Bürker chamber, but this can lead to rather great variations in the number of cells counted. Great care must also be taken as differences in initial cell seeding densities can greatly influence the speed of cell cycle progression and hereby lead to rather large variations in cell numbers after several doubling times. A semi-quantitative approach based on the DNA incorporation of thymidine analogs such as bromodeoxyuridine or 5-ethynyl-2′-deoxyuridine (EdU) offers an interesting alternative to manual cell counting (Fig. 10.2). The following procedure is based on the Molecular Probes Click-iT® EdU Imaging kits.

a. Cells are seeded in 12-well plates containing sterile, collagen-coated glass coverslips at a density of 1×10^4 cells/well and allowed to settle overnight in a humidified atmosphere at 37 °C and 5% CO_2. [Seeding cells directly on the glass coverslips is only useful when cell proliferation is checked within the next few days. If proliferation must be checked at a later time (e.g., after 1 week post-IONP incubation), the cells can first be seeded in 25-cm^2 tissue culture falcons at a density of 5×10^5 cells/falcon, be incubated with the IONPs, and then kept in culture in IONP-free medium up to 48 h prior to the assay, where the cells are then trypsinized and reseeded in the 12-well plate at a density of 1×10^4 cells/well, allowed to settle overnight, and then treated similarly as the other cells described below. The density of the cells must be kept rather low to allow further proliferation without any hindrance or cell contact inhibition.] Then, the medium is removed and cells are given fresh medium containing the particles at the desired concentrations and incubated for the appropriate time. [As cell proliferation is not an acute effect but more

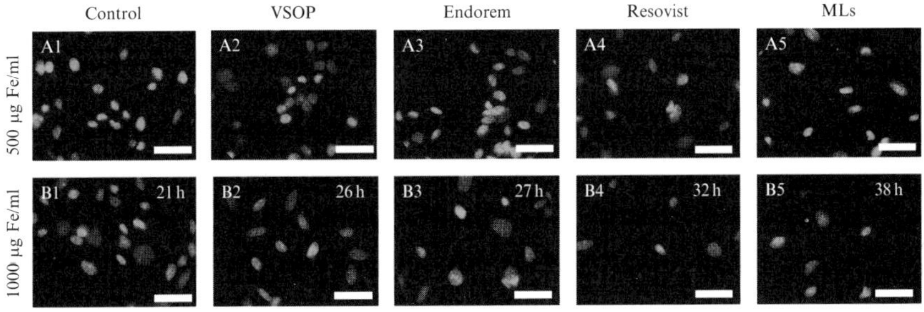

Figure 10.2 (A,B) Cellular proliferation of C17.2 NPCs as assessed by EdU-staining (green) for cells not exposed to any particles (Control) or cells incubated with the indicated particles (VSOP: citrate-coated 4 nm diameter IONPs; Endorem: dextran-coated IONPs with hydrodynamic diameter between 60 and 150 nm; Resovist: carboxydextran-coated IONPs with hydrodynamic diameter of about 75 nm; MLs: lipid-coated 14 nm diameter IONPs) at 500 μg Fe/ml (A) and 1000 μg Fe/ml (B) at 3 days postparticle incubation; nuclei of non-proliferative cells are colored blue (DAPI). Scale bars: 75 μm. Reproduced with permission from Soenen *et al.* (2010b), © Wiley VCH. (See Color Insert.)

a secondary effect, different IONP incubation times are generally not necessary. Normally, one can use only a single incubation time (where maximal IONP uptake was described).]

b. The IONP-containing medium is then removed, and cells are washed three times with sterile PBS and fresh medium not containing any IONPs is added after which the cells are kept in culture in a humidified atmosphere at 37 °C and 5% CO_2 for approximately 48 h.

c. Next, half the medium is removed (0.5 ml/well), after which 0.5 ml of fresh medium is added, supplemented with 20 μM of EdU (resulting in 10 μM of final EdU concentration), after which the cells are kept in culture in a humidified atmosphere at 37 °C and 5% CO_2 for the desired time. [The optimal incubation time with EdU greatly depends on the cell type used. In general, an incubation time of approximately 60–75% of the normal cell doubling time is recommended. During this time, the majority of the cells will have doubled. It is important to keep the incubation time beneath the cell doubling time as otherwise, even cells with a slowed down cell cycle progression could still undergo cell division and stain positively. The average cell doubling times (please note that this depends greatly on cell density and medium composition) of many cell types can be found on the ATCC and ETACC web pages. This can serve as a good indicator to calculate the incubation time, but it is better to measure this yourself under the conditions where your cells are kept, e.g., by means of manual counting using a Bürker chamber.]

d. Next, the medium is removed, and the cells are washed three times with sterile PBS (1 ml/well) after which cells can be fixed in 2% PFA for 15 min at room temperature [for fixating the cells, it is best to use relatively fresh PFA (e.g., maximal 1 week old). For ease of preparation, PFA powder can be weighted (e.g., 4 g), after which an appropriate volume of PBS is added (e.g., if 200 ml of total volume is needed, add 50 ml). Then, a few drops of concentrated NaOH or KOH are added and the mixture can be stirred at elevated temperature (e.g., 50 °C; it is important to keep the temperature below 60° as otherwise the PFA might start to decompose into formaldehyde). After a clear solution is obtained, the pH can be re-adjusted to 7 by adding HCl. After that, the required volume of PBS can further be added to reach a total volume of 200 ml], washed three times with PBS (1 ml/well), permeabilized by incubation with 1% Triton X-100 in PBS for 15 min at room temperature, and washed again three times with PBS (1 ml/well). The cells can then be blocked by incubation with 5% serum-containing PBS, further supplemented with 2% bovine serum albumin for half an hour at room temperature. Then, the cells can be stained for EdU incorporation into the cellular DNA by incubating them with fluorescently tagged azides (making use of the incubation mixture provided in the Click-iT® EdU Cell Proliferation Assay Kit offered by Molecular Probes).

e. Following incubation, the incubation mixture can be removed, and cells are washed three times with PBS (1 ml/well) and then counterstained with a nuclear stain (e.g., 4′,6-diamidino-2-phenylindole (DAPI) which can be incubated at 0.5 μg/ml for 5 min at ambient temperature) and washed three times with PBS (1 ml/well) and then the coverslips can be mounted onto glass microscopy slides as described in Section 3.3.2.d.
f. Cells can then be viewed by fluorescence microscopy, where all cell nuclei stained with DAPI will give a blue color, whereas the cells that have undergone active mitosis will show EdU incorporation and will have a second signal, the color of which depends on the fluorophore-coupled azide which was used for detection. [The fluorescence images taken from the cells can then be further analyzed to get semi-quantitative data on cell cycle progression. By counting the number of blue (= all) nuclei and the green or red (= actively dividing cells), the ratio of the green (or red) nuclei over blue nuclei gives the percentage of cells that have undergone mitosis during the incubation time with EdU. These numbers can be further validated by means of manual cell counting. The visual confirmation allows to investigate whether cells which do appear to still be alive show a reduced proliferation rate, whereas loss of cell numbers observed by manual counting may also signify cell death.]

3.6. Investigation of cell morphology

Cellular stress as, for instance, induced by NP labeling can result in clear morphological alterations (Gupta and Gupta, 2005). Especially when higher intracellular NP levels are achieved, cell spreading was greatly affected, resulting in a decrease in total cell areas and a polarization of the cell from a typical round shaped to a lengthened cell body (Soenen *et al.*, 2010b). Also, the cellular cytoskeleton (the actin and tubulin networks) was shown to be affected, which could impede actin-mediated signaling mechanisms and affect cell mobility, proliferation capacity, and even directly induce cell death (Soenen *et al.*, 2010b). Analyzing cell morphology and cytoskeleton can be based on F-actin and α-tubulin staining and confocal microscopy-based analysis (Fig. 10.3) as described below:

a. Cells (e.g., BOECs) [For these studies, primary endothelial cells, such as BOECs (HUVECs or cells from other species), offer great advantages, as these cells are typically rather large and well spread, with a near-round morphology. This allows a rather simple and straightforward analysis compared to, for example, the neural cell types, which often display cytoplasmic extensions or complete neurites that make it hard to calculate the precise cell surface area and cell aspect ratio.] are seeded at 2×10^4 cells/well in 12-well plates containing sterile, collagen-coated glass coverslips and allowed to settle overnight in a humidified atmosphere

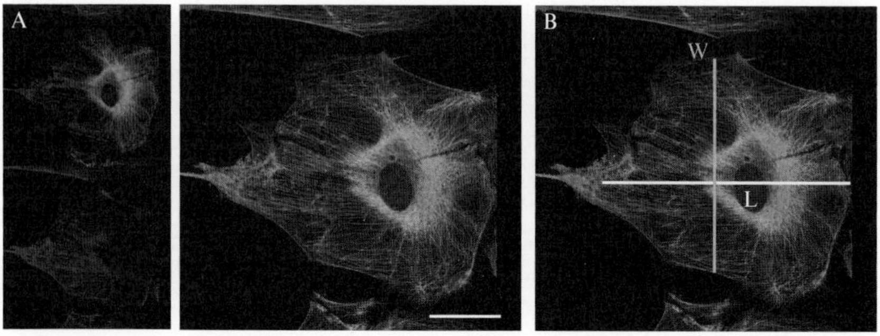

Figure 10.3 (A) Representative confocal image of a hBOEC depicting α-tubulin (green), F-actin (red), and a merged image showing α-tubulin, F-actin, and DAPI nuclear counterstaining (blue). Scale bar: 50 μm. (B) For measuring cell polarity, the cellular length (white bar) and width (light blue bar) must be determined. The ratio of cellular width over length then gives information on the cellular spreading. Please note that cytoplasmic extensions should be excluded for a proper determination of cellular length and width. (See Color Insert.)

at 37 °C and 5% CO_2. [Cell seeding density must be kept rather low to avoid cells getting into contact with neighboring cells, which could by itself lead to morphological changes due to the interaction forces between cells and the limited space provided for the cell to attach on.] Then, the medium is removed, and cells are given fresh medium containing the particles at the desired concentrations and incubated for the appropriate time. [Similar as for the proliferation studies, there is no real need for different incubation times. Generally, 24 h (if maximal uptake is achieved at a faster time, then this time can be taken) is sufficient. As these effects can be transient, cells can be kept in culture for several days postincubation with the IONPs, where samples of the cells can be collected, seeded in 12-well plates as described above, and prepared for analysis as described further in the protocol.]

b. Cell medium is aspirated, and cells are washed three times with PBS, fixed in 2% PFA for 15 min at ambient temperature; washed three times with PBS, permeabilized by incubation with 1% Triton X-100 for 15 min at room temperature; and washed three times with PBS (as described in Section 3.5.d) and then blocked by incubating with PBS containing 10% serum. [The serum should preferably be of the same origin as the secondary antibody which is used further on for tubulin detection. In case a goat anti-murine antibody is used for detection, PBS containing 10% goat serum can be used.]
c. Cells can then be incubated with primary antibody against α-tubulin by exposing the cells to 1 ml/well of 10% serum blocking medium containing the antibody at the desired concentration. [It is important to verify

whether the antibody that is selected is reactive against the origin type of cells. If human or bovine cells are used, the antibody should detect human or bovine α-tubulin, respectively. Antibodies against α-tubulin are commonly available from most antibody suppliers, where the final concentration used may depend on the distributor, but generally, 1–2 μg/ml of antibody is an appropriate concentration.] Incubation can then occur at room temperature for about 2 h. [If desired, incubation can also occur overnight at 4 °C. Here, it is important to nicely cover the cells with the antibody suspension to prevent cells from dehydrating.]

d. Cells are washed three times with blocking solution, and fresh blocking solution with a fluorescently coupled secondary antibody and fluorescent phalloidin can be added and incubated for 1 h at room temperature. [As mentioned in Section 2.3, it is important to verify whether both types of fluorescent conjugates can be used for simultaneous detection. A wide variety of secondary antibodies or phalloidin can be obtained, where the optimal concentration (generally: 1/250–1/500 dilution for the antibody and 1/200–1/500 dilution for phalloidin) will depend on the distributor and should be checked upon purchase.]
e. Next, the incubation medium is removed, and cells are washed three times with 10% serum-containing PBS and the coverslips are then mounted onto glass microscopy slides for further storage as described in Section 3.3.2.d. [Cells can also be stained with DAPI for nuclear counterstaining as described in Section 3.5.e.]

3.6.1. Calculation of cell areas and cell polarity

For calculating cell areas, images can be taken at relatively smaller magnifications (e.g., 20 ×), where preferentially multiple cells are completely visible per image. Using image analysis software, such as ImageJ, the size of the cells can be calculated as follows:

a. The image file is opened in ImageJ.
b. The multicolor image is converted to grayscale 8-bit image.
c. At Image → Adjust → Threshold, the threshold is set for the image. This value should be chosen so that the background is not selected, but every cell is.
d. Then, at Analyze → Measure, the total area of all regions that were above the selected threshold is obtained. By dividing this value by the number of cells on the image, the average area of the cells, expressed in pixels, is obtained. By means of the scale bar (or using Photoshop® or similar tools), this value can be converted to the μm scale. [It is important to measure the complete cells only. If any cells are only partly in the image whereas the rest is outside of the field of view, you can use Analyze → Analyze particles and then select "exclude on edges" so that any cells which are on the border of the image file are not part of the measurement.]

e. For a good analysis, a rather large number of cells are required. Generally, 200–300 cells can give a rather nice idea on the global spreading of the cells. Taking lower magnification images are therefore recommended to increase the number of cells per image and hereby decrease the number of images which need to be analyzed.

Calculating the cell aspect ratio can also be done using ImageJ or the original software associated with the confocal microscopy setup. To this end, higher magnification images (e.g., 60 ×) can be taken, where the main cell axis (longest straight line across the cell body) and the short axis (longest straight line perpendicular to the main cell axis) must be drawn (Fig. 10.3B).

a. The length of the main cell axis is measured.
b. The length of the short cell axis is measured.
c. The length of the short axis is divided by the length of the main cell axis to get the cell aspect ratio. This value lies between 0 (only theoretically possible for a thin line) and 1 (for a perfectly round cell). The closer this value is to 0, the less well-spread the cell is.
d. At least 20 cells/condition (randomly selected) should be analyzed in this way to be able to perform statistically significant analyses.

3.7. Rapid and quantitative cell functionality assay using PC12 cells

The assessment of cell functionality after IONP labeling is usually based on stem cell differentiation, which can take up several weeks and is frequently accompanied by relatively high numbers of cell death (Soenen and De Cuyper, 2009) and is often detected by visual confirmation of the expression of certain marker proteins (Walczak *et al.*, 2007). These aspects have led to contradicting observations with respect to the possible inhibition of stem cell differentiation by IONP labeling (Arbab *et al.*, 2005b; Chen *et al.*, 2010). A semi-quantitative and rapid approach of cell responsiveness, without any associated cell death during the differentiation process itself, would offer a lot of benefits for evaluating the effect of IONPs on stem cell differentiation. Pisanic *et al.* (2007) suggested the use of PC12 cells, a transformed cell line which, under controlled stimulation with NGF, can reproduce most of the aspects of the neuronal differentiation process (Greene and Tischler, 1976). Using this cell model, we have previously shown that it is a sensitive method to evaluate cell responsiveness after labeling with a variety of IONPs (Soenen *et al.*, 2010a, 2011a) (Fig. 10.4).

a. Cells (PC12 cells) are seeded in 12-well plates containing collagen-coated glass coverslips at a density of 1×10^4 in full medium and allowed

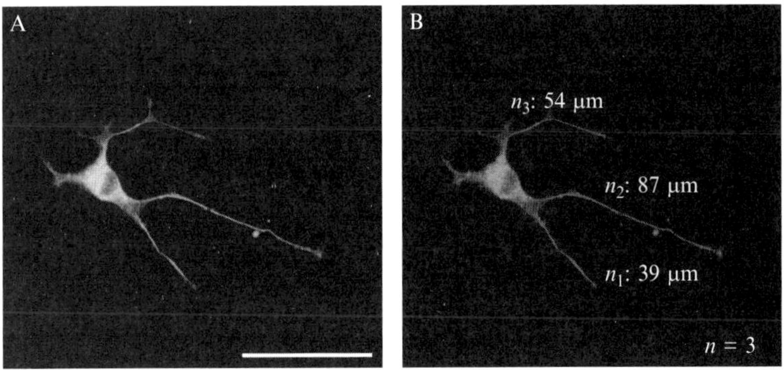

Figure 10.4 (A) Representative confocal image of a PC12 cell depicting both α-tubulin (green) and F-actin (red) staining. Scale bar: 50 μm. (B) The results obtained from calculating neurite lengths and number of neurites using the NeuriteJ plugin for ImageJ. A grayscale version of the previous image is shown along with the calculated trajectories for the neurites, their individual lengths, and the total number of neurites extending from this cell. (See Color Insert.)

to settle overnight in a humidified atmosphere at 37 °C and 5% CO_2. [Cells should be seeded at a rather low density to allow further cultivation and neurite outgrowth. A cell density of about 1000–2000 cells/cm^2 is recommended. At higher densities, cells will also grow more in clumps, which will hinder an efficient analysis of cellular responsiveness.] Then, the medium is removed and cells are given fresh medium containing the particles at the desired concentrations and incubated for the appropriate time. [Similar as for the proliferation and morphological assays, this assay also does not require multiple time points of IONP incubation. In general, 24-h incubation (or sooner, depending on when maximal uptake levels are reached) is sufficient.]

b. The medium is aspirated, and cells are washed three times with PBS and fresh growth medium is added after which the cells are further cultured for 12 h in a humidified atmosphere at 37 °C and 5% CO_2.
c. Medium is removed and fresh induction medium containing only 1% serum and 100 ng/ml NGF (see Section 3.1.b for a full description of the medium) is added and cells are further maintained in this medium for 48 h, where at 24 h half of the medium is removed and replaced with fresh NGF induction medium.
d. The medium can be removed, and cells are washed three times with PBS after which they can be fixed permeabilized, and blocked with 10% serum-containing medium as described in Section 3.6.b. Cells can then be stained for α-tubulin (and possibly also for F-actin, but this is not necessary) as described in Section 3.6.

e. Then, the coverslips can be mounted on glass microscopy slides as described in Section 3.3.2.d and stored prior to viewing the samples by fluorescence microscopy.

3.7.1. Calculating neurite length and number of neurites

The average number of neurites per cell can be easily counted manually, but determining the average neurite length requires image postprocessing software tools, such as ImageJ, which allow to use the plug-in NeuronJ (please see following site for terms and conditions of use and a short tutorial: http://www.imagescience.org/meijering/software/neuronj/), which was specifically developed to facilitate such calculations (Meijering *et al.*, 2004) (Fig. 10.4B). [NeuronJ is free to use software, originally developed by Erik Meijering and can be used for non-profit purposes, taking into account the terms and conditions mentioned with the software. Because the calculation is based on a personal selection of the "optimal paths" and the figures to be analyzed, it is best to let the analysis be done by someone different than the person who took the images and who has no knowledge of sample treatment history.]

a. The image file is opened with ImageJ.
b. The image is converted to a grayscale 8-bit image.
c. The image file is saved as a 8-bit image. Then go to ImageJ → Plugins → NeuronJ (after installation of the two required files into the plugins folder of ImageJ) to open the NeuronJ plugin.
d. The 8-bit grayscale image is opened using NeuronJ.
e. To add tracings which indicate the presence of the neurites, click on "add tracings" (5th button from the left) and then click at the end of the neurite to indicate where the neurite tracing should begin. Then move your mouse along the neurite and a path will be drawn which traces the neurite itself. See if the drawn tracing corresponds well with the actual neurite and then double-click at the opposite end of the neurite to finish tracing. [Neurites can be quite branched, so sometimes the tracing which will be drawn will not correspond well with the actual neurite location. To improve this, it is best to move the mouse back along the neurite path until a shorter tracing is drawn which does colocalize well with the actual neurite. Then click once and continue moving your mouse along the neurite path to add short tracks of the tracing which all colocalize well with the neurite itself. Double-click only when reaching the actual end point of the neurite and NeuronJ will automatically convert all the shorter tracing pieces into one track, covering the whole neurite.]
f. This process is repeated for every neurite.
g. To get the number and size of the neurites, click "measure tracings" (8th button from the left). A pop-up window will appear where everything can be left indicated. Then click "run" and three new windows will

appear with the numerical data. Of interest are the windows with "Tracings" which gives a list with the length of every tracing drawn and the window "Groups" which gives the average length of all the tracings and the total number of tracings. [The length is indicated in pixels and must be converted to μm which can be done by measuring the length of the scale bar in pixels and then divide the length of the tracing (in pixels) by the length of the scale bar (in pixels) and multiply the outcome with the μm scale with which the scale bar correlates.]

h. To obtain statistically relevant information, the number of images and cells analyzed should be as high as possible. In general, 100–150 cells/condition is surely sufficient to allow statistical analysis. Furthermore, images should be taken from multiple samples which were treated identically (e.g., 5 different samples/condition) to verify whether the obtained results are generally representative.

3.8. Investigating possible IONP degradation

When IONPs are used in cell labeling studies, any cell-internalized particles are generally found back in endosomal structures. During internalization, the particles will be subjected to the different pH values of their surroundings, being 7.0–7.4 for the extracellular medium and 5.5 for early and late endosomes, after which the IONPs will remain in lysosomes where the local pH can be as low as 4.5 (Tycko and Maxfield, 1982). Arbab *et al.* (2005a) first described the use of a lysosomal model system to evaluate the effect of this low pH on particle stability and MRI contrast. Later on, we further elaborated this model, combining quantitative iron determination, MRI monitoring, and in-cell IONP degradation (Soenen *et al.*, 2010a, 2011a).

3.8.1. pH-dependent IONP degradation

a. IONPs can be diluted to 200 μg Fe/ml in the different buffer systems as described in Section 2.5, reaching a total volume of 5 ml. These stock solutions are then placed in sterile tissue culture falcons (25 cm^2).
b. The falcons are then incubated in a humidified atmosphere at 37 °C and 5% CO_2 where samples can be collected (200 μl for iron determination and 100 μl for MRI) at different time points (e.g., 0, 3, 6, 12, 24, 36, 48, 120, and 240 h). [During incubation and sample collection, it is important to keep the mixtures sterile as bacteria may interfere with the outcome of both assays.]
c. For iron determination, the 200 μl samples can all be diluted 10 times with PBS (pH 7.0) after which the amount of free Fe^{3+} can be measured using Tiron, as described in Section 3.2. [In contrast to the cellular iron determination, no HCl or HNO_3 is added here. These concentrated acids lead to direct acid hydrolysis of the IONPs which is necessary to get

the total iron content for cell-based work, but which would greatly hinder the current application. Without using the acids, only the free Fe^{3+} is available for binding to Tiron (IONPs which are still complete will not form complexes with Tiron) and thus, only the extent of particles which have been degraded is indicated.] In short, 0.4 ml of diluted sample is taken, to which 0.1 ml of Tiron (0.25 *M*), 0.5 ml KOH (4 N), and 1 ml of phosphate buffer (0.2 *M*, pH 9.5) are added. Following 15 min stabilization, the absorbance can be measured spectrophotometrically at 480 nm (Fig. 10.5A–C). [The first sample which is collected immediately after suspending the particles in the appropriate buffer solutions can be used as an internal reference to indicate the amount of the labile iron pool, which is always associated with IONPs. For any other samples taken from the same pH buffers at later time

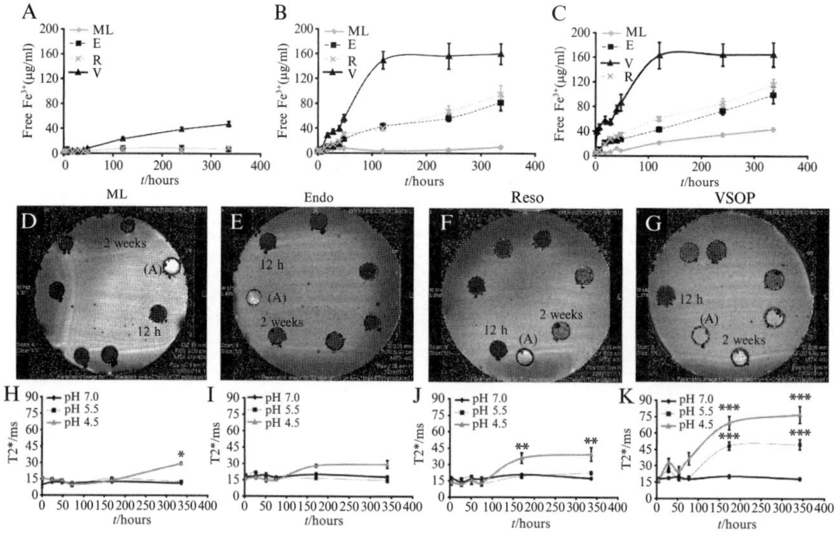

Figure 10.5 pH effect on particle degradation and MR signal intensities. (A–C) The amount of free ferric iron in function of time for four NPs (MLs, Endorem, VSOP, and Resovist) incubated at 200 µg Fe/ml in 20 m*M* sodium citrate containing cell culture medium at different pH (A: pH 7.0; B: pH 5.5; C: pH 4.5; $n=4$). (D–G) Representative T_2* maps obtained for the various particles (D: ML; E: Endorem; F: VSOP; G: Resovist) in the above-described medium at pH 4.5. The samples were collected at different time points after addition of the NPs to the acidic culture medium and are represented clockwise in terms of increasing incubation times, going from (A): pure agar to 12 h; 24 h; 48 h; 72 h; 1 week, and 2 weeks. (H–K) T_2* values obtained when calculating the respective T_2* maps of the four NPs (H: ML; I: Endorem; J: VSOP; K: Resovist) at pH 7.0, 5.5, and 4.5 in function of different incubation times. Significant increases of T_2* relaxation times of NPs treated at pH 5.5 or 4.5 compared with the values obtained at pH 7.0 are indicated (*$p<0.05$; **$p<0.01$; ***$p<0.001$). Reproduced with permission from Soenen *et al.* (2010a), © Wiley VCH.

points, the last value (at time point x) should be subtracted with the original iron value (at time point 0) to get the actual amount of pH-dependent Fe^{3+} secretion.]

d. For MRI, the 100 μl sample can be diluted 10× in PBS after which 250 μl was transferred into an Eppendorf tube and solidified by adding 750 μl agarose gel (1.5% in distilled water) (Fig. 10.5D–G). [The amount of sample which should be taken depends greatly on the intrinsic contrast generated by the particles. It would be optimal to first take a few random samples of the stock solution of IONPs and then perform MR measurements to find at which IONP concentration low T_2 and T_2★ values can be obtained, which are still below the saturation level of the MR machine to be able to pick up slight differences.]

e. MRI should then be performed on the samples by placing all the agar tubes inside one bigger agar block and measuring the samples with different sequences to enable the construction of T_1, T_2, and T_2★ maps and calculate the respective T_1, T_2, and T_2★ values (Fig. 10.5H–K).

3.8.2. Intracellular IONP degradation

The major limitation in the above-described procedures is that only the effect of the pH can be easily studied in this manner. Although this is a crucial parameter which needs to be studied carefully, the IONPs which are present in lysosomes will also encounter degradative lysosomal proteins, such as Cathepsin-L (See *et al.*, 2009). Especially for biologically relevant IONP coatings (less for synthetic and non-degradable polymer coatings), the intracellular degradation should also be studied. This may be done by isolating lysosomes and exposing the particles to lysosomal fractions. Alternatively, the intracellular degradation of IONPs affects the cellular Fe^{3+} pool, which can also be verified by monitoring regulation of proteins which are key components in iron homeostasis (Pawelczyk *et al.*, 2006). Such proteins would include the iron storage protein ferritin or the iron transport molecule transferrin receptor (TfR) which have been described to show aberrant expression levels upon cellular IONP labeling (Schafer *et al.*, 2007; Soenen *et al.*, 2010a). The expression of surface-located TfR levels can be monitored by flow cytometry (Fig. 10.6):

a. Cells (e.g., C17.2 cells) are seeded at 2×10^5 cells/falcon in 25-cm^2 tissue culture falcons (5 ml of cell medium) and allowed to settle overnight after which fresh medium containing the IONPs at various concentrations is added and allowed to incubate at 37 °C and 5% CO_2 for the time required to reach maximal cellular iron levels (generally: 24 h). For controls, cells are incubated with IONP-free medium.

b. Medium is removed and cells are washed twice with PBS. Then, fresh medium (5 ml/dish) without any IONPs is added and the cells are kept in culture for 24 h.

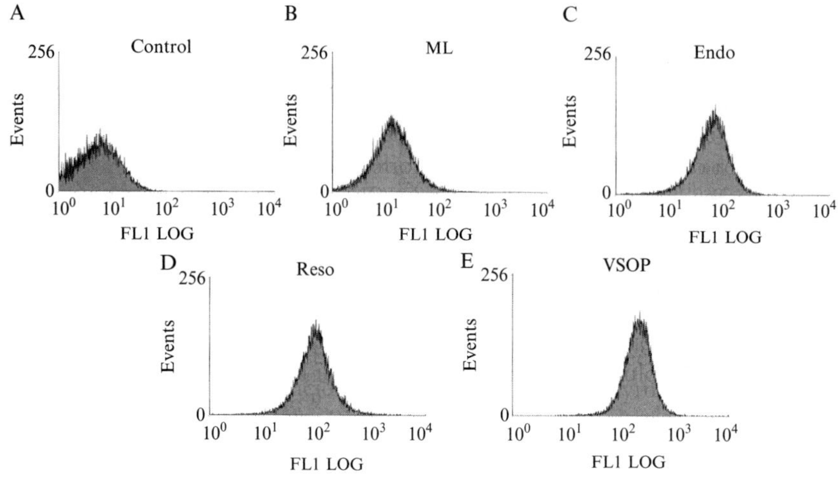

Figure 10.6 Flow cytometric analysis of TfR1 expression in (A) untreated control cells or cells incubated with (B) MLs, (C) Endorem, (D) Resovist, or (F) VSOP at 1 week of continuous culture post-NP-exposure, where there were no differences in proliferation rate or cell densities between the different groups. Expression levels were normalized using isotype control antibody. Reproduced with permission from Soenen *et al.* (2010a), © Wiley VCH. (For color version of this figure, the reader is referred to the Web version of this chapter.)

c. Then, the cells are passaged and reseeded in new 25-cm^2 tissue culture dishes at a density of 2×10^5 cells/dish, after which the cells are kept in culture for 48 h. [As iron plays an important role in cell cycle progression, it is important to control the cell density during seeding as this should be the same for all samples. Also, for the same reason, it is best not to use IONP concentrations which induce great toxic effects (>30% cell death) or which impede cell proliferation (>30% reduction in cell proliferation) as this may also affect TfR expression levels.]
d. Cells are then lifted by scraping [as the purpose of this assay is to assess the expression level of surface-located receptors, the use of enzymatic digestion such as with trypsin should be avoided to minimize receptor degradation], and reseeded at 2×10^5 cells/dish in 25-cm^2 dishes. The remaining cells are split in two aliquots [the number of cells required for analysis should be sufficient. As part of the cells have to be reseeded and the remaining cells need to be divided into two equal parts, the number of cells should be carefully controlled. In case of the fast-proliferating C17.2 cells, the conditions described above result in 300,000 cells/sample (150,000/type of antibody). If the cells proliferate more slowly, longer cultivation times can be taken (e.g., only after 72 or 96 h) or cells can be seeded at higher original densities] and centrifuged for 5 min at

1.2 rcf, washed with 0.5 ml PBS/sample and centrifuged again. These steps are then repeated twice. Per sample, one tube was incubated with fluorophore-conjugated anti-TfR1 antibody at 1 μg/ml in flow cytometry buffer (PBS supplemented with 1% BSA and 0.1% fetal bovine serum), and the remaining tube was incubated with a fluorophore-conjugated isotype control antibody under identical conditions. Incubation takes 30 min and occurs at ambient temperature.

e. Cells are washed for 5 min with flow cytometry buffer (0.5 ml/tube) and centrifuged for 5 min at 1.2 rcf. In total, cells are washed and centrifuged three times. Cells are then resuspended in 0.4 ml flow cytometry buffer and analyzed using a fluorescence-activated cell sorter setup.

f. The cells which were reseeded in Step d in this section can be kept in culture for another 48 h after which they can also be lifted by scraping, reseeded and analyzed by flow cytometry as described in Steps d and e in this section. By keeping the cells in culture, the levels of TfR1 can be analyzed at multiple time points after IONP incubation. As NP degradation may be somewhat slow and the effect on TfR is a secondary effect which requires some time, it is important to include multiple time points (e.g., 3 different time points up to 1–2 weeks after IONP labeling).

4. Concluding Remarks

Although the amount of data on IONP-induced cytotoxic effects has steadily increased over the years, the possible mechanisms involved have not been fully elucidated and the high variations between different studies have made it nearly impossible to compare different particles with regard to uptake efficiency, cytotoxicity, and MR-based efficiency. To allow researchers to address the possible cytotoxicity of IONPs, there is a big need for more uniformity in terms of incubation conditions, viability assays used, and cell types studied. Here, several protocols are described which could be used as a basic model from which to start investigating IONP toxicity. These studies are highly important to allow the careful investigation of the possible contribution of any novel type of IONPs and comparison of the already existing ones to find the optimal formulation in terms of low toxicity, high cellular uptake, and efficient MRI contrast, which will undoubtedly enhance the fields of cell transplantation and non-invasive imaging.

Acknowledgments

We thank Uwe Himmelreich (KULeuven, MoSAIC, Biomedical Imaging, Department of Radiology, KULeuven, Belgium) for his excellent assistance in MRI. S. J. S. is a postdoctoral researcher of the FWO.

REFERENCES

Arbab, A. S., Bashaw, L. A., Miller, B. R., Jordan, E. K., Lewis, B. K., Kalish, H., and Frank, J. A. (2003). Characterization of biophysical and metabolic properties of cells labeled with superparamagnetic iron oxide nanoparticles and transfection agent for cellular MR imaging. *Radiology* **229,** 838–846.

Arbab, A. S., Wilson, L. B., Ashari, P., Jordan, E. K., Lewis, B. K., and Frank, J. A. (2005a). A model of lysosomal metabolism of dextran coated superparamagnetic iron oxide (SPIO) nanoparticles: Implications for cellular magnetic resonance imaging. *NMR Biomed.* **18,** 383–389.

Arbab, A. S., Yocum, G. T., Rad, A. M., Khakoo, A. Y., Fellowes, V., Read, E. J., and Frank, J. A. (2005b). Labeling of cells with ferumoxides-protamine sulfate complexes does not inhibit function or differentiation capacity of hematopoietic or mesenchymal stem cells. *NMR Biomed.* **18,** 553–559.

Chan, W. C. W., Maxwell, D. J., Gao, X., Bailey, R. E., Han, M., and Nie, S. (2002). Luminescent quantum dots for multiplexed biological detection and imaging. *Curr. Opin. Biotechnol.* **13,** 40–46.

Chen, Y. C., Hsiao, J. K., Liu, H. M., Lai, I. Y., Yao, M., Hsu, S. C., Ko, B. S., Chen, Y. C., Yang, C. S., and Huang, D. M. (2010). The inhibitory effect of superparamagnetic iron oxide nanoparticle (Ferucarbotran) on osteogenic differentiation and its signaling mechanism in human mesenchymal stem cells. *Toxicol. Appl. Pharmacol.* **245,** 272–279.

De Cuyper, M., and Soenen, S. J. (2010). Cationic magnetoliposomes. *Methods Mol. Biol.* **605,** 97–111.

De Meyer, S. F., Vanhoorelbeke, K., Chuah, M. K., Pareyn, I., Gillijns, V., Hebbel, R. P., Collen, D., Deckmyn, H., and Vandendriessche, T. (2006). Phenotypic correction of von Willebrand disease type 3 blood-derived endothelial cells with lentiviral vectors expressing von Willebrand factor. *Blood* **107,** 4728–4736.

Greene, L. A., and Tischler, A. S. (1976). Establishment of a noradrenergic clonal line of rat adrenal pheochromocytoma cells which respond to nerve growth-factor. *Proc. Natl. Acad. Sci. USA* **73,** 2424–2428.

Gupta, A. K., and Gupta, M. (2005). Cytotoxicity suppression and cellular uptake enhancement of surface modified magnetic nanoparticles. *Biomaterials* **26,** 1565–1573.

Han, G., Ghosh, P., and Rotello, V. M. (2007). Functionalized gold nanoparticles for drug delivery. *Nanomedicine (Lond.)* **2,** 113–123.

Himmelreich, U., and Dresselaers, T. (2009). Cell labeling and tracking for experimental models using magnetic resonance imaging. *Methods* **48,** 112–124.

Hu, F. X., Neoh, K. G., Cen, L., and Kang, E. T. (2006). Cellular response to magnetic nanoparticles "PEGylated" via surface-initiated atom transfer radical polymerization. *Biomacromolecules* **7,** 809–816.

Hussain, S. M., Braydich-Stolle, L. K., Schrand, A. M., Murdock, R. C., Yu, K. O., Mattie, D. M., Schlager, J. J., and Terrones, M. (2009). Toxicity evaluation for safe use of nanomaterials: Recent achievements and technical challenges. *Adv. Mater.* **21,** 1549–1559.

Lewinski, N., Colvin, V., and Drezek, R. (2008). Cytotoxicity of nanoparticles. *Small* **4,** 26–49.

Maurer-Jones, M. A., Bantz, K. C., Love, S. A., Marquis, B. J., and Haynes, C. L. (2009). Toxicity of therapeutic nanoparticles. *Nanomedicine (Lond.)* **4,** 219–241.

Meijering, E., Jacob, M., Sarria, J. C., Steiner, P., Hirling, H., and Unser, M. (2004). Design and validation of a tool for neurite tracing and analysis in fluorescence microscopy images. *Cytometry A* **58,** 167–176.

Nel, A. E., Madler, L., Velegol, D., Xia, T., Hoek, E. M., Somasundaran, P., Klaessig, F., Castranova, V., and Thompson, M. (2009). Understanding biophysicochemical interactions at the nano-bio interface. *Nat. Mater.* **8,** 543–557.

Pawelczyk, E., Arbab, A. S., Pandit, S., Hu, E., and Frank, J. A. (2006). Expression of transferrin receptor and ferritin following ferumoxides-protamine sulfate labeling of cells: Implications for cellular magnetic resonance imaging. *NMR Biomed.* **19,** 581–592.

Pisanic, T. R., Blackwell, J. D., Shubayev, V. I., Finones, R. R., and Jin, S. (2007). Nanotoxicity of iron oxide nanoparticle internalization in growing neurons. *Biomaterials* **28,** 2572–2581.

Raynal, I., Prigent, P., Peyramaure, S., Najid, A., Rebuzzi, C., and Corot, C. (2004). Macrophage endocytosis of superparamagnetic iron oxide nanoparticles—Mechanisms and comparison of Ferumoxides and Ferumoxtran-10. *Invest. Radiol.* **39,** 56–63.

Razzaq, M. Y., Anhalt, M., Frormann, L., and Weidenfeller, B. (2007). Thermal, electrical and magnetic studies of magnetite filled polyurethane shape memory polymers. *Mater. Sci. Eng. A* **444,** 227–235.

Rivera Gil, P., Oberdorster, G., Elder, A., Puntes, V., and Parak, W. J. (2010). Correlating physico-chemical with toxicological properties of nanoparticles: The present and the future. *ACS Nano* **4,** 5527–5531.

Schafer, R., Kehlbach, R., Wiskirchen, J., Bantleon, R., Pintaske, J., Brehm, B. R., Gerber, A., Wolburg, H., Claussen, C. D., and Northoff, H. (2007). Transferrin receptor upregulation: In vitro labeling of rat mesenchymal stem cells with superparamagnetic iron oxide. *Radiology* **244**(2), 514–523.

See, V., Free, P., Cesbron, Y., Nativo, P., Shaheen, U., Rigden, D. J., Spiller, D. G., Fernig, D. G., White, M. R., Prior, I. A., Brust, M., Lounis, B., *et al.* (2009). Cathepsin L digestion of nanobioconjugates upon endocytosis. *ACS Nano* **3,** 2461–2468.

Shapiro, E. M., Skrtic, S., and Koretsky, A. P. (2005). Sizing it up: Cellular MRI using micron-sized iron oxide particles. *Magn. Reson. Med.* **53,** 329–338.

Soenen, S. J., and De Cuyper, M. (2009). Assessing cytotoxicity of (iron oxide-based) nanoparticles: An overview of different methods exemplified with cationic magnetoliposomes. *Contrast Media Mol. Imaging* **4,** 207–219.

Soenen, S. J., and De Cuyper, M. (2010). Assessing iron oxide nanoparticle toxicity in vitro: Current status and future prospects. *Nanomedicine (Lond.)* **5,** 1261–1275.

Soenen, S. J., and De Cuyper, M. (2011). How to assess cytotoxicity of (iron oxide-based) nanoparticles. A technical note using cationic magnetoliposomes. *Contrast Media Mol. Imaging* **6,** 153–164.

Soenen, S. J., Baert, J., and De Cuyper, M. (2007). Optimal conditions for labelling of 3T3 fibroblasts with magnetoliposomes without affecting cellular viability. *Chembiochem* **8,** 2067–2077.

Soenen, S. J., Brisson, A. R., and De Cuyper, M. (2009a). Addressing the problem of cationic lipid-mediated toxicity: The magnetoliposome model. *Biomaterials* **30,** 3691–3701.

Soenen, S. J., Hodenius, M., and De Cuyper, M. (2009b). Magnetoliposomes: Versatile innovative nanocolloids for use in biotechnology and biomedicine. *Nanomedicine (Lond.)* **4,** 177–191.

Soenen, S. J., Illyes, E., Vercauteren, D., Braeckmans, K., Majer, Z., De Smedt, S. C., and De Cuyper, M. (2009c). The role of nanoparticle concentration-dependent induction of cellular stress in the internalization of non-toxic cationic magnetoliposomes. *Biomaterials* **30,** 6803–6813.

Soenen, S. J., Himmelreich, U., Nuytten, N., Pisanic, T. R., Ferrari, A., and De Cuyper, M. (2010a). Intracellular nanoparticle coating stability determines nanoparticle diagnostics efficacy and cell functionality. *Small* **6,** 2136–2145.

Soenen, S. J., Nuytten, N., De Meyer, S. F., De Smedt, S. C., and De Cuyper, M. (2010b). High intracellular iron oxide nanoparticle concentrations affect cellular cytoskeleton and focal adhesion kinase-mediated signaling. *Small* **6,** 832–842.

Soenen, S. J., Himmelreich, U., Nuytten, N., and De Cuyper, M. (2011a). Cytotoxic effects of iron oxide nanoparticles and implications for safety in cell labelling. *Biomaterials* **32,** 195–205.

Soenen, S. J., Rivera Gil, P., Montenegro, J. M., Parak, W. J., De Smedt, S. C., and Braeckmans, K. (2011b). Cellular toxicity of inorganic nanoparticles: Common aspects and guidelines for improved nanotoxicity evaluation. *Nano Today* **6,** 446–465.

Stroh, A., Zimmer, C., Gutzeit, C., Jakstadt, M., Marschinke, F., Jung, T., Pilgrimm, H., and Grune, T. (2004). Iron oxide particles for molecular magnetic resonance imaging cause transient oxidative stress in rat macrophages. *Free Radic. Biol. Med.* **36,** 976–984.

Sun, C., Lee, J. S., and Zhang, M. (2008). Magnetic nanoparticles in MR imaging and drug delivery. *Adv. Drug Deliv. Rev.* **60,** 1252–1265.

Tycko, B., and Maxfield, F. R. (1982). Rapid acidification of endocytic vesicles containing alpha-2-macroglobulin. *Cell* **28,** 643–651.

Walczak, P., Kedziorek, D. A., Gilad, A. A., Barnett, B. P., and Bulte, J. W. M. (2007). Applicability and limitations of MR tracking of neural stem cells with asymmetric cell division and rapid turnover: The case of the shiverer dysmyelinated mouse brain. *Magn. Reson. Med.* **58,** 261–269.

CHAPTER ELEVEN

Cytotoxicity of Gold Nanoparticles

Yu Pan, Matthias Bartneck, *and* Willi Jahnen-Dechent

Contents

1. Introduction 225
2. Dosage and Quantification of Gold Nanoparticles 227
3. Aggregation State of Nanoparticles in Fluids 228
4. Cell-Based Nanotoxicity Studies 230
    4.1. Cytotoxicity 230
    4.2. Cell cycle arrest and proliferation inhibition 232
    4.3. Cell death 233
    4.4. Oxidative stress 236
References 239

Abstract

Nanomaterials are now routinely used in technical as well as medical applications. The very physicochemical properties that favor nanomaterial application are the prime cause that these materials cannot be considered "generally safe." We are still far from predicting the toxicological profile of new nanoparticles, despite continuous attempts to establish a structure–function relation between the physical and chemical properties of nanoparticles and their interactions with biological systems. Herein, we summarize some basic concept to assess nanoparticle toxicity, death pathways, cell cycle, and oxidative stress in response to nanoparticle exposure of cells.

1. Introduction

Nanotechnology is an expanding branch of material sciences. Nanomaterials have a high surface to volume ratio, possess quantum size effects, and are different from their bulk form in many respects (Schmid *et al.*, 1999). Favorable optical, mechanical, and electronic properties of materials in the nanoscale allow novel applications in high-technology and biomedical

Helmholtz Institute for Biomedical Engineering, RWTH Aachen University, Aachen, Germany

Methods in Enzymology, Volume 509
ISSN 0076-6879, DOI: 10.1016/B978-0-12-391858-1.00012-5

science. These altered properties together with their small size, which is on a similar scale like biological macromolecules, may cause nanomaterials to directly affect biological systems, whereas the same compound in bulk form may be inert and nontoxic. Therefore, the risk assessment of nanomaterials is indispensable for safe and sustainable nanotechnology development.

Nanoparticles have been shown to influence biological systems in many ways. The size of nanoparticles determines their cellular uptake, endocytosis, cytotoxicity, biodistribution, and clearance pathway (Chithrani *et al.*, 2006; De Jong *et al.*, 2008; Hirn *et al.*, 2011; Pan *et al.*, 2007; Semmler-Behnke *et al.*, 2008; Sonavane *et al.*, 2008). The hydrodynamic diameter of nanoparticles is particularly important in determining the nano–bio interaction (Choi *et al.*, 2007a, 2011). Quantum dots with identical metallic core size, but with anionic or cationic charges (dihydrolipoic acid, cysteamine) had a higher tendency to bind serum proteins, thus increasing the hydrodynamic diameter compared to those with zwitterionic and neutral surface layers. An increased hydrodynamic diameter hinders renal clearance. Prolonged retention of nanoparticles in organisms can thus cause adverse effects. Apart from the size effect, the surface properties of nanoparticles, including charge, ligand density, and hydrophobicity have been shown to control the bio–nano interactions as well (Chompoosor *et al.*, 2010; Harush-Frenkel *et al.*, 2008; Lipka *et al.*, 2010). Nanoparticles with cationic surfaces have a high cellular uptake, high toxicity, and renal clearance rate compared to those with an anionic surface. The most intensively studied cases are the increased uptake rate and toxicity of nanoparticles modified with positively charged polyethyleneimine (PEI) and cell-penetrating peptides. These modifications were met with enhanced cytotoxicity because of membrane damage. The bio–nano interface is crucial in nanotoxicity (Nel *et al.*, 2009). Bio–nano interface reactions affect the size and surface properties of the nanoparticles after binding with solute components in biological fluids (proteins, glycans, and ions) and thus alter the uptake pathways (nonspecific and receptor-mediated). Conversely, bio–nano interactions may change the protein composition in biological fluids, the activity and distribution of the biological molecules in the organism, and they may constitute or reveal hidden immune epitopes.

Improved understanding of nanoparticle toxicity will help to avoid adverse effects and may also assist in designing nanoparticles. In many cases, a suitable surface modification may render toxic nanoparticles less toxic to allow application in medical therapy. The situation we are currently confronting, however, is an enormous amount of newly emerging nanoparticles. To meet this demand, high-throughput and cost-efficiency, and animal welfare-friendly methods are required to obtain integral toxicity data from thousands of different nanoparticles. Cell-based tests are undoubtedly the most widely applied screening method. The toxicity of nanoparticles is commonly expressed as the concentration causing 50% of growth inhibition in cell culture (IC_{50}). Results derived from cell tests are used to derive a reasonable dose for the initial animal experiments.

A growing body of toxicity studies using a wide variety of nanoparticles (TiO_2 (Johnston *et al.*, 2009; Long *et al.*, 2007), Au (Pan *et al.*, 2009), ZnO (Heng *et al.*, 2011; Xia *et al.*, 2008), Ag (AshaRani *et al.*, 2009; Xu *et al.*, 2012), SiO_2 (Ye *et al.*, 2010), Fe_3O_4 (Naqvi *et al.*, 2010), carbon nanotubes (Srivastava *et al.*, 2011), Al_2O_3 (Dey *et al.*, 2008)) has shown that the production of reactive oxygen species (ROS) is a common mechanism causing nanoparticle toxicity. It has been pointed out that the toxicity of nanoparticles is determined by their potency to produce ROS, which is balanced by the antioxidant capacity of an organism to prevent oxidative damage (Nel *et al.*, 2006). A modest increase of oxidative stress can be usually rescued by a corresponding increase in the cellular reducing capacity. Continuous accumulation of oxidative stress, however, triggers irreversible cell death. The addition of glutathione and *N*-acetyl-cysteine, which enhances the antioxidant capacity of the organism reduce greatly the toxicity of, for example, ultrasmall gold nanoparticles (AuNPs; Pan *et al.*, 2009). To this end, methods to measure and quantify cellular oxidative stress in response to nanoparticles will be described.

Cell-based toxicity tests are a mere starting point for toxicity studies. The interaction between different cell types and the dynamic translocation, distribution, sedimentation, and clearance of nanoparticles in living animals (Lasagna-Reeves *et al.*, 2010; Minchin, 2008; Schleh *et al.*, 2012; Semmler-Behnke *et al.*, 2008; Sun *et al.*, 2005), and the organ-specific toxicity together with the metabolic response in the presence of nanoparticles can only be revealed in animal tests. Therefore, it is of great importance to test nanoparticle toxicity in rodent models as early in the study as possible. Many nanoparticles under study are designed specifically for *in vivo* applications. The size and surface are optimized to allow tissue and cell entry, and even to bind specific targets (Choi *et al.*, 2007b; Felsenfeld *et al.*, 1996; Giordano *et al.*, 2009; Hainfeld *et al.*, 2006; Pissuwan *et al.*, 2006; Sperling *et al.*, 2008). Therefore, *in vivo* tests should be done in a vertebrate model capable of basic human organ functions.

2. Dosage and Quantification of Gold Nanoparticles

The dose makes the poison. This principle of toxicology was expressed five centuries ago by Paracelsus and equally pertains to nanomaterials. The dosage of nanoparticles in toxicology research, therefore, should cover the full concentration range from nontoxic to cytotoxic. The toxicity should then be reported as the concentration at which half-maximum toxicity is observed—the inhibitory concentration 50 (IC_{50}). We strongly encourage readers to experimentally determine the molar concentration of gold contained in their nanoparticles and to use equimolar doses to compare the toxicity of various AuNPs. The common practice of using particle concentration is fraught with error, because polydispersity of the particles and their tendency to aggregate renders particle number virtually meaningless. By contrast, [Au] can be verified with great accuracy, both before and after compound addition to cells. Methods to quantify AuNPs in stock solution, in biological fluids, and in organ extracts include high-sensitivity methods including neutron activation analysis (NAA) and inductively coupled plasma mass spectrometry (ICP-MS). Table 11.1 lists several published methods to quantify AuNPs and to study their biodistribution.

3. Aggregation State of Nanoparticles in Fluids

Nanoparticles will only behave as nanomaterials as long as they maintain their small size. Aggregating nanoparticles will progressively behave like bulk material. Close attention should therefore be given to the interaction of nanoparticles and biological fluids, which might greatly influence toxicity. The composition of biological fluids was shown to alter the particle size, surface characters, endocytosis pathway, intracellular trajectory, and the toxicity profile of the nanoparticles. On the other hand, nanoparticles can alter the nature of bioactive proteins depending on size, shape, surface charge, and curvature (Nel *et al.*, 2009). Commonly used media include cell culture medium (with or without serum), phosphate buffered saline (PBS), 0.9% sodium chloride, plasma, and whole blood. Many nanoparticles will aggregate in saline and serum free medium because of the salt content of these solutions. Addition of serum can, to some extent, improve the stability of the nanoparticles in biological fluids by forming a protein halo or corona (Nel *et al.*, 2009). Serum protein binding may alter the biological activity of nanoparticles in more ways than aggregation. Many proteins will facilitate AuNP endocytosis by a process called opsonization. In addition, AuNPs may inhibit coagulation factors and decrease the free fibrinogen content of serum. Binding may be pH-dependent by design as in the "intelligent nanomaterials" meant for drug release at sites of inflammation, which have a lower pH than healthy tissue (Tsai *et al.*, 2011). Nanoparticle binding works both ways in that the particles collect molecules from the surrounding fluid but also bind to macromolecular assemblies or cell surfaces. Ultrasmall AuNPs approach the size of typical protein ligands and may inadvertently activate cell surface receptors. AuNPs are particularly notorious for binding thiol ligands because of the high electronegativity of Au and the electropositivity of sulfur. Thus, AuNP will preferentially bind sulfur containing compounds and may deplete the glutathione-based redox buffering capacity of biological fluids.

Methods to measure the actual size of nanoparticles in fluids after contact with the protein are dynamic light scattering (DLS), small angle X-ray

Table 11.1 Methods to quantify nanoparticles in biological samples

Methods	Sample material	Sensitivity	Data	References
TEM	Cell, tissue	High	Verify the uptake of AuNPs	Goel *et al.* (2009)
ICP-MS	Cell, tissue	High	Quantitative	Lasagna-Reeves *et al.* (2010)
ICP-AES	Cell, tissue	High	Quantitative	Chithrani *et al.* (2006)
NAA	Cell, tissue	10–100 μg/kg	Quantitative	Lipka *et al.* (2010), Semmler-Behnke *et al.* (2008)
UV/vis	Nanoparticles	Low	Semiquantitative	Cho *et al.* (2011)
AAS	Cell, tissue	0.8–1.88 mg/kg	Quantitative	Kattumuri *et al.* (2007)
Silver enhancement	Cell, tissue	400 μg Au/mouse	Semiquantitative, localization	Kim *et al.* (2011)
Spectroscopic photoacoustic	Living mouse	400 μg Au/mouse	Semiquantitative, localization	Kim *et al.* (2011)
Photoacoustic tomography	Living rat	0.8×10^9 nanocages/g body weight	Semiquantitative, localization	Yang *et al.* (2007)
CT	Living swine	86–99 mg/kg	Semiquantitative, localization	Boote *et al.* (2010)
Liquid scanning transmission electron microscopy	Living cells	1.8 n*M*, 2 h	Semiquantitative, localization	Peckys and de Jonge (2011)

TEM, transmission electron microscopy; ICP-MS, inductively coupled plasma mass spectrometry; ICP-AES, inductively coupled plasma atomic emission spectroscopy; NAA, neutron activation analysis; AAS, atomic absorption spectroscopy; UV–vis, ultraviolet–visible spectroscopy; CT, X-ray computed tomography.

scattering (SAXS), size-exclusion chromatography (SEC), isothermal titration calorimetry (ITC), and surface plasmon resonance (SPR). In addition, fluorescence quenching measurement may be used to study nanoparticle–protein interaction. The emission properties of the aromatic residues in proteins (tryptophan, tyrosine, and phenylalanine) and fluorescence quenching by AuNPs of fluorophore probes were employed to study nanoparticle–protein interaction. Applying this method, Lacerda and colleagues showed that the binding kinetics between AuNPs and serum proteins is strictly size-dependent. Bigger particles had a higher tendency to bind proteins (Lacerda *et al.*, 2010). Protein conformation changes in contact with nanoparticles were recently reviewed (Fei and Perrett, 2009).

4. Cell-Based Nanotoxicity Studies

Cell-based assays measure cell morphology, proliferation, viability, toxicity, motility, and production of metabolites. Compared to toxicity testing in animals, cell-based assays are relatively easy and low cost, and the dose of testing compound can be precisely defined. Immortalized cell lines like HeLa cells are commonly used to compare the cytotoxicity of nanoparticles varying in size and surface chemistry (Pan *et al.*, 2007). One disadvantage of immortal cell lines is that their genome and the proliferation pattern deviate from normal healthy cells. This drawback is remedied by using primary cells, which are, however, generally harder to obtain and to grow and should therefore be used for specific purposes only. In addition, the role of nanoparticles in the differentiation of stem cells may be studied by *in vitro* cell tests (Ferreira *et al.*, 2008; Yi *et al.*, 2010).

4.1. Cytotoxicity

Cell morphology is an excellent parameter reflecting the status of cells in response to nanoparticles. Thus, time-lapse movies of cells recorded at low light phase-contrast greatly help in determining appropriate time points for further detailed investigations. Viability and membrane integrity are measured using the cell impermeable stain, trypan blue, or Hoechst 33258, or propidium iodide (PI), or measuring the mitochondrial reducing capacity with 3-(4,5-dimethylthiazol-2-yl)-2,5-diphenyltetrazolium bromide (MTT). Furthermore, assessment of the intracellular ATP level and the release of the intracellular enzyme lactate dehydrogenase (LDH) into the culture medium are also measured as proxies of cell integrity. Commercial assays employ colorimetric or fluorogenic substrates to measure metabolic activity and cell integrity. For medium to high throughput, these assays may be run in a multiplex format using robotic pipetting platforms

(Ferreira *et al.*, 2011). Fluorescence assays may, however, be unsuitable to assay the toxicity of nanoparticles, because many particles, especially small AuNPs, quench fluorescence, and thus cause false negative readings. We have found the MTT assay presented in Protocol I to be simple, cheap, fast, and reliable for toxicity testing of metallic nanoparticles with a wide range of ligands.

4.1.1. Protocol I

4.1.1.1. Cell viability tests using MTT The cells are commonly seeded in 96-well plates. For immortal cell lines like HeLa cells, the seeding density relies on the cell proliferation rate. As a rule of thumb, 90% confluence (30,000–40,000 cells) in untreated controls should be reached at the end of the incubation. A proliferation curve should be recorded for every cell line to identify the growth stages and the proliferation rate of the cells. We have noted earlier that cells differ in their sensitivity to toxic compounds depending on the growth phase of the cell culture. Cells in the logarithmic growth phase are more sensitive than those in the stationary phase (Pan *et al.*, 2007). Our routine method for measuring AuNP toxicity in HeLa cells is as follows:

(1) Two thousand cells are seeded in each well and are incubated for 72 h in 96-well plates.
(2) The supernatant is replaced with 100 μl fresh medium containing different concentrations of nanoparticles. The cells are incubated for another 48 h.
(3) Ten microliters of PBS containing 5 mg/ml MTT is dispensed to each well. The plates are incubated for 2 h. After incubation, the yellowish water-soluble tetrazolium is chemically reduced to a water-insoluble purple formazan product by viable cells engaged in oxidative metabolism. Thus, metabolic activity serves as a proxy of cell number and viability.
(4) The water-insoluble formazan is dissolved in a solvent mixture (100 μl) consisting of isopropanol (80 μl) with hydrochloric acid (0.04 m*M*) and 3% sodium dodecyl sulfate (20 μl).
(5) Absorption of the samples is measured with a plate reader at 595 nm. Triplicate wells are set in each plate and three independent experiments are required to determine the IC_{50}.
(6) IC_{50} values are calculated using a four parameter logistic equation. Data are plotted as a sigmoidal dose–response curve with variable slope using statistics software, for example, GraphPad Prism.

For each material, the IC_{50} values are determined from triplicate wells. IC_{50} values are routinely repeated in three independent experiments with almost identical results. The osmolality, pH, and the reducing capacity of the nanoparticles at the highest testing concentration should be recorded prior to the cytotoxicity measurement. The exposure length should be longer than one proliferation cycle of the cells. We routinely expose the cells to the nanoparticles for 48 h before the MTT test. A positive MTT test can be confirmed by observing the formazan accumulation inside the mitochondria. The cells will appear speckled in bright field microscopy. Interference of AuNPs on the colorimetric reading in the MTT test is prevented, because the supernatant containing AuNPs is replaced by an organic solvent. This washing step minimizes the false-positive photometric reading derived from AuNPs in MTT-based viability tests (Fig. 11.1).

4.2. Cell cycle arrest and proliferation inhibition

Most viability assays are end point tests. A low number of viable cells can be due either to increased cell death or to a low proliferation rate. Nanoparticles can affect cell proliferation in both directions, inhibiting (Kalishwaralal *et al.*, 2011; Karthikeyan *et al.*, 2010) and enhancing (Unfried *et al.*, 2008). Cell-cycle control is critical for normal growth and development. AuNP-triggered loss of cell cycle control in somatic cells may potentially cause inflammation and even tumor formation. This is especially important because many nanoparticles under study are designed as antitumor agents. The assessment of cell cycle arrest in the presence of nanoparticles is therefore indispensable to understand their antitumor mechanisms. Staining of cells with carboxyfluorescein succinimidyl ester (CFSE), 5-bromo-2′-deoxyuridine (BrdU) incorporation, and DNA content measurements are commonly used to study the proliferation of cells treated with nanoparticles.

The cell cycle comprises four distinct phases (G1 phase, S phase, G2 phase, and mitosis), and two checkpoints (G0/G1 and G2/M checkpoints), which assure that no DNA damage is transmitted to daughter cells. Nuclear DNA is duplicated during the synthetic or S phase. The cell cycle stage of a given cell can thus be determined by measuring the DNA amount.

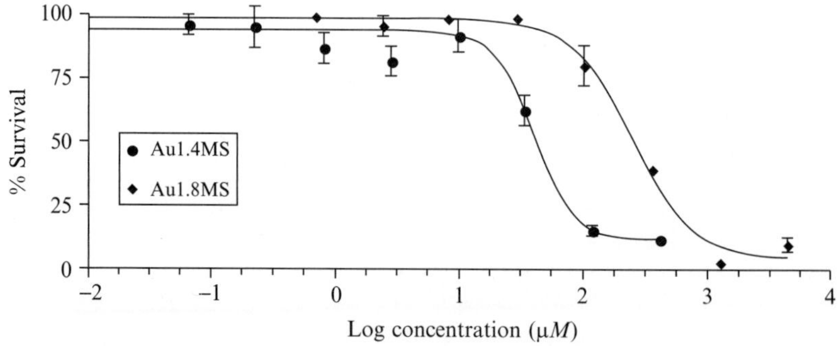

Figure 11.1 Representative sigmoidal dose-response curves showing the cytotoxicity of ultrasmall AuNPs. The cytotoxicity of AuNP of 1.4 and 1.8 nm gold core diameter, respectively, were tested during the logarithmic growth phase of HeLa cells. The IC_{50s} of Au1.4MS and Au1.8MS are 46 and 230 μ*M*, respectively.

Cell proliferation arrest of the cells will be reflected in the DNA content and a changed G1 and G2 ratio. Glucose capped AuNPs have been shown to sensitize prostate cancer cells toward radiotherapy by inducing a G2/M growth arrest (Roa *et al.*, 2009). PI is a fluorescent molecule that binds to DNA in a stoichiometric manner and is commonly used to analyze the cellular DNA content or ploidy (Suzuki *et al.*, 1997; Watson *et al.*, 1987). Hoechst 33342, Hoechst 33258, and 4′,6-diamidino-2-phenylindole (DAPI) are alternative DNA stains. After excitation at 488 nm, PI emits red fluorescence that can be detected with a 562–588 nm band pass filter. Fluorescence intensity is directly proportional to the amount of DNA. Cells in the G1 and M phases have half of the DNA content compared to cells in the S and G2 phases. Thus, cells in G1 phase have a weaker fluorescence intensity compared to those in G2 phase (Bedner *et al.*, 1999; Moore *et al.*, 1998). An additional advantage of PI-based DNA content measurement is that apoptosis can be simultaneously detected as an additional sub-G1 peak and the broadening of the G1 peak, both reflecting apoptosis-related DNA fragmentation.

4.2.1. Protocol II

4.2.1.1. Measuring DNA content to determine the cell cycle phase

(1) Cells are seeded in 6-well plates and incubated for 72 h prior to the addition of nanoparticles.
(2) The supernatant is replaced with 100 μl fresh medium containing nanoparticles. Cells are further incubated for the required period. All cell-material combinations are set up in triplicates.
(3) After the nanoparticle incubation, cells are trypsinized and rinsed with PBS.
(4) Washed cell pellets are fixed in cold 70% ethanol at 4 °C for 60 min.
(5) After fixation, cells are washed with PBS and resuspended in 250 μl PBS. RNAse (250 μl, 1 mg/ml) and PI (500 μl, 0.1 mg/ml) are added at room temperature for 15 min or overnight at 4 °C in the dark.
(6) 20,000 cells are analyzed using a FACSCalibur or FACSCanto or a comparable flow cytometer and the CELLQuest software (Becton-Dickinson) (Fig. 11.2).

4.3. Cell death

Apoptosis, autophagy, necroptosis, aponecrosis (Formigli *et al.*, 2000), and necrosis are major cell death pathways that have been described in great detail (Peter, 2011). They differ with regards to trigger, timing, degree of regulation, and key regulators. Nevertheless, there is considerable overlap, and for practical purposes we confine ourselves to discriminate between apoptosis (slow, regulated, energy dependent—regulated cell death), and necrosis (fast, not energy dependent—unregulated cell death/cell lysis). Cell

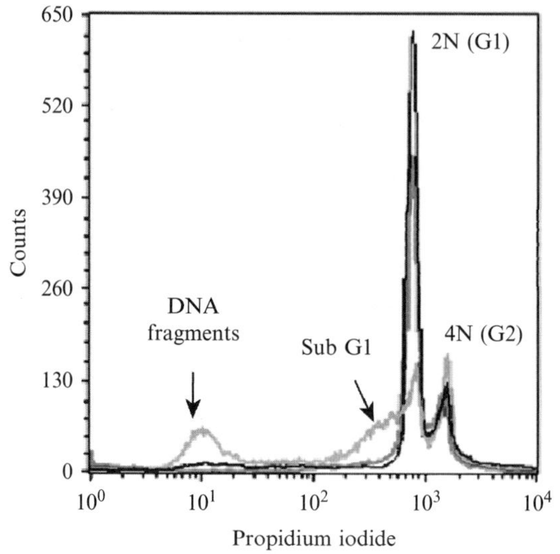

Figure 11.2 Representative flow cytometry diagram of DNA content measurement with propidium iodide staining. Untreated cells (black curve) show a minor peak of propidium iodide (PI) fluorescence at 1.05×10^3 depicting 4N cellular DNA content (G2 phase, about 20% of cells) and a major peak of fluorescence at 8×10^2 indicating 2N cellular DNA content (G1 phase, 70% of the cells). Cells staining intermediary reside in S phase. When the HeLa cells are treated with staurosporine (light gray line) to trigger apoptosis, the relative proportion of cells in G2 phase increases to 50% indicating a G2/M block of the cells. Together with the sub G1 peak and the low fluorescent peak indicating DNA fragments this pattern is typical of apoptosis.

death is a dynamic process (Loos and Engelbrecht, 2009), and time-lapse video microscopy using low light phase contrast illumination once again is very helpful to estimate morphological changes in response to nanoparticles. The microscope used for time-lapse video capture is equipped with a heat plate and CO_2 supply. Apoptotic cells will round up and will show shrinkage of the cytoplasm, blebbing of the plasma membrane, and nuclear condensation (Ziegler and Groscurth, 2004). A rapid cytoplasmic swelling is usually seen in necrotic cells. The membrane lipid phosphatidylserine externalizes from the inner leaflet of the cell membrane to the outer leaflet in the early stage of apoptosis. By using fluorescent annexin V, which specifically binds phosphatidylserine, the number of cells undergoing apoptosis can be quantified. This experiment is usually performed in combination with PI staining to estimate the percentage of necrotic and secondary necrotic cells. PI is membrane impermeable and thus will only enter cells with a damaged cell membrane to stain their nucleus. Caspases and B-cell lymphoma 2 (BCL2) family proteins are major executors of apoptosis. Measuring the activity of these enzymes is useful to detect apoptosis, and to identify the death mechanism. Furthermore, attenuation of caspase enzymes by inhibitors is known to prevent apoptosis but not necrosis. The general caspase inhibitor, Z-VAD-fmk, which binds to the catalytic site of caspase proteases efficiently prevents the induction of apoptosis, and thus may be employed to verify that death pathway. DNA fragmentation is another hallmark of apoptosis. The activation of caspase-activated DNase (CAD) causes cleavage of nuclear DNA into fragments of 180 bp a late event in apoptosis. Therefore, "DNA laddering" is diagnostic of apoptotic cells. The DNA fragmentation can be detected either by gel electrophoresis or by flow cytometry-based DNA content measurement, as described above (Herrmann *et al.*, 1994). The terminal deoxynucleotidyl transferase dUTP nick end labeling (TUNEL) assay detects DNA cleavage sites that are abound in apoptotic cells. The TUNEL assay is thus also commonly used to detect apoptosis.

4.3.1. Protocol III

4.3.1.1. Determination of apoptosis versus necrosis Annexin V and PI are double-staining probes for apoptosis by detecting the externalization of phosphatidylserine and membrane integrity. The apoptotic cells externalize their phosphatidylserine early in apoptosis when the cell membrane is still intact. Therefore, the early apoptotic cells have a positive annexin V, but a negative PI signal. In contrast to apoptotic cells, necrotic cells lose membrane integrity and both annexin V and PI will penetrate the leaky cell membrane to stain intracellular phosphatidylserine (annexin V) and nuclear DNA (PI). The test cannot discriminate between apoptotic cells at the late stage (secondary necrosis) and necrotic cells. Thus, time course measurements are required to determine the cell death pathway.

(1) Cells are seeded into 6-well plates (HeLa 40,000 cells/well) and incubated for 72 h at 37 °C with 5% CO_2 prior to the addition of nanoparticles. The seeding density is adjusted depending on the growth rate of the cell lines used and on the incubation duration. At least 20,000 cells are required at the end of the experiment to allow cell analysis by flow cytometry in the untreated samples.
(2) After 72 h of incubation, nanoparticles at the desired concentrations are applied.
(3) After the exposure to the nanoparticles, the supernatant containing detached apoptotic or necrotic cells together with the trypsinized cells are collected and washed twice with binding buffer followed with the addition of fluorescein isothiocyanate (FITC)-labeled annexin V.
(4) Annexin V binding to phosphatidylserine is calcium dependent. Therefore, the binding buffer must contain 2.5 m*M* calcium chloride throughout all steps including washing.

(5) The cells are incubated at room temperature for 15 min. Thereafter, an aliquot of the PI stock solution (10 μl, 50 μg/ml) is added to each well, and cells are further incubated for 5 min before one final wash in binding buffer. For each experiment, untreated cells serve as a negative control and cells incubated for 24 h with staurosporine (0.2 μ*M*) serve as a positive control for apoptosis. Twenty thousand cells are counted for each sample by flow cytometry. Results are analyzed by CELLQuest software (Becton-Dickinson) (Fig. 11.3).

4.4. Oxidative stress

Oxidative stress is considered one of the most salient features of nanoparticle toxicity (Nel *et al.*, 2006). A low level of oxidative stress can be counteracted by the cellular antioxidant defense mechanisms. A moderate level of ROS can initiate inflammatory pathways by activating the NF-κB pathway. Excessive production of ROS will oxidize cellular components including the cell membrane lipids, protein, and DNA and will lead to cell death (Murphy *et al.*, 2011). Current mechanisms of ROS generation include protein conformation changes caused by nanoparticle binding that entail the so-called unfolded protein response (Wiseman *et al.*, 2010), mechanical cell damage induced by nanoparticles, and direct interaction between the nanoparticle metal and oxygen species (Fenton and Haber-Weiss reaction). Little is known about the direct production of ROS by nanoparticles. Methods commonly used to detect oxidative stress can be grouped as follows:

1. The first class of methods measures the accumulation of intracellular ROS (hydroxyl radicals, superoxide anion, hydrogen peroxide, peroxyl radical) by oxidation of the fluorescent probe, 5-(and-6)-chloromethyl-2′,7′-dichlorodihydrofluorescein diacetate, acetyl ester (CM-H_2DCFDA). CM-H_2DCFDA is cell permeable and becomes fluorescent when the dihydrofluorescein is oxidized to fluorescein by intracellular ROS.
2. The increase of intracellular ROS is accompanied by the structural and functional changes of mitochondria, commonly summarized as permeability transition (PT) (Murphy and Steenbergen, 2011). Analysis of mitochondrial integrity is therefore well suited to estimate the cellular energy production states, and the extent of the damage. Membrane permeable cationic fluorescent probes (e.g., JC-1) are commonly used for this purpose. JC-1 accumulates along the mitochondrial potential ($\Delta\psi$) in mitochondria of healthy cells. At high concentrations, JC-1 dimerizes and emits red fluorescence. Upon PT, the mitochondria become leaky and release JC-1 into the cytoplasm where it fluoresces green as a monomer.

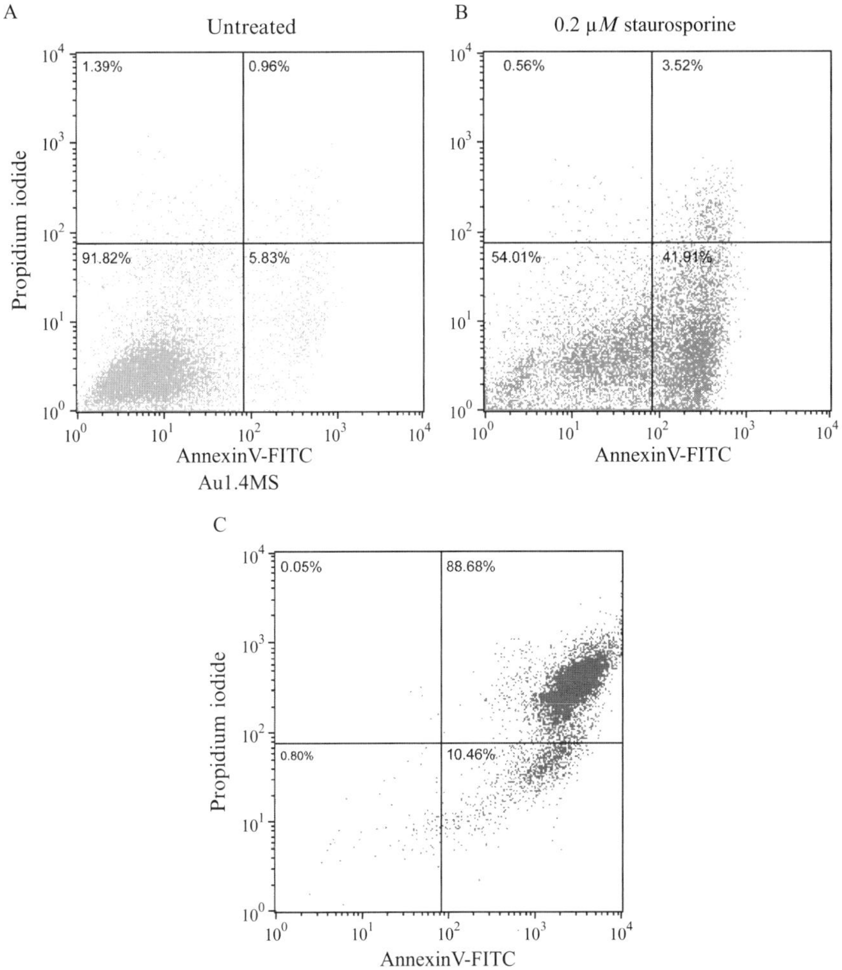

Figure 11.3 Flow cytometric determination of living, apoptotic, and necrotic HeLa cells treated with test compounds. Cells are scored for annexin V/PI double staining to estimate the relative amounts of live cells (annexin V/PI double negative, bottom left quadrant), apoptotic cells (annexin V-positive/PI-negative, bottom right quadrant), and necrotic cells (annexin V/PI-double positive, top right quadrant), respectively. The percentages of cells are given for each quadrant. (A) HeLa untreated, (B) HeLa 0.2 μ*M* staurosporine, 24 h, (C) HeLa 110 μ*M* Au1.4MS, 48 h.

3. Measuring oxidized biomolecules is an alternative method to detect oxidative stress. Commonly measured oxidized biomolecules include peroxidized lipids (malondialdehyde, 4-hydroxynonenal, and 8-iso-prostaglandin; Sevanian and Hochstein, 1985), peroxynitrite (Pryor and Squadrito, 1995), and oxidized DNA derivatives (8-hydroxydeoxyguanosine, 8-oxo-7, 8-dihydroguanine; Dizdaroglu *et al.*, 2002).

4. The fourth commonly used method for oxidative stress measurement, which adapts well to both *in vitro* and *in vivo* models measures the antioxidant capacity of the cell or organism. The best-studied antioxidant measurements include superoxide dismutase, catalase, glutathione peroxidase, glutathione, and ascorbic acid.

In addition to the four methods listed above, few studies have directly measured ROS, and thus oxidative stress, by electron paramagnetic resonance (EPR; Shulaev and Oliver, 2006).

4.4.1. Protocol IV

4.4.1.1. Measuring oxidative stress We measure ROS using CM-H_2DCFDA, because it is both inexpensive and reliable.

(1) HeLa cells are seeded in 6-well plates at initial densities of 40,000 cells in 2 ml and further incubate for 72 h.

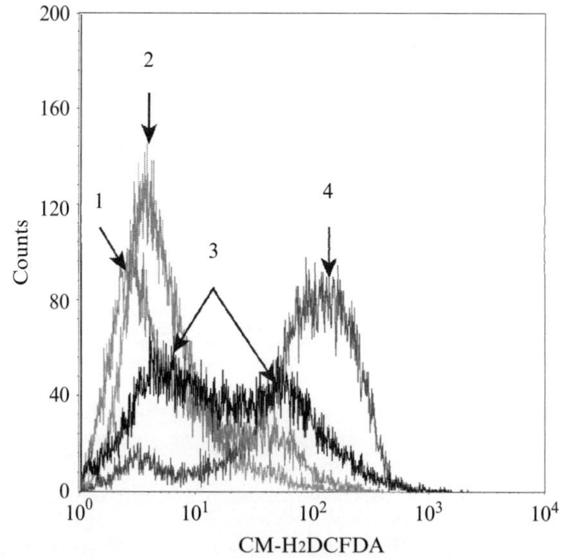

Figure 11.4 CM-H_2DCFDA revealing the accumulation of reactive oxygen species. Cells growing under the normal condition had no reactive oxygen species and the major population settle between 10^0 and 10^1 (curve 2). Treatment with 0.3% H_2O_2 for 30 min induced strong accumulation of intracellular ROS and the cell population shifted to much strong signal around 10^1–10^3 (curve 4). The Au1.4MS induced cellular oxidative stress. The number of unstressed cells (10^0–10^1) decreased while a second peak occurred between 10^1 and 10^3 (curve 3), which indicates the accumulation of the reactive oxygen species. *N*-Acetyl cysteine reduced the ROS (curve 1) induced by Au1.4MS and the majority of cells shifted back to the level of untreated cells (10^0–10^1).

(2) Fresh medium containing nanoparticles are added to the cells. All cell-material combinations are set up in triplicates.
(3) At the end of the incubation, cells are trypsinized and rinsed with PBS.
(4) After rinsing, cells are suspended in a buffer containing CM-H_2DCFDA. For cell assays, the stock solution is diluted in PBS to a final working concentration of 2.5 μM. Treatment of cells with 0.3% H_2O_2 for 30 min serves as a positive control for oxidative stress.
(5) Twenty thousand cells are analyzed using FACSCalibur or FACSCanto flow cytometers and the CELLQuest software (Becton-Dickinson) (Fig. 11.4).

REFERENCES

AshaRani, P. V., Low Kah Mun, G., Hande, M. P., and Valiyaveettil, S. (2009). Cytotoxicity and genotoxicity of silver nanoparticles in human cells. *ACS Nano* **3,** 279–290.

Bedner, E., Li, X., Gorczyca, W., Melamed, M. R., and Darzynkiewicz, Z. (1999). Analysis of apoptosis by laser scanning cytometry. *Cytometry* **35,** 181–195.

Boote, E., Fent, G., Kattumuri, V., Casteel, S., Katti, K., Chanda, N., Kannan, R., and Churchill, R. (2010). Gold nanoparticle contrast in a phantom and juvenile swine: Models for molecular imaging of human organs using X-ray computed tomography. *Acad. Radiol.* **17,** 410–417.

Chithrani, B. D., Ghazani, A. A., and Chan, W. C. (2006). Determining the size and shape dependence of gold nanoparticle uptake into mammalian cells. *Nano Lett.* **6,** 662–668.

Cho, E. C., Zhang, Q., and Xia, Y. (2011). The effect of sedimentation and diffusion on cellular uptake of gold nanoparticles. *Nat. Nanotechnol.* **6,** 385–391.

Choi, H. S., Liu, W., Misra, P., Tanaka, E., Zimmer, J. P., Itty Ipe, B., Bawendi, M. G., and Frangioni, J. V. (2007a). Renal clearance of quantum dots. *Nat. Biotechnol.* **25,** 1165–1170.

Choi, M. R., Stanton-Maxey, K. J., Stanley, J. K., Levin, C. S., Bardhan, R., Akin, D., Badve, S., Sturgis, J., Robinson, J. P., Bashir, R., Halas, N. J., and Clare, S. E. (2007b). A cellular Trojan Horse for delivery of therapeutic nanoparticles into tumors. *Nano Lett.* **7,** 3759–3765.

Choi, C. H., Zuckerman, J. E., Webster, P., and Davis, M. E. (2011). Targeting kidney mesangium by nanoparticles of defined size. *Proc. Natl. Acad. Sci. USA* **108,** 6656–6661.

Chompoosor, A., Saha, K., Ghosh, P. S., Macarthy, D. J., Miranda, O. R., Zhu, Z. J., Arcaro, K. F., and Rotello, V. M. (2010). The role of surface functionality on acute cytotoxicity, ROS generation and DNA damage by cationic gold nanoparticles. *Small* **6,** 2246–2249.

De Jong, W. H., Hagens, W. I., Krystek, P., Burger, M. C., Sips, A. J., and Geertsma, R. E. (2008). Particle size-dependent organ distribution of gold nanoparticles after intravenous administration. *Biomaterials* **29,** 1912–1919.

Dey, S., Bakthavatchalu, V., Tseng, M. T., Wu, P., Florence, R. L., Grulke, E. A., Yokel, R. A., Dhar, S. K., Yang, H. S., Chen, Y., and St Clair, D. K. (2008). Interactions between SIRT1 and AP-1 reveal a mechanistic insight into the growth promoting properties of alumina (Al2O3) nanoparticles in mouse skin epithelial cells. *Carcinogenesis* **29,** 1920–1929.

Dizdaroglu, M., Jaruga, P., Birincioglu, M., and Rodriguez, H. (2002). Free radical-induced damage to DNA: Mechanisms and measurement. *Free Radic. Biol. Med.* **32,** 1102–1115.

Fei, L., and Perrett, S. (2009). Effect of nanoparticles on protein folding and fibrillogenesis. *Int. J. Mol. Sci.* **10,** 646–655.

Felsenfeld, D. P., Choquet, D., and Sheetz, M. P. (1996). Ligand binding regulates the directed movement of beta1 integrins on fibroblasts. *Nature* **383,** 438–440.

Ferreira, L., Karp, J. M., Nobre, L., and Langer, R. (2008). New opportunities: The use of nanotechnologies to manipulate and track stem cells. *Cell Stem Cell* **3,** 136–146.

Ferreira, M. V., Jahnen-Dechent, W., and Neuss, S. (2011). Standardization of automated cell-based protocols for toxicity testing of biomaterials. *J. Biomol. Screen.* **16,** 647–654.

Formigli, L., Papucci, L., Tani, A., Schiavone, N., Tempestini, A., Orlandini, G. E., Capaccioli, S., and Orlandini, S. Z. (2000). Aponecrosis: Morphological and biochemical exploration of a syncretic process of cell death sharing apoptosis and necrosis. *J. Cell. Physiol.* **182,** 41–49.

Giordano, R. J., Edwards, J. K., Tuder, R. M., Arap, W., and Pasqualini, R. (2009). Combinatorial ligand-directed lung targeting. *Proc. Am. Thorac. Soc.* **6,** 411–415.

Goel, R., Shah, N., Visaria, R., Paciotti, G. F., and Bischof, J. C. (2009). Biodistribution of TNF-alpha-coated gold nanoparticles in an *in vivo* model system. *Nanomedicine (Lond.)* **4,** 401–410.

Hainfeld, J. F., Slatkin, D. N., Focella, T. M., and Smilowitz, H. M. (2006). Gold nanoparticles: A new X-ray contrast agent. *Br. J. Radiol.* **79,** 248–253.

Harush-Frenkel, O., Rozentur, E., Benita, S., and Altschuler, Y. (2008). Surface charge of nanoparticles determines their endocytic and transcytotic pathway in polarized MDCK cells. *Biomacromolecules* **9,** 435–443.

Heng, B. C., Zhao, X., Tan, E. C., Khamis, N., Assodani, A., Xiong, S., Ruedl, C., Ng, K. W., and Loo, J. S. (2011). Evaluation of the cytotoxic and inflammatory potential of differentially shaped zinc oxide nanoparticles. *Arch. Toxicol.* **85**(12), 1517–1528.

Herrmann, M., Lorenz, H. M., Voll, R., Grunke, M., Woith, W., and Kalden, J. R. (1994). A rapid and simple method for the isolation of apoptotic DNA fragments. *Nucleic Acids Res.* **22,** 5506–5507.

Hirn, S., Semmler-Behnke, M., Schleh, C., Wenk, A., Lipka, J., Schaffler, M., Takenaka, S., Moller, W., Schmid, G., Simon, U., and Kreyling, W. G. (2011). Particle size-dependent and surface charge-dependent biodistribution of gold nanoparticles after intravenous administration. *Eur. J. Pharm. Biopharm.* **77,** 407–416.

Johnston, H. J., Hutchison, G. R., Christensen, F. M., Peters, S., Hankin, S., and Stone, V. (2009). Identification of the mechanisms that drive the toxicity of TiO_2 particulates: The contribution of physicochemical characteristics. *Part. Fibre Toxicol.* **6,** 33.

Kalishwaralal, K., Sheikpranbabu, S., BarathManiKanth, S., Haribalaganesh, R., Ramkumarpandian, S., and Gurunathan, S. (2011). Gold nanoparticles inhibit vascular endothelial growth factor-induced angiogenesis and vascular permeability via Src dependent pathway in retinal endothelial cells. *Angiogenesis* **14,** 29–45.

Karthikeyan, B., Kalishwaralal, K., Sheikpranbabu, S., Deepak, V., Haribalaganesh, R., and Gurunathan, S. (2010). Gold nanoparticles downregulate VEGF-and IL-1β-induced cell proliferation through Src kinase in retinal pigment epithelial cells. *Exp. Eye Res.* **91,** 769–778.

Kattumuri, V., Katti, K., Bhaskaran, S., Boote, E. J., Casteel, S. W., Fent, G. M., Robertson, D. J., Chandrasekhar, M., Kannan, R., and Katti, K. V. (2007). Gum arabic as a phytochemical construct for the stabilization of gold nanoparticles: *In vivo* pharmacokinetics and X-ray-contrast-imaging studies. *Small* **3,** 333–341.

Kim, S., Chen, Y., Luke, G., and Emelianov, S. (2011). *In vivo* three-dimensional spectroscopic photoacoustic imaging for monitoring nanoparticle delivery. *Biomed. Opt. Express* **2,** 2540–2550.

Lacerda, S. H., Park, J. J., Meuse, C., Pristinski, D., Becker, M. L., Karim, A., and Douglas, J. F. (2010). Interaction of gold nanoparticles with common human blood proteins. *ACS Nano* **4,** 365–379.

Lasagna-Reeves, C., Gonzalez-Romero, D., Barria, M. A., Olmedo, I., Clos, A., Sadagopa Ramanujam, V. M., Urayama, A., Vergara, L., Kogan, M. J., and Soto, C. (2010). Bioaccumulation and toxicity of gold nanoparticles after repeated administration in mice. *Biochem. Biophys. Res. Commun.* **393,** 649–655.

Lipka, J., Semmler-Behnke, M., Sperling, R. A., Wenk, A., Takenaka, S., Schleh, C., Kissel, T., Parak, W. J., and Kreyling, W. G. (2010). Biodistribution of PEG-modified gold nanoparticles following intratracheal instillation and intravenous injection. *Biomaterials* **31,** 6574–6581.

Long, T. C., Tajuba, J., Sama, P., Saleh, N., Swartz, C., Parker, J., Hester, S., Lowry, G. V., and Veronesi, B. (2007). Nanosize titanium dioxide stimulates reactive oxygen species in brain microglia and damages neurons in vitro. *Environ. Health Perspect.* **115,** 1631–1637.

Loos, B., and Engelbrecht, A. M. (2009). Cell death: A dynamic response concept. *Autophagy* **5,** 590–603.

Minchin, R. (2008). Nanomedicine: Sizing up targets with nanoparticles. *Nat. Nanotechnol.* **3,** 12–13.

Moore, A., Donahue, C. J., Bauer, K. D., and Mather, J. P. (1998). Simultaneous measurement of cell cycle and apoptotic cell death. *Methods Cell Biol.* **57,** 265–278.

Murphy, E., and Steenbergen, C. (2011). What makes the mitochondria a killer? Can we condition them to be less destructive? *Biochim. Biophys. Acta* **1813,** 1302–1308.

Murphy, M. P., Holmgren, A., Larsson, N. G., Halliwell, B., Chang, C. J., Kalyanaraman, B., Rhee, S. G., Thornalley, P. J., Partridge, L., Gems, D., Nystrom, T., Belousov, V., *et al.* (2011). Unraveling the biological roles of reactive oxygen species. *Cell Metab.* **13,** 361–366.

Naqvi, S., Samim, M., Abdin, M., Ahmed, F. J., Maitra, A., Prashant, C., and Dinda, A. K. (2010). Concentration-dependent toxicity of iron oxide nanoparticles mediated by increased oxidative stress. *Int. J. Nanomedicine* **5,** 983–989.

Nel, A., Xia, T., Madler, L., and Li, N. (2006). Toxic potential of materials at the nanolevel. *Science* **311,** 622–627.

Nel, A., Madler, L., Velegol, D., Xia, T., Hoek, E. M., Somasundaran, P., Klaessig, F., Castranova, V., and Thompson, M. (2009). Understanding biophysicochemical interactions at the nano-bio interface. *Nat. Mater.* **8,** 543–557.

Pan, Y., Neuss, S., Leifert, A., Fischler, M., Wen, F., Simon, U., Schmid, G., Brandau, W., and Jahnen-Dechent, W. (2007). Size-dependent cytotoxicity of gold nanoparticles. *Small* **3,** 1941–1949.

Pan, Y., Leifert, A., Ruau, D., Neuss, S., Bornemann, J., Schmid, G., Brandau, W., Simon, U., and Jahnen-Dechent, W. (2009). Gold nanoparticles of diameter 1.4 nm trigger necrosis by oxidative stress and mitochondrial damage. *Small* **5,** 2067–2076.

Peckys, D. B., and de Jonge, N. (2011). Visualizing gold nanoparticle uptake in live cells with liquid scanning transmission electron microscopy. *Nano Lett.* **11,** 1733–1738.

Peter, M. E. (2011). Programmed cell death: Apoptosis meets necrosis. *Nature* **471,** 310–312.

Pissuwan, D., Valenzuela, S. M., and Cortie, M. B. (2006). Therapeutic possibilities of plasmonically heated gold nanoparticles. *Trends Biotechnol.* **24,** 62–67.

Pryor, W. A., and Squadrito, G. L. (1995). The chemistry of peroxynitrite: A product from the reaction of nitric oxide with superoxide. *Am. J. Physiol.* **268,** L699–L722.

Roa, W., Zhang, X., Guo, L., Shaw, A., Hu, X., Xiong, Y., Gulavita, S., Patel, S., Sun, X., Chen, J., Moore, R., and Xing, J. Z. (2009). Gold nanoparticle sensitize radiotherapy of prostate cancer cells by regulation of the cell cycle. *Nanotechnology* **20,** 375101.

Schleh, C., Semmler-Behnke, M., Lipka, J., Wenk, A., Hirn, S., Schaffler, M., Schmid, G. N., Simon, U., and Kreyling, W. G. (2012). Size and surface charge of gold nanoparticles determine absorption across intestinal barriers and accumulation in secondary target organs after oral administration. *Nanotoxicology* **6,** 36–46.

Schmid, G., Baumle, M., Geerkens, M., Helm, I., Osemann, C., and Sawitowski, T. (1999). Current and future applications of nanoclusters. *Chem. Soc. Rev.* **28,** 179–185.

Semmler-Behnke, M., Kreyling, W. G., Lipka, J., Fertsch, S., Wenk, A., Takenaka, S., Schmid, G., and Brandau, W. (2008). Biodistribution of 1.4- and 18-nm gold particles in rats. *Small* **4,** 2108–2111.

Sevanian, A., and Hochstein, P. (1985). Mechanisms and consequences of lipid peroxidation in biological systems. *Annu. Rev. Nutr.* **5,** 365–390.

Shulaev, V., and Oliver, D. J. (2006). Metabolic and proteomic markers for oxidative stress. New tools for reactive oxygen species research. *Plant Physiol.* **141,** 367–372.

Sonavane, G., Tomoda, K., and Makino, K. (2008). Biodistribution of colloidal gold nanoparticles after intravenous administration: Effect of particle size. *Colloids Surf. B Biointerfaces* **66,** 274–280.

Sperling, R. A., Rivera Gil, P., Zhang, F., Zanella, M., and Parak, W. J. (2008). Biological applications of gold nanoparticles. *Chem. Soc. Rev.* **37,** 1896–1908.

Srivastava, R. K., Pant, A. B., Kashyap, M. P., Kumar, V., Lohani, M., Jonas, L., and Rahman, Q. (2011). Multi-walled carbon nanotubes induce oxidative stress and apoptosis in human lung cancer cell line-A549. *Nanotoxicology* **5,** 195–207.

Sun, X., Rossin, R., Turner, J. L., Becker, M. L., Joralemon, M. J., Welch, M. J., and Wooley, K. L. (2005). An assessment of the effects of shell cross-linked nanoparticle size, core composition, and surface PEGylation on *in vivo* biodistribution. *Biomacromolecules* **6,** 2541–2554.

Suzuki, T., Fujikura, K., Higashiyama, T., and Takata, K. (1997). DNA staining for fluorescence and laser confocal microscopy. *J. Histochem. Cytochem.* **45,** 49–53.

Tsai, D. H., Delrio, F. W., Keene, A. M., Tyner, K. M., Maccuspie, R. I., Cho, T. J., Zachariah, M. R., and Hackley, V. A. (2011). Adsorption and conformation of serum albumin protein on gold nanoparticles investigated using dimensional measurements and in situ spectroscopic methods. *Langmuir* **27,** 2464–2477.

Unfried, K., Sydlik, U., Bierhals, K., Weissenberg, A., and Abel, J. (2008). Carbon nanoparticle-induced lung epithelial cell proliferation is mediated by receptor-dependent Akt activation. *Am. J. Physiol. Lung Cell. Mol. Physiol.* **294,** L358–L367.

Watson, J. V., Chambers, S. H., and Smith, P. J. (1987). A pragmatic approach to the analysis of DNA histograms with a definable G1 peak. *Cytometry* **8,** 1–8.

Wiseman, R. L., Haynes, C. M., and Ron, D. (2010). Snapshot: The unfolded protein response. *Cell* **140**590–590.e592.

Xia, T., Kovochich, M., Liong, M., Madler, L., Gilbert, B., Shi, H., Yeh, J. I., Zink, J. I., and Nel, A. E. (2008). Comparison of the mechanism of toxicity of zinc oxide and cerium oxide nanoparticles based on dissolution and oxidative stress properties. *ACS Nano* **2,** 2121–2134.

Xu, H., Qu, F., Lai, W., Andrew Wang, Y., Aguilar, Z. P., and Wei, H. (2012). Role of reactive oxygen species in the antibacterial mechanism of silver nanoparticles on Escherichia coli O157:H7. *Biometals* **25**(1), 45–53.

Yang, X., Skrabalak, S. E., Li, Z. Y., Xia, Y., and Wang, L. V. (2007). Photoacoustic tomography of a rat cerebral cortex *in vivo* with au nanocages as an optical contrast agent. *Nano Lett.* **7,** 3798–3802.

Ye, Y., Liu, J., Xu, J., Sun, L., Chen, M., and Lan, M. (2010). Nano-SiO_2 induces apoptosis via activation of p53 and Bax mediated by oxidative stress in human hepatic cell line. *Toxicol. In Vitro* **24,** 751–758.

Yi, C., Liu, D., Fong, C. C., Zhang, J., and Yang, M. (2010). Gold nanoparticles promote osteogenic differentiation of mesenchymal stem cells through p38 MAPK pathway. *ACS Nano* **4,** 6439–6448.

Ziegler, U., and Groscurth, P. (2004). Morphological features of cell death. *News Physiol. Sci.* **19,** 124–128.

CHAPTER TWELVE

Design of Target-Seeking Antifibrotic Compounds

Tero A. H. Järvinen*,†,‡

Contents

1. Introduction 244
2. *In Vivo* Phage Display 246
   2.1. Generation of phage library 246
   2.2. *In vivo* screening 248
   2.3. Phage rescue and amplification 249
   2.4. Individual phage clones 250
   2.5. PCR and sequencing of the insert coding region 251
3. Homing Peptides 251
   3.1. Peptides and homing experiments 251
   3.2. Immunohistochemistry 251
   3.3. Cell culture experiments with peptides and phage 252
4. Multifunctional Fusion Proteins 254
   4.1. Cloning 254
   4.2. Production of recombinant proteins—Mammalian expression 254
   4.3. Production of recombinant proteins—Baculovirus expression 254
   4.4. Characterization of recombinant decorins 256
5. Function of Multifunctional Fusion Protein 257
   5.1. Cell proliferation and cell binding assays 257
   5.2. Wound healing model 258
   5.3. Histology 258
   5.4. Quantitative analysis of immunostaining and histochemical staining 259
6. Concluding Remarks 259
Acknowledgments 259
References 260

* Vascular Mapping Laboratory, Center for Nanomedicine, Sanford-Burnham Medical Research Institute at UCSB, University of California, Santa Barbara, California, USA
† Cancer Center, Sanford-Burnham Medical Research Institute, La Jolla, California, USA
‡ Department of Orthopedic Surgery, University of Tampere and Tampere City Hospital, Tampere, Finland

Methods in Enzymology, Volume 509
ISSN 0076-6879, DOI: 10.1016/B978-0-12-391858-1.00013-7

Abstract

Selective delivery of drugs and biotherapeutics to the site of disease (synaphic targeting) has a number of advantages. First, the enhanced accumulation of the therapeutic compound at the target tissue increases drug efficacy without increasing side effects. Alternatively, the dose of the drug can be lowered to reduce the side effects. On the practical level, when a drug is difficult or expensive to make, being able to lower the dose may be the key to commercial viability. Certain targeting systems can change the distribution of the drug in a beneficial way. Examples include wider distribution and deeper penetration of the drug in the target tissue, active intracellular targeting when desirable, and even targeting to a particular subcellular organelle. In this chapter, we illustrate these principles by describing the development of a targeting system for an antifibrotic biotherapeutic, decorin. The system is based on vascular homing peptide (sequence: CARSKNKDC; referred to as CAR) that specifically recognizes angiogenic blood vessels in injured (regenerating) and inflammatory tissues and can deliver a payload to such tissues with high selectivity. So far, the CAR-targeted decorin has been shown to promote tissue repair with reduced scarring in a skin wound model, but this biotherapeutic can potentially be used in other injuries and in various fibrotic diseases.

1. Introduction

The delivery of a payload to a specific location in the body commonly uses antibodies or peptides as the targeting element. Peptides have several advantages: they are small, simple to synthesize, cheap, and generally non-immunogenic. Tissue-penetrating and cell-internalizing properties of some peptides can also be major advantages in certain applications. A disadvantage of peptides is their relatively low ligand binding affinity. The low affinity can be compensated for by multivalent presentation of a peptide on the surface of a nanoparticle, hence the popularity of peptides as targeting elements in nanomedical applications (Ruoslahti *et al.*, 2010). However, one-to-one peptide-payload conjugates can be quite effective, particularly when the peptide is used to augment an inherent affinity of the payload to the target. A prime example is tumor necrosis factor α targeted to tumors with a tumor-homing peptide from our laboratory (Curnis *et al.*, 2000); such a conjugate is now in phase 3 clinical trials (Corti and Curnis, 2011). The CAR-decorin fusion protein we describe here also falls in this category of compounds because decorin has a selective, albeit not very effective, affinity for injured and fibrotic tissues.

In vivo screening of phage display peptide libraries is an effective tool for identifying homing peptides for specific targets (Pasqualini and Ruoslahti, 1996). Because the phage is a nanoparticle, these screens almost exclusively probe the vasculature. The extensive specialization of the endothelium among normal tissue and in pathological lesions revealed by this method has led to the concept of "vascular zip codes" (Ruoslahti, 2002, 2004). The CAR peptide we used in constructing the targeted decorin fusion protein was isolated using *in vivo* phage display with skin and tendon wounds as the target tissues (Järvinen and Ruoslahti, 2007, 2010). The blood vessels are an advantageous primary target in synaphic drug delivery: they are available for circulating compounds to bind to and serve as a gateway to the parenchymal tissue. We find that a vascular homing peptide for a tissue or lesion often also binds to the corresponding parenchymal cells, and this is the case with CAR, which also recognizes the granulation tissue in wounds (Järvinen and Ruoslahti, 2007). This dual expression of the peptide receptor is an advantage because it drives the translocation of the blood vessel-bound peptide into the injured tissue, where the decorin payload can suppress scar formation.

Permanent scars form upon healing of tissue injuries in response to a wide variety of insults ranging from ischemia, trauma, burn injury, surgery, or inflammation in almost any other organ. Tissue fibrosis, that is not the result of any obvious injury, is the final outcome of a group of common and serious diseases, as well as in aging. The basis of scarring is thought to be that response to tissue injury in mammals is focused on quick sealing of an injury. Rapid proliferation of fibroblasts and extracellular matrix produced by these cells replaces the damaged tissue with fibrotic tissue, the scar (Aarabi *et al.*, 2007). A culprit in this process is the transforming growth factor-β1 (TGF-β1), hence inhibiting TGF-β1 activity is a major focus of preventing scar formation and fibrosis (Aarabi *et al.*, 2007; Border and Ruoslahti, 1992).

Decorin is a natural inhibitor of TGF-β that shuts down the TGF-β responses to injury and inflammation (Border and Ruoslahti, 1992; Hildebrand *et al.*, 1994). In addition to blocking TGF-b, decorin also inhibits other important inducers of fibrosis/scarring; epidermal growth factor receptors (ERBBs), myostatin and connective tissue growth factor (CTGF/CCN2) (see for review Järvinen and Ruoslahti, 2012). The ability of decorin to prevent scar formation and fibrosis and to promote tissue regeneration has been demonstrated in a large number of tissue injuries and diseases (Border and Ruoslahti, 1992; Border *et al.*, 1992; Faust *et al.*, 2009; Järvinen and Ruoslahti, 2010; Li *et al.*, 2004, 2007; Zhu *et al.*, 2007). No serious side effects have been encountered in all these studies. Despite this favorable profile assembled over two decades, decorin has not progressed to the clinic. One apparent reason for this is that decorin is a proteoglycan that is manufactured in mammalian cell expression systems and is therefore difficult and expensive to produce. We have found CAR-decorin to be more than 100 times more active than the native decorin against the scar-inducing isoforms of TGF-β. This high potency, the selective homing to wounds, and *in vivo* activity in wound healing make CAR-decorin a stronger candidate than decorin for clinical development (Järvinen and Ruoslahti, 2010, 2011). Here, we describe the design and development of this targeted decorin as an example of how peptide-based synaphic targeting can be applied to a medical problem.

2. *In Vivo* Phage Display

To provide new tools for systemic treatment of tissue injuries, we have identified peptides that, when injected systemically in to the circulation, seek out (home to) regenerating, injured tissue (Järvinen and Ruoslahti, 2007). For that purpose, we used *in vivo* phage display to identify vascular homing peptides that target the angiogenic vasculature forming at the healing skin and tendon wounds/injuries (Järvinen and Ruoslahti, 2007). We cloned and expressed random CX7C (where X denotes for any random amino acid) cyclic peptide library (diversity of 1.0×10^9 peptides) at the end of the coat protein of T7 bacteriophage. *In vivo* phage display offers a unique opportunity to screen for an almost unlimited (potentially billions) number of potential peptidic drug candidates simultaneously in an *in vivo* setting (Järvinen and Ruoslahti, 2007; Ruoslahti, 2002; Teesalu *et al.*, 2012).

Having identified the potent wound homing peptide CARSKNKDC (CAR), we showed that the CAR peptide homes to angiogenic blood vessels and delivers payloads up to 200-fold more than controls to regenerating skin, skeletal muscle, and tendons (Järvinen and Ruoslahti, 2007). More recently, it has been shown that CAR selectively also homes to the inflammatory/remodeled vasculature of lungs in pulmonary arterial hypertension (Urakami *et al.*, 2011) as well as the myocardial infarction site at the heart (Kean *et al.*, 2011). These results expand the use of CAR peptide from solely targeting angiogenesis to diseases with an inflammatory component.

Then we designed, cloned, and expressed in a mammalian expression system, a multifunctional recombinant protein consisting of the CAR peptide fused with decorin (Järvinen and Ruoslahti, 2010) (Fig. 12.1).

2.1. Generation of phage library

The peptide libraries are prepared by using NNK-oligonucleotides (5′- AAT TCT TGC NNK NNK NNK NNK NNK NNK NNK TGC GGA-3′and 5′-A GCT TCC GCA NNM NNM NNM NNM NNM NNM NNM GCA AG-3′, where the 5′ end is phosphorylated) encoding a random library of cyclic peptides of the general structure CX7C (C, cysteine; X, any amino acid) (Fig. 12.2). Libraries with CX7C-structure have yielded the best vascular homing peptides in our hands (Järvinen and Ruoslahti, 2007; Teesalu *et al.*, 2009, 2012; Sugahara *et al.*, 2009). The oligonucleotide mixture was heated to 95 °C, allowed to cool down slowly at room temperature, and then cloned into the T7Select 415-1b vector according to the manufacturer's instructions (T7Select, Merck Millipore, Nottingham, UK). This vector displays peptides in all 415 copies of the phage capsid protein as a C-terminal fusion (Fig. 12.2). We use either T7Select 415-1b or T7Select10-3b vectors for our peptide library screens and T7Select1-1b and T7Select10-3b vectors for displaying larger proteins and cDNA libraries (Brown and Ruoslahti, 2004).

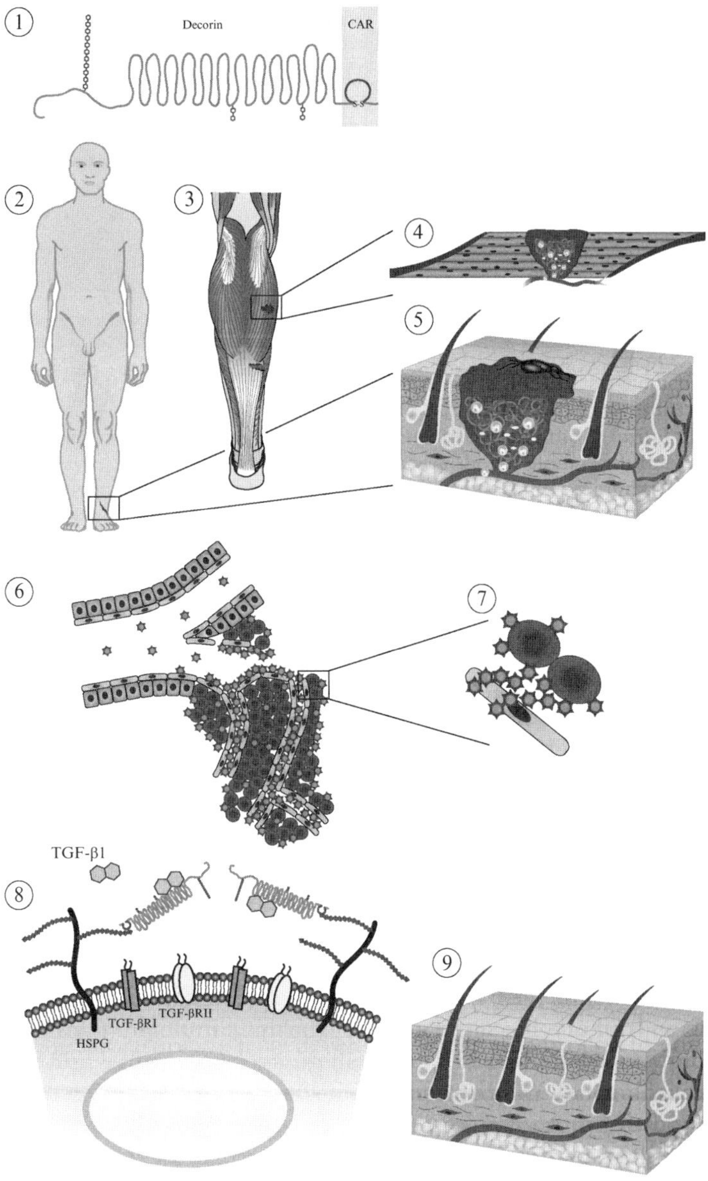

Figure 12.1 Schematic representation of the mechanism of action of the multifunctional therapeutic molecule, CAR-decorin (Järvinen and Ruoslahti, 2010). CAR-decorin (1) is a systemically administered, target-seeking, multifunctional biotherapeutic that inhibits scar formation. The molecule can be targeted to injury taking place at any organ of the body (2, 3) (or multiple organs simultaneously). The CAR homing peptide targets angiogenic vasculature, which forms at the site of the injury (4, 5). The peptide (and any payload attached to it; Blue stars) then extravasates into surrounding tissue (6), where it binds to its receptor(s) on the cell surface of the scar-producing fibroblasts (7). CAR

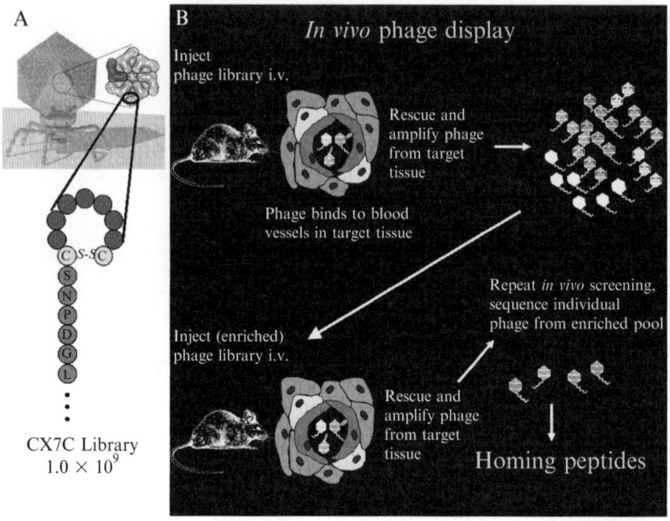

Figure 12.2 Schematic presentation of constructing the CX7C-peptide library in phage and the principle of *in vivo* phage display. (A) A cyclic CX7C-peptide library has been cloned to the C-terminus of the phage coat protein and expressed in 415 copies in T7 Select 415-1b phage. (B) Phage libraries will be injected to circulation. As the homing peptides on the phage surface bind to endothelium in the tissues, there is an enrichment of phage that binds to the endothelium of the target tissue. Cell suspensions will be prepared, and the bound phage will be rescued and amplified by adding *E. coli*. The amplified phage pool recovered from the target tissue will be reinjected into mice at a similar disease stage, and the cycle will be repeated several times to ensure that phage clones that specifically bind (i.e., home) toward target will be recovered. A set of phage clones will randomly be collected from homing phage population that shows enriched homing toward target tissue. The peptide-encoding DNA inserts will be amplified by PCR, and the PCR products sequenced. *Ex vivo* screening can be performed to prescreen the library for clones that potentially home to the target. Modified from Ruoslahti (2004). (For the color version of this figure, the reader is referred to the Web version of this chapter.)

2.2. *In vivo* screening

Our laboratory first reported the use of phage display *in vivo* to discover peptide and protein ligands for macromolecules that are expressed in an organ- and tissue-specific manner (Pasqualini and Ruoslahti, 1996; Ruoslahti, 2002, 2004). The unique feature of *in vivo* phage display is that the screen is both

binding to heparan sulfate proteoglycans (HSPGs) provides docking sites in the proximity of the main scar-inducing growth factors TGF-β1 and TGF-β2 (8) facilitating the neutralization of these growth factors by the therapeutic part of the molecule, decorin (9). This mechanism results in therapeutic response seen as reduced scar formation in the skin (Järvinen and Ruoslahti, 2010, 2011). Picture by Helena Schmidt; reproduced with permission from Finnish Medical Journal Duodecim (originally published in Duodecim 2011;127:50–51). (For interpretation of the references to color in this figure legend, the reader is referred to the Web version of this chapter.)

positive and subtractive; clones that bind to the vasculature of organs and tissues other than the target organ are removed while phage specific for the target organ gets enriched (Fig. 12.2).

We sometimes use *ex vivo* (homogenized suspension of the target organ) as a preselection method; one good example of an *ex vivo* screen is the targeting of intravascular blood clots that form in tumors (Pilch *et al.*, 2006). The preselective *ex vivo* screen toward blood clots was carried out before the actual *in vivo* screening on tumor bearing mice took place (Pilch *et al.*, 2006). The difference between the *ex vivo* and *in vivo* selections is that the phage clones are only exposed to the vasculature *in vivo* whereas they are exposed to all cell types in the *ex vivo* variation. No matter how successful the *ex vivo* screen might have been, we still want to emphasize that the primary outcome is always the phage clones derived from the *in vivo* screens, that is, clones capable of homing to the target organ through the vasculature.

The screening process (Järvinen and Ruoslahti, 2007) involves usually three (sometimes four) *in vivo* selection rounds (that can be preceded by the *ex vivo* screen) (Fig. 12.2). Eight-week-old Sprague-Dawley rats are injected with the library [$1.0–5.0 \times 10^{11}$ phage with the diversity of the library around 1.0×10^9, that is, 100–500 copies of each individual phage clone, in 1.5 ml of phosphate-buffered saline (PBS)] through the tail vein (or intracardially) and perfused 10–12′ later through the heart (by placing the needle of the syringe into the heart's left ventricle just to the apex and by punching a hole to the right atrium by scissors while applying positive pressure from the syringe) with 1% bovine serum albumin (BSA) in Dulbecco's modified Eagle's medium (DMEM, total perfusion amount 50–100 ml) to remove unbound intravascular phage. The short circulation time focuses the screening on the blood vessels because the phage, being a nanoparticle, cannot readily exit from the blood vessels.

Once target tissue samples have been collected, each sample is gently disrupted by a hand-held homogenizer (Omni TH-2, Omni Int., Kennesaw, GA) in large volume of wash media and then the cell and tissue suspension is washed and centrifuged 3–5 times (1500–2000 rpm × 5′) in a large volume of 1% BSA-DMEM (15–50 ml). The number of washes can be reduced by successful and thorough perfusion.

2.3. Phage rescue and amplification

For the rescue of the phage particles that have homed to the target organ, we added 100 μl of 1% NP-40 to each cell/tissue pellet after the final wash and allowed the cells to lyse on ice for 5′. The overnight BL21 (415-1b) or BLT5615 (10-3b) cultures should be diluted to an OD at 600 nm of 1 using M9LB (10-3b vector requires isopropyl thiogalactoside (IPTG) to the bacterial culture at a final concentration of 10 m*M*) and 900 μl added to the lysed cells. The phage particles were rescued by incubating for 5′ at room temperature while rotating them after which the samples were titered. In all,

1.0–100 μl from each sample was added to a 14-ml polystyrene tube along with 300 μl of a BL21 or BLT5615 and mixed with 4 ml of melted (45–50 °C) top agar. Top agar containing the phage and the bacteria is poured onto LG agar (BL21) or 50 μg/ml carbenicillin LB agar (BLT5615) plates. The bacterial plates were then incubated either at 37 °C or at room temperature after which the titer could easily be determined by calculating plaques from the plate or the individual phage clones isolated by punching the plaques from the bacterial cell culture plate with a sterile pipette tip.

The rescued phage should be amplified and purified for the next round of screening by adding the samples remaining for the titering (i.e the rescued library/individual phage clone) to 1.0–10 ml of an OD at 600 nm of 0.5 culture of BL21 or BLT5615 (plus 10 m*M* IPTG). The culture (including the phages) is then incubated and shaken at 37 °C for 2 h until the culture is completely lysed and clarified. Concerning the peptide library *in vivo* screening, the amplification step should not be performed before the actual sample titers are known. The amplification step should be reduced in volume to the minimum needed for the next round of screening. This is recommended due to that the nonrecombinant phage particles proliferate slightly faster than the clones with the full-length peptide insert and can take over the library if the amplification step is excessive for the needs of the next round. The culture is then centrifuged at 7670 × *g* for 10 min. The supernatant is filtered using a 0.2 μm low protein binding filter (PVDF membrane) and stored at 4 °C. A typical titer of the T7 culture supernatant is 10^{11} pfu/ml.

The first *in vivo* round should include numerous animals because most of the selection process takes place during the first round of screening. After the first round of screening, the obtained libraries should be pooled and further rounds of selection performed with the pooled library.

2.4. Individual phage clones

To rule out potential bias, such as other random mutations (favoring its survival in the circulation) taking place in the phage clone during the *in vivo*—phage screen, it is recommended that the selected individual clones to be re-cloned to the T7Select vector arms before testing them individually. The following primers, expressing CAR and control peptides, were used as described above to prepare phage: CAR, 5′-AATTCCTGCGCGC GTTCGAAGAATAAGGATTGCTA-3′ and 5′-AGCTTAGCAATCC TTATTCTTCGAACGCGCGCAGG-3′; mutant-CAR (*m*CAR; CA**Q**S **N**NKDC), 5′-AATTCCTGCGCGCAGTCGAACAATAAGGATTGC TA-3′ and 5′-AGCTTAGCAATCCTTATTGTTCGACTGCGCGCA GG-3′; The cloning process was carried out as described above. To test single phage clones *in vivo*, 1×10^{10} phage in 1.5 ml of PBS were injected through the tail vein of a rat. After the circulation period and perfusion performed as described above, tissue samples were harvested and weighed.

Phage titers per milligram of wet tissue were then determined (tissues placed in preweighted tubes with 1% BSA + DMEM), and the results were expressed as the ratio between the titer of the test clone and nonrecombinant control phage in the same tissue.

2.5. PCR and sequencing of the insert coding region

Plaques formed on the bacterial culture plate can be used to isolate and amplify the recombinant phage's insert coding region by PCR. A single plaque is punched by a sterile pipette tip and placed in 10 μl of 1 × TBS on 96-well plate. The following primers were used for the PCR reaction (10 pmol for each reaction): T7 "Phage Up" 5′-AGCGGACCAGATTATCGCTA-3′ and T7 "Phage Down" 5′-AACCCCTCAAGAC CCGTTTA-3′. PCR can be carried out by Phusion® Flash High-Fidelity PCR Master Mix (Finnzymes, Vantaa, Finland) using the following PCR conditions: hold at 72 °C for 10′, 35 cycles of 94 °C for 50 s, 50 °C for 1′, and 72 °C for 1′; hold at 72 °C for 10; and hold at 4 °C until the samples are taken for sequencing. PCR reaction can be checked by standard agarose gel electrophoresis.

3. Homing Peptides

3.1. Peptides and homing experiments

The procedures follow essentially that described in Järvinen and Ruoslahti (2007). Peptides were synthesized with an automated peptide synthesizer by using standard solid-phase fluorenylmethoxycarbonyl chemistry. During synthesis, the peptides were labeled with either fluorescein (FITC) or 5-carboxyfluorescein (FAM) using an amino-hexanoic acid spacer (Laakkonen *et al.*, 2002, Urakami *et al.*, 2011). To test for *in vivo* homing, each fluorescein-conjugated peptide is injected intravenously into the tail vein of rats (250 – 500 μg of peptide) or mice (50–100 μg of peptide) under anesthesia. The isoflurane anesthesia is recommend; The animals recover from it immediately and can move freely during the entire circulation time. The peptides were allowed to circulate for different periods of time (1–4 h), followed by perfusion through heart (50 ml of 1 × PBS with 1% BSA) as described above. Tissues were embedded into Optimal Cutting Temperature compound OCT Tissue-Tek; Sakura Finetek, Torrance, CA) either unfixed or after the overnight fixation with 4% paraformaldehyde, frozen, and then processed for microscopy. The samples should be shielded from the room light by foil throughout the procedure.

3.2. Immunohistochemistry

In our procedure (Järvinen and Ruoslahti, 2007) frozen, unfixed tissue sections were fixed in acetone for 10 min and incubated with 0.5% blocking reagent for 1 h (NEN Life Sciences, Boston, MA). Tissue sections were incubated with the primary antibody overnight at 4 °C. The following monoclonal antibodies (mAbs) and polyclonal antibodies (pAbs) were used: rabbit anti-T7-phage affinity-purified pAb (1:100) (Laakkonen *et al.*, 2002), rat anti-mouse CD31 mAb (1:200; BD Pharmingen, San Diego, CA), and rabbit antifluorescein isothiocyanate (FITC) pAb (1:200; Invitrogen, Carlsbad, CA). The primary antibodies were detected with labeled (either with peroxidase or fluorescent label) secondary antibodies, and each staining experiment includes sections stained with species-matched immunoglobulins as negative controls. The sections were washed several times with PBS, mounted in Vectashield mounting medium with 4,6-diamidino-2-phenylindole (DAPI) (Vector Laboratories, Burlingame, CA), and visualized under an inverted fluorescent or light microscope.

Although the homing peptides were labeled with FITC, some of them, especially the tissue penetrating peptides like CAR, can be difficult detect, because they efficiently extravasate into surrounding tissue and the signal from the peptide can become insufficiently distinct from the background autofluorescence (Järvinen and Ruoslahti, 2007; Urakami *et al.*, 2011). To circumvent this problem, we usually detect the FITC-labeled peptides by immunohistochemistry using anti-FITC antibody converting the fluorescent signal to light microscopy by peroxidase-labeled secondary antibody.

3.3. Cell culture experiments with peptides and phage

We have tested cell binding and internalization of CAR peptide by using chinese hamster ovary cells (CHO-K) (American Type Culture Collection Rockville, MD) and a glycosaminoglycan mutant variant of CHO-K cells; pgsA-745 (kindly provided by Dr. J. Esko, University of California at San Diego, La Jolla, CA) (Esko *et al.*, 1985). Cells are maintained in α-minimal essential medium and Earle's salt supplemented with 10% fetal bovine serum, 100 μg/ml streptomycin sulfate, 100 U of penicillin G/ml, and 292 μg/ml L-glutamine (Invitrogen). For the phage binding assay, the cells are detached with 0.5 mmol/l ethylenediaminetetraacetic acid solution (Irvine Scientific, Santa Ana, CA). In all, 1×10^{10} phages are added to 15 ml of culture media containing 1×10^{6} cells in a test tube. The samples are slowly rotated for 2 h at +4 °C. The cells are then washed six times and transferred to a new tube each time. After a final wash, the cells are counted and cell-bound phage titers are determined as described above. Heparinase pretreatment of the CHO-K cells is performed using a combination of 1.5 IU/ml heparinase I (*Flavobacterium heparinum*; heparin lyase, EC 4.2.2.7, Sigma-Aldrich, St. Louis, MO) and 1.25 IU/ml heparinase III (*F. heparinum*; heparin-sulfate lyase, EC 4.2.2.8, Sigma-Aldrich) in serum-free

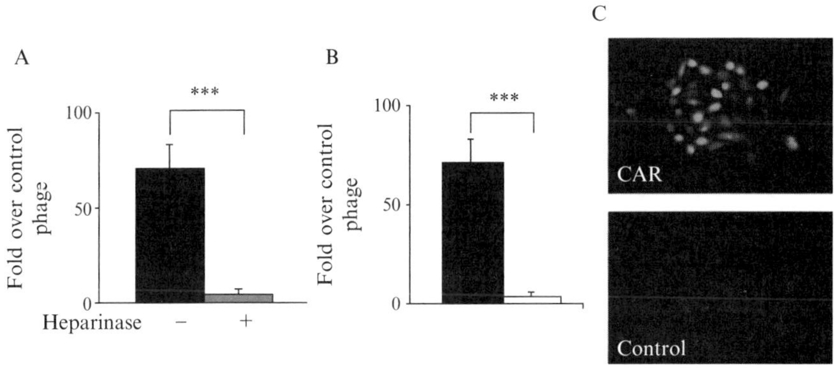

Figure 12.3 Binding of CAR phage and peptide to cell surface heparan sulfate, heparin, and CHO-K cells. (A) Heparinase pretreatment of the CHO-K cells (gray bar) suppresses the binding of the CAR phage to these cells. (B) CAR phage, but not a control phage, binds to heparin-coated beads. (C) Fluorescein-conjugated peptides were incubated at 5 μmol/l concentration with CHO-K cells for 4 h. The cells were washed, fixed, stained with the nuclear stain DAPI (blue), and examined for green fluorescence from the labeled peptides. The CAR peptide produces strong green fluorescence in the CHO-K cells (Top), but a control peptide gives no detectable cellular fluorescence. Error bars (A and B) represent mean ± SEM for three or more separate experiments performed in duplicate. $^{\star}P < 0.05$; $^{\star\star\star}P < 0.001$. (A) Two-way analysis of variance; (B) unpaired Student's *t*-test. (originally published in Järvinen and Ruoslahti, 2007.) (For interpretation of the references to color in this figure legend, the reader is referred to the Web version of this chapter.)

culture media for 2 h on the cell culture dish at 37 °C (Fig. 12.3) (Järvinen and Ruoslahti, 2007).

To study peptide internalization, we incubated cells seeded for 2 days on plastic coverslips with 10 μmol/l fluorescein-conjugated peptides for 30 min to 72 h; the cells are washed three times with PBS, and fixed with 4% paraformaldehyde for 20 min at room temperature (Fig. 12.3). After several washes with PBS, the nuclei are visualized by staining with DAPI, and the slides are mounted with ProLong Gold antifade reagent (Invitrogen). The images are acquired using an inverted and confocal microscopes (Olympus IX81 and Olympus Fluoview FV1000, Olympus, Melville, NY). Z-stack images are taken by confocal microscope every 1 μm through the cells.

To measure phage binding to heparin as shown in Figure 12.3, heparin-coated acrylic beads (Sigma-Aldrich) 10% (v/v) are suspended in 20 mmol/l Na_2HPO_4 buffer, pH 7.2, containing 0.2 mol/l NaCl (El-Sheikh *et al.*, 2002). A total of 5.0×10^9 phage particles are incubated with the beads for 1 h at room temperature. The beads are washed with 1% BSA + 1 × PBS and transferred to a new tube five times, and the bound phage is eluted with 1.2 mol/l NaCl (pH 7.2) and titrated as described above (Fig. 12.3).

4. Multifunctional Fusion Proteins

4.1. Cloning

A full-length human decorin cDNA (Krusius and Ruoslahti, 1986) together with a Kozak sequence to the N-terminus, *Eco*RI and *Sal*I restriction sites, thrombin cleavage site, and his-tag all into the C-terminus of the decorin are subcloned into the mammalian expression vector pcDNA3.1/myc-his-C (Invitrogen) by PCR (Fig. 12.4). The wound homing peptides are cloned between the C-terminus of the decorin and the his-tag placing the (thrombin cleavage site being between peptide and the his-tag). This was accomplished using *Eco*RI and *Sal*I restriction sites regenerated by PCR during the first step of cloning process. Human serum albumin (HSA) targeted with CAR homing peptide is cloned by replacing decorin with HSA cDNA (Sheffield *et al.*, 2009) in the expression vector, and *m*CAR-decorin by replacing the CAR sequence with that of *m*CAR (provided above). A map of the C-terminus of the decorin fusion proteins is shown in Fig. 12.4 (Järvinen and Ruoslahti, 2010).

4.2. Production of recombinant proteins—Mammalian expression

The procedures follow essentially that described in Järvinen and Ruoslahti (2010). We have produced recombinant decorin and HSA constructs in 293-F cells using the FreeStyle 293 expression system from Invitrogen, according to the manufacturer's instructions (Fig. 12.4) (Järvinen and Ruoslahti, 2010). For the transfection, the pcDNA3.1/myc-his-C plasmid DNA is mixed with the Opti-MEM I media and 293fectin transfection reagent according to the instructions. The cells are cultured for 48 h after the transfection and the decorins were isolated from the media on Ni-NTA agarose beads (Qiagen, GmbH, Germany) using 5 ml of beads per 500 ml of media. After an overnight incubation at +4 °C, the beads are washed with PBS, and decorin was eluted with PBS containing 300 m*M* imidazole, dialyzed against PBS, and stored at −80 °C.

4.3. Production of recombinant proteins—Baculovirus expression

Alternatively, recombinant decorins can be produced in a baculovirus expression system (Bac-to-Bac system, Invitrogen) (Fig. 12.4). The above-described decorin constructs were cloned into pFasBac1-vector. The native signal

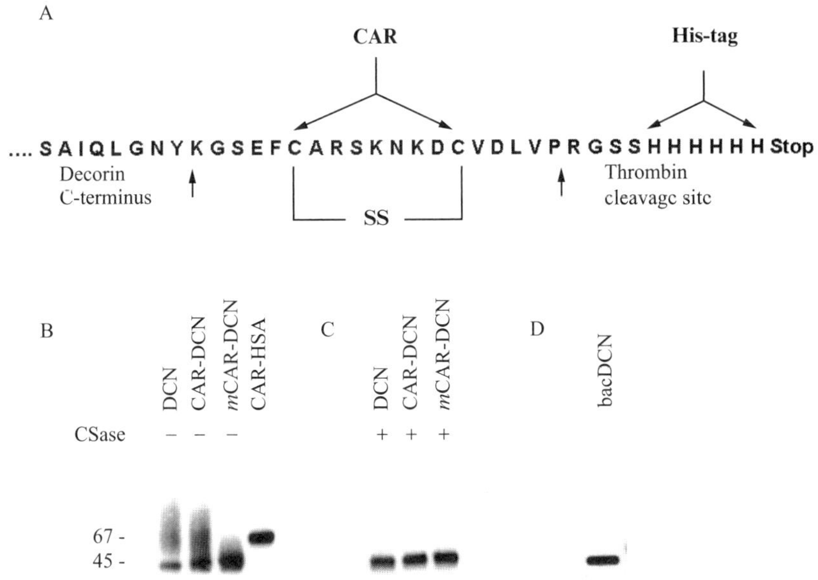

Figure 12.4 Cloning and production of recombinant fusion proteins. (A) A schematic representation showing the insertion of the CAR peptide sequence, thrombin cleavage site, and a polyhistidine-tag C-terminal of full-length decorin. (B) Gel electrophoretic analysis of recombinant decorins and HSA modified with the CAR peptide. The recombinant proteins were expressed in mammalian cells, purified on a Ni-column, separated on gradient SDS-PAGE gels, and detected with a monoclonal anti-6-histidine tag antibody. (C) The recombinant decorins were also digested with chondroitinase ABC prior to electrophoresis. The decorins migrate as sharp bands at 45 kDa with a smear above it. The sharp bands correspond to the core proteins, and the smear is caused by heterogeneity in the chondroitin sulfate chain attached to most of the decorin molecules (Border *et al.*, 1992; Yamaguchi and Ruoslahti, 1988; Yamaguchi *et al.*, 1990). The treatment of decorins with chondroitinase ABC removes the glycosaminoglycan side chain and the smear above the sharp bands disappears after the chondroitinase treatment. (D) Decorin produced in a baculovirus (bacDCN) expression system was run as a control. It runs as a sharp band because this expression system is not capable of producing glycosaminoglycans. (Modified from Järvinen and Ruoslahti, 2010)

peptide of decorin is replaced with that from the honeybee melittin gene for efficient secretion using the following primer: 5′-GATCTATGAAATTCT-TAGTCAACGTTGCCCTTGTTTTTATGGTCGTATACATTTCTT ACATCTATGCG GGATCC-3′. Dh10Bac *Escherichia coli* cells are transfected with pFasBac1-decorin DNA and plated. Six recombinant bacmid

DNA batches are purified from individual colonies selected using blue/white screening. Presence of the decorin insert in the bacmids is verified by PCR. SF9 cells are transfected with recombinant bacmides, and six low titer virus stocks are collected. They are amplified three times to produce high titer stocks. The high titer virus stocks are tested for decorin expression, and one with the highest yield is selected for protein production. One liter of SF9 cells is infected with decorin-containing virus at MOI of 5 and incubated for 60 h. Cells are isolated from conditioned media by centrifugation and the decorin is isolated from the media with Ni-NTA agarose beads.

4.4. Characterization of recombinant decorins

The procedures follow essentially that described in Järvinen and Ruoslahti (2010). Briefly, the recombinant fusion proteins can be analyzed on SDS-PAGE on 4–20% acrylamide gradient gels (Fig. 12.4). Some of the gels are stained with Coomassie Blue, while others were used to transfer the proteins to PVDF membrane, and immunoblots can be performed with monoclonal anti-6-histidine tag antibody (1:1 000, clone 18184, Novus Biologicals, Littleton, CO) (Fig. 12.3) or with monoclonal antidecorin (1: 500, clone 115 402, R&D Systems), and detected with goat anti-mouse IgG-HRP (diluted 1:25,000; Bio-Rad, Hercules, CA), and then developed using ECL + plus chemiluminescence reagent (Amersham Biosciences, Piscataway, NJ), according to the manufacturer's instructions.

For mass spectrometry analysis, recombinant proteins were separated on 10% SDS-PAGE. Protein bands were detected by silver staining, extracted, and subjected to in-gel trypsin digestion and peptide mass fingerprinting in MALDI-TOF by standard procedures (Bruker Daltronics AutoflexII, Bremen, Germany).

Protein folding is examined by differential scanning calorimetry using N-DSC II differential calorimeter (Calorimetry Sciences Corp., Provo, UT) at a scanning rate of 1.0 K/min (0–120 °C) under 3.0 atm of pressure. Protein samples are dialyzed against PBS (pH 7.4) and the analyses were carried out at 1.0mg/ml of protein with PBS as reference.

The recombinant decorins with glycosaminoglycan side chain (chondroitin sulfate) attached to the core protein (1.0 mg/ml) are digested with chondroitinase ABC (EC 4.2.2.4., Sigma-Aldrich, St. Louis, MO) in chondroitinase buffer (0.3 *M* NaCl, 10 m*M* Tris acetate (Tris + glacial acetic acid), 0.01 *M* EDTA, pH 7.4) containing protease inhibitors (Carrino *et al.*, 2003, 2010). Chondroitinase ABC is added at 0.8 unit/mg of recombinant protein; the samples are incubated at 37 °C for 1 h, frozen, and examined by SDS-PAGE.

5. Function of Multifunctional Fusion Protein

5.1. Cell proliferation and cell binding assays

The effect of decorin preparations on TGF-β-induced cell proliferation is determined on CHO-K cells as described previously (Järvinen and Ruoslahti, 2010; Yamaguchi and Ruoslahti, 1988; Yamaguchi *et al.*, 1990) (Fig. 12.5). The cells are grown either in medium containing dialyzed fetal bovine serum (Invitrogen further dialyzed with the Slide-A-Lyzer dialysis cassette with the cut-off value of 25 k MWCO from Pierce, Rockford, IL) or in medium supplemented only with TGF-β1, -β2, or -β3 (7.5 ng/ml, Biovision, Mountain View, CA). The cells are plated at a density of 1×10^4 cells per well in 24-well plates and cultured in 1 ml of culture medium. For the TGF-β experiment, the cells were allowed to adhere and spread out by growing them under full cell culture media for the 1st 24 h, after which the media was changed to one where the serum was replaced by

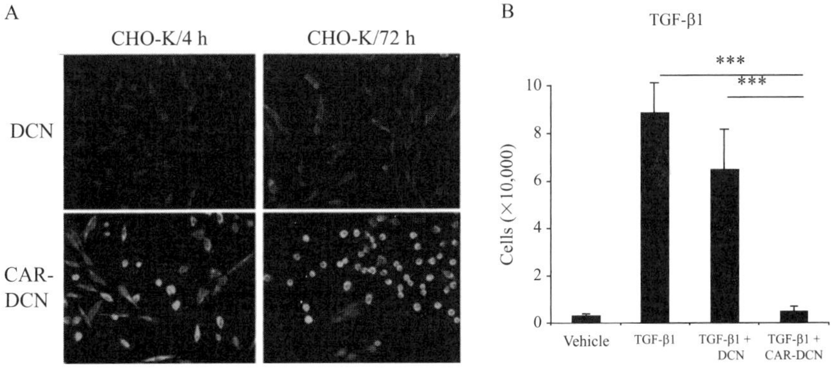

Figure 12.5 Cell binding and inhibition of cell spreading and TGF-β-dependent cell proliferation by decorins. (A) The culture media of CHO-K cells were supplemented with 0.3 μg/ml of decorins daily for 4 h or 3 days. After washing, the decorins were detected with anti-his-tag antibody and FITC-conjugated secondary antibody (green). The CAR-targeted decorin bound to CHO-K cells more (4 h) and inhibited cell spreading more strongly (3 days) than unmodified decorin. Cell nuclei were stained by DAPI (Blue). (B) CHO-K cells were grown in the presence of TGF-β1 (7.5 ng/ml) and 0.3 μg/ml of decorins. Half of the medium was changed daily for 4 days (7). Error bars represent mean ± standard deviation (SD). Two to four separate experiments were performed in triplicate. CAR targeted decorin was particularly potent in inhibiting TGF-β1-driven cell proliferation (B; (***) $P \leq 0.001$ both compared to control and decorin; ANOVA). (Modified from Järvinen and Ruoslahti, 2010). (For interpretation of the references to color in this figure legend, the reader is referred to the Web version of this chapter.)

one of the TGF-β isoforms listed above. Half of the medium was replaced daily with fresh medium containing recombinant decorins or HSA as well as TGF-β1/-β2/-β3 (7.5 ng/ml). Cells are collected by trypsinization, washed, and resuspended in 1 ml of PBS containing 2 μg/ml of propidium iodide (PI) and 20,000 CountBright counting beads (Invitrogen), and analyzed by counting viable cells in FACS (Fig. 12.5).

To study decorin binding and internalization, CHO-K and pgsA-745 cells seeded on plastic coverslips are incubated with different decorins (0.3 μg/24 h) for 4–72 h, washed three times with PBS, and fixed with 4% paraformaldehyde for 20 min at room temperature. After several washes with PBS and blocking of nonspecific background staining by CAS-Block reagent (Invitrogen), the immunocytochemistry is carried out as described above (Fig. 12.5) (Järvinen and Ruoslahti, 2010).

5.2. Wound healing model

Eight- to twelve-week-old male BALB/c mice (weighting 23–25 g) are induced into the anesthesia with 4% isoflurane and 1.5 l/min of oxygen and the anesthesia was maintained at approximately 1.5% isoflurane at 1.0 l/min of oxygen. The skin is shaved, cleaned, and disinfected with betadine and 70% alcohol. Treatment trials are conducted on mice that have circular, 6- or 8-mm-diameter, full thickness (including panniculus carnosus muscle) excision wounds in the dorsal skin. The wounds are first marked by a biopsy bunch and then cut with scissors along the markings made by a biopsy punch. All skin wounds are left uncovered without a dressing. The animals recovered from the isoflurane-induced anesthesia immediately and were able to move without noticeable limp (Järvinen and Ruoslahti, 2007).

5.3. Histology

Wound tissues are removed by cutting the entire back skin covering the wounds as a single large unit that also includes the underlying muscle. This will avoid any procedure-induced breakage to the wounds and to the underlying granulation tissue. The skin block is placed on filter membrane and fixed in 10% buffered zinc formalin (Statlab Medical Products, Lewisville, TX) by overnight incubation. After the fixation, the wounds are bisected longitudinally (along the spine) at the middle of the wounds for mounding. The tissues are then dehydrated in graded alcohol series and embedded in paraffin. Six-micrometer sections from the middle of the wound are stained with hematoxylin/eosin. Alternatively, the Masson trichrome procedure can be used, or the tissues can be processed for immunohistochemistry. Each side of the wound is evaluated using one section. The analysis is performed using the average of these two values.

5.4. Quantitative analysis of immunostaining and histochemical staining

Microscopic slides were scanned at an absolute magnification of 400 × (resolution of 0.25 μm/pixel [100,000 pixel/in.]) using the Aperio ScanScope® CS and XT systems (Aperio Technologies, CA, USA) (Krajewska *et al.*, 2009). The background illumination levels are calibrated using a prescan procedure. The acquired digital images representing whole tissue sections were evaluated for image quality. Slides are viewed and analyzed remotely using desktop personal computers employing the Web-based ImageScope™ viewer. The Spectrum Analysis algorithm package and Image Scope analysis software (version 9; Aperio Technologies, Inc.) are applied to quantify immunohistochemical signal. These algorithms make use of a color deconvolution method to separate stains so that quantification of individual stains avoids cross-contamination (Krajewska *et al.*, 2009). Using the software, each stain is individually calibrated by analyzing single-stained sections and recording the average RGB optical density vectors (Krajewska *et al.*, 2009). The algorithms calculate the area of positive staining, the average positive intensity (optical density), as well as the percentage of weak (1+), medium (2+), and strong (3+) positive staining.

6. Concluding Remarks

Systemic treatment of injured tissues would be important since a systemically administered therapeutic agent will reach the site of injury regardless of the location. We have used vascular targeting technology to devise an improved version of a physiological antiscarring agent, decorin, that has been extensively validated in preclinical testing but has not reached the clinic. The increased efficacy of our targeted decorin may make systemic treatment of tissue injuries and fibrotic diseases a clinically feasible option. This opens up new possibilities not only for the treatment of skin wounds we used as the model, but also in traumatology, surgery, the treatment of tissue injuries that result in permanent scar formation (myocardial infarction, stroke), and fibrotic diseases.

Acknowledgments

T. A. H. J. received support from the Academy of Finland, Sigrid Juselius Foundation, Instrumentarium Research Foundation (both Helsinki, Finland), and Competitive Research Funding of the Pirkanmaa Hospital District, Tampere University Hospital.

REFERENCES

Aarabi, S., Longaker, M. T., and Gurtner, G. C. (2007). Hypertrophic scar formation following burns and trauma: New approaches to treatment. *PLoS Med.* **4,** e234.

Border, W. A., and Ruoslahti, E. (1992). Transforming growth factor-β in disease: The dark side of tissue repair. *J. Clin. Invest.* **90,** 1–7.

Border, W. A., Noble, N. A., Yamamoto, T., Harper, J. R., Yamaguchi, Y., Pierschbacher, M. D., and Ruoslahti, E. (1992). Natural inhibitor of transforming growth factor-β protects against scarring in experimental kidney disease. *Nature* **360,** 361–364.

Brown, D. M., and Ruoslahti, E. (2004). Metadherin, a cell surface protein in breast tumors that mediates lung metastasis. *Cancer Cell* **5,** 365–374.

Carrino, D. A., Onnerfjord, P., Sandy, J. D., Cs-Szabo, G., Scott, P. G., Sorrell, J. M., Heinegård, D., and Caplan, A. I. (2003). Age-related changes in the proteoglycans of human skin. Specific cleavage of decorin to yield a major catabolic fragment in adult skin. *J. Biol. Chem.* **278,** 17566–17572.

Carrino, D. A., Calabro, A., Darr, A. B., Dours-Zimmermann, M. T., Sandy, J. D., Zimmermann, D. R., Sorrell, J. M., Hascall, V. C., and Caplan, A. I. (2010). Age-related differences in human skin proteoglycans. *Glycobiology* **21,** 257–268.

Corti, A., and Curnis, F. (2011). Tumor vasculature targeting through NGR peptide-based drug delivery systems. *Curr. Pharm. Biotechnol.* **12,** 1128–1134.

Curnis, F., Sacchi, A., Borgna, L., Magni, F., Gasparri, A., and Corti, A. (2000). Enhancement of tumor necrosis factor α antitumor immunotherapeutic properties by targeted delivery to aminopeptidase N (CD13). *Nat. Biotechnol.* **18,** 1185–1190.

El-Sheikh, A., Liu, C., Huang, H., and Edgington, T. S. (2002). A novel vascular endothelial growth factor heparin-binding domain substructure binds to glycosaminoglycans in vivo and localizes to tumor microvascular endothelium. *Cancer Res.* **62,** 7118–7123.

Esko, J. D., Stewart, T. E., and Taylor, W. H. (1985). Animal cell mutants defective in glycosaminoglycan biosynthesis. *Proc. Natl. Acad. Sci. USA* **82,** 3197–3201.

Faust, S. M., Lu, G., Wood, S. C., and Bishop, D. K. (2009). TGFβ neutralization within cardiac allografts by decorin gene transfer attenuates chronic rejection. *J. Immunol.* **183,** 7307–7313.

Hildebrand, A., Romaris, M., Rasmussen, L. M., Heinegård, D., Twardzik, D. R., Border, W. A., and Ruoslahti, E. (1994). Interaction of the small interstitial proteoglycans biglycan, decorin and fibromodulin with transforming growth factor β. *Biochem. J.* **302,** 527–534.

Järvinen, T. A. H., and Ruoslahti, E. (2007). Molecular profiling of vasculature in injured tissues. *Am. J. Pathol.* **171,** 702–711.

Järvinen, T. A. H., and Ruoslahti, E. (2010). Target seeking anti-fibrotic compound enhances wound healing and suppresses scar formation. *Proc. Natl. Acad. Sci. USA* **107,** 21671–21676.

Järvinen, T. A. H., and Ruoslahti, E. (2011). Uusi lääkeainen estää arven muodostusta. In Press. Duodecim 127, 50–51. (in Finnish).

Järvinen, T. A. H., and Ruoslahti, E. (2012). Targeted antiscarring therapy for tissue injuries. *Adv. Wound Care,* in press.

Kean, T. J., Duesler, L., Young, R. G., Dadabayev, A., Olenyik, A., Penn, M., Wagner, J., Fink, D. J., Caplan, A. I., and Dennis, J. E. (2011). Development of a peptide-targeted, myocardial ischemia-homing, mesenchymal stem cell. *J. Drug Target.* **20,** 23–32.

Krajewska, M., Smith, L. H., Rong, J., Huang, X., Hyer, M. L., Zeps, N., Iacopetta, B., Linke, S. P., Olson, A. H., Reed, J. C., and Krajewski, S. (2009). Image analysis algorithms for immunohistochemical assessment of cell death events and fibrosis in tissue sections. *J. Histochem. Cytochem.* **57,** 649–663.

Krusius, T., and Ruoslahti, E. (1986). Primary structure of an extracellular matrix proteoglycan core protein deduced from cloned cDNA. *Proc. Natl. Acad. Sci. USA* **83,** 7683–7687.

Laakkonen, P., Porkka, K., Hoffman, J. A., and Ruoslahti, E. (2002). A tumor-homing peptide with a targeting specificity related to lymphatic vessels. *Nat. Med.* **8,** 751–755.

Li, Y., Foster, W., Deasy, B. M., Chan, Y., Prisk, V., Tang, Y., Cummins, J., and Huard, J. (2004). Transforming growth factor-β1 induces the differentiation of myogenic cells into fibrotic cells in injured skeletal muscle: A key event in muscle fibrogenesis. *Am. J. Pathol.* **164,** 1007–1019.

Li, Y., Li, J., Zhu, J., Sun, B., Branca, M., Tang, Y., Foster, W., Xiao, X., and Huard, J. (2007). Decorin gene transfer promotes muscle cell differentiation and muscle regeneration. *Mol. Ther.* **15,** 1616–1622.

Pasqualini, R., and Ruoslahti, E. (1996). Organ targeting in vivo using phage display peptide libraries. *Nature* **380,** 364–368.

Pilch, J., Brown, D. M., Komatsu, M., Järvinen, T. A. H., Yang, M., Peters, D., Hoffman, R. M., and Ruoslahti, E. (2006). Peptides selected for binding to clotted plasma accumulate in tumor stroma and wounds. *Proc. Natl. Acad. Sci. USA* **103,** 2800–2804.

Ruoslahti, E. (2002). Specialization of tumour vasculature. *Nat. Rev. Cancer* **2,** 83–90.

Ruoslahti, E. (2004). Vascular zip codes in angiogenesis and metastasis. *Biochem. Soc. Trans.* **32,** 397–402.

Ruoslahti, E., Bhatia, S. N., and Sailor, M. J. (2010). Targeting of drugs and nanoparticles to tumors. *J. Cell Biol.* **188,** 759–768.

Sheffield, W. P., Eltringham-Smith, L. J., Gataiance, S., and Bhakta, V. (2009). A long-lasting, plasmin-activatable thrombin inhibitor aids clot lysis *in vitro* and does not promote bleeding *in vivo*. *Thromb. Haemost.* **101,** 867–877.

Sugahara, K. N., Teesalu, T., Karmali, P. P., Kotamraju, V. R., Agemy, L., Girard, O. M., Hanahan, D., Mattrey, R. F., and Ruoslahti, E. (2009). Tissue-penetrating delivery of compounds and nanoparticles into tumors. *Cancer Cell* **16,** 510–520.

Teesalu, T., Sugahara, K. N., Kotamraju, V. R., and Ruoslahti, E. (2009). C-end rule peptides mediate neuropilin-1-dependent cell, vascular, and tissue penetration. *Proc. Natl. Acad. Sci. USA* **106,** 16157–16162.

Teesalu, T., Sugahara, K. N., and Ruoslahti, E. (2012). Mapping of vascular ZIP codes by phage display. *Meth. Enzymol.* **503,** 35–56.

Urakami, T., Järvinen, T. A. H., Oka, M., Sawada, J., Ambalavanan, N., Mann, D., McMurtry, I., Ruoslahti, E., and Komatsu, M. (2011). Peptide-directed highly selective targeting of pulmonary arterial hypertension. *Am. J. Pathol.* **178,** 2489–2495.

Yamaguchi, Y., and Ruoslahti, E. (1988). Expression of human proteoglycan in Chinese hamster ovary cells inhibits cell proliferation. *Nature* **336,** 244–246.

Yamaguchi, Y., Mann, D. M., and Ruoslahti, E. (1990). Negative regulation of transforming growth factor-β by the proteoglycan decorin. *Nature* **346,** 281–284.

Zhu, J., Li, Y., Shen, W., Qiao, C., Ambrosio, F., Lavasani, M., Nozaki, M., Branca, M. F., and Huard, J. (2007). Relationships between transforming growth factor-β1, myostatin, and decorin: Implications for skeletal muscle fibrosis. *J. Biol. Chem.* **282,** 25852–25863.

CHAPTER THIRTEEN

Design and Fabrication of *N*-Alkyl-Polyethylenimine-Stabilized Iron Oxide Nanoclusters for Gene Delivery

Gang Liu,[*,†,‡] Zhiyong Wang,[§] Seulki Lee,[†] Hua Ai,[§,1] *and* Xiaoyuan Chen[*,†,1]

Contents

1. Introduction 264
2. Materials 265
   2.1. Reagents 265
   2.2. Equipment 265
3. Methods 266
   3.1. Fabrication of *N*-alkyl-polyethylenimine-stabilized iron oxide nanocluster 266
   3.2. N-alkyl-PEI2K-SPIO-siRNA complex formation and properties 267
   3.3. Cell transfection with IO–siRNA complexes 268
   3.4. Cytotoxicity assay 268
   3.5. MRI study of transfected cells 269
4. Notes 269
5. Anticipated Results 271
6. Summary 274
Acknowledgments 274
References 275

* Center for Molecular Imaging and Translational Medicine, School of Public Health, Xiamen University, Xiamen, China
† Laboratory of Molecular Imaging and Nanomedicine, National Institute of Biomedical Imaging and Bioengineering, National Institutes of Health, Bethesda, Maryland, USA
‡ Sichuan Key Laboratory of Medical Imaging, North Sichuan Medical College, Nanchong, China
§ National Engineering Research Center for Biomaterials, Sichuan University, Chengdu, China
1 Corresponding authors. E-mail: huaai@scu.edu.cn; shawn.chen@nih.gov

Methods in Enzymology, Volume 509
ISSN 0076-6879, DOI: 10.1016/B978-0-12-391858-1.00009-5

Abstract

With the rapid development of nanotechnology, inorganic magnetic nanoparticles, especially iron oxide nanoparticles (IOs), have emerged as great vehicles for biomedical diagnostic and therapeutic applications. In order to rationally design IO-based gene delivery nanovectors, surface modification is essential and determines the loading and release of the gene of interest. Here we highlight the basic concepts and applications of nonviral gene delivery vehicles based on low molecular weight *N*-alkyl polyethylenimine-stabilized IOs. The experimental protocols related to these topics are described in this chapter.

1. Introduction

Magnetic nanoparticles, such as iron oxide nanoparticles (IOs), are an important class of nanomaterials with interesting properties (biocompatibility, magnetic properties, physical/chemical stability, and low cost) that have been developed into various functional agents for applications in imaging, cell labeling, hyperthermia, and drug/gene delivery (Liu *et al.*, 2010b, 2011a,c; Xie *et al.*, 2011). To rationally design IO-based gene delivery nanovectors, surface modification processes are necessary to stabilize IOs to create strong interactions for binding of the therapeutic gene and to control the release mechanism (Hao *et al.*, 2010; Liu *et al.*, 2010a; Taratula *et al.*, 2011; Veiseh *et al.*, 2010). In addition, surface engineering can play a central role in determining the internalization rate and toxicity of the particles (Liu *et al.*, 2009; Wang *et al.*, 2009). A common strategy is to convert the IO surface to a polycation layer, onto which the highly negatively charged therapeutic gene can be electrostatically attracted and tethered. This is mostly achieved by introducing one layer of cationic polymers onto IO surfaces to design nonviral gene delivery vehicles (Arsianti *et al.*, 2010; Cheong *et al.*, 2009; Scherer *et al.*, 2002).

Among many nonviral gene carriers, polyethylenimine (PEI) is one of the most effective gene delivery vehicles that can bind negatively charged nucleic acids via electrostatic interactions, protect nucleic acids from degradation by restriction of endonucleases, and enter cells through rapid endocytosis (Demeneix and Behr, 2005; Godbey *et al.*, 1999). In addition, PEI is also able to delay the acidification and fusion of intracellular organelles through "proton sponge" effects, which cause osmotic swelling of endosomes and subsequent release of the trapped nucleic acid into the cytosol (Boussif *et al.*, 1995; Demeneix and Behr, 2005). It is well known that the transfection efficiency and cytotoxicity of PEI correlate strongly with its molecular weight. High molecular weight PEI exhibits much higher transfection efficiency as well as cytotoxicity compared to its lower molecular weight analogues (Godbey *et al.*, 1999). However, recent studies show that cross-linked low molecular weight PEI, or IO/gold nanoparticles "wrapped" in low molecular weight PEI, could enhance gene delivery efficiency without compromising its low cytotoxicity (Hu *et al.*, 2010; Huang *et al.*, 2010; Kievit *et al.*, 2009; Liu *et al.*, 2011d; Thomas and Klibanov, 2003). It is possible that the efficient unpacking of complexes and subsequent siRNA release from IOs, with its unique nanostructure, into the cytoplasm, contribute to an enhanced gene-silencing effect when compared with *N*-alkyl-PEI2k/siRNA complexes (Liu *et al.*, 2011d). Here we describe an approach to form IO–siRNA complexes that can be used to monitor the transfection efficiency, as well as to track the cells in gene therapy applications.

2. Materials

2.1. Reagents

PEI2k (MW: 1.8 kD) (Alfa Asear); 1-iodododecane (Aldrich Chemical Co.); iron(III) acetylacetonate (Aldrich Chemical Co.); 1,2-hexadecanediol (Aldrich Chemical Co.); benzyl ether (Aldrich Chemical Co.); oleic acid (Aldrich Chemical Co.); oleylamine (Aldrich Chemical Co.); chloroform (Sigma); dimethyl sulfoxide (Sigma); heparin (Sigma); phosphate-buffered saline (PBS) (Gibco); RPMI 1640 medium (Invitrogen); fetal bovine serum (FBS) (Hyclone); TAE electrophoresis buffer (Gibco); ethidium bromide (Gibco); agarose gel (Gibco); L-glutamine (Mediatech, Inc.); penicillin–streptomycin (Mediatech, Inc.); streptomycin (Mediatech, Inc.); trypsin–EDTA (Mediatech, Inc.); D-luciferin (Gold Bio Technology, Inc.); 3-(4,5-dimethylthiazol-2-yl)-2,5-diphenyltetrazolium bromide (Sigma); firefly luciferase siRNA (Ambion); Lipofectamine 2000 (Invitrogen, Grand Island, NY, USA); plasmid DNA encoding firefly luciferase protein (System Biosciences); murine 4T1 breast cancer cell line (American Type Culture Collection (ATCC)); the 4T1 cell line stably expressing luciferase gene (4T1-fluc).

2.2. Equipment

1. Reaction vessel, magnetic stirrer, and magnetic rod
2. Thermal analysis instrument
3. Dialysis membranes
4. Freeze dry system
5. Nuclear magnetic resonance (NMR) spectrometer
6. Dynamic light scattering (DLS) and zeta potential instrument
7. Transmission electron microscope (TEM)
8. Atomic force microscopy (AFM)
9. Gel electrophoresis apparatus equipped with an image acquisition system
10. Controlled environment incubator (37 °C, 5% CO_2) for cell culture
11. Pipettes, tips, and cell culture plates
12. Centrifuge and centrifuge tubes
13. Inverted microscope
14. Cell counter

15. Fluorescence microplate reader
16. MR scanner and ImageJ software
17. Xenogen IVIS-100 system and Living Image v.3.1 software

3. Methods

3.1. Fabrication of *N*-alkyl-polyethylenimine-stabilized iron oxide nanocluster

3.1.1. Synthesis of *N*-alkyl-PEI2K

1. 1-Iodododecane (1 mmol) is added to dry PEI (0.25 mmol) and dissolved in absolute ethanol (10 ml) in a 50-ml round-bottom flask.
2. Then, the container is put into liquid nitrogen. After the solution is converted into solid, the container is evacuated, put into argon, and heated to 55 °C for 6 h with magnetic stirring.
3. After cooling to room temperature, the solution is incubated at room temperature overnight to remove the ethanol by rotary evaporation.
4. The crude product is dissolved in water, treated with 11 mol% of NaOH, and dialyzed in a dialysis bag extensively against water (molecular mass cutoff, 1 kDa) at room temperature.
5. The pure products are analyzed using ^{1}H NMR ($CDCl_3$).

3.1.2. Synthesis of IOs

1. $Fe(acac)_3$ (1 mmol) is mixed with 1,2-hexadecanediol (5 mmol), oleic acid (3 mmol), and oleylamine (3 mmol) in benzyl ether (10 ml) under nitrogen.
2. The mixture is heated to reflux (300 °C) for 1 h.
3. After cooling to room temperature (23 °C), ethanol is used to yield a dark-brown precipitate from the solution.
4. The product is redispersed in hexane with oleic acid and oleylamine, and reprecipitated with ethanol to give 6 nm Fe_3O_4 nanoparticles.
5. The 6 nm Fe_3O_4 nanoparticles are used as seeds to synthesize 12 nm Fe_3O_4 nanoparticles.
   - 5.1. $Fe(acac)_3$ (1 mmol) is mixed with 1,2-hexadecanediol (5 mmol), benzyl ether (10 ml), oleic acid (1 mmol), oleylamine (1 mmol), and 6 nm Fe_3O_4 nanoparticles (41 mg) dissolved in hexane.
   - 5.2. The mixture is heated to 100 °C for 30 min to remove hexane.
   - 5.3. The mixture is heated to 300 °C to reflux for 30 min. After cooling to room temperature (23 °C), ethanol is used to yield a dark-brown precipitate from the solution.
   - 5.4. The product is redispersed in hexane with oleic acid and oleylamine, and reprecipitated with ethanol to give 12 nm Fe_3O_4 nanoparticles.
6. The SPIO nanocrystals are analyzed by DLS and TEM.

3.1.3. Preparation of *N*-alkyl-PEI2K-IO nanoparticles

1. The SPIO nanoparticles (5 mg) are redispersed in 1 ml chloroform, together with *N*-alkyl-PEI2K (3 mg).
2. The above solution is added to 10 ml water with probe sonication (VCX130; Sonic & Materials Inc.) at room temperature.
3. The solution is mixed by shaking for 24 h, and the remaining chloroform is removed by rotary evaporation.
4. The composition of *N*-alkyl-PEI2K-IOs is determined via thermogravimetry analysis (TGA) on a NETZSCH Thermal Analysis Instrument at a heating rate of 10 °C/min from 35 °C to 1000 °C, under a N_2 atmosphere.
5. The size and morphology are established through DLS, zeta potential, AFM, and TEM analysis.
6. The T_2 relaxivity is determined at 3T on a clinical MR scanner (Siemens Sonata): TR = 5000 ms, TE values ranging from 6 to 170 ms.
7. The r_2 relaxivity is calculated through the curve fitting of 1/relaxation time (s^{-1}) versus the iron concentration (mM Fe).

3.2. N-alkyl-PEI2K-SPIO-siRNA complex formation and properties

1. For the binding assay, an appropriate amount of *N*-alkyl-PEI2k-IOs is added to the siRNA solution (6 pmol) with different N/P ratios (i.e., N/P ratio from 5 to 50).
2. The mixed solutions are incubated at room temperature for 20–30 min to allow for the formation of IO–siRNA complexes.
3. The IO–siRNA complexes are characterized by DLS and zeta potential measurements.
4. The IO–siRNA complexes are loaded onto a 2% agarose gel with ethidium bromide (0.5 μg/ml) for visualization.
5. The electrophoresis is run at 100 V for 15 min.
6. The gel is placed on top of a UV transilluminator (Sigma-Aldrich) to observe the migration of the IO–siRNA complexes within the agarose gel.
7. For the siRNA release assay, the IO–siRNA complexes are mixed at the full-binding ratio (N/P ratio 20:1) with varying amounts of heparin and incubated for approximately 10 min.
8. The above solution is loaded into the agarose gel, and the electrophoresis is run at 100 V for 15 min.
9. The IO–siRNA complexes and the dissociated siRNA are detected as described above.
10. For serum stability assay, IO–siRNA complexes (N/P ratio 20:1) are incubated with serum (final 50% concentration) at 37 °C.

11. After incubation for 24 h, heparin (10 μg) is added to the treated IO–siRNA complex samples.
12. The above solution is loaded into the agarose gel, and the electrophoresis is run at 100 V for 15 min.
13. The IO–siRNA complexes and undegraded siRNA are detected as described above.

3.3. Cell transfection with IO–siRNA complexes

1. 4T1 cells stably expressing the luciferase gene (4T1-fluc) are cultured in RPMI 1640 medium supplemented with 10% FBS at 37 °C with 5% CO_2.
2. The cells are plated at a density of 10^4 cells per well in 100 μl of appropriate complete growth medium without antibiotics in a 96-well, sterile microtiter plates.
3. For each well of cells, 6 pmol of siRNA (fluc-siRNA) is diluted into 25 μl RPMI 1640 medium without serum in 96-well, sterile microtiter plates.
4. For each well of cells, appropriate amounts of the *N*-alkyl-PEI2k-IOs or 0.4 μl Lipofectamine 2000 (Invitrogen) is diluted into 25 μl RPMI 1640 medium without serum and incubated for 5 min at room temperature.
5. The transfection agents from Step 4 are added to each well containing diluted siRNA from Step 3, mixed gently, and incubated at room temperature for 20 min to allow IO–siRNA complexes to form.
6. The IO–siRNA complexes (50 μl) are added directly to each well of the plates containing cells and mixed gently.
7. The cells are incubated (37 °C, 5% CO_2) for 3 h to allow the siRNA transfection.
8. The transfection medium is removed and the cells are cultured in fresh RPMI 1640 medium supplemented with 10% FBS at 37 °C with 5% CO_2.
9. Forty-eight hours posttransfection, 20 μl D-luciferin solution (3 mg/ml) is added into each well, and the cells are imaged in the plate using a Xenogen IVIS-100 system.
10. The imaging data are analyzed with Living Image v.3.1 software.

3.4. Cytotoxicity assay

1. After the IVIS imaging experiment, the medium is replaced with fresh serum-enriched media, and 20 μl MTT solution (3-(4,5-dimethylthiazol-2-yl)-2,5-diphenyltetrazolium bromide, 5 mg/ml in PBS) is added into each well. The microplate is placed on a shaker (200 rpm for 5 min) to thoroughly mix the MTT into the media.

2. The cells are incubated (37 °C, 5% CO_2) for 4 h to allow the MTT to be metabolized.
3. The culture media are removed.
4. Formazan (the MTT metabolic product) is resuspended in 200 μl DMSO. The culture plate is placed on a shaker (200 rpm for 5 min) to thoroughly mix the formazan into the solvent.
5. The optical density is read at 560 nm and the background at 670 nm is subtracted.

3.5. MRI study of transfected cells

1. After transfection as described above, the cells are washed three times using PBS with heparin sodium (10 U/ml) and the transfected cells are harvested.
2. The cells (10^6) are suspended in 100 μl agarose gel (2%).
3. A T_2-weighted normal anatomic fast-spin echo MRI study is performed in a 7 T Bruker magnet (Bruker, Rheinstetten, Germany) with a 5-cm volume coil: TR = 4200 ms, TE = 36.0 ms, number of averages = 2, FOV = 30 × 30 mm, slice thickness = 1.0 mm.
4. The T_2-weighted imaging is analyzed using NIH ImageJ (Bethesda, MD, USA).

4. Notes

1. In terms of size data between DLS and AFM, larger diameters of the IOs determined by AFM may be caused by the AFM tip broadening effect and particle-flattening on the mica surface.
2. The N/P ratios can be calculated based on PEI nitrogen per nucleic acid phosphate (1 μg of siRNA has 3 nmol of phosphate and 0.9 μg PEI contains 10 nmol of amine nitrogen).
3. The optimal incubation period for a maximal siRNA effect will need to be tested. Unnecessarily long incubation times may result in decreased transfection activity resulting from degradation.
4. Incubating the cells for a short time at 37 °C before IVIS imaging can increase the signal.
5. Washing transfected cells with PBS with heparin sodium (10 U/ml) will help eliminate IOs from the extracellular surface.
6. Air can induce a hypointense signal on T_2-weighted sequences. To minimize any artifacts, the cells are placed gently into the agarose gel phantom to make sure no air is trapped.
7. MRI parameters should be optimized for different IO concentrations and magnetic field strengths.

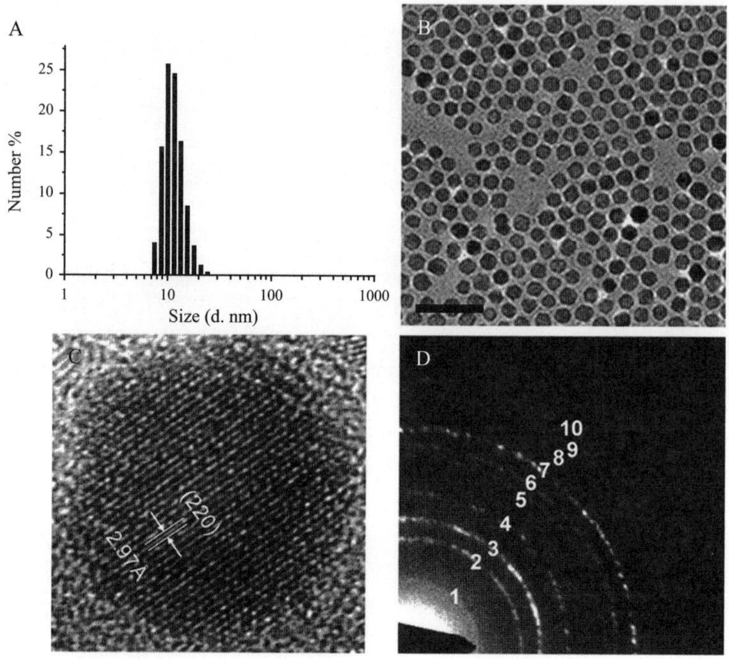

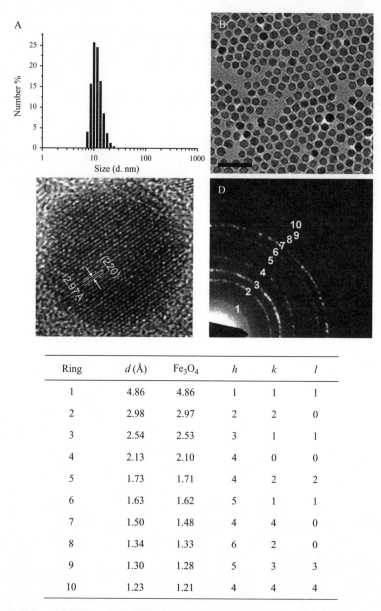

Ring	d (Å)	Fe_3O_4	h	k	l
1	4.86	4.86	1	1	1
2	2.98	2.97	2	2	0
3	2.54	2.53	3	1	1
4	2.13	2.10	4	0	0
5	1.73	1.71	4	2	2
6	1.63	1.62	5	1	1
7	1.50	1.48	4	4	0
8	1.34	1.33	6	2	0
9	1.30	1.28	5	3	3
10	1.23	1.21	4	4	4

Figure 13.1 (A) DLS and (B) TEM images of magnetite nanoparticles in hexane; (C) measured lattice spacing, d (Å), based on the rings in (D) and standard atomic spacing for Fe_3O_4 along with their respective Miller indexes from the PDF database (from Wang *et al.*, 2012).

5. Anticipated Results

The synthesized hydrophobic SPIO nanoparticles are monodisperse with a diameter of 12 nm (Fig. 13.1) (Wang *et al.*, 2012). The structure of the SPIO nanoparticles characterized by high-resolution TEM and selected area electron diffraction indicates that the distance between two adjacent planes is 2.97 Å, corresponding to (220) planes in the spinel-structured Fe_3O_4.

Alkyl-PEI2k can transfer hydrophobic IOs from organic solvent to aqueous phase and hold multiple IOs with a controlled clustering structure, leading to higher r_2 values (>300 Fe mM^{-1} s^{-1}) (Liu *et al.*, 2011b). Both TEM and AFM analyses show that the hydrophilic nanocomposites are well dispersed without obvious aggregation (Fig. 13.2), and TEM studies also show that each nanocomposite is a cluster of a few closely packed SPIO nanoparticles (Fig. 13.2D).

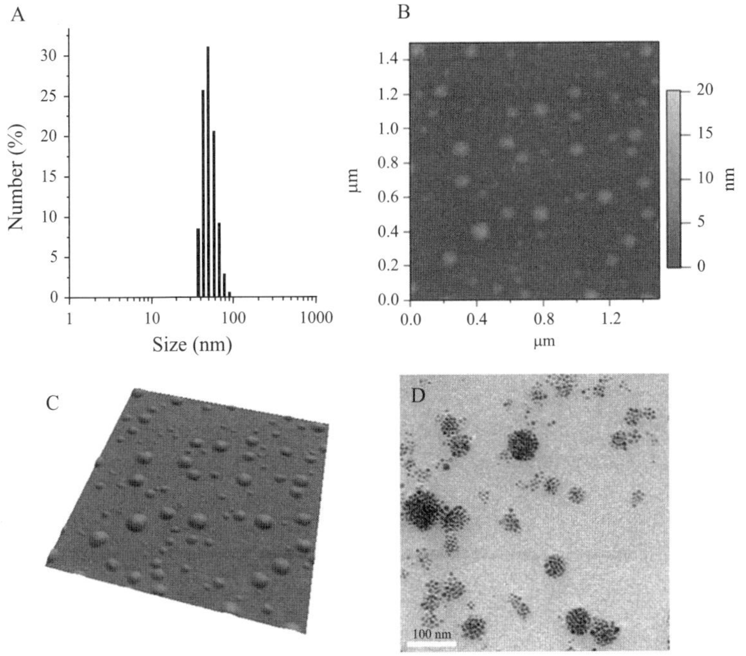

Figure 13.2 Physical characterization of *N*-alkyl-PEI2k-IOs. DLS (A), AFM height image (B), 3D AFM image (C) of the same area in (B) and TEM image (D) of SPIO nanoparticle clusters (from Liu *et al.*, 2011b). (For the color version of this figure, the reader is referred to the Web version of this chapter.)

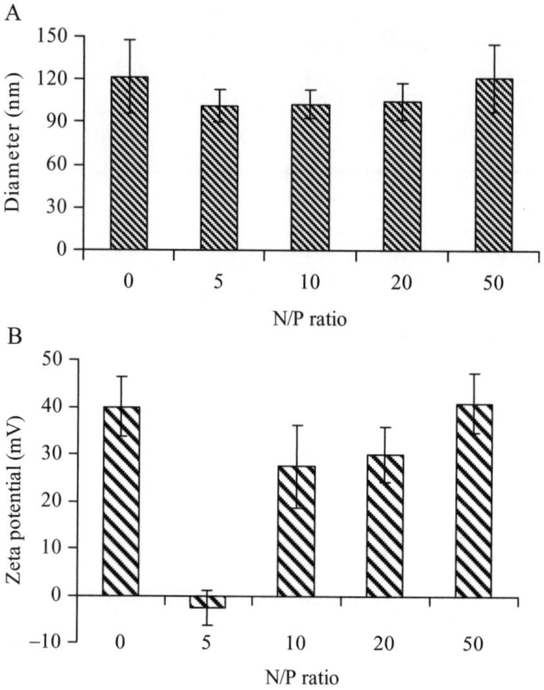

Figure 13.3 Physical characterization of *N*-alkyl-PEI2k-IO/siRNA complexes. (A) Average diameter and (B) zeta potential of *N*-alkyl-PEI2k-IO/siRNA complexes at various N/P ratios (from Liu *et al.*, 2011d).

The formed *N*-alkyl-PEI2k-IO/siRNA complexes in pure water demonstrate an average hydrodynamic diameter of about 100 nm with relatively narrow distribution (Fig. 13.3) (Liu *et al.*, 2011d). The zeta potentials of the complexes decrease when increasing the initial siRNA concentration (Fig. 13.3), suggesting successful loading of siRNA onto the nanoparticle surface.

The results of the retardation assay illustrate that the retardation efficiency increases with decreasing initial siRNA concentration (Fig. 13.4) (Liu *et al.*, 2011d). The naked siRNA can be found to be released from *N*-alkyl-PEI2k-IO/siRNA complexes when heparin is added, presumably due to the stronger association of heparin with the particle surface than that of the siRNA (Fig. 13.4). *N*-alkyl-PEI2k-IO/siRNA complexes are obviously more stable than naked siRNA in the presence of serum, as shown in Fig. 13.4.

N-alkyl-PEI2k-IO/lucsiRNA complexes can induce silencing of specific genes of interest (Fig. 13.5) (Liu *et al.*, 2011d). The improved gene-silencing effects of *N*-alkyl-PEI2k-IO/lucsiRNA complexes may be due to an increased intracellular delivery, as physically stable siRNA polyelectrolyte

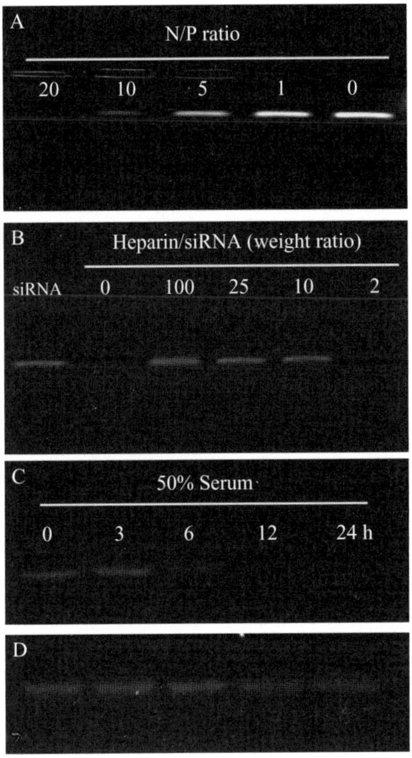

Figure 13.4 Agarose gel electrophoresis of *N*-alkyl-PEI2k-IO/siRNA complexes: (A) electrophoretic retardation analysis of siRNA binding with *N*-alkyl-PEI2k-IOs/siRNA samples; (B) release of siRNA with the addition of heparin at various concentrations; (C, D) serum stability of siRNA when complexed with *N*-alkyl-PEI2k-IOs at an N/P ratio of 20. The study was performed in 50% serum solution for a predetermined incubation time (from Liu *et al.*, 2011d).

complexes with a size around 100 nm are more readily internalized by cells through the endocytotic pathway. The *N*-alkyl-PEI2k-IO/siRNA complexes have no obvious cytotoxicity on 4T1-fluc cells under transfection conditions even at the highest concentration (25 μg Fe/ml) (Liu *et al.*, 2011d), which is fivefold higher than the concentration used in the cell transfection experiments, confirming that the gene silencing is a pure consequence of the RNAi effect.

At a 7 T magnetic field, the *N*-alkyl-PEI2k-IO/siRNA complex-transfected cells exhibit obviously decreased signal intensities on T_2-weighted images, and higher N/P ratios are associated with decreased signal intensities (Liu *et al.*, 2011d). The internalized nanoparticles can shorten the spin–spin relaxation time by dephasing the spins of neighboring water protons, resulting in hypointensities on T_2-weighted images.

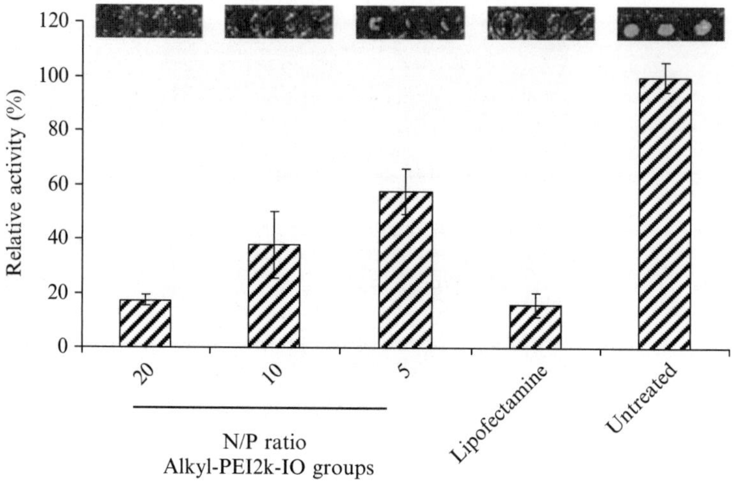

Figure 13.5 Inhibition of fluc gene expression by *N*-alkyl-PEI2k-IOs/siRNA (siRNA = 6 pmol) at various N/P ratios. (See Color Insert.)

6. Summary

N-alkyl-PEI2k-IOs are synthesized using *N*-alkyl-PEI2k as the phase-transfer material, which bind siRNA and result in well-dispersed nanoparticles with uniform structure and narrow size distribution. Such *N*-alkyl-PEI2k-IOs have high siRNA binding capability, protecting siRNA from enzymatic degradation for effective siRNA delivery. With siRNA loading, *N*-alkyl-PEI2k-IOs induce enhanced luciferase gene (fluc) silencing in fluc-4T1 cells in cell culture with good biocompatibility. Meanwhile, the transfected cells display strong signal contrast compared to untreated cells on T_2-weighted MR imaging. This multifunctional nanocomposite system shows great potential for gene delivery with noninvasive imaging monitoring capability.

Acknowledgments

This work was supported by the Intramural Research Program (IRP) of the National Institutes of Biomedical Imaging and Bioengineering (NIBIB), National Institutes of Health (NIH), and the International Cooperative Program of National Science Foundation of China (NSFC) (81028009). The work was also supported by Projects of Sichuan Province (2011JQ0032, 2010SZ0294, and 09ZA036) and National Natural Science Foundation of China (20974065, 50603015, 51173117, and 81101101).

REFERENCES

Arsianti, M., Lim, M., Marquis, C. P., and Amal, R. (2010). Assembly of polyethylenimine-based magnetic iron oxide vectors: Insights into gene delivery. *Langmuir* **26,** 7314–7326.

Boussif, O., Lezoualc'h, F., Zanta, M. A., Mergny, M. D., Scherman, D., Demeneix, B., and Behr, J. P. (1995). A versatile vector for gene and oligonucleotide transfer into cells in culture and in vivo: Polyethylenimine. *Proc. Natl. Acad. Sci. USA* **92,** 7297–7301.

Cheong, S. J., Lee, C. M., Kim, S. L., Jeong, H. J., Kim, E. M., Park, E. H., Kim, D. W., Lim, S. T., and Sohn, M. H. (2009). Superparamagnetic iron oxide nanoparticles-loaded chitosan-linoleic acid nanoparticles as an effective hepatocyte-targeted gene delivery system. *Int. J. Pharm.* **372,** 169–176.

Demeneix, B., and Behr, J. P. (2005). Polyethylenimine (PEI). *Adv. Genet.* **53,** 217–230.

Godbey, W. T., Wu, K. K., and Mikos, A. G. (1999). Poly(ethylenimine) and its role in gene delivery. *J. Control. Release* **60,** 149–160.

Hao, R., Xing, R., Xu, Z., Hou, Y., Gao, S., and Sun, S. (2010). Synthesis, functionalization, and biomedical applications of multifunctional magnetic nanoparticles. *Adv. Mater.* **22,** 2729–2742.

Hu, C., Peng, Q., Chen, F., Zhong, Z., and Zhuo, R. (2010). Low molecular weight polyethylenimine conjugated gold nanoparticles as efficient gene vectors. *Bioconjug. Chem.* **21,** 836–843.

Huang, H., Yu, H., Tang, G., Wang, Q., and Li, J. (2010). Low molecular weight polyethylenimine cross-linked by 2-hydroxypropyl-gamma-cyclodextrin coupled to peptide targeting HER2 as a gene delivery vector. *Biomaterials* **31,** 1830–1838.

Kievit, F. M., Veiseh, O., Bhattarai, N., Fang, C., Gunn, J. W., Lee, D., Ellenbogen, R. G., Olson, J. M., and Zhang, M. (2009). PEI-PEG-Chitosan copolymer coated iron oxide nanoparticles for safe gene delivery: Synthesis, complexation, and transfection. *Adv. Funct. Mater.* **19,** 2244–2251.

Liu, G., Swierczewska, M., Lee, S., and Chen, X. (2010a). Functional nanoparticles for molecular imaging guided gene delivery. *Nano Today* **5,** 524–539.

Liu, G., Swierczewska, M., Niu, G., Zhang, X., and Chen, X. (2011a). Molecular imaging of cell-based cancer immunotherapy. *Mol. Biosyst.* **7,** 993–1003.

Liu, G., Tian, J., Liu, C., Ai, H., Gu, Z., Gou, J., and Mo, X. (2009). Cell labeling efficiency of layer-by-layer self-assembly modified silica nanoparticles. *J. Mater. Res.* **24,** 1317–1321.

Liu, G., Wang, Z., Lu, J., Xia, C., Gao, F., Gong, Q., Song, B., Zhao, X., Shuai, X., Chen, X., Ai, H., and Gu, Z. (2011b). Low molecular weight alkyl-polycation wrapped magnetite nanoparticle clusters as MRI probes for stem cell labeling and in vivo imaging. *Biomaterials* **32,** 528–537.

Liu, G., Xia, C., Wang, Z., Lv, F., Gao, F., Gong, Q., Song, B., Ai, H., and Gu, Z. (2011c). Magnetic resonance imaging probes for labeling of chondrocyte cells. *J. Mater. Sci. Mater. Med.* **22,** 601–606.

Liu, G., Xie, J., Zhang, F., Wang, Z., Luo, K., Zhu, L., Quan, Q., Niu, G., Lee, S., Ai, H., and Chen, X. (2011d). N-alkyl-PEI functional iron oxide nanocluster for efficient siRNA delivery. *Small* **7,** 2742–2749.

Liu, G., Yang, H., Zhang, X. M., Shao, Y., and Jiang, H. (2010b). MR imaging for the longevity of mesenchymal stem cells labeled with poly-L-lysine-resovist complexes. *Contrast Media Mol. Imaging* **5,** 53–58.

Scherer, F., Anton, M., Schillinger, U., Henke, J., Bergemann, C., Kruger, A., Gansbacher, B., and Plank, C. (2002). Magnetofection: Enhancing and targeting gene delivery by magnetic force in vitro and in vivo. *Gene Ther.* **9,** 102–109.

Taratula, O., Garbuzenko, O., Savla, R., Wang, Y. A., He, H., and Minko, T. (2011). Multifunctional nanomedicine platform for cancer specific delivery of siRNA by superparamagnetic iron oxide nanoparticles-dendrimer complexes. *Curr. Drug Deliv.* **8,** 59–69.

Thomas, M., and Klibanov, A. M. (2003). Conjugation to gold nanoparticles enhances polyethylenimine's transfer of plasmid DNA into mammalian cells. *Proc. Natl. Acad. Sci. USA* **100,** 9138–9143.

Veiseh, O., Kievit, F. M., Fang, C., Mu, N., Jana, S., Leung, M. C., Mok, H., Ellenbogen, R. G., Park, J. O., and Zhang, M. (2010). Chlorotoxin bound magnetic nanovector tailored for cancer cell targeting, imaging, and siRNA delivery. *Biomaterials* **31,** 8032–8042.

Wang, Z., Liu, G., Sun, J., Gong, Q., Song, B., Sun, S., Ai, H., and Gu, Z. (2012). N-alkyl-polyethylenimine stabilized iron oxide nanoparticles as MRI visible transfection agents. *J. Nanosci. Nanotechnol.* 10.1166/jnn.2011.5151.

Wang, Z., Liu, G., Sun, J., Wu, B., Gong, Q., Song, B., Ai, H., and Gu, Z. (2009). Self-assembly of magnetite nanocrystals with amphiphilic polyethylenimine: Structures and applications in magnetic resonance imaging. *J. Nanosci. Nanotechnol.* **9,** 378–385.

Xie, J., Liu, G., Eden, H. S., Ai, H., and Chen, X. (2011). Surface-engineered magnetic nanoparticle platforms for cancer imaging and therapy. *Acc. Chem. Res.* **44,** 883–892.

CHAPTER FOURTEEN

Cell-Penetrating Peptide-Based Systems for Nucleic Acid Delivery: A Biological and Biophysical Approach

Sara Trabulo,* Ana L. Cardoso,* Ana M. S. Cardoso,* Nejat Düzgüneş,† Amália S. Jurado,*,‡ *and* Maria C. Pedroso de Lima*,‡

Contents

* CNC-Center for Neuroscience and Cell Biology, University of Coimbra, Coimbra, Portugal
† Department of Biomedical Sciences, Arthur A. Dugoni School of Dentistry, University of the Pacific, San Francisco, California, USA
‡ Department of Life Sciences, University of Coimbra, Coimbra, Portugal

Methods in Enzymology, Volume 509
ISSN 0076-6879, DOI: 10.1016/B978-0-12-391858-1.00014-9

Abstract

The increasing knowledge on the genetic basis of disease provides a platform for the development of promising gene-targeted therapies that can be applied to numerous pathological conditions, including cancer. Such genetic-based approaches involve the use of nucleic acids as therapeutic agents, either for the insertion or for the repair and regulation of specific genes. However, despite the huge pharmacological potential of these molecules, their application remains highly dependent on the development of delivery systems capable of mediating efficient cellular uptake. The discovery of a class of small peptides, the so-called cell-penetrating peptides (CPPs), which are able to very efficiently cross cell membranes through a mechanism that is independent of membrane receptors or transporters and avoids lysosomal enzymatic degradation, has been enthusiastically considered of key interest to improve noninvasive cellular delivery of therapeutic molecules. A large number of CPPs have been applied successfully to mediate the intracellular delivery of nucleic acids, including the $S4_{13}PV$ peptide for which interactions with membranes and resulting biological effects are illustrated in this chapter. Here, we provide a description of the experimental procedures for the preparation of CPP-based nucleic acid complexes and assessment of their formation, the selection of those protocols leading to the most efficient complexes, the biophysical characterization of CPP membrane interactions, and the evaluation of the biological and cytotoxic activity of the complexes.

1. Introduction

Since the first studies describing the ability of a class of small peptides to be translocated across biological membranes in a very efficient, nontoxic process, apparently independent of membrane receptors and energy consumption, cell-penetrating peptides (CPPs), also often referred to as protein transduction domains (PTDs), have been used extensively as nucleic acid delivery systems (Frankel and Pabo, 1988; Green and Loewenstein, 1988; Lindgren *et al.*, 2000). Regardless of the great variability in their amino acid sequence, CPPs are usually short-peptide sequences rich in basic amino acids (lysine and arginine), in some cases exhibiting the ability to be arranged in amphipathic alpha-helical structures. Among all CPPs described to date, which include peptides derived from PTDs, chimeric peptides and peptides of synthetic origin, those derived from the HIV-1 Tat protein (Frankel and Pabo, 1988; Green and Loewenstein, 1988), and from the homeodomain of the Antennapedia protein of *Drosophila* (Derossi *et al.*, 1994; Joliot *et al.*, 1991), Tat and Penetratin peptides, respectively, as well as the synthetic Pep-1 peptide (Morris *et al.*, 2001), the model amphipathic peptide (Oehlke *et al.*, 1998), and transportan (Pooga *et al.*, 1998) are the best characterized. These peptides have been used successfully for the intracellular delivery of different cargoes, including nanoparticles

(Zhao *et al.*, 2002), full-length proteins (Morris *et al.*, 2001; Schwarze *et al.*, 1999), liposomes (Khalil *et al.*, 2007; Kogure *et al.*, 2004, 2007), and nucleic acids (Kumar *et al.*, 2007; Meade and Dowdy, 2007; Morris *et al.*, 1997), both *in vitro* and *in vivo*. Some CPPs, such as Tat (Torchilin, 2008), transportan (Kilk *et al.*, 2005), $S4_{13}PV$ (Trabulo *et al.*, 2008, 2010), and polyarginine peptides (Khalil *et al.*, 2007; Kogure *et al.*, 2004, 2007; Zhang *et al.*, 2006), have been associated with other nonviral vectors. This strategy was shown to improve nucleic acid delivery, while offering the possibility of combining efficient packaging, delivery, and targeting in a single nanocarrier.

Although CPPs constitute one of the most promising tools for delivering nucleic acids into cells, the exact mechanisms underlying their cellular uptake, and that of their conjugates, remain far from being fully understood and are still the object of some controversy. Understanding of such mechanisms should reveal important clues about how to optimize the use of CPPs for delivery purposes, thus contributing to the generation of novel and improved nucleic acid delivery systems.

Several endocytotic and nonendocytotic pathways for CPP internalization have been proposed, depending on the features of the CPP, its associated cargo, the target cell type, and the membrane lipid composition (Lindgren *et al.*, 2000). In this context, the interactions between CPPs and membrane lipids have been suggested to play a major role in CPP membrane translocation (Thoren *et al.*, 2004). Therefore, the understanding of the role of peptide–lipid interactions in the mechanisms of CPP membrane translocation may help the rational design of more efficient CPP-based drug delivery systems.

It is also known that several physicochemical parameters, including size, charge density, and colloidal stability, can affect the biological activity of drug delivery systems. These parameters influence the ability of CPPs to complex with and protect nucleic acids, the ability of the CPP/nucleic acid complexes to bind and be internalized by target cells and, ultimately, the transfection efficiency of these systems.

In this chapter, we describe experimental procedures to characterize CPP-based nucleic acid systems, from both a biological and a biophysical perspective, and illustrate their application in the delivery of nucleic acids with some typical results.

2. Methods for Preparation of CPP-Based Nucleic Acid Complexes

The considerable interest that CPPs have evoked among the scientific community has triggered the synthesis of a large variety of CPPs, the development of diverse methods for the preparation of CPP-based delivery

Table 14.1 Sequences of peptides that can be used for nucleic acid delivery

Penetratin	RQIKIWFQNRRMKWKK
Pep-1	KETWWETWWTEWSQPKKKRKV
Transportan	GWTLNSAGYL LGKINLKALAALAKK IL
Polyarginine (R8)	RRRRRRRR
Tat	GRKKRRQRRRPP
$S4_{13}$-PV	ALWKTLLKKVLKAPKKKRKVC
Reverse NLS	ALWKTLLKKVLKAVKRKKKPC
Scrambled $S4_{13}$-PV	KTLKVAKWLKKAKPLRKLVKC

systems. The $S4_{13}$PV, Penetratin, Pep-1, Polyarginines (e.g., R8), Transportan, and Tat peptides (Table 14.1) are examples of CPPs that can be used to generate nucleic acid delivery systems.

The design of control peptides of each CPP under study constitutes an important step when investigating CPP internalization mechanisms and transfection. Table 14.1 includes the sequences of the reverse NLS (nuclear localization signal) and scrambled $S4_{13}$PV peptides that have been used as controls for examining the structural requirements of the $S4_{13}$PV peptide regarding internalization and transfection activity (Mano *et al.*, 2005, 2006; Trabulo *et al.*, 2008, 2010). The scrambled peptide was generated on the basis of the $S4_{13}$PV peptide sequence, by randomization of the sequential position of amino acid residues. In the reverse NLS peptide, the sequence corresponding to the nuclear localization signal of the SV40 large T antigen (aa 14–20) is inverted. Thus, both resulting peptides have the same amino acid composition and overall charge, but a distinct primary sequence (Table 14.1).

The concentration of the CPPs can be determined by amino acid analysis and light absorption at 280 nm. Amino acid analysis is usually performed in a Beckman 6300 automatic analyzer, following acid hydrolysis of the peptide.

CPP-based nucleic acid complexes are usually prepared through noncovalent association of nucleic acids with CPPs. Although conjugation offers some advantages for *in vivo* applications, including reproducibility of the procedure and control of the stoichiometry of the CPP–cargo conjugates (Heitz *et al.*, 2009), this strategy also has some drawbacks, such as the possibility to compromise the biological activity of the cargo (Heitz *et al.*, 2009) and the need to generate and test a new construct for any given nucleic acid cargo. Noncovalent strategies, therefore, appear to be more promising, particularly in the case of negatively charged nucleic acids, which can readily interact with positively charged CPPs.

CPP-based nucleic acid complexes can be prepared from the simple association of the CPP and nucleic acids (binary complexes) or in combination with other delivery systems, such as cationic liposomes, originating ternary complexes (Trabulo *et al.*, 2008, 2010). The experimental procedures for the preparation of cationic liposomes and CPP-based nucleic acid binary or ternary complexes are outlined below.

2.1. Cationic liposomes

Cationic liposomes can be purchased from several companies or prepared in the laboratory. Cationic liposome formulations for nucleic acid delivery can be easily prepared from a mixture of a cationic lipid (e.g., 1,2-dioleoyl-3-trimethylammonium-propane—DOTAP or dioctadecylamidoglycylspermidine—DOGS) with a "helper" lipid (cholesterol, CHOL or dioleoylphosphatidylethanolamine, DOPE), obtained from stock solutions in chloroform, which is dried under nitrogen to obtain a thin lipid film. The lipid film can be dissolved in ultrapure ethanol and the resulting ethanol solution then injected into HEPES-buffered saline (HBS; 140 m*M* NaCl, 10 m*M* HEPES, pH 7.4), or, alternatively, the lipid film can be hydrated, and the resulting liposomes are sonicated briefly and then extruded through polycarbonate membranes. Both procedures for preparing cationic liposomes have been described in more detail in a previous volume of Methods in Enzymology (Cardoso *et al.*, 2009). While the first procedure allows the rapid preparation of small liposomes (120–150 nm) that can be used for *in vitro* transfection assays, the second is particularly useful to generate highly concentrated lipid formulations, as required for *in vivo* studies, since it enables the production of liposomes with a homogenous size distribution, even when the lipid film is hydrated with very small volumes of water or dextrose solution.

It is expected that a small amount of lipid will be lost after liposome manipulations (e.g., during the extrusion); then the lipid quantification of the final liposome preparation is necessary. Since DOTAP and other cationic lipids do not contain phosphate groups for lipid quantification by a spectroscopic methodology (Bartlett, 1959), total lipid concentration may be estimated by measuring the inorganic phosphate provided by the colipid (in the case of phospholipids, such as DOPE) after hydrolysis by a standard method. When cholesterol is included in the liposomes as a colipid, quantification may be performed using the Infinity™ Cholesterol Reagent (Thermo Fisher Scientific). From the known relative proportion of cationic lipid to colipid, the total lipid concentration can be determined easily. Typically, 70–80% of the initial amount of lipid is recovered.

Liposomes should be stored at 4 °C under nitrogen or argon and used within 3 weeks after preparation to avoid aggregation.

2.2. CPP-based nucleic acid complexes

The CPP/nucleic acid binary complexes are prepared by mixing the nucleic acids in a buffered saline solution (e.g., HBS) with different amounts of peptide (depending on the desired CPP/nucleic acid charge ratio), followed by incubation for 15–30 min at room temperature to allow the formation of the complexes. These complexes should be used immediately after their preparation. Complexes of nucleic acids with Lipofectamine 2000 (Invitrogen, Carlsbad, CA, USA) or other commercial reagents, prepared according to the instructions of the manufacturer, are used as a positive control in experiments aiming at assessing transfection efficiency.

In the case of ternary complexes involving the combination of binary complexes with cationic liposomes, the order of addition of the components influences significantly the biological activity of the resulting complexes. However, the protocol for obtaining optimal ternary complexes depends on the type of nucleic acid molecules used for preparing the complexes. (i) CPPs can be added to the nucleic acids, followed by the addition of the cationic liposomes. Alternatively, (ii) CPPs can be mixed first with the cationic liposomes, followed by the addition of the nucleic acids, or (iii) the nucleic acids can be mixed with the cationic liposomes, followed by the addition of the CPPs.

According to the results obtained in our laboratory with the $S4_{13}PV$ peptide on the delivery of plasmid DNA, the most efficient ternary complexes are those prepared by adding the cationic liposomes to CPP/plasmid DNA binary complexes (Trabulo *et al.*, 2008). On the other hand, the protocol that results in the most efficient formulation for the delivery of splice-switching oligonucleotides (SSOs) involves the addition of the CPP after the precomplexation of the cationic liposomes with the SSOs (Trabulo *et al.*, 2010). Finally, for siRNA delivery, the most favorable protocol consists of the addition of the siRNA molecules to preformed cationic liposome/CPP complexes (personal communication). After the addition of each component, the mixtures are mixed gently and incubated for 15 or 30 min (for plasmid DNA or siRNAs and oligonucleotides (ONs), respectively) at room temperature to allow the formation of the complexes with a variety of charge ratios.

Since the efficiency of the ternary complexes changes according to the type of nucleic acids, and possibly, the different CPPs, the influence of the order of addition of the components on the efficiency of the resulting formulation should be checked when optimizing a ternary CPP-based nucleic acid delivery system.

2.3. Assessment of complex formation

Complexation of nucleic acids by the CPPs can be evaluated by monitoring the electrophoretic mobility of the complexes (binary or ternary), prepared at increasing CPP/nucleic acid charge ratios, as described above. The different complexes can be separated on a 1% agarose gel in Tris–acetate buffer (0.5 μg of nucleic acids per lane). The agarose gel is stained with ethidium bromide to allow the visualization of uncomplexed nucleic acids.

Typical results for the complexation of plasmid DNA with the $S4_{13}PV$ peptide at increasing peptide/DNA charge ratios are shown in Fig. 14.1.

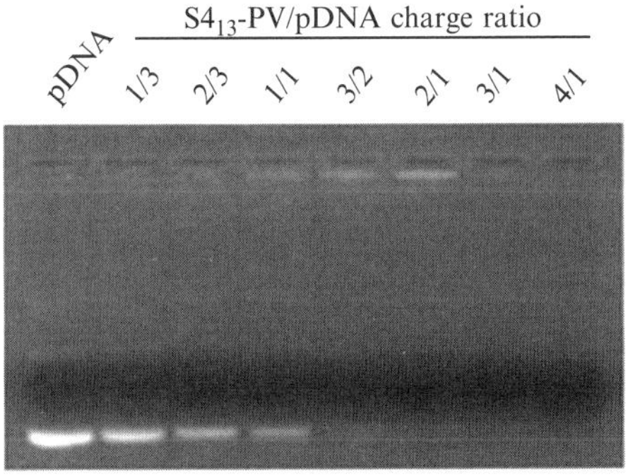

Figure 14.1 Complexation of plasmid DNA by the S4$_{13}$PV peptide. Peptide/DNA complexes were prepared by mixing plasmid DNA with increasing amounts of the peptide, at room temperature for 15 min. Formation of peptide/plasmid DNA complexes was monitored by evaluating the electrophoretic mobility of the different mixtures in a 1% agarose gel (0.5 μg of plasmid DNA per lane was used). Free plasmid DNA is shown for comparison.

Complexes prepared at charge ratios higher than 3:2 did not migrate in the agarose gel, indicating a total neutralization of the DNA negative charges by the peptide. Similar results were obtained for the reverse NLS or scrambled peptides, demonstrating that the S4$_{13}$PV and its analogs are able to form complexes through electrostatic interactions with plasmid DNA, and that an excess of peptide positive charge is required for complete complexation of DNA.

3. Methods for Physical Characterization of CPP-Based Nucleic Acid Complexes and Their Interactions with Membranes

The success in nucleic acid delivery mediated by CPPs is in large part determined by the physical properties, namely, size and surface charge density, of the resulting CPP/nucleic acid complexes. The procedures currently used for this physical characterization can be found in a previous volume of Methods in Enzymology (Cardoso *et al.*, 2009). The interactions of the nanocarriers with membranes constitute another key element for successful nucleic acid delivery. In this context, the study of CPP–lipid interactions is of crucial importance from two perspectives: first, addressing the physical characteristics of nucleic acid complexes prepared from formulations that include liposomes in addition to a CPP; second, unraveling how the physical properties of the membrane lipid bilayer are affected upon interaction with the CPP. Some techniques that can be used toward a biophysical approach of CPP–lipid interactions, requiring neither very specific equipment nor specialized skills, are described below.

3.1. Differential scanning calorimetry assay

Differential scanning calorimetry (DSC) is a very sensitive technique to address the effects of CPP–lipid interactions from the perspective of the lipids. The first requirement for the use of this technique is that the lipids or lipid mixtures undergo a phase transition (or transitions) within the temperature range of operation of the calorimeter and at temperatures distinct from those at which phase transitions for other elements of the mixture (CPP or nucleic acids in the case under study) occur. When the focus is on the effect of CPP on its nucleic acid complexes including the lipids in the formulation, the method of preparation of the complexes should be reproduced. When the objective is to mimic the interaction of CPPs with membranes, the addition of the CPP to preformed liposomes of a specific composition assumes a higher biological relevance. Although large unilamellar vesicles (LUV) have been preferred by several authors as a more realistic model of a membrane lipid bilayer, the multilamellar vesicles (MLV) continue to be the choice of many researchers, due to the advantage of showing much more cooperative transitions (Alves *et al.*, 2008). The liposomes (MLV) are prepared as described above, and the hydration and lipid dispersion by shaking and vortexing should be performed at 7–10 °C above the transition temperature of the lipids. The selected lipid concentration of DSC samples depends on the sensitivity of the calorimeter. In the studies performed in our laboratory with the $S4_{13}PV$, using a Perkin-Elmer Pyris 1 differential scanning calorimeter, liposomal samples are prepared at 75 m*M* in phospholipid. Liposomes are then incubated with different concentrations of the peptide at a temperature above the phase transition and preferentially overnight to achieve a dynamical equilibrium. Afterward, the suspensions are centrifuged at 148,000 × *g* for 45 min at 4 °C to obtain a pellet, this being then carefully transferred into a 50 μl capacity aluminum pan, which is subsequently sealed hermetically by means of a crimper. A void pan or a pan with an adequate volume of buffer is used as reference. When a microcalorimeter system (MCS DSC, MicroCal Inc., North Hampton, MA, USA) is used for DSC measurements, relatively diluted liposome suspensions in buffer can be used and are transferred to the DSC sample holder after being degassed. For data acquisition and subsequent analysis of thermograms, the calorimeter software is normally used. In the DSC profiles, which register the excess of heat capacity (Cp) of the system as a function of temperature, different characteristic temperatures can be defined (Fig. 14.2) after constructing the interpolated baseline (the line connecting the pre- and posttransition baselines): the initial peak temperature, the extrapolated peak onset temperature, the peak maximum temperature,

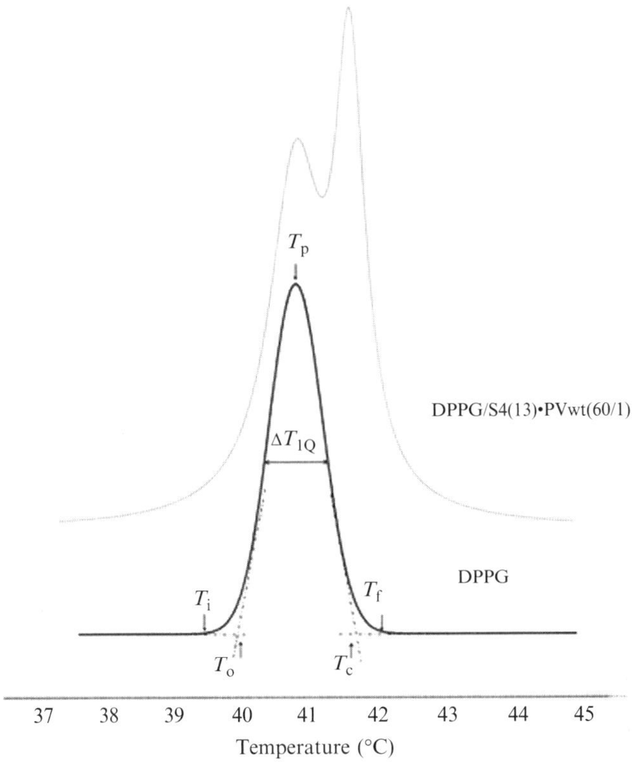

Figure 14.2 DSC profiles (heating scans) of DPPG bilayers in the absence (—) and presence (—) of the $S4_{13}PV$ peptide at a lipid/peptide molar ratio of 60/1. In the DPPG DSC profile, characteristic temperatures are indicated: T_i, when DSC record first deviates from the base line; T_o, determined by extrapolating to the baseline a tangent to the ascending slope of the endothermic peak; T_p, the peak maximum; T_c, determined by extrapolating to the baseline a tangent to the descendent slope of the endothermic peak; and T_f, obtained when the curve reaches again the baseline. $\Delta T_{1/2}$ (peak width at half maximum height) is also indicated.

the extrapolated peak completion temperature, and the final peak temperature (T_i, T_o, T_p, T_c, and T_f, respectively; see Fig. 14.2 for their determination). Heating–cooling cycles over an appropriate temperature range are then obtained at previously selected scanning rates. The DSC operator should be aware that the heating rate is one important parameter that determines the position of the peak maximum (T_p) and the slope of the ascending part of the peak (Hohne *et al.*, 1996). By contrast, the extrapolated peak onset temperature (T_o) defined for a given sample does not vary significantly under different experimental conditions, this being the reason why this temperature is often preferred to characterize lipid phase transitions. To check the reproducibility of the data, at least three scans for each sample should be recorded. For defining the range of the phase transition or lateral phase separation, the difference $T_f - T_i$ or $T_c - T_o$ can be used. If, however, we intend to define the sharpness of the phase transition, the temperature width at half height ($\Delta T_{1/2}$) is normally measured (Fig. 14.2). Values of $\Delta T_{1/2} < 0.1$ °C are obtained with pure lipid systems, but in the case of lipid mixtures such as lipid extracts of some native membranes (Jurado *et al.*, 1991a), values as much as 15–20 °C can be registered. The software also provides the enthalpy change (ΔH) of the transition, which is proportional to the area of the endothermic peak and is usually expressed in kcal/mol or kJ/mol. Since the amount of lipid transferred to the pan can vary, the calorimetric enthalpy change provided by the software of the calorimeter, which is calculated for a preestablished amount of lipid (typically, 1 mg), should be normalized for the exact phospholipid content in each pan. For this purpose, the pans are carefully opened at the end of the DSC assay and the lipid is dissolved in chloroform/methanol (3:1) and dried with N_2 flow. The phospholipid content is determined by measuring the inorganic phosphate, by any adequate methodology (see e.g., Düzgüneş, 2003; Jurado *et al.*, 1991b). Since the change in free energy (ΔG) in a system at the phase transition midpoint temperature equals zero, the entropy change of the transition can be calculated from the equation: $\Delta S = \Delta H / T_m$, where T_m represents the temperature at which the phase transition is half complete and corresponds to the temperature at which the excess specific heat reaches a maximum when the peak is perfectly symmetric. The width of the transition, $\Delta T_{1/2}$, has also a significant physical meaning. The van't Hoff enthalpy, ΔH_{vH}, is related with the T_m and $\Delta T_{1/2}$ values, being approximately determined from the equation: $\Delta H_{vH} \approx 6.9\ T_m^2 / \Delta T_{1/2}$ (Cherry, 2005). The ratio between ΔH_{vH} and the calorimetrically determined enthalpy of the transition (ΔH) defines the size of the cooperative unit (SCU) in molecules, that is, $\Delta H_{vH} / \Delta H = \mathrm{SCU}$ (Ruthaven *et al.*, 2005). Therefore, for a highly cooperative phase transition (e.g., the phase transition of a highly pure phospholipid), SCU approaches infinity and the transition is very sharp; on the contrary, when the transition is noncooperative, SCU approaches unity and the transition is very broad.

The information that can be obtained with this type of assays with respect to CPP–lipid interactions depends on the membrane models employed. Attention should be paid to the following: shifts of the temperature ranges of endothermic events assigned to pretransitions (between two gel phases, $L_{\beta'}$ and $P_{\beta'}$ phases) and main phase transitions (from a gel to a liquid-crystalline phase, L_α) or transitions between a fluid lamellar phase and an inverted Hexagonal phase; broadening ($\Delta T_{1/2}$) of the peaks; occurrence of thermal anomalies (the appearance of new peaks or shoulders); and alterations of ΔH. On the basis of an accurate interpretation of these data, an extensive characterization of CPP–lipid interactions in terms of the consequences for the lipid physical behavior can be performed. Therefore, a decrease or increase of the enthalpy change of the lipid main phase transition is often associated to peptide-induced disruption or enhancement of the packing of lipid acyl chains, respectively. However, a reduction of the transition enthalpy in the presence of peptides may also indicate that a fraction of lipid molecules is somehow "sequestered" by tight interaction with the peptide, being inhibited to participate in the bulk phase transition. The extent of the perturbation should be related with the structural characteristics of CPP (e.g., the presence of charge and the length of the hydrophobic transmembrane segment) and the composition of the host bilayer (e.g., regarding the charge of lipid head groups and the length or structure of the lipid acyl chains), since lipid thermotropic properties are influenced by the occurrence of electrostatic interactions or hydrophobic mismatch (Morein *et al.*, 2000). The pretransition of some saturated phospholipids like PC or L-α-phosphatidylglycerol (PG) is very sensitive to the presence of impurities, and several peptides (e.g., penetratin, melittin, and δ-lisin) promote a decrease of the enthalpy or even abolishment of this phase transition (Alves *et al.*, 2008). Shifts of the phase transition temperatures or split of the endotherm is often associated to perturbations at the level of the lateral distribution of lipid molecules and their interactions at the plane of the bilayer. This type of effects is illustrated in Fig. 14.2, showing the alterations induced by the incorporation of the $S4_{13}PV$ peptide in DPPG liposomes, whose typical DSC profile shows a relatively symmetric peak, centered at 41 °C, and with a T_o at 40 °C. The splitting of the endothermic peak in the presence of the peptide reflects the induction of a lateral phase separation. The lower-temperature component, centered at 41 °C, might correspond to nonperturbed lipid domains with low concentration of the peptide, whereas the component shifted to higher temperatures might represent the peptide-enriched lipid domains, where lipid intermolecular repulsions (due to the negatively charged head groups) are attenuated by the interaction with the cationic peptide.

DSC also allows for monitoring the propensity of the lipid system to adopt nonlamellar phases, which could be favored or, by contrast, counteracted by the presence of peptides. However, one should be aware that the transition from a L_α phase to a H_{II} phase displays a very low enthalpy change, hence requiring high lipid concentrations or high sensitivity calorimeters to be detected. More adequate techniques for this purpose are the ^{31}P NMR or X-ray diffraction, although requiring more sophisticated equipment, which might not be available in cell biology or nanomedicine laboratories.

Although DSC provides important information to study peptide–lipid interactions, it should be complemented with other biophysical approaches. Fluorescence anisotropy represents an interesting possibility, merely requiring a spectrofluorimeter equipped with two polarizers and adequate fluorescent probes, as described below. This technique has the advantage over

DSC of allowing gain insights into the lipid molecular organization and dynamics at different depths of the lipid bilayer, as affected by CPP interaction.

3.2. Fluorescence anisotropy assay

Fluorescence anisotropy is a very sensitive and reproducible technique that provides a spectroscopic parameter that can be interpreted in terms of membrane fluidity. By using a series of different fluorescent probes, this technique offers the possibility of detecting alterations in membrane physical properties at different depths of the lipid bilayer. Several fluorescent probes can be used efficiently for this purpose, including 1,6-diphenyl-1,3,5-hexatriene (DPH) and its anionic and cationic analogs 3-[*p*-(6-phenyl)-1,3,5-hexatrienyl]phenylpropionic acid (DPH-PA) and 1-(4-trimethylaminophenyl)-6-phenyl-1,3,5-hexatriene (DPH-TMA), respectively, which can be purchased from Sigma-Aldrich (St. Louis, MO). These probes have been extensively used in membrane studies (Cardoso *et al.*, 2011; Guillen *et al.*, 2008; Rosa *et al.*, 2000; Zoonens *et al.*, 2008) because of their advantageous spectroscopic properties, such as high molar absorption coefficient ($\varepsilon_{350\text{nm}} \approx 80{,}000\ M^{-1}\ \text{cm}^{-1}$), high quantum yield in hydrophobic environments (0.8 in hexane, at 25 °C), and negligible emission in aqueous media, as well as absorption and emission spectra at distinct wavelength ranges, preventing energy transfer between DPH molecules (Shinitzky and Barenholz, 1974). While DPH embeds into the lipid bilayer, being deeply buried in the hydrocarbon core due to its high hydrophobicity (Andrich and Vanderkooi, 1976), its charged derivatives are anchored at the lipid–water interface, with the DPH moiety intercalated between the upper portions of the fatty acyl chains (Prendergast *et al.*, 1981; Trotter and Storch, 1989).

Steady-state fluorescence anisotropy measurements of membrane fluidity are based on the principle of photoselectivity (Gennis, 1989). When a suspension of liposomes labeled with a fluorescent probe such as DPH is exposed to light of a convenient wavelength (340–360 nm) polarized along the z-axis, the probe molecules preferentially excited are those whose transition dipole moments are also oriented in the vertical direction. The emission at an adequate wavelength (450 nm for the probes mentioned above) is monitored under two different conditions: when the incident and the emitted light are both polarized along the z-axis (I_{II} component) and when the emitted light is polarized in a perpendicular direction with respect to the exciting radiation ($I_{\perp}$). The values of the components I_{II} and $I_{\perp}$ are used to obtain the anisotropy (r) according to Eq. (14.1), where G is the grating correction factor for the optical system, obtained from the ratio of the perpendicular and the parallel polarized emission components when the excitation light is polarized in the horizontal plane (Chen *et al.*, 1977).

$$r = \frac{I_{II} - GI_{\perp}}{I_{II} + 2GI_{\perp}} \tag{14.1}$$

The value of r reported by a fluorescent probe embedded into a membrane indicates the higher or lower constraints to the rotational diffusion and reorientation of probe molecules, which, in turn, reflect the lower or higher free volume existent in the region where the probe resides within the membrane. Therefore, indirectly, r describes the freedom degree and level of mobility (range and rate of motion) of phospholipid hydrocarbon chains. In order to distinguish information regarding dynamics and range of motion, measurements of time-dependent anisotropy $r(t)$ following a flash of exciting light will be required. This technique, however, involves more sophisticated equipment; thus, most fluorescence polarization studies on CPP–lipid interactions are made in the steady state with continuous excitation and emission.

For an adequate interpretation of fluorescence anisotropy data in the context of CPP-induced changes of membrane fluidity, researchers should be aware of some experimental limitations. Thus, in the presence of foreign molecules, fluorescence anisotropy changes may not describe alterations of membrane fluidity but rather reflect alterations in the excited state lifetime of the fluorophore. Then, to prevent misinterpretations, the influence of CPPs on the fluorescence lifetime of fluidity probes should be checked (measurements might be made with a Fluorolog Tau-3 Lifetime system).

In terms of lipid preparations, some precautions should also be taken. The suspension of liposomes should be clear enough to prevent intense light scattering, which might affect fluorescence measurements. Thus, preformed liposomes are subjected to a mild sonication, which allows obtaining smaller liposomes displaying less concentric bilayers, although maintaining the character of MLV. When a homogeneous population is required, LUV, obtained by extrusion through polycarbonate filters, with defined pore size, are a good choice, since they most closely mimic the lipid bilayers of biological membranes. Fluorescent probes are added to liposomes, by subsequently injecting very small amounts, ca. 0.2 μl, of concentrated probe solutions (in dimethylformamide) into the liposome suspension, under vortexing and at a temperature 7–10 °C above lipid phase transition, in order to favor probe incorporation into the lipid bilayers. The concentration of the probe stock solution should be chosen so that the total amount of probe solution added to a liposome suspension (about 2 ml) in the cuvette, to obtain a probe/lipid molar ratio of about 1:400, does not exceed 5–10 μl. The final liposome preparation containing the fluorescent probe and the desired CPP concentration is maintained above the transition temperature for 2 h and then left to equilibrate overnight in the dark. To correct fluorescence measurements for the contribution of light scattering, appropriate blanks with equivalent volumes of the probe solvent should be prepared, although these corrections are often found to be negligible (blank fluorescence intensity lower than 5% of that of the respective sample).

Fluorescence anisotropy data provided by pure liposome preparations (not doped with foreign molecules) plotted as a function of temperature may provide two types of curves, depending on membrane lipid composition: a curve showing a monotonic decrease of anisotropy with increasing temperatures, or a sigmoidal curve showing an abrupt decrease of anisotropy in a more or less broad temperature range (Fig. 14.3). In the first case, often occurring when a complex mixture of lipids with different transition temperatures constitutes the liposome suspension, or a suspension of native membranes is

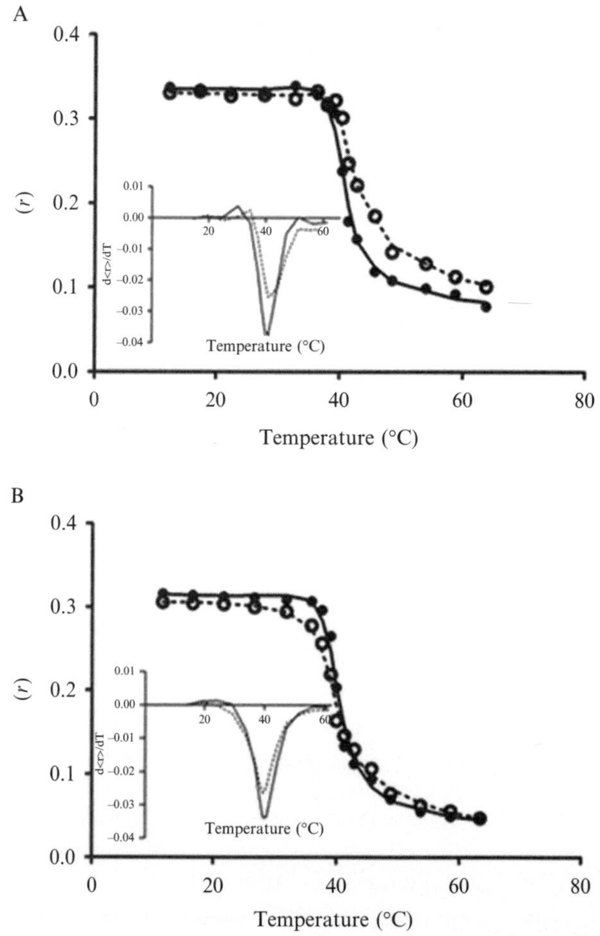

Figure 14.3 Fluorescence anisotropy (r) of DPH-PA (A) or DPH (B) in DPPG bilayers in the absence (-●-) and presence of the $S4_{13}PV$ peptide (¨○¨), at a lipid/peptide molar ratio of 40/1. The insets represent the derivative curves, the peak indicating the transition temperature, corresponding to the main curves. The thermotropic profiles are typical assays of at least three independent experiments.

studied, the progressive decrease of anisotropy describes the continuous temperature-dependent increase of lipid disorder. In the second case, a lipid phase transition is detected centered at a temperature (transition midpoint temperature) equivalent to that at which a DSC profile obtained with the same liposome preparation would show the peak of the corresponding endothermic transition (see e.g., Jurado *et al.*, 1991a). The phase transition midpoint temperature can be evidenced in a fluorescence anisotropy thermogram by plotting the derivative of the respective curve (Fig. 14.3).

The addition of CPPs to liposome preparations may induce more or less pronounced perturbations of the lipid packing at low temperatures (highly ordered lipid bilayers), at high temperatures (fluid membranes) or over the whole transition phase temperature range. They can also differently affect the low- and high temperature regions of the thermogram with an eventual abolishment of the phase transition or promote a redistribution of lateral pressures across the bilayer thickness, which can be evidenced by using probes residing at different depths in the bilayer. Finally, membrane disturbance will depend on CPP structure and the phospholipid composition of the host bilayer, regarding acyl chain length and unsaturation or head group size, polarity and charge, as well as on the cholesterol content of the lipid bilayer.

Figure 14.3 shows the effects of $S4_{13}PV$ peptide on the fluorescence anisotropy (r) of two fluorescent probes (DPH and DPH-PA) incorporated in liposomes of DPPG. The thermogram of pure DPPG liposomes displays an abrupt decrease of r in a temperature range between 40 and 42 °C, which matches the temperature range of the endothermic transition detected in the same liposomes by DSC (see Fig. 14.2). On the other hand, the corresponding derivative curve shows a peak centered at 41 °C, which coincides with T_p of the DSC profile. Regarding CPP-doped liposomes, a remarkable difference is noticed between the thermograms obtained with DPH and its charged derivative. With the former, the thermograms are almost superimposable, whereas with the latter, a significant increase of r is observed in the presence of the CPP, at temperatures above 40 °C (corresponding to T_o in the DSC profile), although no effects are noticed at lower temperatures. These observations indicate that the $S4_{13}PV$ peptide does not perturb fluidity (inversely related with fluorescence anisotropy) in the hydrophobic core of the bilayer (probed by DPH) but promotes an increase of lipid packing, during and above the phase transition, in the most upper regions of the lipid acyl chains, reflecting an increase of van der Waals interactions. This effect, which might be a consequence of the electrostatic interactions established between the positive charges of the peptide and the negative charges of the lipid phosphate, counteracting head group repulsion and lipid molecule separation, is consistent with the splitting of the endotherm detected by DSC, generating a component shifted to higher temperatures. Therefore, with calorimetric and spectrofluorimetric techniques, we may gather complementary information in order to fully characterize CPP–membrane interactions.

4. Methods for Evaluation of Biological Activity of CPP-Based Nucleic Acid Complexes

The ultimate goal when developing a delivery system is to determine the biological activity of the carried nucleic acids in the target cells. In this regard, both gene expression and silencing of target genes can be assessed by methodologies including flow cytometry, luminescence, Western blot, and real-time PCR, which have been described in a previous volume of Methods in Enzymology (Cardoso *et al.*, 2009).

Typical results for gene expression (transfection efficiency) upon delivery of plasmid DNA mediated by the $S4_{13}PV$ peptide and its analogs, as assessed by flow cytometry, are illustrated in Fig. 14.4.

The $S4_{13}PV$ and the reverse NLS peptides mediated transfection more efficiently than the scrambled peptide. Importantly, these observations correlate well with previous studies, which demonstrated that the cellular internalization of the $S4_{13}PV$ and the reverse NLS peptides occurs to a larger extent than that of the scrambled peptide, a fact that was attributed to the presence of the Dermaseptin sequence in the CPP (Mano *et al.*, 2005).

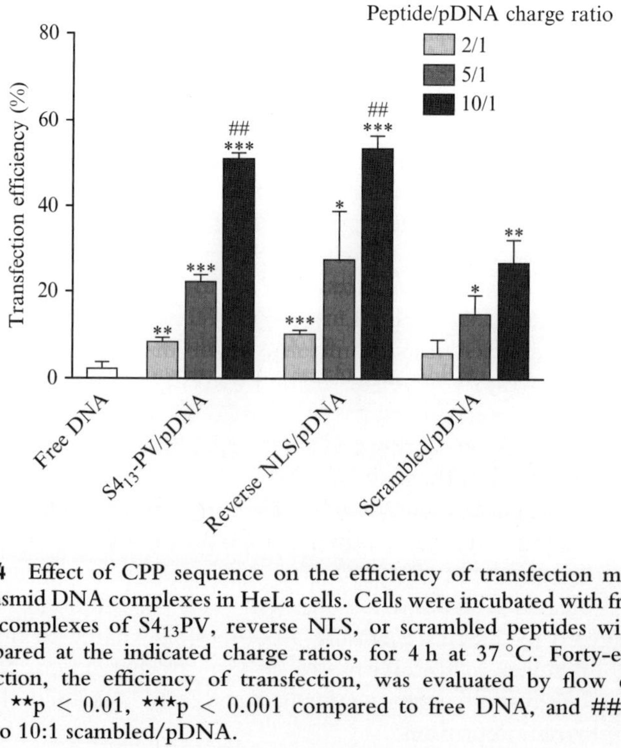

Figure 14.4 Effect of CPP sequence on the efficiency of transfection mediated by peptide/plasmid DNA complexes in HeLa cells. Cells were incubated with free plasmid DNA and complexes of $S4_{13}PV$, reverse NLS, or scrambled peptides with plasmid DNA, prepared at the indicated charge ratios, for 4 h at 37 °C. Forty-eight hours posttransfection, the efficiency of transfection, was evaluated by flow cytometry. *$p < 0.05$, **$p < 0.01$, ***$p < 0.001$ compared to free DNA, and ## $p < 0.01$ compared to 10:1 scrambled/pDNA.

The silencing of target genes through the antisense technology constitutes an approach of paramount pharmacological interest, since it allows modulating a variety of cellular functions. The finding of the high frequency of alternative splicing, with most studies reporting alternative splice forms in more than 70% of human genes (Sazani and Kole, 2003; Tazi *et al.*, 2009), and the fact that aberrant splicing is often related to a large number of human diseases underscore the importance of this approach for modulating pre-mRNA splicing. The control of pre-mRNA splicing pattern can be accomplished by delivering SSOs (Kole *et al.*, 2004; Mercatante *et al.*, 2002; Sazani and Kole, 2003). A reporter splice-switching assay that provides a clear, sequence specific and sensitive read-out was developed by Kole and collaborators to facilitate the assessment of SSO delivery to the nucleus (Kang *et al.*, 1998). The principles of this assay and the experimental procedure employed for determination of the biological activity of SSOs when delivered by the $S4_{13}PV$ peptide-based binary and ternary complexes are described below.

4.1. Splicing correction assay

In this assay, the coding sequence of luciferase is interrupted by a defective intron from a thalassemic human β-globin gene, which carries a point mutation that creates an aberrant 5′ splice site and activates an additional 3′ splice site upstream. The use of this aberrant splice site prevents the production of correctly spliced luciferase mRNA. In the absence of a steric-blocking ON treatment, virtually no luciferase mRNA or protein can be detected, while in the presence of a specific SSO, both the correctly spliced mRNA and luciferase activity can be quantified (Kang *et al.*, 1998). Using this assay, the effect of several parameters on the efficiency of the intracellular delivery of the SSOs, such as the amino acid arrangement of the CPP or the order of addition of the different components of the delivery system, can be easily assessed. In a typical experiment, the delivery systems are incubated for 4 h in OptiMEM medium with exponentially growing HeLa pLuc/705 cells (1.75×10^5 cells/well seeded and cultured overnight in 24 well-plates). Cells are then washed and further incubated for 20 h in complete DMEM medium containing 10% fetal bovine serum. After this incubation period, cells are washed twice with ice-cold PBS and lysed with Reporter Lysis Buffer (Promega, Madison, WI). Luciferase activity can be quantified in a Berthold Centro LB 960 Luminometer (Berthold Technologies, Bad Wildbad, Germany) using the Luciferase Assay System substrate (Promega, Madison, WI). Total protein levels are determined with the BCA™ Protein Assay Kit (Pierce, Rockford, IL), using an ELISA plate reader (Dynatech MR 5000, Dynatech Labs, Chantilly, VA) at 550 nm, to allow normalization of luciferase activity and correction for differences in cell number. Luciferase activity can be expressed as relative luminescence units per microgram protein.

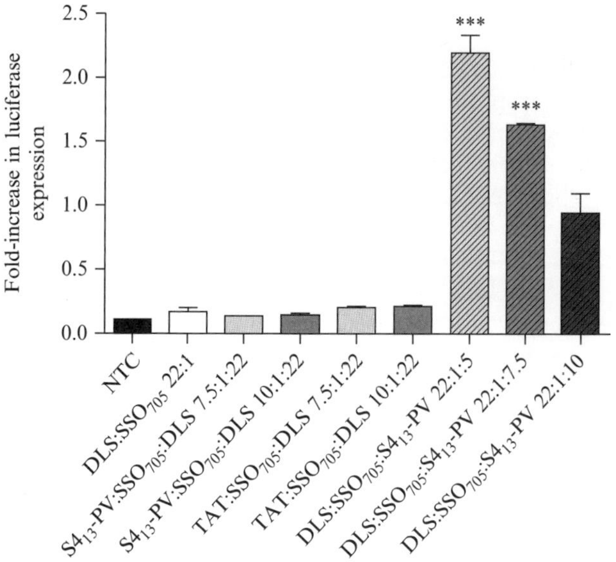

Figure 14.5 Splicing correction in HeLa pLuc/705 cells mediated by CPP-based complexes. HeLa pLuc/705 cells were incubated with the different DLS:SSO_{705}:peptide ternary complexes, for 4 h at 37 °C, at a final SSO_{705} concentration of 100 n*M*. The ternary complexes were obtained by the addition of either the CPP to preformed complexes of DLS with the SSO_{705} (CPP:SSO_{705}:DLS) or DLS to preformed complexes of CPP with the SSO_{705} (DLS:SSO_{705}:CPP). Luciferase activity was evaluated 24 h posttransfection. Results are expressed as fold increase over the DLS:SSO_{705} complexes (mean ± SD) and are representative of three independent experiments. Results for nontreated cells (NTC) are shown as a control. $^{***}p < 0.001$ compared to DLS:SSO_{705} 22:1.

The results presented in Fig. 14.5 show that the $S4_{13}$PV peptide-based SSO delivery systems are more efficient in promoting splice correction when this CPP is added after the precomplexation of DLS (DOGS/DOPE liposome of Thierry) cationic liposomes with the SSOs, than when the CPP is precomplexed with the SSOs before the addition of the DLS liposomes (Trabulo *et al.*, 2010).

5. Methods for Evaluation of Cytotoxicity of CPP-Based Nucleic Acid Complexes

Transfection of certain cell types by CPP-based nucleic acid delivery systems may lead to toxicity, which may limit their application *in vitro* and *in vivo*. Thus, evaluation of the biological activity of nucleic acid delivery systems (such as CPP-based complexes) should include studies of their effects on cell viability. Tumor-targeted nucleic acid delivery systems can also result in cell death through triggering of apoptosis, which, in this case, could represent a desired biological effect.

5.1. Measurement of unspecific cytotoxicity

Different techniques can be used to assess the cytotoxic effects of the nucleic acid delivery vectors, including lactate dehydrogenase, Alamar blue, and nuclear fragmentation assays. These experimental approaches and their applications in biocompatibility studies have been described previously in detail (Cardoso *et al.*, 2009). In our studies with the CPP-based complexes, the Alamar blue assay has been applied to evaluate cell viability following transfection. This assay measures the redox capacity of the cells due to the production of metabolites as a result of cell growth. Besides its simplicity, this assay offers the advantage of allowing determination of viability over the culture period without the detachment of adherent cells, which can thus be used for other experiments or submitted to further analysis (Trabulo *et al.*, 2008, 2010).

5.2. Measurement of apoptosis-triggering effects

Cells may be induced to activate the mechanisms of programmed cell death by the knockdown of an antiapoptotic protein using siRNAs or ONs, or the overexpression of a proapoptotic protein, through the delivery of plasmid DNA, which can be of great relevance to achieve a therapeutic benefit in tumor-targeted approaches. Such activation mechanisms occur following exposure to a cytotoxic stimulus and are usually accompanied by phosphatidylserine (PS) exposure on the outer leaflet of the plasma membrane, nuclear fragmentation, and caspase activation. Studies to evaluate the rate of cell apoptosis and caspase activity can be performed using the protocols described below.

5.2.1. Determination of apoptosis rate

Apoptosis detection can be performed by flow cytometry analysis using the ApopNexin™ FITC Apoptosis detection Kit (Chemicon, Millipore, MA, USA) as follows (Trabulo *et al.*, 2011). Briefly, 48 h postincubation with a drug, like doxorubicin, cells are collected after digestion with trypsin, washed with PBS three times, and resuspended in 500 μl of binding buffer (10 m*M* HEPES/NaOH, 140 m*M* NaCl, 2.5 m*M* $CaCl_2$, pH 7.4). Cells are then stained with Annexin V conjugated with FITC, which binds, in a calcium-dependent manner and with very high affinity, to membranes containing the negatively charged phospholipid PS. The cells are also incubated with propidium iodide, which is used to distinguish apoptotic cells with intact membranes from lysed, necrotic cells. After 15 min incubation in the dark at room temperature, the apoptosis rate should be immediately determined by flow cytometry. Bicolor analysis of cell suspensions allows the separation of the whole cell population into viable cells, early apoptotic cells and late apoptotic cells, or necrotic cells, and the calculation of the percentage of each cell type with respect to the total cell population.

5.2.2. Evaluation of caspase-3/7 activity

Caspase-3/7 activity can be determined using the SensoLyte® Homogeneous AMC Caspase-3/7 Assay Kit for fluorimetric analysis (Ana Spec, Fremont, CA, USA). Cell extracts should be prepared according to the manufacturer's instructions (SensoLyte® Homogeneous AMC Caspase-3/7 Assay Kit, Ana Spec, Fremont, CA, USA). Cells are washed with ice-cold PBS and lysed with 1 × lysis buffer, diluted from the 10 × lysis buffer included in the assay kit (SensoLyte® Homogeneous AMC Caspase-3/7 Assay Kit, Ana Spec, Fremont, CA, USA). Subsequently, cells are scraped off from the plates, and the cell suspension is collected in a microcentrifuge tube. The cell suspension is rotated on a rotating apparatus for 30 min at 4 °C and then centrifuged at 2500 × *g* for 10 min at 4 °C. The supernatant is then collected and stored at − 80 °C. Caspase 3/7 activity is measured 48 h post-transfection as the ability of the cell extract to catalyze the cleavage of Ac-DEVD-AMC and release the AMC fluorochrome. Protein content of cell extracts can be determined using the Bio-Rad protein assay (Bio-Rad). To initiate the enzymatic reaction, 50 μl of the cell lysate is transferred into the wells of a black 96-well plate (Costar, Cambridge, CA, USA), followed by the addition of 50 μl of assay buffer containing DTT and caspase-3/7 substrate. Reagents are mixed by shaking the plate in a shaker for 30–60 s at 100–200 rpm, and the fluorescence signal is continuously recorded every 30 min for 5–7 h. The plate should always be maintained at 37 °C between and during measurements. Caspase-3/7 activity can be determined by first plotting data as relative fluorescence units (RFU)/mg of protein versus time for each sample, and then determining the slope of the linear portion of the data plot.

6. Concluding Remarks

The application of CPPs for nucleic acid delivery is still in its infancy. The coadjuvation of CPPs with other components such as cationic liposomes has been shown to be a promising strategy to be exploited. However, as important as the composition of CPP-based nucleic acid complexes seems to be, how the different components are arranged and interact with each other, as well as how the particle as a whole interacts with cell membranes are also highly significant. Therefore, as emphasized in this chapter, the protocols followed for the preparation of the CPP-based complexes are determinants of success in the delivery of nucleic acids. Different protocols have to be optimized according to the carried cargo, and probably to the features of the CPP itself, the cell type, and membrane lipid composition. Although these protocols have been established in a tentative-error basis, biophysical studies addressing the physical interactions between the different components of the complexes and between these and the lipids of cell membranes could provide a solid platform for the establishment of structure–activity relationships toward a rational design of CPP-based complexes that are increasingly competent for nucleic acid delivery.

ACKNOWLEDGMENTS

This work was supported by the Portuguese Foundation for Science and Technology and COMPETE and FEDER (Grants PTDC/QUI-BIQ/103001/2008 and PTDC/BIO/65627/2006).

REFERENCES

Alves, I. D., Goasdoue, N., Correia, I., Aubry, S., Galanth, C., Sagan, S., Lavielle, S., and Chassaing, G. (2008). Membrane interaction and perturbation mechanisms induced by two cationic cell-penetrating peptides with distinct charge distribution. *Biochim. Biophys. Acta* **1780,** 948–959.

Andrich, M. P., and Vanderkooi, J. M. (1976). Temperature dependence of 1,6-diphenyl-1,3,5-hexatriene fluorescence in phophoslipid artificial membranes. *Biochemistry* **15,** 1257–1261.

Bartlett, G. R. (1959). Colorimetric assay methods for free and phosphorylated glyceric acids. *J. Biol. Chem.* **234,** 469–471.

Cardoso, A., Trabulo, S., Moreira, J. N., Düzgüneş, N., and Pedroso de Lima, M. C. (2009). Targeted lipoplexes for siRNA delivery. *Methods Enzymol.* **465,** 267–287.

Cardoso, A. M., Faneca, H., Almeida, J. A., Pais, A. A., Marques, E. F., Pedroso de Lima, M. C., and Jurado, A. S. (2011). Gemini surfactant dimethylene-1,2-bis(tetradecyldimethylammonium bromide)-based gene vectors: A biophysical approach to transfection efficiency. *Biochim. Biophys. Acta* **1808,** 341–351.

Chen, L. A., Dale, R. E., Roth, S., and Brand, L. (1977). Nanosecond time-dependent fluorescence depolarization of diphenylhexatriene in dimyristoyllecithin vesicles and the determination of "microviscosity". *J. Biol. Chem.* **252,** 2163–2169.

Cherry, R. J. (2005). Membrane protein dynamics: Rotational dynamics. *In* "The Structure of Biological Membranes," (P. L. Yeagle, ed.), pp. 389–410. CRC Press, Boca Raton, Florida, USA.

Derossi, D., Joliot, A. H., Chassaing, G., and Prochiantz, A. (1994). The third helix of the Antennapedia homeodomain translocates through biological membranes. *J. Biol. Chem.* **269,** 10444–10450.

Düzgüneş, N. (2003). Preparation and quantitation of small unilamellar liposomes and large unilamellar reverse-phase evaporation liposomes. *Methods Enzymol.* **367,** 23–27.

Frankel, A. D., and Pabo, C. O. (1988). Cellular uptake of the tat protein from human immunodeficiency virus. *Cell* **55,** 1189–1193.

Gennis, R. B. (1989). *Biomembranes molecular structure and function.* Springer-Verlag, New York.

Green, M., and Loewenstein, P. M. (1988). Autonomous functional domains of chemically synthesized human immunodeficiency virus tat trans-activator protein. *Cell* **55,** 1179–1188.

Guillen, J., Perez-Berna, A. J., Moreno, M. R., and Villalain, J. (2008). A second SARS-CoV S2 glycoprotein internal membrane-active peptide. Biophysical characterization and membrane interaction. *Biochemistry* **47,** 8214–8224.

Heitz, F., Morris, M. C., and Divita, G. (2009). Twenty years of cell-penetrating peptides: From molecular mechanisms to therapeutics. *Br. J. Pharmacol.* **157,** 195–206.

Hohne, G. W., Hemminger, W. F., and Flammershein, H. J. (1996). *Differential Scanning Calorimetry: An Introduction for Practitioners.* Springer-Verlag, New York.

Joliot, A. H., Triller, A., Volovitch, M., Pernelle, C., and Prochiantz, A. (1991). Alpha-2,8-Polysialic acid is the neuronal surface receptor of antennapedia homeobox peptide. *New Biol.* **3,** 1121–1134.

Jurado, A. S., Almeida, L. M., and Madeira, V. M. (1991a). Fluidity of bacterial membrane lipids monitored by intramolecular excimerization of 1.3-di(2-pyrenyl)propane. *Biochem. Biophys. Res. Commun.* **176,** 356–363.

Jurado, A. S., Pinheiro, T. J., and Madeira, V. M. (1991b). Physical studies on membrane lipids of *Bacillus stearothermophilus* temperature and calcium effects. *Arch. Biochem. Biophys.* **289,** 167–179.

Kang, S. H., Cho, M. J., and Kole, R. (1998). Up-regulation of luciferase gene expression with antisense oligonucleotides: Implications and applications in functional assay development. *Biochemistry* **37,** 6235–6239.

Khalil, I. A., Kogure, K., Futaki, S., Hama, S., Akita, H., Ueno, M., Kishida, H., Kudoh, M., Mishina, Y., Kataoka, K., Yamada, M., and Harashima, H. (2007). Octaarginine-modified multifunctional envelope-type nanoparticles for gene delivery. *Gene Ther.* **14,** 682–689.

Kilk, K., El-Andaloussi, S., Jarver, P., Meikas, A., Valkna, A., Bartfai, T., Kogerman, P., Metsis, M., and Langel, U. (2005). Evaluation of transportan 10 in PEI mediated plasmid delivery assay. *J. Control. Release* **103,** 511–523.

Kogure, K., Moriguchi, R., Sasaki, K., Ueno, M., Futaki, S., and Harashima, H. (2004). Development of a non-viral multifunctional envelope-type nano device by a novel lipid film hydration method. *J. Control. Release* **98,** 317–323.

Kogure, K., Akita, H., and Harashima, H. (2007). Multifunctional envelope-type nano device for non-viral gene delivery: Concept and application of Programmed Packaging. *J. Control. Release* **122,** 246–251.

Kole, R., Williams, T., and Cohen, L. (2004). RNA modulation, repair and remodeling by splice switching oligonucleotides. *Acta Biochim. Pol.* **51,** 373–378.

Kumar, P., Wu, H., McBride, J. L., Jung, K. E., Kim, M. H., Davidson, B. L., Lee, S. K., Shankar, P., and Manjunath, N. (2007). Transvascular delivery of small interfering RNA to the central nervous system. *Nature* **448,** 39–43.

Lindgren, M., Hallbrink, M., Prochiantz, A., and Langel, U. (2000). Cell-penetrating peptides. *Trends Pharmacol. Sci.* **21,** 99–103.

Mano, M., Teodosio, C., Paiva, A., Simões, S., and Pedroso de Lima, M. C. (2005). On the mechanisms of the internalization of $S4_{13}PV$ cell-penetrating peptide. *Biochem. J.* **390,** 603–612.

Mano, M., Henriques, A., Paiva, A., Prieto, M., Gavilanes, F., Simões, S., and Pedroso de Lima, M. C. (2006). Cellular uptake of $S4_{13}PV$ peptide occurs upon conformational changes induced by peptide-membrane interactions. *Biochim. Biophys. Acta* **1758,** 336–346.

Meade, B. R., and Dowdy, S. F. (2007). Exogenous siRNA delivery using peptide transduction domains/cell-penetrating peptides. *Adv. Drug Deliv. Rev.* **59,** 134–140.

Mercatante, D. R., Mohler, J. L., and Kole, R. (2002). Cellular response to an antisense-mediated shift of Bcl-x pre-mRNA splicing and antineoplastic agents. *J. Biol. Chem.* **277,** 49374–49382.
Morein, S., Koeppe, I. R., Lindblom, G., de Kruijff, B., and Killian, J. A. (2000). The effect of peptide/lipid hydrophobic mismatch on the phase behavior of model membranes mimicking the lipid composition in Escherichia coli membranes. *Biophys. J.* **78,** 2475–2485.
Morris, M. C., Vidal, P., Chaloin, L., Heitz, F., and Divita, G. (1997). A new peptide vector for efficient delivery of oligonucleotides into mammalian cells. *Nucleic Acids Res.* **25,** 2730–2736.
Morris, M. C., Depollier, J., Mery, J., Heitz, F., and Divita, G. (2001). A peptide carrier for the delivery of biologically active proteins into mammalian cells. *Nat. Biotechnol.* **19,** 1173–1176.
Oehlke, J., Scheller, A., Wiesner, B., Krause, E., Beyermann, M., Klauschenz, E., Melzig, M., and Bienert, M. (1998). Cellular uptake of an alpha-helical amphipathic model peptide with the potential to deliver polar compounds into the cell interior non-endocytically. *Biochim. Biophys. Acta* **1414,** 127–139.
Pooga, M., Hallbrink, M., Zorko, M., and Langel, U. (1998). Cell penetration by transportan. *FASEB J.* **12,** 67–77.
Prendergast, F. G., Haugland, R. P., and Callahan, P. J. (1981). 1-[4-(Trimethylamino) phenyl]-6-phenylhexa-1,3,5-triene: Synthesis, fluorescence properties, and use as a fluorescence probe of lipid bilayers. *Biochemistry* **20,** 7333–7338.
Rosa, S. M., Antunes-Madeira, M. C., Matos, M. J., Jurado, A. S., and Madeira, V. M. (2000). Lipid composition and dynamics of cell membranes of *Bacillus stearothermophilus* adapted to amiodarone. *Biochim. Biophys. Acta* **1487,** 286–295.
Ruthaven, N. A., McElhaney, L., and McElhaney, R. N. (2005). The mesomorphic phase behaviour of lipid bilayers. *In* "The Structure of Biological Membranes," (P. L. Yeagle, ed.), pp. 53–107. CRC Press, Boca Raton, Florida, USA.
Sazani, P., and Kole, R. (2003). Modulation of alternative splicing by antisense oligonucleotides. *Prog. Mol. Subcell. Biol.* **31,** 217–239.
Schwarze, S. R., Ho, A., Vocero-Akbani, A., and Dowdy, S. F. (1999). In vivo protein transduction: Delivery of a biologically active protein into the mouse. *Science* **285,** 1569–1572.
Shinitzky, M., and Barenholz, Y. (1974). Dynamics of the hydrocarbon layer in liposomes of lecithin and sphingomyelin containing dicetylphosphate. *J. Biol. Chem.* **249,** 2652–2657.
Tazi, J., Bakkour, N., and Stamm, S. (2009). Alternative splicing and disease. *Biochim. Biophys. Acta* **1792,** 14–26.
Thoren, P. E., Persson, D., Esbjorner, E. K., Goksor, M., Lincoln, P., and Norden, B. (2004). Membrane binding and translocation of cell-penetrating peptides. *Biochemistry* **43,** 3471–3489.
Torchilin, V. P. (2008). Tat peptide-mediated intracellular delivery of pharmaceutical nanocarriers. *Adv. Drug Deliv. Rev.* **60,** 548–558.
Trabulo, S., Mano, M., Faneca, H., Cardoso, A. L., Duarte, S., Henriques, A., Paiva, A., Gomes, P., Simões, S., and de Lima, M. C. (2008). $S4_{13}PV$ cell-penetrating peptide and cationic liposomes act synergistically to mediate intracellular delivery of plasmid DNA. *J. Gene Med.* **10,** 1210–1222.
Trabulo, S., Resina, S., Simões, S., Lebleu, B., and Pedroso de Lima, M. C. (2010). A non-covalent strategy combining cationic lipids and CPPs to enhance the delivery of splice correcting oligonucleotides. *J. Control. Release* **145,** 149–158.
Trabulo, S., Cardoso, A. M., Santos-Ferreira, T., Cardoso, A. L., Simões, S., and Pedroso de Lima, M. C. (2011). Survivin silencing as a promising strategy to enhance the sensitivity of cancer cells to chemotherapeutic agents. *Mol. Pharm.* **8,** 1120–1131.

Trotter, P. J., and Storch, J. (1989). 3-[p-(6-phenyl)-1,3,5-hexatrienyl]phenylpropionic acid (PA-DPH): Characterization as a fluorescent membrane probe and binding to fatty acid binding proteins. *Biochim. Biophys. Acta* **982,** 131–139.

Zhang, C., Tang, N., Liu, X., Liang, W., Xu, W., and Torchilin, V. P. (2006). siRNA-containing liposomes modified with polyarginine effectively silence the targeted gene. *J. Control. Release* **112,** 229–239.

Zhao, M., Kircher, M. F., Josephson, L., and Weissleder, R. (2002). Differential conjugation of tat peptide to superparamagnetic nanoparticles and its effect on cellular uptake. *Bioconjug. Chem.* **13,** 840–844.

Zoonens, M., Reshetnyak, Y. K., and Engelman, D. M. (2008). Bilayer interactions of pHLIP, a peptide that can deliver drugs and target tumors. *Biophys. J.* **95,** 225–235.

CHAPTER FIFTEEN

Multifunctional Envelope-Type Nano Device (MEND) for Organelle Targeting Via a Stepwise Membrane Fusion Process

Yuma Yamada,[1] Hidetaka Akita,[1] *and* Hideyoshi Harashima

Contents

1. Programmed Packaging Concept and Construction of R8-MEND 302
2. Screening of Lipid Compositions for Their Ability to Fuse with Nuclear and Mitochondrial Membranes 304
   2.1. Liposome preparation for the screening assay by FRET 305
   2.2. The use of FRET in a nuclear membrane fusion assay 305
   2.3. The use of FRET in a mitochondrial membrane fusion assay 307
   2.4. Analysis results regarding fusion activity of various liposomes with nuclear and mitochondrial membranes 308
3. Construction of Tetra-Lamellar MEND (T-MEND) 309
   3.1. Protocol for preparation of a T-MEND 310
   3.2. Characteristics of a T-MEND 311
4. Nuclear Gene Delivery 314
   4.1. The nuclear membrane, an ultimate barrier to gene delivery to the nonmitotic cells 314
   4.2. T-MEND as a nuclear delivery carrier of pDNA to DC 315
   4.3. Upgrading of T-MEND function by controlling intracellular trafficking and intranuclear decondensation 316
5. Mitochondrial Bioactive Molecule Delivery Using a Dual Function-MITO-Porter as a MEND for Mitochondrial Delivery 318
   5.1. Schematic image of mitochondrial genome targeting and construction of a DF-MITO-Porter 318
   5.2. Intracellular observation of DF-MITO-Porter and evaluation of mitochondrial targeting activity 320
   5.3. Evaluation of mtDNA levels after the mitochondrial delivery of DNase I using the DF-MITO-Porter 321

Faculty of Pharmaceutical Sciences, Hokkaido University, Sapporo, Japan
[1] These authors contributed equally.

Methods in Enzymology, Volume 509
ISSN 0076-6879, DOI: 10.1016/B978-0-12-391858-1.00015-0

Abstract

A single cell contains a variety of organelles. Included among these organelles are the nucleus that regulates the central dogma, mitochondria that function as an energy plant, the Golgi apparatus that determines the destination of endogenous protein, and others. If it were possible to prepare a nano craft that could specifically target a specific organelle, this would open a new field of research directed toward therapy for various diseases. We recently developed a new concept of "Programmed Packaging," by which we succeeded in creating a multifunctional envelope-type nano device (MEND) as a nonviral gene-delivery system. Our attempts to target certain organelles (nucleus and mitochondria) are described here, mainly focusing on the construction of a tetra-lamellar MEND (T-MEND), and on methods for screening the organelle-specific fusogenic envelope. The critical structural elements of the T-MEND include an organelle-specific membrane-fusogenic inner envelope and a cellular membrane-fusogenic outer envelope. The resulting T-MEND can be utilized to overcome intracellular membrane barriers, since it involves stepwise membrane fusion. To deliver cargos into a target organelle in our strategy, the carriers must fuse with the organelle membrane. Therefore, we screened a series of lipid envelopes that have the potential for fusing with an organelle membrane by monitoring the inhibition of fluorescence resonance energy transfer and identified the optimal lipid conditions for nuclear and mitochondrial membrane fusion. Finally, we describe the delivery of a bioactive molecule targeted to the nucleus and mitochondria in living cells, demonstrating that this system can be useful for targeting various organelles.

1. Programmed Packaging Concept and Construction of R8-MEND

We recently developed a multifunctional envelope-type nano device (MEND) based on a new packaging concept called "Programmed Packaging" (Kogure *et al.*, 2004, 2008), in which various functional devices that control intracellular trafficking are packaged into single nanoparticles so as to permit them to function at the appropriate place and time. This concept consists of three components: (1) a program to overcome all barriers, (2) design of functional devices and their three-dimensional assignments, and (3) a nanotechnology for assembling all the devices into a nano-sized structure (Moriguchi *et al.*, 2005).

The ideal MEND, as shown in Fig. 15.1, consists of a condensed plasmid DNA (pDNA) core and a lipid envelope structure equipped with the

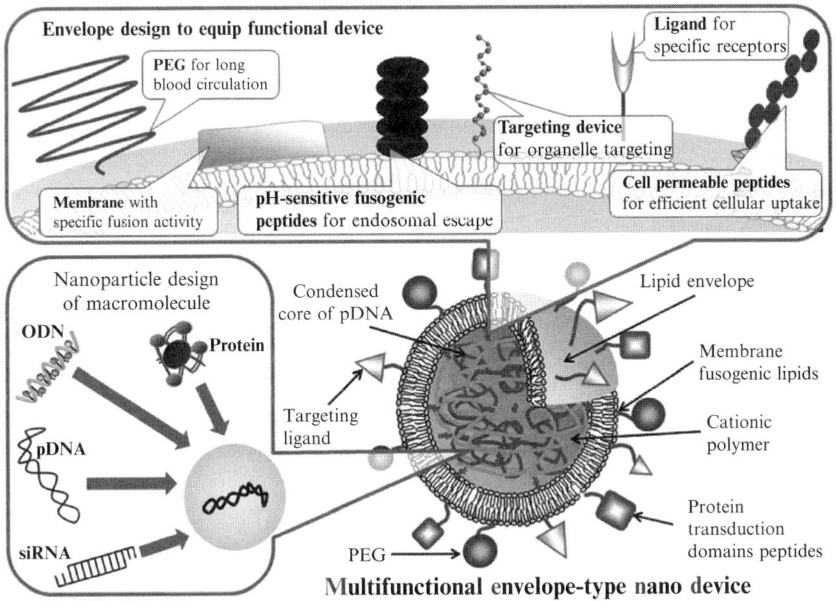

Figure 15.1 Schematic representation of multifunctional envelope-type nano device (MEND). The MEND consists of condensed pDNA molecules, coated with a lipid envelope modified with functional devices such as PEG for long blood circulation, ligand for specific targeting, a cell permeable peptide to increase intracellular availability, and pH-sensitive fusogenic peptides to enhance endosomal escape. Various cargos (not only pDNA but also oligo-nucleic acids, siRNA, proteins, and others) can be efficiently encapsulated in MEND.

various functional devices. The condensation of pDNA into a compact core prior to its inclusion in a lipid envelope has several advantages as follows: protection of pDNA from DNase; size control; and improved packaging efficiency, as the result of electrostatic interactions between the condensed core and the lipid envelope. To date, we have been successful in efficiently packaging not only pDNA (Kogure *et al.*, 2004; Masuda *et al.*, 2005; Moriguchi *et al.*, 2005; Suzuki *et al.*, 2008; Yamada *et al.*, 2010) but also oligo-nucleic acids (Nakamura *et al.*, 2006; Yamada *et al.*, 2005), siRNA (Akita *et al.*, 2010; Nakamura *et al.*, 2007), and proteins (Suzuki *et al.*, 2007) into MEND. The MEND was constructed based on a packaging process that involves three consecutive steps: (i) pDNA condensation with polycations, (ii) hydration of the lipid film for electrostatic binding of the condensed pDNA, and (iii) sonication to package the condensed pDNA with lipids.

Octaarginine (R8), an artificially designed cell permeable peptide, can be introduced on the surface of the MEND (R8-MEND). The high-density R8-MEND can induce macropinocytosis and escape lysosomal degradation

(Khalil *et al.*, 2006). The R8-MEND, with a high endosomal escape efficiency, showed transfection activities as high as that for adenovirus in dividing cells (Khalil *et al.*, 2007). Moreover, it was reported that the R8-MEND can be topically applied to the skin of a 4-week-old ICR mouse for *in vivo* delivering a gene to hair follicles (Khalil *et al.*, 2007). The R8-MEND containing pDNA-encoding bone morphogenetic protein receptor type 1 A (Bmpr1a), which is involved in the hair growth cycle, was applied to the mouse skin. The R8-MEND-formulated Bmpr1a gene was found to extend the hair growth period.

2. Screening of Lipid Compositions for Their Ability to Fuse with Nuclear and Mitochondrial Membranes

Targeted delivery of an engineered gene or gene product to the nucleus or a mitochondrion is an essential first step toward the therapeutic restoration of a missing cellular function. The nuclear localization signal (NLS) peptide can be used to guide a protein to the nucleus (Yoneda *et al.*, 1992). However, the same NLS does not function as such when attached to a pDNA (Nagasaki *et al.*, 2003; Tanimoto *et al.*, 2003): the positively charged NLS can be neutralized by the anionic pDNA, and it is difficult to control the dimensions of the carrier, thus, to maintain a sufficiently small size to permit it to pass through the nuclear pore (<39 nm). We also faced this problem in the case of the delivery of macromolecules to mitochondria: the conjugation of a mitochondrial targeting signal peptide to exogenous proteins and small linear DNAs was found to aid their delivery to mitochondria (Flierl *et al.*, 2003; Schatz, 1987; Seibel *et al.*, 1999; Vestweber and Schatz, 1989), but this strategy was not viable for delivering pDNA. Because large molecules such as pDNA and folded proteins do not readily pass through the mitochondrial membrane, they cannot be delivered easily to mitochondria (Endo *et al.*, 1995; Wiedemann *et al.*, 2004).

To overcome these problems, we recently proposed an original and innovative strategy for overcoming the intracellular membrane via membrane fusion. To deliver cargos to target organelles using our strategy, the liposomes must fuse with the organelle membrane. As described above, the R8-MEND composed of 1,2-dioleoyl-*sn*-glycero-3-phosphatidylethanolamine (DOPE) and cholesteryl hemisuccinate (CHEMS) showed high transfection activities in dividing cells (Khalil *et al.*, 2007); however, these lipids may not be the best lipid composition for the nucleus of nondividing cells and mitochondria. Therefore, screening was initiated for lipid compositions that are adequate for fusion to the nuclear and mitochondrial membranes. This was done by monitoring the cancelation of fluorescence resonance energy transfer (FRET) between donor and acceptor fluorophores, modified on the surface of liposomes.

2.1. Liposome preparation for the screening assay by FRET

Dually labeled liposomes for screening the fusogenic membrane with a target organelle are prepared with various compositions of lipids, as described below. DOPE is purchased from Avanti Polar lipids (Alabaster, AL, USA). CHEMS, cardiolipin (CL), phosphatidic acid (PA), phosphatidylglycerol (PG), phosphatidylinositol (PI), phosphatidylserine (PS), and sphingomyelin (SM) are purchased from Sigma (St. Louis, MO, USA). Stearyl octaarginine (STR-R8) (Futaki *et al.*, 2001) is obtained from KURABO INDUSTRIES LTD (Osaka, Japan).

1. Lipid films are formed on the bottom of a glass tube by evaporating the organic solvent containing 125 nmol lipids [DOPE/X = 1:1, molar ratio] or 137.5 nmol lipids [DOPE/X = 9:2, molar ratio]. For the FRET analysis, both 1 mol% of 7-nitrobenz-2-oxa-1, 3-diazole labeled DOPE (NBD-DOPE; Avanti Polar lipids) and 0.5 mol% of rhodamine-DOPE (Avanti Polar lipids) were incorporated into the lipid films.
2. To generate empty vesicles, 0.25 mL of buffer was applied to the dried lipid film. As a buffer for the liposome preparation, nuclear transport buffer (20 m*M* HEPES, 110 m*M* KOAc, 3 m*M* NaOAc, 2 m*M* MgOAc, 0.5 m*M* EGTA 4Na, and 2 m*M* DTT) and mitochondrial isolation buffer (MIB: 250 m*M* sucrose, 2 m*M* Tris–HCl, pH 7.4) were used for the screening of nuclear and mitochondrial fusogenic lipid composition, respectively.
3. After incubation for 10 min for hydration, the suspensions were sonicated using a bath-type sonicator (85 W, Aiwa Co., Tokyo, Japan) for 15 s.
4. An STR-R8 solution (10 mol% of lipids) was then added to the suspension, followed by incubation for 30 min at room temperature to permit the R8 to become attached to the surface of the liposome.

2.2. The use of FRET in a nuclear membrane fusion assay

We investigated the membrane fusion activity of the liposomes with the nucleus using FRET (Akita *et al.*, 2009; Maier *et al.*, 2002; Struck *et al.*, 1981). Liposomal membranes are labeled with both NBD-DOPE (excitation at 460 nm and emission at 534 nm) and rhodamine-DOPE (excitation at 550 nm and emission at 590 nm) so that energy transfer will occur from NBD to rhodamine. Membrane fusion between the labeled liposomes and the nucleus will lead to the diffusion of NBD and rhodamine into the lipid membranes, which causes a reduction in energy transfer, resulting in an increase in the fluorescence intensity at 530 nm.

2.2.1. Isolation of nucleus from HeLa cells

1. HeLa human cervix carcinoma cells (RIKEN Cell Bank, Tsukuba, Japan) are cultured with Dulbecco's modified Eagle medium (DMEM, Invitrogen Corp., Carlsbad, CA) containing 10% fetal bovine serum (FBS, Invitrogen Corp.), under 5% CO_2//air at 37 °C.
2. After removing the medium, the cells are washed twice with wash buffer (0.5% bovine serum albumin (BSA) and 0.1% NaN_3 in PBS (−)).
3. Then, 500 μL of wash buffer is added to the cells in 1.5-mL sample tubes, and the cells are isolated by centrifugation (1000 × *g*, 4 °C, 5 min). All subsequent steps are carried out on ice.
4. After removing the supernatant, the pellet is suspended in 100 μL of ice-cold Hank's Buffered Salt Solution (HBSS)/0.1% BSA.
5. After the addition of 100 μL of ice-cold HBSS/0.1% BSA/0.2% NP-40, the resulting solution is incubated on ice for 5 min.
6. After incubation, 800 μL of ice-cold HBSS/0.1% BSA is added to the solution, and pelleted by centrifugation (1400 × *g*, 4 °C, 1 min).
7. After removing the supernatant, the pellet is suspended in 1 mL of ice-cold HBSS/0.02% NP-40, pelleted by centrifugation (1400 × *g*, 4 °C, 1 min). This procedure is repeated one more time.
8. After removing the supernatant, the pellet is suspended in nuclear transport buffer to give the isolated nuclear suspension.

2.2.2. FRET analysis using isolated nucleus

1. A 10-μL aliquot of dually labeled liposome (lipid concentration, 550 μ*M*) is added to the isolated nuclei (corresponding to the 0.075 m*M* of phosphatidylcholine, as determined by a Test Kit, Wako) in 90 μL of nuclear transport buffer, which is then incubated for 30 min at 37 °C.
2. After incubation, energy transfer is assessed by measuring the fluorescence intensity (excitation at 470 nm and emission at 530 nm).
3. The maximum fluorescence is defined as the fluorescence of liposomes when dissolved in Triton X-100 (final concentration, 0.5%, v/v).
4. Fusion activity (%) is estimated by the reduction in the level of energy transfer in accordance with membrane fusion and is calculated as follows:

$$\text{Fusion activity } (\%) = (F - F_0)/(F_{\text{max}} - F_0) \times 100$$

where F, F_0, and F_{max} represent the fluorescence intensity of labeled liposome after incubation with nucleus, the fluorescence intensity of labeled liposome after incubation without nucleus, and the maximum fluorescence intensity after the Triton X-100 treatment, respectively.

2.3. The use of FRET in a mitochondrial membrane fusion assay

2.3.1. Isolation of mitochondria from rat liver

Mitochondria are isolated from rat liver essentially as described previously (Shinohara *et al.*, 1997). All animal protocols were approved by the institutional animal care and research advisory committee at the Faculty of Pharmaceutical Sciences, Hokkaido University, Sapporo, Japan.

1. Adult male Wistar rats (6–8 weeks old; Sankyo Labo Service (Sapporo, Japan)) are sacrificed and the livers are removed after bleeding had largely subsided and then placed in approximately 20 mL of ice-cold MIB containing EDTA (MIB [+EDTA]) per 10 g of liver. All subsequent steps are carried out on ice.
2. Livers were chopped into small pieces and the suspension is homogenized in a glass homogenizer (50 mL capacity) with a pestle. Three complete up and down cycles with the pestle are made. The pestle is motor driven and is operated at approximately 550 rpm.
3. The homogenate is diluted approximately 1:3 with MIB [+EDTA] and centrifuged at 800 × *g* for 5 min.
4. The supernatant is transferred into ice-cold tubes and centrifuged at 7500 × *g* for 10 min.
5. The pellets are washed twice with MIB [+EDTA], and once further with EDTA-free MIB to obtain the isolated mitochondrial suspension.

The purity of the mitochondrial preparation was greater than 90%, as judged by electron microscopy observations (Shinohara *et al.*, 2002), and the respiration control index of the mitochondria was determined to be 4.5–6 using a Clark oxygen electrode (YSI 5331; Yellow Springs Instrument Co., Yellow Springs, OH, USA), indicating that the mitochondria were intact and respiring.

2.3.2. FRET analysis using isolated mitochondria

The membrane fusion activity of the liposomes with mitochondria is assessed by FRET, as previously reported (Yamada *et al.*, 2008).

1. A 10-μL aliquot of dually labeled liposomes (lipid concentration, 550 μ*M*) is added to a mitochondrial suspension (corresponding to the 0.9 mg of mitochondrial protein/mL, determined by a BCA protein assay kit (Pierce; Rockford, IL, USA)) in 90 μL of MIB, and incubated for 30 min at 25 °C.
2. Subsequent manipulations to evaluate the mitochondrial fusogenic activities of various liposomes are carried out as described in Section 2.2.2.

2.4. Analysis results regarding fusion activity of various liposomes with nuclear and mitochondrial membranes

To select the lipid compositions of the MEND for nuclear targeting via membrane fusion, we screened for fusogenic activities directed at the nucleus by monitoring the cancelation of FRET. As a result, FRET inhibition was observed for all of the tested liposome compositions (Akita *et al.*, 2009). We also evaluated the fusion activity between the R8-unmodifeid liposomes and mitochondria, and compared their fusion activities between nucleus and mitochondria (Fig. 15.2A). The graph shows the relationship between mitochondrial membrane fusion activity (x-axis) and nuclear membrane fusion activity (y-axis). The R8-unmodified liposomes showed a high fusogenic activity with the nucleus but did not effectively fuse with

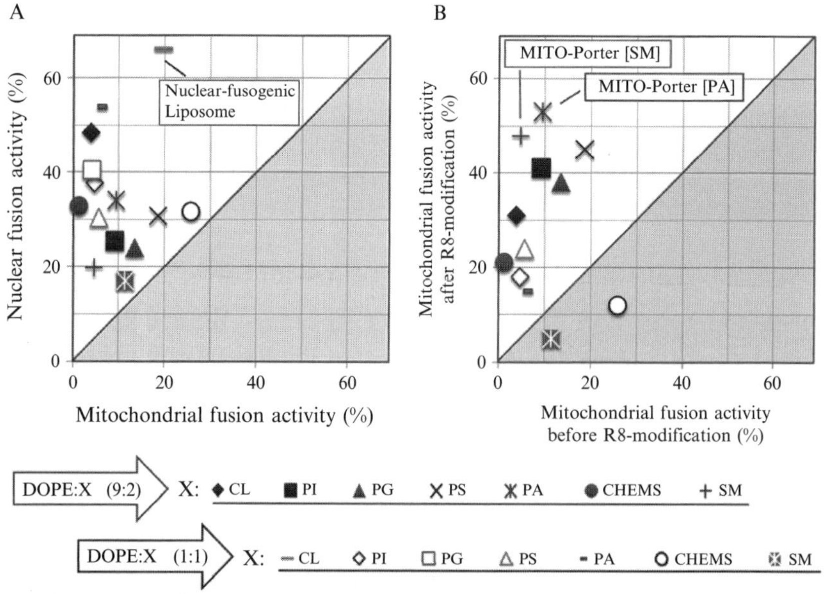

Figure 15.2 Fusion activity of various liposomes with nuclear and mitochondrial membranes. Lipid compositions with fusogenic activity toward the nucleus and mitochondria were identified by monitoring the cancelation of FRET between NBD and rhodamine incorporated in the candidate liposomes as a donor–acceptor pair. Since 1,2-dioleoyl-*sn*-glycero-3-phosphatidylethanolamine (DOPE) is a known fusogenic lipid, it was included in all of the candidate lipid compositions. Cancelation of FRET was monitored after incubating dual-labeled liposomes with an isolated nucleus or isolated mitochondria. (A) Relationship between mitochondrial membrane fusion activity (x-axis) and nuclear membrane fusion activity (y-axis). (B) Mitochondrial membrane fusion activity of various liposomes before (x-axis) and after R8-modification (y-axis). CL, cardiolipin; PI, phosphatidyl inositol; PG, phosphatidyl glycerol; PS, phosphatidyl serine; PA, phosphatidic acid; CHEMS, cholesteryl hemisuccinate; SM, sphingomyelin.

mitochondria. The most negatively charged lipid composition of DOPE:CL (=1:1, molar ratio) gave the highest fusion activity with the nucleus (66% of nuclear fusion activity, Fig. 15.2A) (Akita *et al.*, 2009) and showed a low fusogenic activity with mitochondria. This finding suggests that this specific membrane fusion might be useful for achieving selective nuclear delivery.

In our strategy for delivering cargo to mitochondria, the liposomes must fuse with the mitochondrial membrane after mitochondrial binding. A previous investigation involving binding assays between liposomes and mitochondria showed that R8-unmodified liposomes had a low binding activity, but liposome-binding activity increased significantly after R8-modification (Yamada and Harashima, 2008). This result indicates that the addition of a cationic peptide, R8, to the liposome surface enhances mitochondrial binding, probably because mitochondria maintain a high negative potential. We investigated the mitochondrial fusion activity of various liposomes with different lipid compositions before and after R8-modification (Fig. 15.2B). The graph shows the relationship between the mitochondrial membrane fusion activity before R8-modification (*x*-axis) and the values after R8-modification (*y*-axis). R8-modified liposomes showed a higher fusogenic activity than R8-unmodifeid liposomes, suggesting that strong electrostatic binding between the R8-modified liposomes and mitochondria stimulates liposomal fusogenic activity. It is noteworthy that R8-modified liposomes, which are composed of DOPE/SM/STR-R8 (9:2:1) or DOPE/PA/STR-R8 (9:2:1), showed a high fusogenic activity with mitochondria. Moreover, lipid compositions containing SM had a lower cytotoxicity than that of PA, indicating that the lipid composition, DOPE/SM/STR-R8 (9:2:1), was optimal for the carriers for mitochondrial delivery (Yamada *et al.*, 2011).

3. Construction of Tetra-Lamellar MEND (T-MEND)

As described above, the R8-MEND is capable of inducing macropinocytosis and can escape lysosomal degradation, leading to transfection activities as high as that for adenovirus in dividing cells (Khalil *et al.*, 2007). However, in nondividing cells, the ultimate barrier, namely the nuclear membrane, must also be overcome. The R8-MEND cannot in corporate two kinds of envelopes with different compositions for different membrane fusions. To solve this problem, an innovative nanotechnology was developed in which a tetra-lamellar MEND (T-MEND) was constructed. In this particle, a condensed nanoparticle of pDNA was coated with nuclear and endosomal membrane-fusogenic lipid envelopes in a stepwise manner to overcome these intracellular barriers via serial membrane fusion (Akita *et al.*, 2009). This technology has recently been used for mitochondrial delivery (Yamada *et al.*, 2011).

We succeeded in the identification of a highly fusogenic lipid composition for the nucleus (DOPE/CL = 1:1, molar ratio) and mitochondria (DOPE/SM/STR-R8 = 9:2:1, molar ratio), as described in Section 2.4. Lipids having fusogenic activity toward the endosome were also screened by incubating dually labeled liposomes with cultured cells at pH 6.5, which mimics the acidic environment of the endosome. As a result, R8-modified liposomes composed of DOPE/PA/STR-R8 (7:2:1, molar ratio) exhibited a high fusion activity and also showed the highest transfection activity (> 10-fold that of conventional R8-MEND prepared with DOPE/CHEMS/STR-R8 (9:2:0.5, molar ratio)). The endosome-fusogenic activity of this liposome in living cells has been demonstrated previously by spectral imaging (El-Sayed *et al.*, 2008). In this section, the protocol for the preparation of T-MENDs with an optimal lipid composition for nuclear targeting and their characteristics are described.

3.1. Protocol for preparation of a T-MEND

To construct the T-MEND, we developed a new packaging technology that involves a multilayering procedure consisting of three steps (Fig. 15.3). First, the DNA is packaged into a condensed core via electrostatic

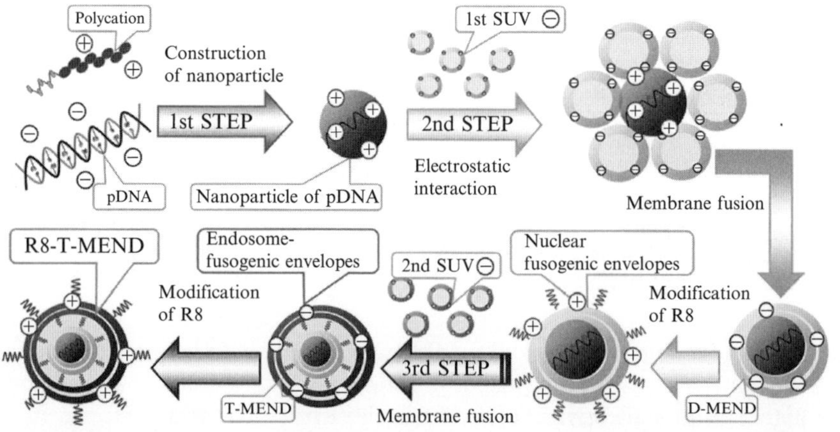

Figure 15.3 Schematic diagram illustrating a method for preparation of T-MEND. Plasmid DNA (pDNA) is first condensed with a polycation via electrostatic interactions. The condensed particle is then coated with the first two-layered envelope by assembling a negatively charged small unilamellar vesicle (SUV) around the positively charged DNA/polycation (D-MEND). The D-MEND is then modified by stearyl octaarginine (STR-R8) to reverse the surface charge and then coated with the second envelope by the fusion of a negatively charged SUV triggered by the assembly of SUV around the positively charged D-MEND. As a result, the tetra-lamellar MEND (T-MEND) is prepared. Finally, an STR-R8 solution was added to the suspension of the T-MEND to modify the outer envelope with R8.

interactions with a polycation. Second, a lipid double-lamellar MEND (D-MEND) is prepared through the membrane fusion of neighboring small unilamellar vesicles (SUVs), triggered by the assembly of negatively charged SUVs around the positively charged core. The surface of the resulting particles is then modified with STR-R8. Third, a second lipid coating is performed, again through the fusion of negatively charged SUV around the positive surface charges of the D-MEND. Finally, the carrier surface is modified with STR-R8 to give the final R8-T-MEND, which is a tetra-lamellar structured nanoparticle.

1. Lipid films composed of DOPE/CL (1:1, molar ratio, total lipid content: 0.55 μmol) are hydrated with 1 mL of 10 m*M* HEPES buffer (pH 7.4) for 10 min at room temperature. The hydrated lipid film is then sonicated using a probe-type sonicator to form SUV.
2. Condensed DNA particles are prepared by mixing 50 μL of pDNA solution (pEGFPLuc; BD Bioscience Clontech, Palo Alto, CA, USA, 0.1 mg/mL) with 75 μL of protamine solution (Calbiochem; Darmstadt, Germany, 0.1 mg/mL) with vortexing (final charge ratio (±) = 2.2).
3. The suspended nuclear membrane-fusogenic liposomes and condensed pDNA particles are then mixed at a ratio of 2:1 (v/v) to coat the condensed pDNA particles with a double-lipid envelope, as described previously (Lee and Huang, 1996).
4. The STR-R8 solution (20 mol% of total lipid) is added to the suspension of double-layered nuclear membrane-fusogenic particles.
5. This suspension of nuclear membrane-fusogenic particles is then mixed with endosome-fusogenic SUVs composed of DOPE/PA (7:2, molar ratio, total lipid content: 0.55 μmol) at a ratio of 1:2 (v/v) to generate particles with a double endosome-fusogenic envelope, which we refer to as a T-MEND.
6. The STR-R8 solution (10 mol% of endosome-fusogenic lipid) is added to the suspension of T-MEND to modify the outer envelope with R8.

3.2. Characteristics of a T-MEND

3.2.1. The physiochemical properties of the T-MEND and its intermediate particles

The sizes and ζ-potentials of the T-MEND and its intermediate particles are summarized in Table 15.1. Particle diameters are measured using a quasi-elastic light scattering method, and ζ-potentials are determined electrophoretically using laser Doppler velocimetry (Zetasizer Nano ZS; Malvern Instruments, Herrenberg, Germany). The pDNA is first packaged into a condensed core via electrostatic interactions with a polycation (+18 mV of ζ-potential and 65 nm in size). Second, the D-MEND is prepared through membrane fusion of neighboring negatively charged

Table 15.1 Physicochemical properties of T-MEND and its intermediates

	Diameter (nm)	ζ-Potential (mV)
Nanoparticle of pDNA	65	18
D-MEND R8 (−)	112	−84
D-MEND R8 (+)	134	40
T-MEND R8 (−)	142	−57
T-MEND R8 (+)	164	55

SUVs around the positively charged core (−84 mV, 112 nm). The surface of the resulting particles is then modified with STR-R8 (+40 mV, 134 nm). Third, a second lipid coating is performed, again through the fusion of negatively charged SUV around the positive surface charges of R8-modified D-MEND (−57 mV, 142 nm). If the second SUVs (approximately 80 nm) only bind to the D-MEND (~130 nm), the mean diameter would be expected to be approximately 300 nm. However, the resulting particle had a diameter of 164 nm, suggesting that the second SUVs fuse with each other, forming a T-MEND. The conversion of ζ-potential in each coating process strongly suggests that the DNA core was subject to stepwise encapsulation. Thus, the resulting T-MEND particle consists of two types of lipid layers.

3.2.2. Discontinuous sucrose density gradient ultracentrifugation

The encapsulation of pDNA in two different lipid envelopes is confirmed by discontinuous sucrose density gradient ultracentrifugation (Fig. 15.4A), as reported previously (Kogure *et al.*, 2004). T-MENDs prepared with 3 mol% NBD-DOPE (inner envelope) and 0.5 mol% rhodamine-DOPE (outer envelope) are layered on a discontinuous sucrose density gradient (0–60%) and centrifuged at 160000 × *g* for 2 h at 20 °C. A 1-mL aliquot is collected from the top, and the fluorescence intensities of NBD and rhodamine are measured (Fig. 15.4A(a)), and pDNA was visualized by staining with ethidium bromide after agarose gel electrophoresis (Fig. 15.4A(b)). Empty liposomes were enriched in fraction #5–8. Furthermore, the three principal components, inner lipid; outer lipid; and DNA, were contained in one fraction (fraction #10), indicating that the DNA core was successfully coated with the two different lipid layers.

3.2.3. Phase-contrast transmission electron microscopy

The structures of the conventional MEND and T-MEND are visualized by phase-contrast transmission electron microscopy (Fig. 15.4B). A suspension of the T-MEND is dropped onto a copper grid coated with a carbon film. After carefully removing excess liquid with the tip of

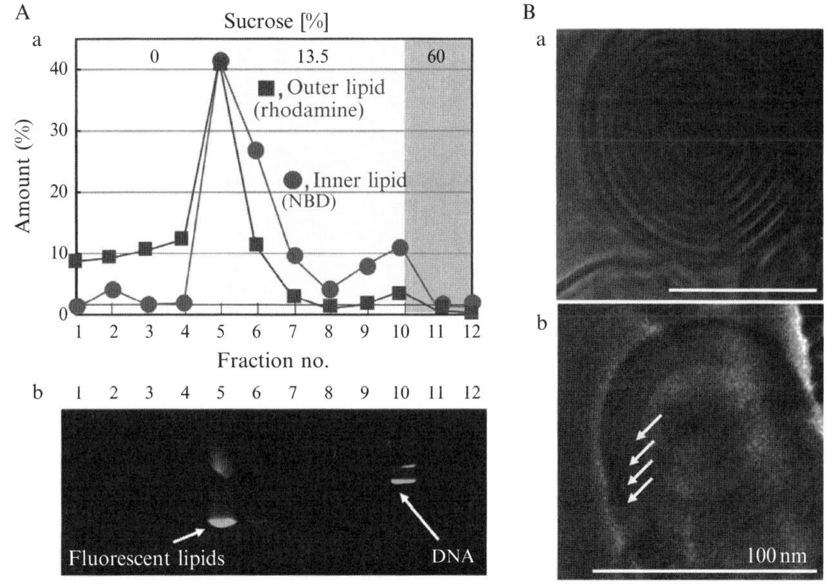

Figure 15.4 Characteristics of the T-MEND. (A) Distribution of the inner and outer lipids and plasmid DNA (pDNA) after discontinuous sucrose density gradient ultracentrifugation of the T-MEND, which consisted of NBD-DOPE (first lipid) and rhodamine-DOPE (second lipid). The amount of lipids was estimated by measuring the fluorescence intensities of NBD and rhodamine (a). pDNA was visualized by staining with ethidium bromide after agarose gel electrophoresis of a sample from each fraction (b). (B) Electron microscopic images of conventional MEND (a) and T-MEND (b). Morphological images of MENDs were obtained by phase-contrast transmission electron microscopy.

a filter paper, the sample is rapidly frozen in liquid ethane using a Leica rapid-freezing device (Leica EM CPC system, Wetzlar, Germany). The grid, with the ice-embedded samples, is transferred to the specimen chamber of a transmission electron microscope using a cryo-transfer system. The specimen chamber is cooled with liquid helium to reduce specimen damage caused by the electron beam. For observation, a JEOL JEM-3100FFC electron microscope (Tokyo, Japan) with an HDC phase plate inserted into the back focal plane of the objective lens is operated at 300 kV (Danev and Nagayama, 2001; Ohta *et al.*, 2006). While approximately 10 lipid layers were observed for the conventional MEND produced by lipid hydration (Fig. 15.4B(a)), the number of lipid layers for the T-MEND was controlled at 4 (Fig. 15.4B(b)). Therefore, our multilayering method can be used to construct a T-MEND.

4. Nuclear Gene Delivery

4.1. The nuclear membrane, an ultimate barrier to gene delivery to the nonmitotic cells

For a successful gene-delivery system, the nuclear membrane is the ultimate barrier that must be overcome. One of the typical results revealing its barrier function has been observed in the cell-to-cell variation in transgene expression. Flow cytometry analyses revealed that the percent of marker gene expression-positive cells to all cells analyzed was sharply enhanced when the cell cycle progressed through the M-phase (Tseng *et al.*, 1999). Many investigators believe that the M-phase-specific enhancement in transgene expression is the result of the temporal breakdown of the nuclear membrane, which allows pDNA to gain access to the nucleoplasm (Brunner *et al.*, 2000, 2002; James and Giorgio, 2000; Mortimer *et al.*, 1999).

Other examples have been obtained by more direct evaluations of the nuclear transport of pDNA. The mutual transport of various types of molecules between the cytosol and the nucleus was carried out via the nuclear pore complex (NPC), which allows the passive transport of gold nanoparticles with sizes <9 nm (Allen *et al.*, 2002). Wolff and coworkers investigated the effect of pDNA size on nuclear transport activity using digitonin-permeabilized cells (Hagstrom *et al.*, 1997; Ludtke *et al.*, 1999). Short DNA (<200 bp) fragments were effectively imported into the nucleus, whereas nuclear import decreased when the size of the pDNA increased and was negligible when the size reached >1500 bp. The typically used pDNA (>5 kbp) is too large to pass through the NPCs. In fact, microinjection studies demonstrated that more than a 100-fold larger amount of pDNA is required to achieve transgene expression by cytoplasmic microinjection, compared to that by nuclear microinjection. This suggests that less than 1% of the cytoplasm-microinjected pDNA reaches the nucleus (Pollard *et al.*, 1998). Therefore, efficient systems for the nuclear delivery of pDNA would be highly desirable in terms of developing an efficient gene-delivery system. In mitotic cells, pDNA primarily enters the nucleus when the nuclear membrane structure breaks down during the M-phase (Marenzi *et al.*, 1999; Mortimer *et al.*, 1999; Tseng *et al.*, 1999; Wilke *et al.*, 1996). However, in nondividing cells, the nuclear membrane structure remains intact. Therefore, the nuclear membrane is the ultimate barrier to be overcome for successful nuclear delivery in nondividing cells.

The mechanism for the NLS-dependent nuclear import of proteins has been extensively studied (Stewart, 2007). One of the most well-defined NLS signal is a peptide (PKKKRKV) derived from the SV40 T-antigen (NLS_{SV40}). Endogenous or chemically modified NLSs bind to importin α and are subsequently recognized by importin β that facilitates the nuclear import of the cargo via its efficient interaction between hydrophobic phenylalanine–glycine (FG)-rich domains inside the nuclear pore (Ben-Efraim and Gerace, 2001). In the light of these findings, many attempts have been made to improve the nuclear delivery processes by using NLS peptides. NLS peptides or importin proteins were conjugated to pDNA itself (Nagasaki *et al.*, 2003; Sebestyen *et al.*, 1998; Tanimoto *et al.*, 2003) and/or pDNA-condensing polycations (Chan and Jans, 1999; Chan *et al.*, 2000; Subramanian *et al.*, 1999). Alternatively, specific DNA sequences, such as the NF-κB-binding sequence (Breuzard *et al.*, 2008; Mesika *et al.*, 2001, 2005), are introduced to the pDNA, which allow the binding of nuclear transcription factors in the cytosol, and support nuclear import.

One of the targets of this system is dendritic cells (DCs), which play a crucial role in the initiation of an immunoresponse and the regulation of cell-mediated immunoreactions. Whereas the R8-MEND can facilitate a high gene expression in dividing HeLa cells (Khalil *et al.*, 2006, 2007), the transfection activity in JAWS II cells, a dendrite-derived cell line, was negligible. Additional modification of NLSs with polycations and/or lipid envelopes failed to improve transfection activity, which was not achieved in JAWS II cells. The most important drawback in the NLS-dependent transport is that the threshold size of the NPC would be limited to >39 nm in size (Dworetzky *et al.*, 1988; Pante and Kann, 2002), a size that is much smaller than that of the pDNA/polycation core particle. Thus, the NPC-independent pathway represents an alternative pathway for nuclear entry.

4.2. T-MEND as a nuclear delivery carrier of pDNA to DC

We recently proposed an NPC-independent nuclear transport approach, in which intracellular membrane barriers are overcome via stepwise membrane fusion using a T-MEND. The critical structural elements of the T-MEND are a DNA–polycation condensed core coated with two nuclear membrane-fusogenic inner envelopes and endosome-fusogenic outer envelopes, which are shed in a stepwise fashion as the core makes its way through the cell (Fig. 15.5) (Akita *et al.*, 2009). When a MEND was prepared using a single endosome-fusogenic or nuclear membrane-fusogenic lipids, stimulation of transfection activity was rarely observed in JAWS II cells. By contrast, when a T-MEND was prepared with both double nuclear (inner) and endosomal-fusogenic membranes (outer), the transfection activity was dramatically increased by several hundred-fold (Fig. 15.6a and b). Replacement of the nuclear- and/or endosomal-fusogenic envelopes with nonfusogenic ones resulted in a drastic loss of function. Thus, a multilayered structure with an adequate lipid combination is an essential requirement for successful gene expression.

The intracellular fusion events in live cells were also confirmed by means of spectral imaging using a donor- and acceptor dually labeled inner

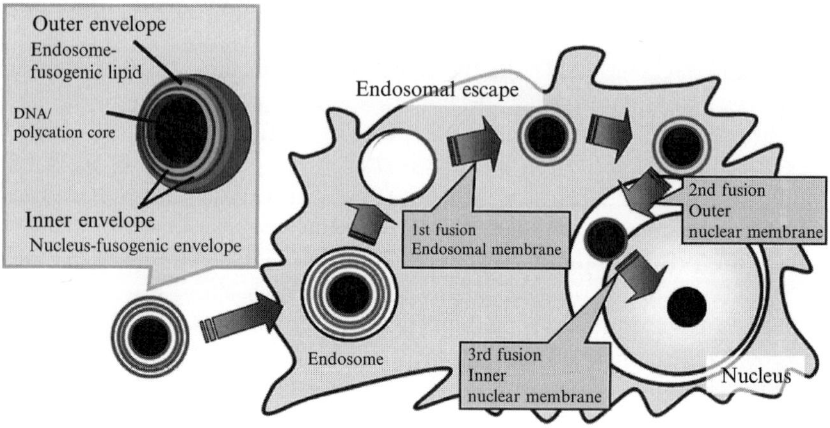

Figure 15.5 Schematic diagram illustrating a multicoational envelope-type nano device for a nuclear delivery of pDNA via membrane fusion. pDNA is condensed with a polycation and then coated with a nucleus-fusogenic lipid membrane (inner) and an endosome-fusogenic lipid membrane (outer). The carrier is designed to escape from the endosome into the cytosol with a fusogenic function of outer envelope (1st fusion). In the cytoplasmic, the carrier fuses with the nuclear membrane by virtue of surface-modified R8. The encapsulated pDNA cargo is delivered to the interior of the nucleus through successive fusions of the liposomal inner nucleus-fusogenic membrane with the two membranes of the nucleus (second and third fusion).

envelope as a probe. In the cells, inner envelope-derived fluorescence signals were detected as punctate forms both on the nucleus and in other intracellular regions. For an objective evaluation, clusters were randomly selected from both regions and the relative fluorescence intensity at 530 nm to that at 590 nm (F530/590) was calculated. As a result, F530/590 was significantly higher at the nucleus, suggesting that FRET was canceled preferentially at the nuclear membrane (Akita *et al.*, 2009).

4.3. Upgrading of T-MEND function by controlling intracellular trafficking and intranuclear decondensation

Further improvements in transfection activity were achieved by (i) modifying it with functional devices that further induce membrane fusion and (ii) controlling the status of pDNA condensation. As a fusion inducer, we have frequently used cholesteryl-GALA (Akita *et al.*, 2010; Kakudo *et al.*, 2004), which changes its conformation from a random coil to an α-helix in an acidic pH environment (i.e., endosomes and lysosomes) (Li *et al.*, 2004; Nicol *et al.*, 1996). Thus, this peptide is designed to overcome the endosomal membrane barrier. However, considering the environment in the cytoplasm, peptides that possess fusogenic activity in a neutral pH environment are essential for inducing fusion with the nuclear membrane. Thus, the KALA

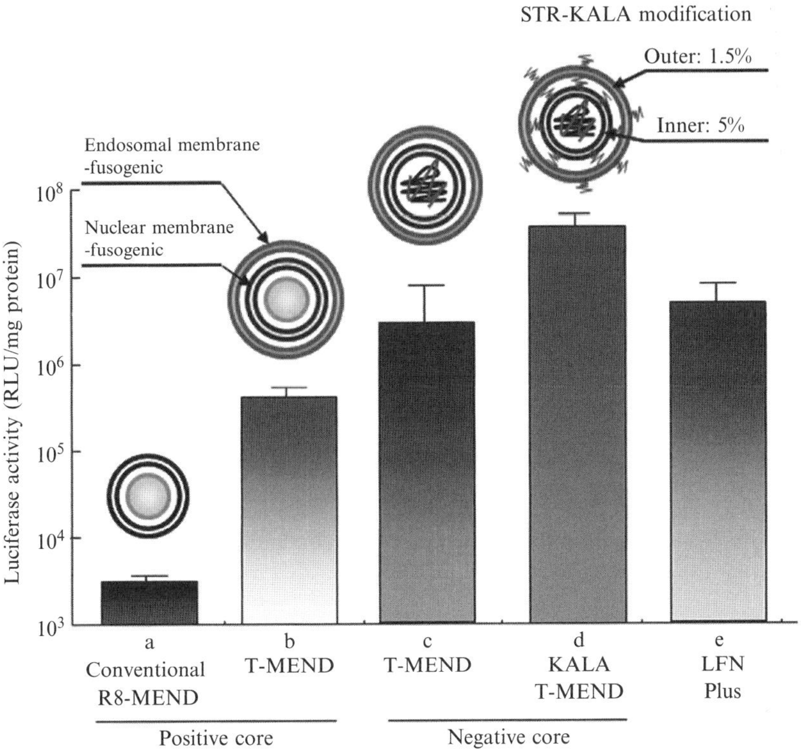

Figure 15.6 Function of T-MENDs as a gene-delivery carrier to the dendritic cells. JAWSII cells were transfected with MENDs and T-MENDs containing luciferase-encoding pDNA. Luciferase activity is expressed as relative light units (RLU) per milligram protein. Error bars represent the standard deviations for three independent experiments performed in triplicate. (a) Conventional R8-MEND composed of pDNA/protamine and DOPE/CHEMS as a lipid envelope component; (b) T-MEND prepared with positively charged pDNA/protamine core particle with nucleus-fusogenic inner lipid membrane (DOPE/CL) and an endosome-fusogenic outer lipid membrane (DOPE/PA); (c) T-MEND prepared with negatively charged pDNA/protamine core particle; (d) T-MEND prepared with negatively charged pDNA/protamine core particle and KALA-modified lipid envelopes at optimized density; (e) Lipofectamine PLUS (commercially available transfection reagent).

peptide was used as an alternative pH-independent fusion inducer (Wyman *et al.*, 1997). Cholesterol can be a candidate for anchoring, but it is also known to function as a lipid membrane stabilizer and could potentially inhibit the membrane fusion process. We alternatively employed a stearyl moiety for this purpose. In addition, based on the recent finding that the nuclear decondensation process is a key rate-limiting process for nonviral gene vectors (Hama *et al.*, 2006, 2007), we also altered the charge ratio (+/−), in an attempt to enhance the nuclear release of pDNA (Fig. 15.6b and c).

For anchoring the KALA peptide to the surface of liposomes to permit the KALA peptide to spontaneously orient themselves outward from the lipid envelopes, we synthesized KALA modified with a hydrophobic residue. KALA-modification further improved the transfection efficiency by one order of magnitude (Fig. 15.6c and d), and it exceeded Lipofectamine PLUS, a commercially available transfection reagent (Fig. 15.6d and e). These modifications acted synergistically to improve the transfection activities of the T-MENDs and reached expression levels that were sufficient for successful antigen presentation to MHC-Class I molecules (Shaheen *et al.*, 2011).

5. Mitochondrial Bioactive Molecule Delivery Using a Dual Function-MITO-Porter as a MEND for Mitochondrial Delivery

Mitochondrial dysfunction has recently been implicated in a variety of diseases (Chen and Chan, 2009; Kyriakouli *et al.*, 2008; Reeve *et al.*, 2008; Schapira, 2006; Tuppen *et al.*, 2010; Wallace, 2005). Mutations and defects of mitochondrial DNA (mtDNA) are thought to be causes of mitochondrial diseases. Therefore, mitochondrial gene therapy would be expected to be useful and productive for the treatment of various diseases. To achieve such an innovative therapy, it will be necessary to deliver therapeutic agents into mitochondria in living cells. In previous studies, we reported on the development of a MITO-Porter, a liposome-based carrier that is capable of introducing macromolecular cargos into mitochondria via membrane fusion (Yamada *et al.*, 2008). Using the green fluorescence protein as a model macromolecule and analysis by confocal laser scanning microscopy (CLSM), we were able to confirm mitochondrial macromolecule delivery by the MITO-Porter. In this section, we discuss our approaches regarding mitochondrial delivery of bioactive molecules using a dual function (DF)-MITO-Porter, which possesses mitochondria-fusogenic inner and endosome-fusogenic outer envelopes, aimed at mtDNA.

5.1. Schematic image of mitochondrial genome targeting and construction of a DF-MITO-Porter

The critical structural elements of the DF-MITO-Porter include a complexed particle of cargos that are coated with mitochondria-fusogenic inner membranes and endosome-fusogenic outer membranes (Fig. 15.7A). Modification of the outer envelope surface with a high density of R8 greatly assists in the efficient internalization of the carriers into cells (Khalil *et al.*, 2006). Inside the cell, the carrier escapes from the endosome into the cytosol via membrane fusion, a process that is mediated by the outer

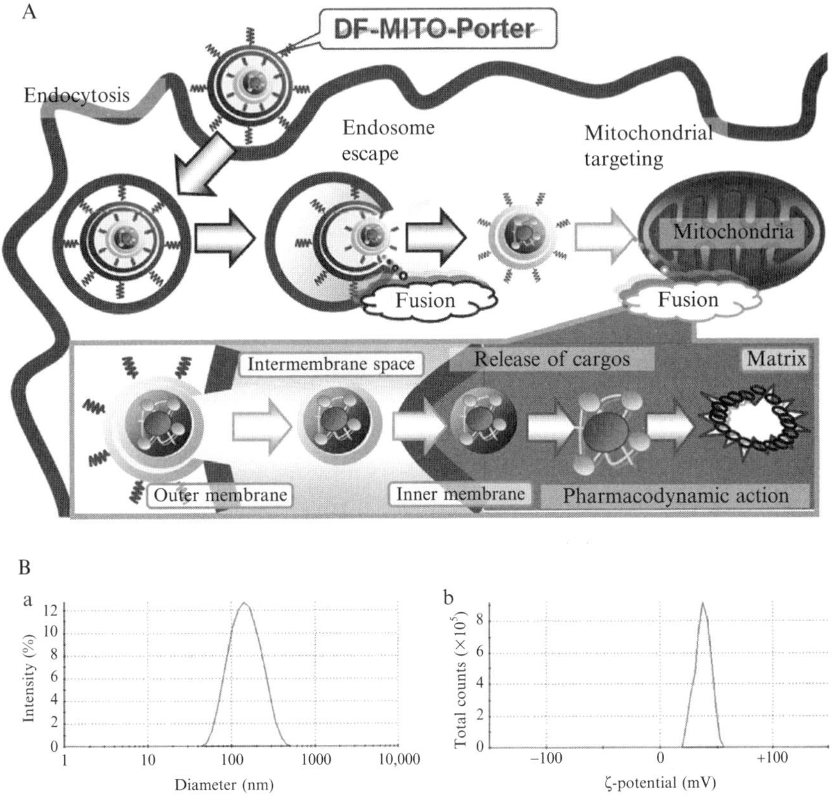

Figure 15.7 Schematic image of mitochondrial genome targeting using DF-MITO-Porter and the physicochemical property of the carrier. (A) Schematic image of mitochondrial delivery of DNase I protein via a series of membrane fusions using dual function (DF)-MITO-Porter. Octaarginine functions as a cell-uptake device. Once inside the cell, the carrier escapes from the endosome into the cytosol via membrane fusion, a process that is mediated by the outer endosome-fusogenic lipid membranes. The carrier then binds and fuses with the mitochondrial membrane to deliver the cargos to mitochondria. It was expected that mtDNA would be digested when the mitochondrial delivery of DNase I progressed. (B) Distribution of particle size (a) and ζ-potential (b) of the DF-MITO-Porter were measured.

endosome-fusogenic lipid membranes. The carrier then binds and fuses with the mitochondrial membrane. In this study, DNase I protein was chosen as a model bioactive macromolecule, which permitted us to estimate the mitochondrial gene targeting of the carrier. It was expected that mtDNA would be digested, as the mitochondrial matrix delivery of DNase I progressed.

The DF-MITO-Porter encapsulating DNase I was prepared based on the multilayering method, as shown in Fig. 15.3 (Akita *et al.*, 2009; Yamada *et al.*, 2011). The nanoparticle with a positive charge was constructed by mixing DNase I protein and STR-R8 at a complex-inducer/protein molar ratio of 10. The outer envelope had an endosome-fusogenic composition [DOPE/PA/STR-R8 (7:2:1, molar ratio)] (Akita *et al.*, 2009; El-Sayed *et al.*, 2008), while inner envelope had a mitochondria-fusogenic composition [DOPE/SM/CHEMS/STR-R8 (9:2:1:1, molar ratio)] (Yamada *et al.*, 2008). As shown in Fig. 15.7B, the DF-MITO-Porter was a positively charged nanoparticle with a homogeneous structure. A dynamic light scattering analysis indicated a single population that was small in size (Fig. 15.7B(a)). Furthermore, the population of the carrier also exhibited single peak in ζ-potential (Fig. 15.7B(b)).

5.2. Intracellular observation of DF-MITO-Porter and evaluation of mitochondrial targeting activity

Intracellular observation using CLSM permitted us to compare the mitochondrial targeting activity between DF-MITO-Porter and conventional MITO-Porter. The carriers labeled with 0.5 mol% NBD-DOPE are incubated with HeLa cells (final lipid concentration, 13.5 μmol/L). Mitochondria are stained with MitoFluor Red 589 (Molecular Probes, Eugene, OR) prior to observation by CLSM (LSM510; Carl Zeiss, Jena, Germany). A series of images are obtained using an LSM510 microscope with a water immersion objective lens (Achroplan 63×/NA=0.95). In the case of the DF-MITO-Porter, numerous yellow clusters were observed, indicating that the carriers (green) were colocalized with red-stained mitochondria (Yamada *et al.*, 2011).

We also evaluated the fraction of mitochondrial targeted positive cells based on images observed by CLSM. The fraction of mitochondrial targeted positive cells of the total number of observed cells is calculated as follows:

$$\text{Fraction of mitochondrial targeted positive cells (\%)} = \frac{N_{\text{mt}}}{N_{\text{tot}}} \times 100$$

where N_{mt} and N_{tot} represent the number of mitochondrial targeted positive cells and the total number of cells observed, respectively.

Z-series images of the intracellular location of the carriers are obtained. The yellow pixel areas where the carriers (green) are colocalized with mitochondria (red) are marked in each *xy* plane. We define cells where carriers are colocalized with mitochondria in at least one Z-series image as mitochondrial targeted positive cells. Consequently, the fraction of mitochondrial targeted positive cells of the DF-MITO-Porter (more than 80%) was considerably higher than the corresponding value for the conventional MITO-Porter (Fig. 15.8A) (Yamada *et al.*, 2011). This suggests that a nanostructure in which two types of envelopes with different lipid compositions

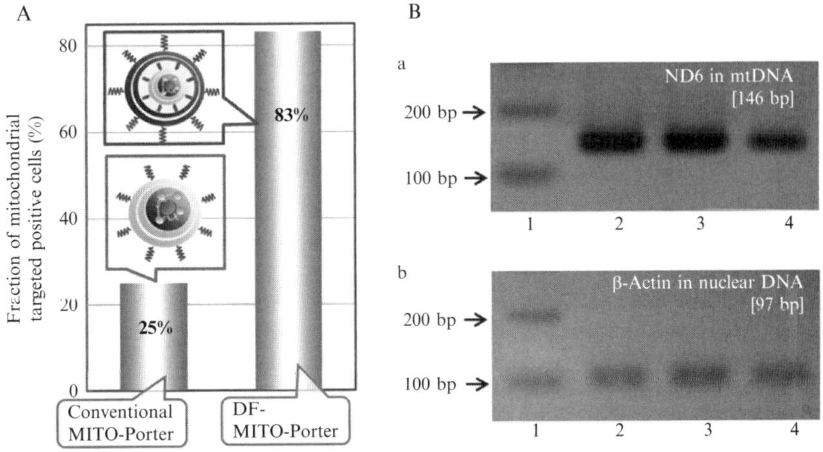

Figure 15.8 Evaluation of mitochondrial targeting activity and mtDNA levels after the mitochondrial delivery. (A) Fraction of mitochondrial targeted positive cells was calculated from the intracellular observation images. (B) Evaluation of mtDNA levels after the mitochondrial delivery of DNase I. After mitochondrial delivery of DNase I protein, cellular DNA was purified and subjected to the PCR, where ND6 and β-actin genes were assayed in order to detect mtDNA (a) and nuclear DNA (b). Lane 1, 100 bp DNA ladder; lane 2, nontreatment; lane 3, control carrier with low mitochondrial fusion activity; lane 4, DF-MITO-Porter.

(mitochondrial- and endosome-fusogenic lipids) are integrated can be very useful for mitochondrial delivery in living cells.

5.3. Evaluation of mtDNA levels after the mitochondrial delivery of DNase I using the DF-MITO-Porter

After the mitochondrial delivery of DNase I using the DF-MITO-Porter, we evaluated the mtDNA levels within cells using PCR. DNase I protein that was encapsulated in carriers is incubated with HeLa cells. The cellular DNA is then purified and subjected to PCR. PCR assays to detect the ND6 and β-actin genes are performed to evaluate both mtDNA and nuclear DNA levels. Figure 15.8B shows agarose gel electrophoresis data for the PCR products derived from mtDNA (Fig. 15.8B(a)) and nuclear DNA (Fig. 15.8B(b)). In the case of the DF-MITO-Porter, a decrease in the mtDNA level was observed (lanes 4 in Fig. 15.8B(a)), whereas carriers with a low mitochondrial fusion activity had a negligible effect on mtDNA levels (lane 3 in Fig. 15.8B(a)). This result suggests that DNase I is delivered to mitochondria by the DF-MITO-Porter, as indicated by the digestion of mtDNA by DNase I to some extent. On the other hand, no decrease in nuclear DNA levels was detected for any of the carriers (Fig. 15.8B(b)). Based on these results, it is presumed that the mitochondrial specific fusion activity of the DF-MITO-Porter might be involved in a pathway related to the selective digestion of mtDNA (Yamada *et al.*, 2011), demonstrating its potential for use in therapies aimed at mtDNA.

6. Conclusions

In summary, we propose a novel strategy for the targeted delivery of macromolecules to a specific organelle (i.e., nucleus and mitochondria) based on the concept of stepwise membrane fusion, which is achieved using multi-coated nanoparticles with endosome- and organelle-fusogenic lipid envelopes. At this time, a double-layered coating is added in a stepwise manner, based on the fusion of SUVs. Thus, in principle, only a particle with even numbers of lipid envelopes can be prepared. In the future, we hope to establish methods for coating with a single lamellar lipid envelope (lipid bilayers) and studies in the area are currently ongoing. This would be an essential technology for the development of a Triple-lamellar MEND, which is the most appropriate system for overcoming the one endosomal membrane and double nuclear/mitochondrial membranes. Also, additional customizing of the lipid composition and/or modification of functional devices with reference to the targeted cells and its application may permit this particle to be used for *in vivo* gene delivery.

ACKNOWLEDGMENTS

This work was supported, in part by, the Program for Promotion of Fundamental Studies in Health Sciences of the National Institute of Biomedical Innovation, Japan (NIBIO), Funding Program for Next Generation World-Leading Researchers (NEXT Program), a Grant-in-Aid for Young Scientists (A) and a Grant-in-Aid for Scientific Research (S) from the Ministry of Education, Culture, Sports, Science and Technology of Japanese Government (MEXT). H.A. is also supported by the Asahi Glass Foundation. We also thank Dr. Milton Feather for his helpful advice in writing the chapter.

REFERENCES

Akita, H., Kudo, A., Minoura, A., Yamaguti, M., Khalil, I. A., Moriguchi, R., Masuda, T., Danev, R., Nagayama, K., Kogure, K., and Harashima, H. (2009). Multi-layered nanoparticles for penetrating the endosome and nuclear membrane via a step-wise membrane fusion process. *Biomaterials* **30,** 2940–2949.

Akita, H., Kogure, K., Moriguchi, R., Nakamura, Y., Higashi, T., Nakamura, T., Serada, S., Fujimoto, M., Naka, T., Futaki, S., and Harashima, H. (2010). Nanoparticles for ex vivo siRNA delivery to dendritic cells for cancer vaccines: Programmed endosomal escape and dissociation. *J. Control. Release* **143,** 311–317.

Allen, N. P., Patel, S. S., Huang, L., Chalkley, R. J., Burlingame, A., Lutzmann, M., Hurt, E. C., and Rexach, M. (2002). Deciphering networks of protein interactions at the nuclear pore complex. *Mol. Cell. Proteomics* **1,** 930–946.

Ben-Efraim, I., and Gerace, L. (2001). Gradient of increasing affinity of importin beta for nucleoporins along the pathway of nuclear import. *J. Cell Biol.* **152,** 411–417.

Breuzard, G., Tertil, M., Goncalves, C., Cheradame, H., Geguan, P., Pichon, C., and Midoux, P. (2008). Nuclear delivery of NFκB-assisted DNA/polymer complexes: Plasmid DNA quantitation by confocal laser scanning microscopy and evidence of nuclear polyplexes by FRET imaging. *Nucleic Acids Res.* **36,** e71.

Brunner, S., Sauer, T., Carotta, S., Cotten, M., Saltik, M., and Wagner, E. (2000). Cell cycle dependence of gene transfer by lipoplex, polyplex and recombinant adenovirus. *Gene Ther.* **7,** 401–407.

Brunner, S., Furtbauer, E., Sauer, T., Kursa, M., and Wagner, E. (2002). Overcoming the nuclear barrier: Cell cycle independent nonviral gene transfer with linear polyethylenimine or electroporation. *Mol. Ther.* **5,** 80–86.

Chan, C. K., and Jans, D. A. (1999). Enhancement of polylysine-mediated transferrinfection by nuclear localization sequences: Polylysine does not function as a nuclear localization sequence. *Hum. Gene Ther.* **10,** 1695–1702.

Chan, C. K., Senden, T., and Jans, D. A. (2000). Supramolecular structure and nuclear targeting efficiency determine the enhancement of transfection by modified polylysines. *Gene Ther.* **7,** 1690–1697.

Chen, H., and Chan, D. C. (2009). Mitochondrial dynamics—Fusion, fission, movement, and mitophagy—In neurodegenerative diseases. *Hum. Mol. Genet.* **18,** R169–R176.

Danev, R., and Nagayama, K. (2001). Transmission electron microscopy with Zernike phase plate. *Ultramicroscopy* **88,** 243–252.

Dworetzky, S. I., Lanford, R. E., and Feldherr, C. M. (1988). The effects of variations in the number and sequence of targeting signals on nuclear uptake. *J. Cell Biol.* **107,** 1279–1287.

El-Sayed, A., Khalil, I. A., Kogure, K., Futaki, S., and Harashima, H. (2008). Octaarginine- and octalysine-modified nanoparticles have different modes of endosomal escape. *J. Biol. Chem.* **283,** 23450–23461.

Endo, T., Nakayama, Y., and Nakai, M. (1995). Avidin fusion protein as a tool to generate a stable translocation intermediate spanning the mitochondrial membranes. *J. Biochem.* **118,** 753–759.

Flierl, A., Jackson, C., Cottrell, B., Murdock, D., Seibel, P., and Wallace, D. C. (2003). Targeted delivery of DNA to the mitochondrial compartment via import sequence-conjugated peptide nucleic acid. *Mol. Ther.* **7,** 550–557.

Futaki, S., Ohashi, W., Suzuki, T., Niwa, M., Tanaka, S., Ueda, K., Harashima, H., and Sugiura, Y. (2001). Stearylated arginine-rich peptides: A new class of transfection systems. *Bioconjug. Chem.* **12,** 1005–1011.

Hagstrom, J. E., Ludtke, J. J., Bassik, M. C., Sebestyen, M. G., Adam, S. A., and Wolff, J. A. (1997). Nuclear import of DNA in digitonin-permeabilized cells. *J. Cell Sci.* **110**(Pt. 18), 2323 2331.

Hama, S., Akita, H., Ito, R., Mizuguchi, H., Hayakawa, T., and Harashima, H. (2006). Quantitative comparison of intracellular trafficking and nuclear transcription between adenoviral and lipoplex systems. *Mol. Ther.* **13,** 786–794.

Hama, S., Akita, H., Iida, S., Mizuguchi, H., and Harashima, H. (2007). Quantitative and mechanism-based investigation of post-nuclear delivery events between adenovirus and lipoplex. *Nucleic Acids Res.* **35,** 1533–1543.

James, M. B., and Giorgio, T. D. (2000). Nuclear-associated plasmid, but not cell-associated plasmid, is correlated with transgene expression in cultured mammalian cells. *Mol. Ther.* **1,** 339–346.

Kakudo, T., Chaki, S., Futaki, S., Nakase, I., Akaji, K., Kawakami, T., Maruyama, K., Kamiya, H., and Harashima, H. (2004). Transferrin-modified liposomes equipped with a pH-sensitive fusogenic peptide: An artificial viral-like delivery system. *Biochemistry* **43,** 5618–5628.

Khalil, I. A., Kogure, K., Futaki, S., and Harashima, H. (2006). High density of octaarginine stimulates macropinocytosis leading to efficient intracellular trafficking for gene expression. *J. Biol. Chem.* **281,** 3544–3551.

Khalil, I. A., Kogure, K., Futaki, S., Hama, S., Akita, H., Ueno, M., Kishida, H., Kudoh, M., Mishina, Y., Kataoka, K., Yamada, M., and Harashima, H. (2007). Octaarginine-modified multifunctional envelope-type nanoparticles for gene delivery. *Gene Ther.* **14,** 682–689.

Kogure, K., Moriguchi, R., Sasaki, K., Ueno, M., Futaki, S., and Harashima, H. (2004). Development of a non-viral multifunctional envelope-type nano device by a novel lipid film hydration method. *J. Control. Release* **98,** 317–323.

Kogure, K., Akita, H., Yamada, Y., and Harashima, H. (2008). Multifunctional envelope-type nano device (MEND) as a non-viral gene delivery system. *Adv. Drug Deliv. Rev.* **60,** 559–571.

Kyriakouli, D. S., Boesch, P., Taylor, R. W., and Lightowlers, R. N. (2008). Progress and prospects: Gene therapy for mitochondrial DNA disease. *Gene Ther.* **15,** 1017–1023.

Lee, R. J., and Huang, L. (1996). Folate-targeted, anionic liposome-entrapped polylysine-condensed DNA for tumor cell-specific gene transfer. *J. Biol. Chem.* **271,** 8481–8487.

Li, W., Nicol, F., and Szoka, F. C., Jr. (2004). GALA: A designed synthetic pH-responsive amphipathic peptide with applications in drug and gene delivery. *Adv. Drug Deliv. Rev.* **56,** 967–985.

Ludtke, J. J., Zhang, G., Sebestyen, M. G., and Wolff, J. A. (1999). A nuclear localization signal can enhance both the nuclear transport and expression of 1 kb DNA. *J. Cell Sci.* **112**(Pt. 12), 2033–2041.

Maier, O., Oberle, V., and Hoekstra, D. (2002). Fluorescent lipid probes: Some properties and applications (a review). *Chem. Phys. Lipids* **116,** 3–18.

Marenzi, S., Adams, R. L., Zardo, G., Lenti, L., Reale, A., and Caiafa, P. (1999). Efficiency of expression of transfected genes depends on the cell cycle. *Mol. Biol. Rep.* **26,** 261–267.

Masuda, T., Akita, H., and Harashima, H. (2005). Evaluation of nuclear transfer and transcription of plasmid DNA condensed with protamine by microinjection: The use of a nuclear transfer score. *FEBS Lett.* **579,** 2143–2148.

Mesika, A., Grigoreva, I., Zohar, M., and Reich, Z. (2001). A regulated, NFkappaB-assisted import of plasmid DNA into mammalian cell nuclei. *Mol. Ther.* **3,** 653–657.

Mesika, A., Kiss, V., Brumfeld, V., Ghosh, G., and Reich, Z. (2005). Enhanced intracellular mobility and nuclear accumulation of DNA plasmids associated with a karyophilic protein. *Hum. Gene Ther.* **16,** 200–208.

Moriguchi, R., Kogure, K., Akita, H., Futaki, S., Miyagishi, M., Taira, K., and Harashima, H. (2005). A multifunctional envelope-type nano device for novel gene delivery of siRNA plasmids. *Int. J. Pharm.* **301,** 277–285.

Mortimer, I., Tam, P., MacLachlan, I., Graham, R. W., Saravolac, E. G., and Joshi, P. B. (1999). Cationic lipid-mediated transfection of cells in culture requires mitotic activity. *Gene Ther.* **6,** 403–411.

Nagasaki, T., Myohoji, T., Tachibana, T., Futaki, S., and Tamagaki, S. (2003). Can nuclear localization signals enhance nuclear localization of plasmid DNA? *Bioconjug. Chem.* **14,** 282–286.

Nakamura, Y., Kogure, K., Yamada, Y., Futaki, S., and Harashima, H. (2006). Significant and prolonged antisense effect of a multifunctional envelope-type nano device encapsulating antisense oligodeoxynucleotide. *J. Pharm. Pharmacol.* **58,** 431–437.

Nakamura, Y., Kogure, K., Futaki, S., and Harashima, H. (2007). Octaarginine-modified multifunctional envelope-type nano device for siRNA. *J. Control. Release* **119,** 360–367.

Nicol, F., Nir, S., and Szoka, F. C., Jr. (1996). Effect of cholesterol and charge on pore formation in bilayer vesicles by a pH-sensitive peptide. *Biophys. J.* **71,** 3288–3301.

Ohta, A., Danev, R., Nagayama, K., Mita, T., Asakawa, T., and Miyagishi, S. (2006). Transition from nanotubes to micelles with increasing concentration in dilute aqueous solution of potassium N-acyl phenylalaninate. *Langmuir* **22,** 8472–8477.

Pante, N., and Kann, M. (2002). Nuclear pore complex is able to transport macromolecules with diameters of about 39 nm. *Mol. Biol. Cell* **13,** 425–434.

Pollard, H., Remy, J. S., Loussouarn, G., Demolombe, S., Behr, J. P., and Escande, D. (1998). Polyethylenimine but not cationic lipids promotes transgene delivery to the nucleus in mammalian cells. *J. Biol. Chem.* **273,** 7507–7511.

Reeve, A. K., Krishnan, K. J., and Turnbull, D. (2008). Mitochondrial DNA mutations in disease, aging, and neurodegeneration. *Ann. NY Acad. Sci.* **1147,** 21–29.

Schapira, A. H. (2006). Mitochondrial disease. *Lancet* **368,** 70–82.

Schatz, G. (1987). 17th Sir Hans Krebs lecture. Signals guiding proteins to their correct locations in mitochondria. *Eur. J. Biochem.* **165,** 1–6.

Sebestyen, M. G., Ludtke, J. J., Bassik, M. C., Zhang, G., Budker, V., Lukhtanov, E. A., Hagstrom, J. E., and Wolff, J. A. (1998). DNA vector chemistry: The covalent attachment of signal peptides to plasmid DNA. *Nat. Biotechnol.* **16,** 80–85.

Seibel, M., Bachmann, C., Schmiedel, J., Wilken, N., Wilde, F., Reichmann, H., Isaya, G., Seibel, P., and Pfeiler, D. (1999). Processing of artificial peptide-DNA-conjugates by the mitochondrial intermediate peptidase (MIP). *Biol. Chem.* **380,** 961–967.

Shaheen, S. M., Akita, H., Nakamura, T., Takayama, S., Futaki, S., Yamashita, A., Katoono, R., Yui, N., and Harashima, H. (2011). KALA-modified multi-layered nanoparticles as gene carriers for MHC class-I mediated antigen presentation for a DNA vaccine. *Biomaterials* **32,** 6342–6350.

Shinohara, Y., Sagawa, I., Ichihara, J., Yamamoto, K., Terao, K., and Terada, H. (1997). Source of ATP for hexokinase-catalyzed glucose phosphorylation in tumor cells: Dependence on the rate of oxidative phosphorylation relative to that of extramitochondrial ATP generation. *Biochim. Biophys. Acta* **1319,** 319–330.

Shinohara, Y., Almofti, M. R., Yamamoto, T., Ishida, T., Kita, F., Kanzaki, H., Ohnishi, M., Yamashita, K., Shimizu, S., and Terada, H. (2002). Permeability transition-independent release of mitochondrial cytochrome c induced by valinomycin. *Eur. J. Biochem.* **269,** 5224–5230.

Stewart, M. (2007). Molecular mechanism of the nuclear protein import cycle. *Nat. Rev. Mol. Cell Biol.* **8,** 195–208.

Struck, D. K., Hoekstra, D., and Pagano, R. E. (1981). Use of resonance energy transfer to monitor membrane fusion. *Biochemistry* **20,** 4093–4099.

Subramanian, A., Ranganathan, P., and Diamond, S. L. (1999). Nuclear targeting peptide scaffolds for lipofection of nondividing mammalian cells. *Nat. Biotechnol.* **17,** 873–877.

Suzuki, R., Yamada, Y., and Harashima, H. (2007). Efficient cytoplasmic protein delivery by means of a multifunctional envelope-type nano device. *Biol. Pharm. Bull.* **30,** 758–762.

Suzuki, R., Yamada, Y., and Harashima, H. (2008). Development of small, homogeneous pDNA particles condensed with mono-cationic detergents and encapsulated in a multifunctional envelope-type nano device. *Biol. Pharm. Bull.* **31,** 1237–1243.

Tanimoto, M., Kamiya, H., Minakawa, N., Matsuda, A., and Harashima, H. (2003). No enhancement of nuclear entry by direct conjugation of a nuclear localization signal peptide to linearized DNA. *Bioconjug. Chem.* **14,** 1197–1202.

Tseng, W. C., Haselton, F. R., and Giorgio, T. D. (1999). Mitosis enhances transgene expression of plasmid delivered by cationic liposomes. *Biochim. Biophys. Acta* **1445,** 53–64.

Tuppen, H. A., Blakely, E. L., Turnbull, D. M., and Taylor, R. W. (2010). Mitochondrial DNA mutations and human disease. *Biochim. Biophys. Acta* **1797,** 113–128.

Vestweber, D., and Schatz, G. (1989). DNA-protein conjugates can enter mitochondria via the protein import pathway. *Nature* **338,** 170–172.

Wallace, D. C. (2005). The mitochondrial genome in human adaptive radiation and disease: On the road to therapeutics and performance enhancement. *Gene* **354,** 169–180.

Wiedemann, N., Frazier, A. E., and Pfanner, N. (2004). The protein import machinery of mitochondria. *J. Biol. Chem.* **279,** 14473–14476.

Wilke, M., Fortunati, E., van den Broek, M., Hoogeveen, A. T., and Scholte, B. J. (1996). Efficacy of a peptide-based gene delivery system depends on mitotic activity. *Gene Ther.* **3,** 1133–1142.

Wyman, T. B., Nicol, F., Zelphati, O., Scaria, P. V., Plank, C., and Szoka, F. C., Jr. (1997). Design, synthesis, and characterization of a cationic peptide that binds to nucleic acids and permeabilizes bilayers. *Biochemistry* **36,** 3008–3017.

Yamada, Y., and Harashima, H. (2008). Mitochondrial drug delivery systems for macromolecule and their therapeutic application to mitochondrial diseases. *Adv. Drug Deliv. Rev.* **60,** 1439–1462.

Yamada, Y., Kogure, K., Nakamura, Y., Inoue, K., Akita, H., Nagatsugi, F., Sasaki, S., Suhara, T., and Harashima, H. (2005). Development of efficient packaging method of oligodeoxynucleotides by a condensed nano particle in lipid envelope structure. *Biol. Pharm. Bull.* **28,** 1939–1942.

Yamada, Y., Akita, H., Kamiya, H., Kogure, K., Yamamoto, T., Shinohara, Y., Yamashita, K., Kobayashi, H., Kikuchi, H., and Harashima, H. (2008). MITO-Porter: A liposome-based carrier system for delivery of macromolecules into mitochondria via membrane fusion. *Biochim. Biophys. Acta* **1778,** 423–432.

Yamada, Y., Nomura, T., Harashima, H., Yamashita, A., Katoono, R., and Yui, N. (2010). Intranuclear DNA release is a determinant of transfection activity for a non-viral vector: Biocleavable polyrotaxane as a supramolecularly dissociative condenser for efficient intranuclear DNA release. *Biol. Pharm. Bull.* **33,** 1218–1222.

Yamada, Y., Furukawa, R., Yasuzaki, Y., and Harashima, H. (2011). Dual function MITO-Porter, a nano carrier integrating both efficient cytoplasmic delivery and mitochondrial macromolecule delivery. *Mol. Ther.* **19,** 1449–1456.

Yoneda, Y., Semba, T., Kaneda, Y., Noble, R. L., Matsuoka, Y., Kurihara, T., Okada, Y., and Imamoto, N. (1992). A long synthetic peptide containing a nuclear localization signal and its flanking sequences of SV40 T-antigen directs the transport of IgM into the nucleus efficiently. *Exp. Cell Res.* **201,** 313–320.

CHAPTER SIXTEEN

Lipopolyplexes as Nanomedicines for Therapeutic Gene Delivery

Leire García,* Koldo Urbiola,* Nejat Düzgüneş,† *and* Conchita Tros de Ilarduya*

Contents

1. Introduction	328
2. Principle of the Method	329
3. Experimental Procedures	329
3.1. Materials	329
3.2. Cell culture	330
3.3. Preparation of lipopolyplexes	330
3.4. DNA/PEI condensation assay	331
3.5. Particle size and zeta potential measurements	331
3.6. *In vitro* transfection activity	331
4. Application of Lipopolyplexes	332
4.1. Characterization of complexes	332
4.2. *In vitro* transfection activity measurements	334
5. Concluding Remarks	336
Acknowledgments	336
References	337

Abstract

We describe an efficient, nonviral gene transfer system that employs polyethylenimine (PEI 800, 25, 22 kDa), and 1,2-dioleoyl-3-(trimethylammonium) propane (DOTAP) and cholesterol (Chol) as lipids (lipopolyplex), at three different lipid/DNA molar ratios (2/1, 5/1, and 17/1), employing five different formulation strategies. PEIs of 800, 25, and 22 kDa are highly effective in condensing plasmid DNA, leading to a complete condensation at N/P (+/−) ratios above 4. Increasing the molar ratio lipid/DNA in the complex results in higher positive values of the zeta potential, while the particle size increases in some protocols, but not in others. PEI of molecular weight 800 kDa used in the formulation of

* Department of Pharmacy and Pharmaceutical Technology, School of Pharmacy, University of Navarra, Pamplona, Spain

† Department of Biomedical Sciences, University of the Pacific Arthur A. Dugoni School of Dentistry, San Francisco, California, USA

Methods in Enzymology, Volume 509
ISSN 0076-6879, DOI: 10.1016/B978-0-12-391858-1.00016-2

lipopolyplexes results in bigger particles compared to that obtained with the smaller PEI species. Transfection activity is measured using pCMVLuc expressing luciferase is maximal by using strategies 3 and 4 and an N/P molar ratio of 17/1. These complexes have a high efficiency of gene delivery to liver cancer cells, even in the presence of a high serum concentration. Complexes formed with linear PEI are more effective than lipopolyplexes containing branched PEI. The ternary complexes are much more efficient than conventional lipoplexes (cationic lipid and DNA) and polyplexes (cationic polymer and DNA). The same behavior is observed for complexes prepared with the therapeutic gene pCMVIL-12 expressing interleukin-12.

1. Introduction

Gene therapy focuses on the therapeutic use of genes delivered to cells, and may lead to advances in the treatment of numerous diseases including cancer and genetic diseases (Drew and Martin, 1999). Although progress has been made in identifying target structures for cancer gene therapy, actual therapy is limited by the lack of a safe and efficient gene delivery system. Currently, one of the primary objectives of gene therapy is the development of efficient, nontoxic gene carriers that can effectively deliver therapeutic genetic material into specific cell types, including cancer cells.

The two main gene carrier systems that have been utilized in gene therapy are viral vectors and nonviral delivery systems (Pedroso de Lima *et al.*, 2001; Tros de Ilarduya *et al.*, 2010). Viral vectors, including retroviruses, adenoviruses, and adeno-associated viruses, have a high efficiency of gene delivery. Because of serious safety risks with viral vectors that have become apparent in the past few years, however, their utility is being reappraised. Furthermore, the addition of targeting ligands on the surface of viral vectors to transfect specific cell types is problematic. Because of these concerns, nonviral vectors are emerging as a viable alternative. Nonviral systems present significantly lower safety risks, and can be produced easily and inexpensively in large quantities. A major disadvantage of nonviral gene carriers is their low transfection efficiency, especially in the presence of serum.

Strategies such as particle bombardment (Mahato *et al.*, 1999), ultrasound transfection (Newman *et al.*, 2001), or the application of naked DNA (Shi *et al.*, 2002) have been used to deliver genes. Cationic polymers and lipids are the most widely used vectors in nonviral gene and oligonucleotide delivery. Some liposomal formulations have serious limitations because of their low transfection efficiency and cytotoxicity (Gao and Huang, 1995; Meyer *et al.*, 1998). Polycationic polymers can condense DNA, which is an advantage in gene transfer (Goldman *et al.*, 1997; Sorgi *et al.*, 1997). Polyethylenimine (PEI) has been used successfully for *in vitro* and *in vivo* gene delivery (Boussif *et al.*, 1995; Lungwitz *et al.*, 2005). It can condense and protect DNA, facilitate binding to the cell surface, trigger endocytosis, and mediate the release DNA/lipid complexes from endosomes into the cytoplasm as a result of the "proton-sponge" effect (Behr, 1994). It can enter the nucleus (Godbey *et al.*, 1999) and thus accelerate DNA entry into the nucleus (Pollard *et al.*, 1998). Although cationic liposomes can deliver DNA into the cytosol following endocytosis, the entry of the DNA into the nucleus is inefficient (Fisher *et al.*, 2000; Harashima *et al.*, 2001).

2. Principle of the Method

Lipopolyplexes (i.e., a ternary complexes of cationic liposomes, cationic polymer, and DNA) constitute a second generation of nonviral gene carriers that can improve gene transfer compared to the first generation cationic liposome-DNA complexes (lipoplexes) (Gao and Huang, 1996; Lampleta *et al.*, 2003; Matsura *et al.*, 2003; Pelisek *et al.*, 2006; Sorgi *et al.*, 1997; Whitmore *et al.*, 1999; Yamazaki *et al.*, 2000).

Because of our interest in developing new methods to prepare lipopolyplexes as nanomedicines for therapeutic gene delivery, we have evaluated different parameters to optimize the formulation to achieve high transfection activity, including the protocol of preparation, the lipid/DNA molar ratio, and the molecular weight and type of PEI. We hypothesized that the association of PEI with cationic liposomes (lipopolyplexes) would increase luciferase expression compared to lipoplexes (cationic lipid and DNA) and polyplexes (cationic polymer and DNA) alone (García *et al.*, 2007).

3. Experimental Procedures

3.1. Materials

The cationic lipid, 1,2-dioleoyl-3-(trimethylammonium) propane (DOTAP), and cholesterol (Chol) are obtained from Avanti Polar Lipids (Alabaster, AL, USA). PEI 800 (MW 800 kDa, branched) is available from Fluka (Steinheim, Germany), PEI 25 (MW 25 kDa, branched) from Aldrich (Madrid, Spain), and linear PEI (MW 22 kDa, ExGen® 500) from Quimigranel (Madrid, Spain). The plasmids, pCMVLuc (VR-1216) (6934 bps)

(Clontech, Palo Alto, CA, USA) and pCMV100-IL-12 (5500 bps) (kindly provided by Dr. Chen Qian, University of Navarra), encoding luciferase and interleukin-12 (IL-12), respectively, are used for these studies.

3.2. Cell culture

HepG2 human hepatoblastoma cells (American Type Culture Collection, Rockville, MD, USA) are maintained at 37 °C under a 5% CO_2 atmosphere in Dulbecco's modified Eagle's medium high glucose, supplemented with 10% (v/v) heat-inactivated fetal bovine serum (FBS), penicillin (100 units/ml), streptomycin (100 μg/ml), and L-glutamine (4 m*M*; Gibco BRL Life Technologies). Cells are passaged 1:10 following trypsinization once a week.

3.3. Preparation of lipopolyplexes

Lipopolyplexes are prepared with plasmid DNA and B-PEI (branched, 800 or 25 kDa) or L-PEI (linear, 22 kDa) at an N/P ratio of 4. Different amounts of lipids are added to prepare complexes at molar ratios of total lipid/DNA of 2/1, 5/1, and 17/1. The final DNA concentration in the lipopolyplex suspension is 25 μg/ml in a total volume of 3 ml.

Lipopolyplexes are formulated using the following strategies:

Strategy 1 involves drying a chloroform solution of the lipids, DOTAP/Chol (1:0.9 molar ratio), by rotary evaporation, and then hydrating the film with the polyplexes (PEI/DNA) (N/P ratio of 4).

Strategies 2 and 3 are carried out by drying a chloroform solution of the lipids (DOTAP/Chol) and PEI by rotary evaporation. In *Strategy 2*, the film is hydrated with water, and then the plasmid is added. In *Strategy 3*, the mixture of lipids and PEI are hydrated with the plasmid solution.

In strategies 4 and 5, the polyplexes, PEI/DNA, are generated at an N/P ratio of 4 and, after a 15 min incubation, different amounts of preformed cationic liposomes are added, and to obtain complexes at various lipid/DNA N/P ratios (2/1, 5/1, and 17/1). These strategies differ from each other only in the order of addition of plasmid to obtain the polyplexes: PEI added to DNA (*Strategy 4*) or DNA added to PEI (*Strategy 5*).

The prepared complexes are extruded through polycarbonate membranes (100 nm pore diameter), using a Liposofast device (Avestin, Toronto, ON, Canada) to obtain particles with a uniform size distribution.

3.4. DNA/PEI condensation assay

The binding of PEI to DNA is investigated using ethidium bromide, whose fluorescence is enhanced greatly upon binding to DNA, and quenched when it is displaced by the condensation of the DNA structure (Tros de Ilarduya *et al.*, 2002). The assays are carried out in 96-well plates in 10% (w/v) glucose, 10 m*M* HEPES buffer (pH 7.4). DNA (0.6 μg) is mixed with 1.2 μg of ethidium bromide, then increasing amounts of PEI are added to the wells and incubated for 10 min in the dark. Fluorescence is read in an LS 50 spectrofluorimeter (Perkin-Elmer, Mountain View, CA, USA), at excitation and emission wavelengths of 520 and 600 nm, respectively. The relative fluorescence values are evaluated as follows: $F_r = (F_{obs} - F_e) \times 100/(F_0 - F_e)$, where F_r is the relative fluorescence, F_{obs} is the measured fluorescence, F_e is the fluorescence of ethidium bromide in the absence of DNA, and F_0 is the initial fluorescence in the absence of the polycation.

3.5. Particle size and zeta potential measurements

The particle size of the complexes is measured by dynamic light scattering, and the overall charge of the particles by zeta potential measurements, using a Zeta Nano Series particle analyzer (Malvern Instruments, Spain). Aliquots of the complexes are diluted in distilled water, and are measured at least three times immediately after the preparation of the complexes.

3.6. *In vitro* transfection activity

The cells are seeded in medium in 48-well culture plates (Iwaki Microplate, Japan) and incubated for 24 h at 37 °C in 5% CO_2. The medium is removed and 0.2 ml of the complexes (containing 1 μg of plasmid) and 0.3 ml of FBS are added to each well. The complexes are thus incubated with the cells for 4 h in the presence of 60% FBS. The complexes are then removed, and medium containing 10% FBS is added. The cells are further incubated for 48 h; then they are washed with phosphate-buffered saline and lysed with 100 μl of Reporter Lysis Buffer (Promega, Madison, WI, USA). The mixture is incubated at room temperature for 10 min, and then frozen and thawed twice. The cell lysate is centrifuged for 2 min at 12000 × *g* to pellet cellular debris. Twenty microliters of the supernatant are assayed for total luciferase activity, using the Luciferase Assay Reagent (Promega), according to the manufacturer's instructions. A luminometer (Sirius-2, Berthold Detection Systems, Innogenetics, Diagnóstica y Terapéutica, Barcelona, Spain) is used to measure luciferase activity. The protein content of the lysates is measured by the DC protein Assay Reagent (Bio-Rad, Hercules, CA, USA), with bovine serum albumin as the standard. The data are expressed as nanograms of luciferase (based on a standard curve for luciferase activity) per milligram of protein. IL-12 levels are determined by an ELISA kit for murine IL-12p70 (BD OptEIA ELISA sets, Pharmingen, San Diego, CA, USA), according to the manufacturer's instructions.

4. Application of Lipopolyplexes

4.1. Characterization of complexes

We prepared lipopolyplexes according to each of the five strategies by using PEI 800, 25, and 22 kDa, and DOTAP and cholesterol at lipid/DNA N/P ratios of 2/1, 5/1, and 17/1, as described above.

To prepare complexes in which DNA was condensed completely by PEI, the corresponding condensation studies were performed, by measuring the decrease in the ethidium bromide fluorescence. As shown in Fig. 16.1, the fluorescence decreased with increasing N/P ratio (García *et al.*, 2007). The fluorescence reached a plateau at N/P ratios of 4–10, with the DNA being condensed to less than 10% of the uncomplexed DNA, suggesting that DNA is condensed completely at these ratios. The same behavior was observed with all three polymers of PEI. Increasing amounts of lipid added to polyplexes at an N/P ratio of 4 did not show any additional effect on the condensation of the plasmid.

Complexes were characterized in terms of particle size and zeta potential. At a lipid/DNA ratio of 2/1, the largest lipopolyplexes were those prepared using *Strategy 2* (281 nm), whereas at the higher ratios, *Strategy 3*

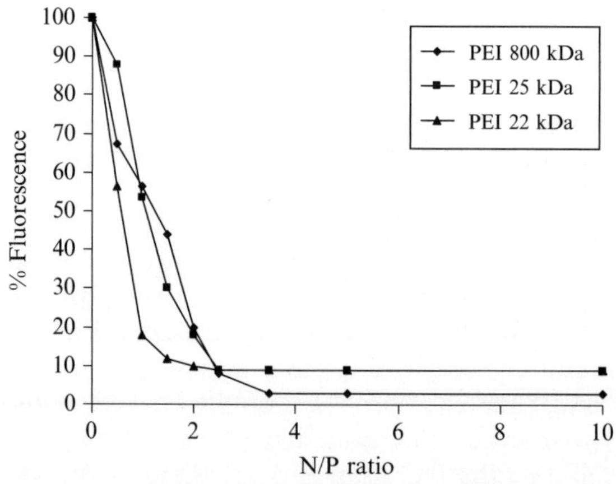

Figure 16.1 PEI/DNA condensation assay. DNA condensation measured as a decrease in the fluorescence of EtBr added to polyplexes (From García *et al.*, 2007).

Table 16.1 Particle size of lipopolyplexes, prepared with PEI 800 kDa, as a function of the protocol of formulation at the indicated lipid/DNA molar ratios (From García *et al.*, 2007)

	Particle size (nm)				
Molar ratio (lipid/DNA)	*Strategy 1*	*Strategy 2*	*Strategy 3*	*Strategy 4*	*Strategy 5*
2/1	188 ± 8	281 ± 10	229 ± 14	157 ± 3	124 ± 2
5/1	187 ± 6	232 ± 8	356 ± 11	181 ± 4	183 ± 1
17/1	205 ± 4	211 ± 20	295 ± 2	214 ± 8	201 ± 8

Table 16.2 Zeta potential of lipopolyplexes, prepared with PEI 800 kDa, as a function of the protocol of formulation at the indicated lipid/DNA molar ratios (From García *et al.*, 2007)

	Zeta potential (mV)				
Molar ratio (lipid/DNA)	*Strategy 1*	*Strategy 2*	*Strategy 3*	*Strategy 4*	*Strategy 5*
2/1	55 ± 3	34 ± 4	35 ± 1	57 ± 2	68 ± 1
5/1	57 ± 1	48 ± 1	54 ± 3	68 ± 1	71 ± 3
17/1	69 ± 2	49 ± 2	59 ± 2	69 ± 1	72 ± 1

produced the largest lipopolyplexes (356 and 295 nm for 5/1 and 17/1 ratios, respectively; Table 16.1) (García *et al.*, 2007). At the 2/1 lipid/DNA ratio, the smallest particle size (124 nm) was obtained using *Strategy 5*. Most of the other complexes had average particle sizes of about 200 nm. Regarding the zeta potential measurements, at the 2/1 lipid/DNA ratio, the smallest lipopolyplexes prepared by *Strategy 5* also had the highest net positive charge (zeta potential = 68 mV), whereas the largest ones (*Strategy 2*) had the lowest zeta potential (34 mV; Table 16.2). All the preparation strategies resulted in net positively charged complexes. Increasing the lipid/DNA ratio appeared to increase the zeta potential for lipopolyplexes prepared by each strategy (Table 16.2).

Complexes were prepared by strategy 4 at an N/P ratio of 17/1 to study the influence of the molecular weight and type of PEI on the particle size and zeta potential of lipopolyplexes. Table 16.3 shows that complexes prepared with PEI 800 of high molecular weight were bigger in particle size compared to lipopolyplexes containing PEI 25 or PEI 22. No differences in the particle size were observed between lipopolyplexes prepared with branched (25 kDa) or linear (22 kDa) PEI of similar molecular weight (García *et al.*, 2007). Complexes prepared with the branched PEIs had a slightly higher zeta potential compared to the linear PEI.

Table 16.3 Influence of molecular weight and type of PEI on the particle size and zeta potential of lipopolyplexes (From García *et al.*, 2007)

	Particle size (nm)	Zeta potential (mV)
PEI 800 branched	214 ± 8	69 ± 1
PEI 25 branched	134 ± 4	60 ± 4
PEI 22 linear	147 ± 2	51 ± 2

Complexes were prepared by protocol 4 at a lipid/DNA molar ratio of 17/1. Results are expressed as the mean ± SD of three independent experiments.

4.2. *In vitro* transfection activity measurements

We examined gene expression mediated by lipopolyplexes as a function of the strategy of formulation and the lipid/DNA molar ratio of the complexes, in the presence of 60% FBS. As shown in Fig. 16.2, lipopolyplexes formulated with strategies 3 and 4 were more effective in transfecting HepG2 cells than complexes generated with the other strategies, 1, 2, and 5, independently of the lipid/DNA molar ratio used to prepare them (García *et al.*, 2007). The highest expression was achieved with complexes prepared by *Strategy 4* at a molar ratio lipid/DNA of 17/1, which showed a 16-, 28-, 1.3-, and 25-fold increased transfection compared to lipopolyplexes formulated by the strategies 1, 2, 3, and 5, respectively, prepared at the same molar ratio. Transfection activity increased by increasing the

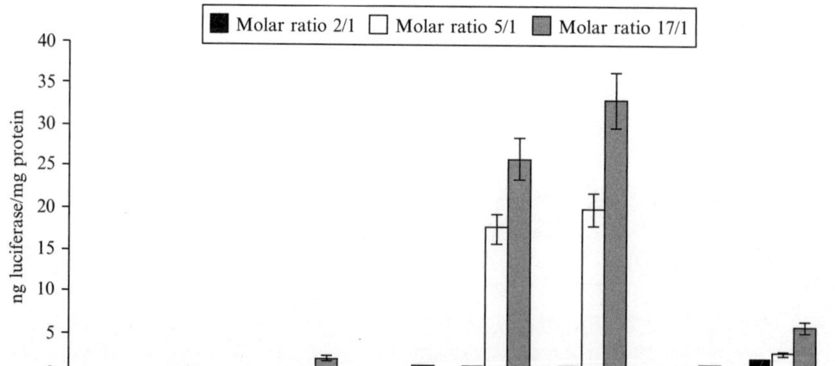

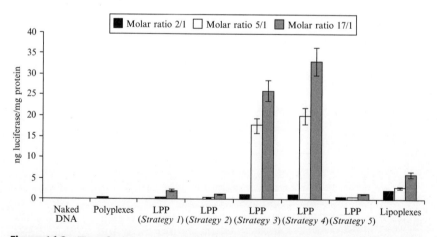

Figure 16.2 Transfection activity in HepG2 cells by polyplexes, lipoplexes, and lipopolyplexes (LPP) in the presence of 60% FBS. Complexes were formulated with branched PEI of 800 kDa with 1 μg of pCMVLuc. The data represent the mean ± SD of three wells and are representative of three independent experiments (From García *et al.*, 2007).

lipid/DNA molar ratio in the lipopolyplexes, independently of the protocol used to prepare the complexes. The transfection activity increased by 1.7- and 33-fold with complexes prepared at the 17/1 molar ratio compared to lipopolyplexes at the 5/1 and 2/1 molar ratios, respectively, by using strategy 4. Lipopolyplexes prepared with strategies 3 and 4 showed much higher transfection activity than conventional polyplexes or lipoplexes, particularly at lipid/DNA ratios of 5/1 and 17/1. Naked DNA did not induce any measurable luciferase expression.

We investigated the effect of the molecular weight and the type of polymer on polyplex- and lipopolyplex-mediated transfection of HepG2 cells. For these experiments, lipopolyplexes were formulated using *Strategy 4* at a lipid/DNA molar ratio of 17/1. No significant differences were found in the levels of gene expression by using PEI 800 or 25 kDa (Fig. 16.3). Complexes formed with linear PEI (22 kDa) were slightly more effective than lipopolyplexes containing branched PEI (800 or 25 kDa). Lipopolyplexes mediated much higher transfection activity than conventional polyplexes, with lipopolyplexes prepared with PEI 22 kDa resulting in 26 times higher luciferase expression than the polyplexes prepared with the same polymer.

We also prepared lipopolyplexes were prepared in the presence of pCMVIL-12 to examine the ability of liver cancer cells to express a potentially therapeutic gene. Lipopolyplexes prepared with linear PEI 22 kDa

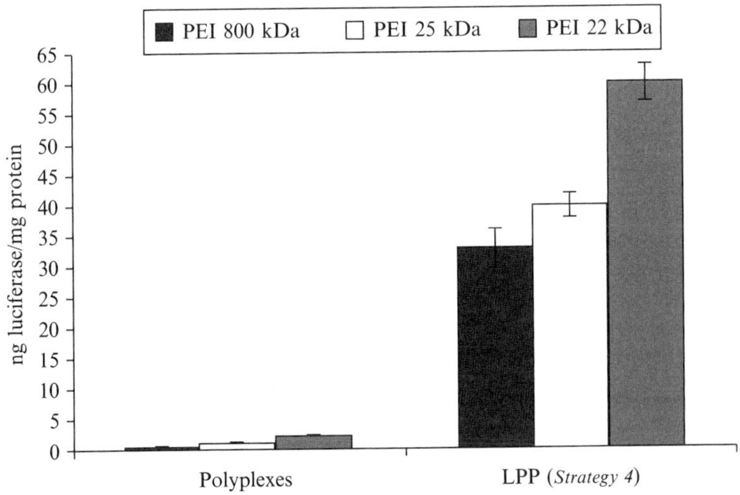

Figure 16.3 Transfection activity by polyplexes and lipopolyplexes (LPP) in HepG2 cells in the presence of 60% FBS. Complexes were prepared with different types of PEI, at a lipid/DNA molar ratio of 17/1, and contained 1 μg of pCMVLuc. The data represent the mean ± SD of three wells and are representative of three independent experiments (From García *et al.*, 2007).

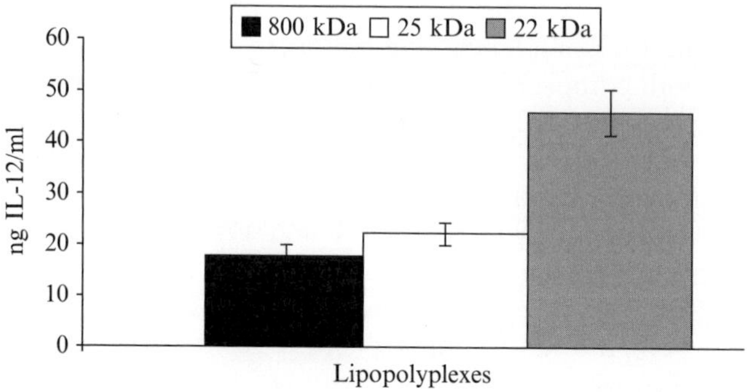

Figure 16.4 IL-12 gene expression following transfection of HepG2 cells by lipopolyplexes (LPP), in the presence of 60% FBS. Complexes were formulated by protocol 4 at a lipid/DNA molar ratio of 17/1 and contained 1 μg of pCMVIL-12. The data represent the mean ± SD of three wells and are representative of three independent experiments (From García *et al.*, 2007).

mediated higher IL-12 expression than lipopolyplexes containing PEI 800 or 25 kDa (Fig. 16.4), in general agreement with the results obtained with the plasmid encoding luciferase (García *et al.*, 2007). Transfection activity of lipoplexes and polyplexes mediated lower levels of transgene expression compared to that observed with lipopolyplexes.

5. Concluding Remarks

We have developed an efficient gene delivery vehicle by combining PEI and DOTAP/Chol liposomes (García *et al.*, 2007). The advantages of these complexes include (i) small particle size to potentially improve transfection efficiency *in vivo*, (ii) reproducibility of transfection efficacy, (iii) efficient transfection of liver cancer cells in the presence of a high concentration of serum, and (iv) stability of the complexes. Since both PEI and liposomes can be modified easily with various ligands, these lipopolyplexes might be useful in the design of targeted carriers specific for certain cell types and the delivery of therapeutic genes. Future studies will investigate the suitability of this nonviral vector for gene transfer *in vivo*.

Acknowledgments

This work was financially supported by the University of Navarra Foundation (FUN) and the Caja de Ahorros de Navarra (CAN).

REFERENCES

Behr, J. P. (1994). Gene transfer with synthetic cationic amphiphiles: prospects for gene therapy. *Bioconjug. Chem.* **5,** 382–389.

Boussif, O., Lezoualch, F., Zanta, M. A., Mergny, M. D., Scherman, D., Demeneix, B., and Behr, J. P. (1995). A versatile vector for gene and oligonucleotide transfer into cells in culture and *in vivo*: Polyethylenimine. *Proc. Natl. Acad. Sci. USA* **92,** 7297–7301.

Drew, J., and Martin, L. A. (1999). What is gene therapy? *In* "Understanding Gene Therapy," (N. R. Lemoine, ed.), pp. 1–10. BIOS Scientific Publishers, Oxford.

Fisher, K. D., Ulbrich, K., Subr, V., Ward, C. M., Mautner, V., Blakey, D., and Seymour, L. W. (2000). A versatile system for receptor-mediated gene delivery permits increased entry of DNA into target dells, enhanced delivery to the nucleus and elevated rates of transgene expression. *Gene Ther.* **7,** 1337–1343.

Gao, X., and Huang, L. (1995). Cationic liposome-mediated gene transfer. *Gene Ther.* **2,** 710–722.

Gao, X., and Huang, L. (1996). Potentiation of cationic liposome-mediated gene delivery by polycations. *Biochemistry* **35,** 1027–1036.

García, L., Buñuales, M., Düzgüneş, N., and Tros de Ilarduya, C. (2007). Serum-resistant lipopolyplexes for gene delivery to liver tumor cells. *Eur. J. Pharm. Bioharm.* **67,** 58–66.

Godbey, W. T., Wu, K. K., and Mikos, A. G. (1999). Tracking the intracellular path of poly (ethylenimine)/DNA complexes for gene delivery. *Proc. Natl. Acad. Sci. USA* **96,** 5177–5181.

Goldman, C. K., Soroceanu, L., Smith, N., Gillespie, G. Y., Shaw, W., Burgess, S., Bilbao, G., and Curiel, D. T. (1997). *In vitro* and *in vivo* gene delivery mediated by a synthetic polycationic amino polymer. *Nat. Biotechnol.* **15,** 462–466.

Harashima, H., Shinohara, Y., and Kiwada, H. (2001). Intracellular control of gene trafficking using liposomes as drug carriers. *Eur. J. Pharm. Sci.* **13,** 85–89.

Lampleta, P., Elomaa, M., Ruponen, M., Urtti, A., Mansito, P. T., and Raasmaja, A. (2003). Different synergistic roles of small polyethylenimine and Dosper in gene delivery. *J. Control. Release* **88,** 173–183.

Lungwitz, U. L., Breunig, M., Blunk, T., and Gopferich, A. (2005). Polyethylenimine-based non-viral gene delivery systems. *Eur. J. Pharm. Biopharm.* **60,** 247–266.

Mahato, R. I., Smith, L. C., and Rolland, A. (1999). Pharmaceutical perspectives of nonviral gene therapy. *Adv. Genet.* **41,** 95–156.

Matsura, M., Yamazaki, Y., Sugiyama, M., Kondo, M., Ori, H., Nango, M., and Oku, N. (2003). Polycation liposome-mediated gene transfer *in vivo*. *Biochim. Biophys. Acta* **2,** 136–143.

Meyer, O., Kirpotin, D., Hong, K., Sternberg, B., Park, J. W., Woodle, M. C., and Papahadjopoulos, D. (1998). Cationic liposomes coated with polyethyleneglycol as carriers for oligonucleotides. *J. Biol. Chem.* **273,** 15621–15627.

Newman, C. M., Lawrie, A., Brisken, A. F., and Cumberland, D. C. (2001). Ultrasound gene therapy: On the road from concept to reality. *Echocardiography* **18,** 339–347.

Pedroso de Lima, M. C., Simões, S., Pires, P., Faneca, H., and Düzgüneş, N. (2001). Cationic lipid-DNA complexes in gene delivery: From biophysics to biological applications. *Adv. Drug Deliv. Rev.* **47,** 277–294.

Pelisek, J., Gaedtke, L., DeRouchey, J., Walker, G. F., Nikol, S., and Wagner, E. (2006). Optimized lipopolyplex formulations for gene transfer to human colon carcinoma cells under *in vitro* conditions. *J. Gene Med.* **8,** 186–197.

Pollard, H., Remy, J. S., Loussouarn, G., Demolombe, S., Behr, J. P., and Escande, D. (1998). Polyethylenimine but not cationic lipids promotes transgene delivery to the nucleus in mammalian cells. *J. Biol. Chem.* **273,** 7507–7511.

Shi, F., Rakhmilevich, A. L., Heise, C. P., Oshikawa, K., Sondel, P. M., and Yang, N. S. (2002). Intratumoral injection of interleukin-12 plasmid DNA, either naked or in complex with cationic lipid, results in similar tumor regression in a murine model. *Mol. Cancer Ther.* **1,** 949–957.

Sorgi, F. L., Bhattacharya, S., and Huang, L. (1997). Protamine sulphate enhances lipid-mediated gene transfer. *Gene Ther.* **4,** 961–968.

Tros de Ilarduya, C., Arangoa, M. A., Moreno-Aliaga, M. J., and Düzgüneş, N. (2002). Enhanced gene delivery *in vitro* and *in vivo* by improved transferrin-lipoplexes. *Biochim. Biophys. Acta* **1561,** 209–221.

Tros de Ilarduya, C., Sun, Y., and Düzgüneş, N. (2010). Gene delivery by lipoplexes and polyplexes. *Eur. J. Pharm. Sci.* **40,** 159–170.

Whitmore, M., Li, S., and Huang, L. (1999). LPD lipopolyplex initiates a potent cytokine response and inhibits tumor growth. *Gene Ther.* **6,** 1867–1875.

Yamazaki, Y., Nango, M., Matsuura, M., Hasegawa, Y., and Oku, N. (2000). Polycation liposomes, a novel nonviral gene transfer system, constructed from cetylated polyethylenimine. *Gene Ther.* **7,** 1148–1155.

CHAPTER SEVENTEEN

Interfering Nanoparticles for Silencing MicroRNAs

Huricha Baigude *and* Tariq M. Rana

Contents

1. Introduction 340
2. Delivery of Short Therapeutic RNAs 341
   2.1. Nanoparticles for short interfering RNA delivery 341
   2.2. Delivery of miRNA inhibitors 342
3. Protocols 344
   3.1. Determining the silencing efficiency of anti-miRs 344
   3.2. Oligonucleotides 346
   3.3. *In vitro* silencing of miRNA by iNOP-7 346
   3.4. *In vivo* silencing of miRNA by iNOP-7 348
4. Concluding Remarks 350
Acknowledgments 350
References 350

Abstract

MicroRNAs (miRNAs) are single-stranded noncoding RNAs ~21-nucleotide (nt) in length and regulate gene expression at the posttranscriptional level. miRNAs are involved in almost every area of biology, including developmental processes, disease pathogenesis, and host–pathogen interactions. Dysregulation of miRNAs in various disease states makes them potential targets for therapeutic intervention. Specific miRNAs can be silenced by anti-microRNAs (anti-miRs) that are chemically modified antisense oligonucleotides complementary to mature miRNA sequences. *In vivo* delivery of anti-miRs is the main barrier in achieving efficient silencing of target miRNAs. A new systemic delivery agent, interfering nanoparticles (iNOPs), was designed and prepared from lipid-functionalized poly-L-lysine dendrimer. iNOPs can efficiently deliver small RNAs, including short interfering RNAs, miRNA mimics, and anti-miRs. Systemic delivery of a chemically stabilized anti-miR-122 by iNOPs effectively silences miR-122 in mouse liver. Intravenous administration of 2 mg/kg anti-miR-122 complexed with iNOP-7 results in 83% specific silencing of target miRNA. The specific silencing of miR-122 by iNOP-7 is long lasting and does not induce an immune response.

Program for RNA Biology, Sanford-Burnham Medical Research Institute, La Jolla, California, USA

Methods in Enzymology, Volume 509
ISSN 0076-6879, DOI: 10.1016/B978-0-12-391858-1.00017-4

1. Introduction

MicroRNAs (miRNAs) are single-stranded RNAs ~21-nt in length that are involved in developmental processes, disease pathogenesis, and host–pathogen interactions (Ambros, 2011; Kim *et al.*, 2009; Krol *et al.*, 2010). The biogenesis of mature miRNAs depends on cleavage of the precursor RNA hairpin structure by two members of the RNase III family, Drosha and Dicer, while other miRNAs can be generated through splicing of miR-coding introns (Carthew and Sontheimer, 2009; Kim *et al.*, 2009). For functional assemblies, miRNAs are loaded into a ribonucleoprotein assembly called the RNA-induced silencing complex (RISC), which serves as the catalytic engine for miRNA-mediated posttranscriptional regulation. Although some studies have suggested a potential role for miRNAs in translational activation (Henke *et al.*, 2008; Orom *et al.*, 2008; Vasudevan *et al.*, 2007), the more common mechanism of miRNA-mediated gene regulation involves repression. In general, miRNAs bind imperfectly to the 3′UTR of target mRNA and block their expression by directly inhibiting the translational steps and/or by enhancing mRNA destabilization (Bagga *et al.*, 2005; Fabian *et al.*, 2010; Guo *et al.*, 2010). Recent studies have identified the role of GW182 protein in the molecular mechanism of miRNA-mediated mRNA deadenylation (Behm-Ansmant *et al.*, 2006; Eulalio *et al.*, 2007; Iwasaki and Tomari, 2009). GW182 directly interacts with all members of the Ago protein family and is localized within P-bodies in the cytoplasm of mammalian cells (Fabian *et al.*, 2010). Another P-body protein, RCK/p54, a DEAD-box helicase, has been shown to interact with the argonaute proteins, Ago1 and Ago2, and modulate miRNA function (Chu and Rana, 2006). RCK/p54 facilitates formation of P-bodies and is a general repressor of translation, suggesting that miRNAs are transferred to P-bodies for further decay or storage (Chu and Rana, 2006).

Many disease-affected tissues have characteristic miRNA expression patterns (Chang *et al.*, 2008; Munker and Calin, 2011; Suzuki *et al.*, 2011). Selective elimination of upregulated miRNAs in disease-affected tissue could provide a potential therapeutic approach (Krutzfeldt *et al.*, 2005; Montgomery and van Rooij, 2011). Specific disease-related endogenous miRNAs can be silenced by their complimentary antisense sequence, anti-miRs. However, development of anti-miR-based therapies faces many challenges including stabilization and optimization of anti-miR sequences and efficient delivery of these sequences to the tissues. While many anti-miR sequences and chemical modifications have been successfully designed to target miRNAs (Lennox and Behlke, 2010), developing efficient *in vivo* delivery agents is essential for successful therapeutic development. An efficient delivery agent should protect anti-miR from degradation by endonuclease in circulation and in tissue, facilitate cellular uptake, and release the cargo in a tissue specific manner (Baigude and Rana, 2009; Baigude *et al.*, 2007; Su *et al.*, 2011).

2. Delivery of Short Therapeutic RNAs

2.1. Nanoparticles for short interfering RNA delivery

Nanoparticles have attracted much attention as nonviral carriers for *in vivo* delivery of short synthetic RNAs. Traditional drug delivery method involving liposome formulation has been readily adopted and optimized for small interfering RNA (siRNA) delivery. Highly efficient *in vivo* siRNA delivery based on liposome has been reported. By optimizing structure and components of lipid molecules together with formulation methods, Love *et al.* were able to knock down 80% a clinically relevant gene transthyretin at a dose as low as 0.03 mg/kg (Love *et al.*, 2010; Semple *et al.*, 2010). Although several approaches have been reported to attenuate liposome for tissue specific siRNA delivery (Peer *et al.*, 2008), main target tissue of lipid-based siRNA delivery has been limited to liver. Cationic polymer-based delivery is another highly attractive approach for systemic delivery of short therapeutic RNAs. Both synthetic polymers such as polyethylenimine (PEI) (Nimesh and Chandra, 2009; Urban-Klein *et al.*, 2005) and naturally occurring polycations such as Chitosan (Howard *et al.*, 2006; Liu *et al.*, 2007; Pille *et al.*, 2006) have been reported to be useful for *in vitro* and *in vivo* siRNA delivery. siRNA can also be conjugated to cationic polymers for more efficient systemic delivery. A polyconjugates was prepared by covalently attaching siRNA to an amphipathic poly (vinyl ether) and was functionalized with hepatocyte targeting ligand *N*-acetylgalactosamine and polyethylene glycol (PEG) (Rozema *et al.*, 2007). The 10 nm polyconjugate nanoparticle successfully delivered siRNA to mouse liver, significantly knocking down the target gene. Cationic peptides have also been used for nanoparticle formulation and siRNA delivery. Apart from popular ones such as Tat peptide (Chiu *et al.*, 2004) and polyarginine (Kim *et al.*, 2006), peptide transduction domains (PTDs) have been reported for targeted siRNA delivery (Eguchi and Dowdy, 2010; Palm-Apergi *et al.*, 2011). Remarkably, a polyarginine functionalized 29-amino acid peptide derived from rabies virus glycoprotein (RVG) enables the transvascular delivery of siRNA to mouse brain by specifically binding to the acetylcholine receptor expressed by neuronal cells (Kumar *et al.*, 2007). Nanoparticles or nanomaterials have also been reported to deliver anti-miRNA *in vitro* (Dong *et al.*, 2011; Kim *et al.*, 2011).

2.2. Delivery of miRNA inhibitors

Mature miRNAs are single-stranded ~21-nt-long RNA molecules. An anti-miR is a single-stranded oligonucleotides containing a complementary sequence to a mature miRNA. Single-stranded 2′-*O*-methyl modified sequences that are complementary to endogenous miRNA can inhibit miRNA function by interacting with the miRNA–RISC nucleoprotein complex (Hutvagner *et al.*, 2004; Meister *et al.*, 2004). Transfection reagents that are used for siRNA delivery can also be used for anti-miRs delivery. Commercially available transfection reagents such as lipofectamine 2000 can deliver anti-miRs to the cells *in vitro*. Chemically stabilized cholesterol-conjugated anti-miRs (antagomirs) have been used for *in vivo* silencing of endogenous miRNA (Krutzfeldt *et al.*, 2005). However, a relatively higher dose (80 mg/kg) had to be administered to achieve significant knockdown of miRNA. Chemically stabilized anti-miR sequences with 2′-fluoro and/or methoxyethyl groups have increased inhibitory activities (Rayner *et al.*, 2011). The anti-miR efficiency of a complementary sequence of miRNA can be further enhanced by incorporating locked nucleic acids nucleotides in the sequence (Chan *et al.*, 2005; Elmen *et al.*, 2008; Lecellier *et al.*, 2005; Wang *et al.*, 2011). Incorporation of double-stranded flanking regions significantly increases inhibitor function (Vermeulen *et al.*, 2007).

We designed and created a series of interfering nanoparticles (iNOPs) for siRNA delivery *in vitro* and *in vivo*. iNOP-7 is polymeric nanoparticles chemically synthesized by functionalizing a generation 4 (G4) poly-L-lysine dendrimer (LDG4) with lipids (such as oleic acid) (Baigude *et al.*, 2007). Chemically synthesized LDG4 is highly branched poly-L-lysine bearing 32 amino groups on the surface. The amino groups, which are positively charged at physiological condition, provide excellent binding affinity for any negatively charged macromolecules, such as oligonucleotides, siRNA, miRNA, and plasmid DNA. The amino groups are also viable to various chemical modifications for further functionalization of the nanoparticles such as tissue specific delivery of nucleic acids. For example, to increase the hydrophobicity of the dendrimer to facilitate cellular uptake of the nanoparticles, we chemically conjugated lipid chains with different degrees of saturation and different length (C12–C18), and found that oleic acid is one of the best lipids for this purpose, although linking other lipids to LDG4 can also increase the hydrophobicity of dendrimer/nucleic acid complex and cellular uptake. By carefully adjusting the number of oleic acid molecules (6–13 oleoyl groups) coupled to LDG4, we were able to find the optimal number that makes the resulting dendrimer least toxic and most efficient in siRNA or miRNA delivery *in vitro* and *in vivo*. iNOP-7, which is composed of LDG with seven oleic acid moieties, is nontoxic at a concentration of up to 1.5 μ*M* for *in vitro* delivery of siRNA. More importantly, iNOP-7 can deliver siRNA (against ApoB) in mice and obtained at least 50% of decrease in ApoB mRNA level (Figs. 17.1 and 17.2).

The soluble, 15 nm nanoparticles readily bind to short nucleic acids (siRNA and/or miRNA) at neutral pH and form nanoparticles with diameters less than 200 nm (Fig. 17.3). Since iNOP-7 is made from a naturally occurring amino acid, it is less toxic and more biocompatible in contrast to other nondegradable polymeric nanoparticles and inorganic nanoparticles. Besides, iNOP-7 is more flexible in terms of surface functionalization because of the more accessible surface groups. One of the iNOPs, iNOP-7 was able to deliver siRNA against apoB in mouse liver (Baigude *et al.*, 2007) (Fig. 17.4). Moreover, iNOP-7 was able to deliver chemically stabilized anti-miR-122 to mouse liver and significantly silenced the target miRNA-122 (Su *et al.*, 2011).

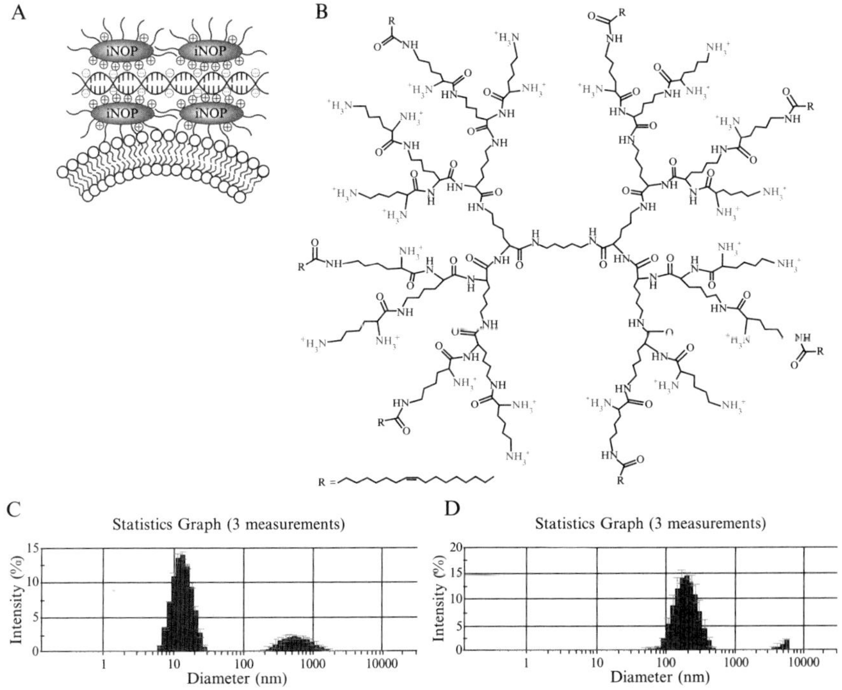

Figure 17.1 (A) Schematic illustration of nanoparticles–siRNA complex (iNOP-7) initiating cellular entry. (B) Chemical structure of the nanoparticles used to construct iNOP-7. Particle size of iNOP-7 before (C) and after (D) complexed with short RNA (in this case, a 21-mer-duplex RNA) measured by dynamic light scattering. (For the color version of this figure, the reader is referred to the Web version of this chapter.)

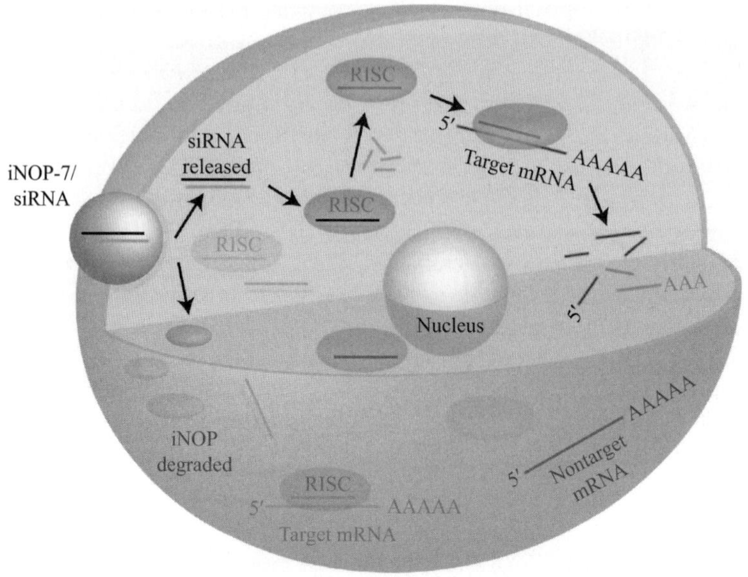

Figure 17.2 Schematic illustration of iNOP-7 bringing siRNA into cytoplasm and knocking down endogenous mRNAs. (For the color version of this figure, the reader is referred to the Web version of this chapter.)

3. Protocols

3.1. Determining the silencing efficiency of anti-miRs

There are two main methods to detect the endogenous miRNA level after transfecting anti-miRs to the cells or tissues: (1) Northern blotting and (2) quantitative PCR. Although there has been no direct proof so far, a duplex of miRNA-anti-miR could potentially degraded by nuclease present in the cell. Northern blotting is a routinely used method for detection of miRNA levels. The silencing effect of anti-miR can also be evaluated by using commercially available quantitative PCR kits.

3.1.1. Determining the silencing efficiency by Northern blotting

RNA from cell culture or tissue can be isolated by homogenizing in TRIZOL according to the manufacturer's instructions. Total RNA can be separated on a 14% acrylamide containing 20% formamide and 8 *M* urea gel, then electroblotted onto Hybond-XL nylon membrane (GE Healthcare, Piscataway, NJ, USA). The probe with γ-^{32}P-labeled oligonucleotides for miRNA or rRNA is hybridized to the membrane at 42 °C. The blots are visualized by scanning in a FLA-5000 scanner (Fujifilm, Stamford, CT, USA).

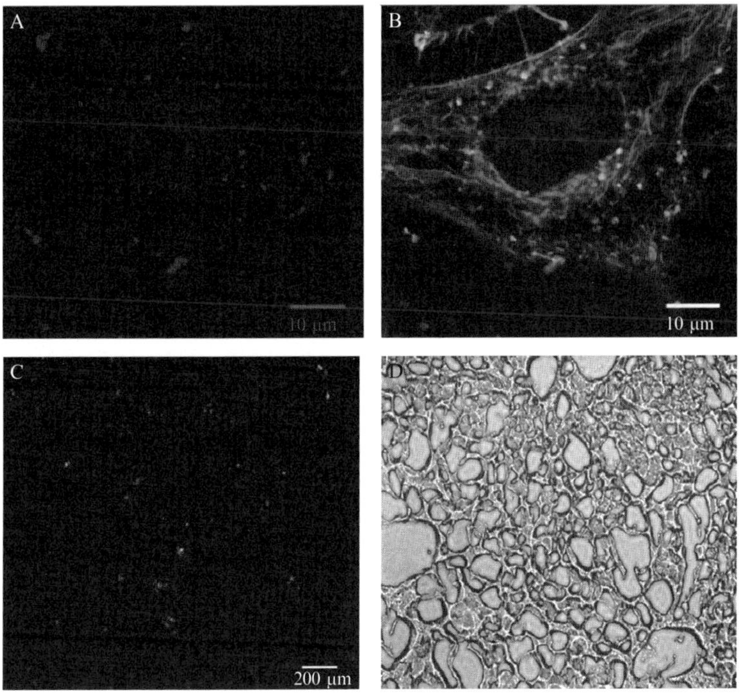

Figure 17.3 Localization of siRNA. (A) Cells were transfected with iNOP-7 containing Cy3-labeled siRNA. Localization of the duplex siRNA after 24 h was monitored by confocal microscopy. Overlay images of siRNA and nuclear DNA (4′,6-diamidino-2-phenylindole [DAPI] stained) are shown. (B) Cells transfected with iNOP-7/Cy3 siRNA complex were stained fluorescein labeled Phalloidin to show cell outlines. (C) Fluorescence microscopy analysis of siRNA distribution *in vivo*. Histology slide of liver shows the distribution of siRNA in liver 4 h after intravenous administration of iNOP-7 was injected. Scale bar: 100 μm. (D) Bright field microscopic image of (C). (See Color Insert.)

3.1.2. Determining the silencing efficiency by quantitative PCR

Silencing efficiency of anti-miR delivered to the cells can be determined by RT-qPCR. The total RNA is used to quantify miR expression level using Mir-X^{TM} miRNA First-Strand Synthesis and SYBR qRT-PCR kit according to the manufacturer's instruction.

3.1.3. Dual luciferase assay

Anti-miR transfection efficiency can be determined by *in vitro* dual luciferase assay. The following is a protocol that we have used for preliminary determination of anti-miR transfection efficiency by iNOP-7.

The miR-122 luciferase constructs are engineered by inserting the full 23-bp sequence complementary to the mature miR-122 into the 3′UTR of pGL3-Control as described previously (Chu and Rana, 2006) (Promega,

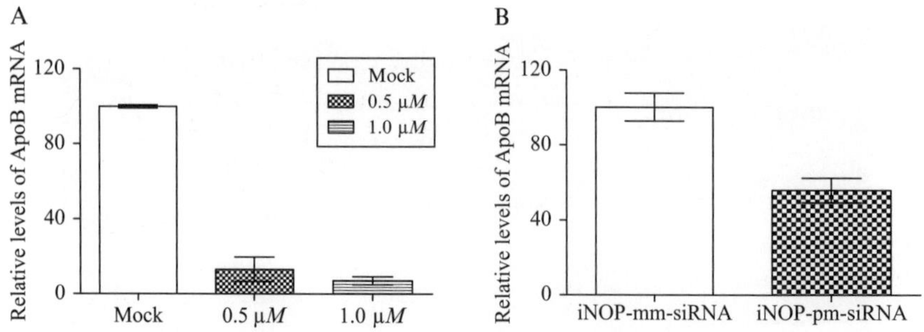

Figure 17.4 (A) *In vitro* delivery of ApoB siRNA to FL83B cells by iNOP-7. More than 90% decrease in ApoB mRNA was achieved by using 1.0 μ*M* concentration of iNOP-7 complexed to ApoB siRNA (50 n*M*). (B) *In vivo* delivery of ApoB siRNA to mouse. A 1.5 mg/kg dose of iNOP-7 complexed to ApoB siRNA was injected to C57BL/6 mice via tail vein. After 48 h, liver mRNA level of ApoB was analyzed by quantitative PCR.

Madison, WI, USA). Huh-7 cells are seeded in 24-well culture plates and transfected with 0.1 μg miR-122 pGL3-control plasmid and 0.015 μg pRL-TK plasmid (Promega) for normalization using iNOP-7. After 4 h of transfection, cells are treated with complete media. Cells are lysed 48 h later, unless otherwise indicated, and luciferase activity is measured using the Dual-Luciferase Reporter Assay System (Promega). These quantitative assays are conducted in multiple replicates for each concentration.

3.2. Oligonucleotides

The anti-miR efficiency of iNOP-7 can be increased further by using chemically modified RNAs (Chiu and Rana, 2002, 2003; Chiu *et al.*, 2004). For example, to knockdown endogenous miRNA-122, we have designed a chemically stabilized 23–24 nucleotides length anti-miR-122 as following (Dharmacon, Lafayette, CO, USA):

anti-miR-122 (CM): 5′-AC^{F}AAAC^{F}AC^{F}C^{F}AUUGUCACACUCCA-3′;
mismatch anti-miR-122 (MM): 5′-UCAC^{F}AAC^{F}CUCC^{F}UAGAAAGA-GUAGA-3′.

The superscript letter F represents 2′-*O*-F modified nucleotides. Chemical modifications are designed based on our previously published rules established in our laboratory (Rana, 2007).

3.3. *In vitro* silencing of miRNA by iNOP-7

3.3.1. Preparation of materials for *in vitro* transfection

iNOP-7 dissolves readily in water. A stock solution of iNOP-7 (0.2 m*M*) is prepared by dissolving 1.16 mg iNOP-7 in 1.0 ml RNase free water. The stock solution is sterilized by filtering through a 0.22 μm spin filter.

To prepare a buffered solution of iNOP-7, it is first dissolved in water at 2 × concentration, and then it is diluted with an equal volume of the buffer of choice. The aqueous solution of iNOP-7 can be preserved at 4 °C for at least 6 months without losing any apparent transfection efficiency.

A 50 μ*M* stock solution of anti-miRs is prepared by suspending the powders provided by the manufacturer in 1 × siRNA buffer.

3.3.2. Preparation of cells for *in vitro* transfection

To achieve the highest transfection efficiency with iNOP-7, actively growing cells are seeded in 6-well plates 15–20 h before transfection. The confluency of the cells at the time of transfection should be about 70%. The state of the cells before transfection can impact anti-miR transfection efficiency. Cells with lower passage number should be used and cellular morphology should be monitored. Cells with similar passage number and appropriate cell density should be used to achieve reproducible transfection results.

For example, Huh-7 cells are maintained at 37 °C with 5% CO_2 in DMEM with High Glucose culture medium (Invitrogen, Carlsbad, CA, USA) supplemented with 10% fetal bovine serum, 100U/ml penicillin and 100 mg/ml streptomycin. Cells are regularly passaged and plated in 6-well culture plates for 16 h before transfection at 70% confluency. The anti-miRs are transfected in Opti-MEM serum-free culture medium for 4 h at 37 °C and then the medium is changed to normal medium with 10% FBS.

3.3.3. Transfection of adherent cells in 6-well plate using iNOP-7

3.3.3.1. Serum-free transfection

1. Check the cells under microscope for optimal confluency.
2. Add 500 μl of serum-free medium (such as Opti-MEM) to a sterile plastic tube and add 5 μl of iNOP-7. Vortex gently to mix.
3. Incubate at room temperature for 5 min.
4. Add anti-miR (1.0 μl of a 50 μ*M* siRNA stock) to another sterile plastic tube containing 500 μl of Opti-MEM.
5. Add (3) to (4) and mix gently. To allow nanoparticle formation, incubate at room temperature for 20 min.
6. Remove cell culture medium and wash the cells twice with PBS, pH 7.4.
7. Add iNOP-7/anti-miR complex dropwise to the cells and gently rock the plate to disperse the complexes.
8. Incubate at 37 °C for 2–3 h.
9. Remove transfection medium and add 2.0 ml complete medium.
10. Incubate for 24 h

3.3.3.2. Transfection in complete medium

1. Check the cells under the microscope for optimal confluency.
2. Add 250 μl of serum-free medium (such as Opti-MEM) to a sterile plastic tube and add 5 μl of iNOP-7. Vortex gently to mix.

3. Incubate at room temperature for 5 min.
4. Add anti-miR (1.0 μl of a 50 μ*M* siRNA stock) to another sterile plastic tube containing 250 μl of Opti-MEM.
5. Add (3) to (4) and mix gently. To allow nanoparticle formation, incubate at room temperature for 20 min.
6. Remove 500 μl of complete medium from each wells of 6-well plate and add iNOP-7/anti-miR complex.
7. Gently rock the plate to disperse the complex.

Different cell lines have different sensitivity to iNOP-7. Therefore, the optimal amount of iNOP-7 appropriate for a certain cell line must be determined. We recommend to test a few different concentrations of iNOP-7 as a preliminary experiment to optimize the conditions. At least three serial concentrations should be tested: 2.5, 5.0, and 7.5 μl for each 6-well plate, which give concentrations of 0.5, 1.0, and 1.5 μ*M*, respectively.

3.4. *In vivo* silencing of miRNA by iNOP-7

3.4.1. Formulation of iNOP-7/anit-miR-122

1. Prepare a stock solution of anti-miRs by resuspending RNAs in 1× siRNA buffer. The concentration of stock RNA solution is 1.0 m*M*, which can be stored at −80 °C. To prevent frequent freeze-thawing, the stock solution should be frozen in small aliquots.
2. Prepare a stock solution of iNOP-7 in RNase free water at 3.0–5.0 mg/ml concentration. Sterilize by filtering. The stock solution of iNOP-7 can be preserved at +4 °C for at least 6 months without losing any apparent activity.
3. Assemble the iNOP-7/anti-miR complex in a sterile plastic tube. Add the anti-miR to 1× HEPES buffer. Amount of RNA varies depending on the dosage. For example, 50 μg anti-miR is added to 100 μl 1× HEPES buffer for 2.0 mg/kg dose for a 25 g mouse. Mix well and keep on ice.
4. Dilute 250 μg stock iNOP-7 in 1× HEPES buffer to make 100 μl solution. Mix well.
5. Carefully add 100 μl anti-miR solution to 100 μl iNOP-7. Mix by pipetting up and down for 5–10 times.
6. Incubate at room temperature for 20 min to form iNOP-7.

3.4.2. *In vivo* delivery of anti-miR-122

1. Animal: 6- to 8-week-old male C57BL/6 mice (Charles River Laboratories, Wilmington, MA) are used for *in vivo* anti-miR (or siRNA) delivery experiment. Mice should be maintained under a 12 h dark cycle in a pathogen-free animal facility.

2. Inject iNOP-7 via the lateral tail vein on three consecutive days with iNOP-7 complexes of anti-miR or its MM anti-miR. Daily dosages of 0.20–5 mg/kg RNA can be delivered in a final volume of 0.2 ml.

3.4.3. Determining the delivery efficiency

The delivery efficiency of iNOP-7 can be determined 24 h after the final injection. In case of siRNA delivery, tissue levels of target mRNA can be determined by RT-qPCR, and the level of protein can be determined by Western blotting, providing that an antibody against target protein is available. In case of anti-miR delivery, endogenous level of miRNA can be determined by either RT-qPCR or Northern blotting (Fig. 17.5).

To determine mRNA levels in tissues, small uniform samples are collected from three regions of the tissue. Total RNA is extracted with Trizol and treated with DNaseI before quantification. mRNA level is determined

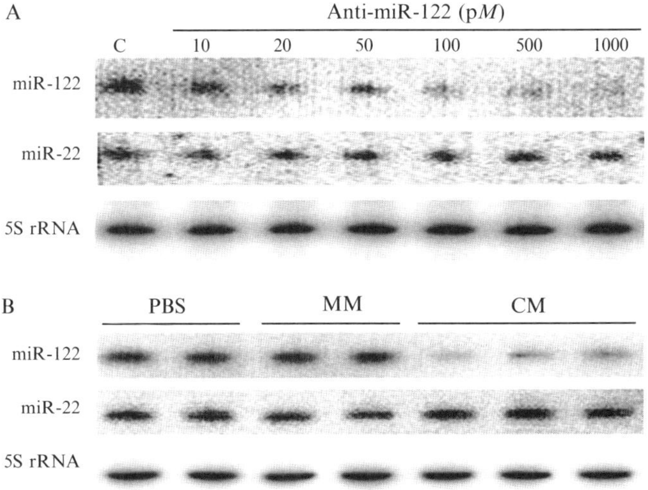

Figure 17.5 (A) Silencing of miR-122 by iNOP-7 *in vitro*. Huh-7 cells were transfected by iNOP-7 containing anti-miR-122 at varying concentrations as indicated for 4 h. Total RNA was isolated from cells 24 h after transfection and separated on 14% polyacrylamide gels. Membrane was probed for miR-122 and miR-22, respectively. 5S rRNA is shown as a loading control. (B) Specific silencing of miR-122 in mice treated with iNOP-7 assembled with chemically modified anti-miR-122. Mice were intravenously injected with 2 mg/kg of iNOP-7 containing chemically modified anti-miR-122, mismatched chemically modified anti-miR-122, or PBS at 0, 12, and 36 h. Tissues were harvested at 24 h after the last injection to measure miRNA or mRNA levels. Total RNA was isolated from mouse liver 24 h after last injection. Samples were separated in 14% polyacrylamide gel and membrane was probed for miR-122, miR-22, and 5S rRNA. 5S rRNA is shown as a loading control. (For the color version of this figure, the reader is referred to the Web version of this chapter.)

by qPCR as described above. After anti-miR treatment, the mRNA levels of genes regulated by miRNA can be determined by RT-qPCR.

4. Concluding Remarks

miRNAs are small endogenous non-coding RNAs that regulate post-transcriptional gene expression and are important in many biological processes. Disease-associated miRNAs have been shown to become potential targets for therapeutic intervention. The newly designed nanoparticles (iNOP-7) can be used to deliver single-stranded anti-miR oligonucleotides to silence disease-related endogenous miRNA. iNOP-7 containing chemically modified anti-miRs specifically silence miRNAs in liver. iNOP-7 treatment is nontoxic and does not induce an immune response. Further, iNOP-7 can be modified to target specific tissues and develop tissue specific RNAi-based therapies.

ACKNOWLEDGMENTS

We thank members of the Rana lab for helpful discussions and critical reading of the chapter. This work was supported in part by grants from the NIH to T. M. R.

REFERENCES

Ambros, V. (2011). MicroRNAs and developmental timing. *Curr. Opin. Genet. Dev.* **21,** 511–517.

Bagga, S., Bracht, J., Hunter, S., Massirer, K., Holtz, J., Eachus, R., and Pasquinelli, A. E. (2005). Regulation by let-7 and lin-4 miRNAs results in target mRNA degradation. *Cell* **122,** 553–563.

Baigude, H., McCarroll, J., Yang, C. S., Swain, P. M., and Rana, T. M. (2007). Design and creation of new nanomaterials for therapeutic RNAi. *ACS Chem. Biol.* **2,** 237–241.

Baigude, H., and Rana, T. M. (2009). Delivery of therapeutic RNAi by nanovehicles. *Chembiochem* **10,** 2449–2454.

Behm-Ansmant, I., Rehwinkel, J., Doerks, T., Stark, A., Bork, P., and Izaurralde, E. (2006). mRNA degradation by miRNAs and GW182 requires both CCR4:NOT deadenylase and DCP1:DCP2 decapping complexes. *Genes Dev.* **20,** 1885–1898.

Carthew, R. W., and Sontheimer, E. J. (2009). Origins and mechanisms of miRNAs and siRNAs. *Cell* **136,** 642–655.

Chan, J. A., Krichevsky, A. M., and Kosik, K. S. (2005). MicroRNA-21 is an antiapoptotic factor in human glioblastoma cells. *Cancer Res.* **65,** 6029–6033.

Chang, T. C., Yu, D., Lee, Y. S., Wentzel, E. A., Arking, D. E., West, K. M., Dang, C. V., Thomas-Tikhonenko, A., and Mendell, J. T. (2008). Widespread microRNA repression by Myc contributes to tumorigenesis. *Nat. Genet.* **40,** 43–50.

Chiu, Y.-L., Ali, A., Chu, C.-Y., Cao, H., and Rana, T. M. (2004). Visualizing a correlation between siRNA localization, cellular uptake, and RNAi in living cells. *Chem. Biol.* **11,** 1165–1175.

Chiu, Y. L., and Rana, T. M. (2002). RNAi in human cells: Basic structural and functional features of small interfering RNA. *Mol. Cell* **10,** 549–561.

Chiu, Y. L., and Rana, T. M. (2003). siRNA function in RNAi: A chemical modification analysis. *RNA* **9,** 1034–1048.

Chu, C. Y., and Rana, T. M. (2006). Translation repression in human cells by microRNA-induced gene silencing requires RCK/p54. *PLoS Biol.* **4,** e210.

Dong, H., Ding, L., Yan, F., Ji, H., and Ju, H. (2011). The use of polyethylenimine-grafted graphene nanoribbon for cellular delivery of locked nucleic acid modified molecular beacon for recognition of microRNA. *Biomaterials* **32,** 3875–3882.

Eguchi, A., and Dowdy, S. F. (2010). Efficient siRNA delivery by novel PTD-DRBD fusion proteins. *Cell Cycle* **9,** 424–425.

Elmen, J., Lindow, M., Silahtaroglu, A., Bak, M., Christensen, M., Lind-Thomsen, A., Hedtjarn, M., Hansen, J. B., Hansen, H. F., Straarup, E. M., McCullagh, K., Kearney, P., and Kauppinen, S. (2008). Antagonism of microRNA-122 in mice by systemically administered LNA-antimiR leads to up-regulation of a large set of predicted target mRNAs in the liver. *Nucleic Acids Res.* **36,** 1153–1162.

Eulalio, A., Rehwinkel, J., Stricker, M., Huntzinger, E., Yang, S. F., Doerks, T., Dorner, S., Bork, P., Boutros, M., and Izaurralde, E. (2007). Target-specific requirements for enhancers of decapping in miRNA-mediated gene silencing. *Genes Dev.* **21,** 2558–2570.

Fabian, M. R., Sonenberg, N., and Filipowicz, W. (2010). Regulation of mRNA translation and stability by microRNAs. *Annu. Rev. Biochem.* **79,** 351–379.

Guo, H., Ingolia, N. T., Weissman, J. S., and Bartel, D. P. (2010). Mammalian microRNAs predominantly act to decrease target mRNA levels. *Nature* **466,** 835–840.

Henke, J. I., Goergen, D., Zheng, J., Song, Y., Schuttler, C. G., Fehr, C., Junemann, C., and Niepmann, M. (2008). microRNA-122 stimulates translation of hepatitis C virus RNA. *EMBO J.* **27,** 3300–3310.

Howard, K. A., Rahbek, U. L., Liu, X., Damgaard, C. K., Glud, S. Z., Andersen, M. O., Hovgaard, M. B., Schmitz, A., Nyengaard, J. R., Besenbacher, F., and Kjems, J. (2006). RNA interference *in vitro* and *in vivo* using a novel chitosan/siRNA nanoparticle system. *Mol. Ther.* **14,** 476–484.

Hutvagner, G., Simard, M. J., Mello, C. C., and Zamore, P. D. (2004). Sequence-specific inhibition of small RNA function. *PLoS Biol.* **2,** E98.

Iwasaki, S., and Tomari, Y. (2009). Argonaute-mediated translational repression (and activation). *Fly (Austin)* **3,** 204–206.

Kim, J. H., Yeom, J. H., Ko, J. J., Han, M. S., Lee, K., Na, S. Y., and Bae, J. (2011). Effective delivery of anti-miRNA DNA oligonucleotides by functionalized gold nanoparticles. *J. Biotechnol.* **155,** 287–292.

Kim, V. N., Han, J., and Siomi, M. C. (2009). Biogenesis of small RNAs in animals. *Nat. Rev. Mol. Cell Biol.* **10,** 126–139.

Kim, W. J., Christensen, L. V., Jo, S., Yockman, J. W., Jeong, J. H., Kim, Y.-H., and Kim, S. W. (2006). Cholesteryl oligoarginine delivering vascular endothelial growth factor siRNA effectively inhibits tumor growth in colon adenocarcinoma. *Mol. Ther.* **14,** 343–350.

Krol, J., Loedige, I., and Filipowicz, W. (2010). The widespread regulation of microRNA biogenesis, function and decay. *Nat. Rev. Genet.* **11,** 597–610.

Krutzfeldt, J., Rajewsky, N., Braich, R., Rajeev, K. G., Tuschl, T., Manoharan, M., and Stoffel, M. (2005). Silencing of microRNAs *in vivo* with 'antagomirs'. *Nature* **438,** 685–689.

Kumar, P., Wu, H., McBride, J. L., Jung, K.-E., Kim, M. H., Davidson, B. L., Lee, S. K., Shankar, P., and Manjunath, N. (2007). Transvascular delivery of small interfering RNA to the central nervous system. *Nature* **448,** 39–43.

Lecellier, C. H., Dunoyer, P., Arar, K., Lehmann-Che, J., Eyquem, S., Himber, C., Saib, A., and Voinnet, O. (2005). A cellular microRNA mediates antiviral defense in human cells. *Science* **308,** 557–560.

Lennox, K. A., and Behlke, M. A. (2010). A direct comparison of anti-microRNA oligonucleotide potency. *Pharm. Res.* **27,** 1788–1799.

Liu, X., Howard, K. A., Dong, M., Andersen, M. O., Rahbek, U. L., Johnsen, M. G., Hansen, O. C., Besenbacher, F., and Kjems, J. (2007). The influence of polymeric properties on chitosan/siRNA nanoparticle formulation and gene silencing. *Biomaterials* **28,** 1280–1288.

Love, K. T., Mahon, K. P., Levins, C. G., Whitehead, K. A., Querbes, W., Dorkin, J. R., Qin, J., Cantley, W., Qin, L. L., Racie, T., Frank-Kamenetsky, M., Yip, K. N., *et al.* (2010). Lipid-like materials for low-dose, *in vivo* gene silencing. *Proc. Natl. Acad. Sci. USA* **107,** 1864–1869.

Meister, G., Landthaler, M., Dorsett, Y., and Tuschl, T. (2004). Sequence-specific inhibition of microRNA- and siRNA-induced RNA silencing. *RNA* **10,** 544–550.

Montgomery, R. L., and van Rooij, E. (2011). Therapeutic advances in microRNA targeting. *J. Cardiovasc. Pharmacol.* **57,** 1–7.

Munker, R., and Calin, G. A. (2011). MicroRNA profiling in cancer. *Clin. Sci. (Lond.)* **121,** 141–158.

Nimesh, S., and Chandra, R. (2009). Polyethylenimine nanoparticles as an efficient *in vitro* siRNA delivery system. *Eur. J. Pharm. Biopharm.* **73,** 43–49.

Orom, U. A., Nielsen, F. C., and Lund, A. H. (2008). MicroRNA-10a binds the 5′UTR of ribosomal protein mRNAs and enhances their translation. *Mol. Cell* **30,** 460–471.

Palm-Apergi, C., Eguchi, A., and Dowdy, S. F. (2011). PTD-DRBD siRNA delivery. *Methods Mol. Biol.* **683,** 339–347.

Peer, D., Park, E. J., Morishita, Y., Carman, C. V., and Shimaoka, M. (2008). Systemic leukocyte-directed siRNA delivery revealing cyclin D1 as an anti-inflammatory target. *Science* **319,** 627–630.

Pille, J. Y., Li, H., Blot, E., Bertrand, J. R., Pritchard, L. L., Opolon, P., Maksimenko, A., Lu, H., Vannier, J. P., Soria, J., Malvy, C., and Soria, C. (2006). Intravenous delivery of anti-RhoA small interfering RNA loaded in nanoparticles of chitosan in mice: Safety and efficacy in xenografted aggressive breast cancer. *Hum. Gene Ther.* **17,** 1019–1026.

Rana, T. M. (2007). Illuminating the silence: Understanding the structure and function of small RNAs. *Nat. Rev. Mol. Cell Biol.* **8,** 23–36.

Rayner, K. J., Sheedy, F. J., Esau, C. C., Hussain, F. N., Temel, R. E., Parathath, S., van Gils, J. M., Rayner, A. J., Chang, A. N., Suarez, Y., *et al.* (2011). Antagonism of miR-33 in mice promotes reverse cholesterol transport and regression of atherosclerosis. *J. Clin. Invest.* **121,** 2921–2931.

Rozema, D. B., Lewis, D. L., Wakefield, D. H., Wong, S. C., Klein, J. J., Roesch, P. L., Bertin, S. L., Reppen, T. W., Chu, Q., Blokhin, A. V., *et al.* (2007). Dynamic polyconjugates for targeted *in vivo* delivery of siRNA to hepatocytes. *Proc. Natl. Acad. Sci. USA* **104,** 12982–12987.

Semple, S. C., Akinc, A., Chen, J., Sandhu, A. P., Mui, B. L., Cho, C. K., Sah, D. W. Y., Stebbing, D., Crosley, E. J., Yaworski, E., *et al.* (2010). Rational design of cationic lipids for siRNA delivery. *Nat. Biotechnol.* **28,** 172–176.

Su, J., Baigude, H., McCarroll, J., and Rana, T. M. (2011). Silencing microRNA by interfering nanoparticles in mice. *Nucleic Acids Res.* **39,** e38.

Suzuki, H., Takatsuka, S., Akashi, H., Yamamoto, E., Nojima, M., Maruyama, R., Kai, M., Yamano, H. O., Sasaki, Y., Tokino, T., Shinomura, Y., Imai, K., and Toyota, M. (2011). Genome-wide profiling of chromatin signatures reveals epigenetic regulation of microRNA genes in colorectal cancer. *Cancer Res.* **71,** 5646–5658.

Urban-Klein, B., Werth, S., Abuharbeid, S., Czubayko, F., and Aigner, A. (2005). RNAi-mediated gene-targeting through systemic application of polyethylenimine (PEI)-complexed siRNA *in vivo*. *Gene Ther.* **12,** 461–466.

Vasudevan, S., Tong, Y., and Steitz, J. A. (2007). Switching from repression to activation: MicroRNAs can up-regulate translation. *Science* **318,** 1931–1934.

Vermeulen, A., Robertson, B., Dalby, A. B., Marshall, W. S., Karpilow, J., Leake, D., Khvorova, A., and Baskerville, S. (2007). Double-stranded regions are essential design components of potent inhibitors of RISC function. *RNA* **13,** 723–730.

Wang, W. C., Juan, A. H., Panebra, A., and Liggett, S. B. (2011). MicroRNA let-7 establishes expression of beta2-adrenergic receptors and dynamically down-regulates agonist-promoted down-regulation. *Proc. Natl. Acad. Sci. USA* **108,** 6246–6251.

CHAPTER EIGHTEEN

Genetic Nanomedicine: Gene Delivery by Targeted Lipoplexes

Nejat Düzgüneş* *and* Conchita Tros de Ilarduya†

Contents

1. Introduction 356
2. Preparation of Protein–Cationic Lipid–DNA Ternary Complexes 357
   2.1. Liposome preparation 357
   2.2. Plasmids 358
   2.3. Lipoplex preparation 358
3. Gene Delivery *In Vitro* 359
4. Cell Viability Following Transfection 361
5. Enhancement of Transfection *In Vitro* 362
6. Targeted Lipoplexes *In Vivo* 364
7. Concluding Remarks 365
Acknowledgments 365
References 366

Abstract

Cationic liposome–DNA complexes (lipoplexes) are used for the delivery of plasmid DNA to cultured cells and various tissues *in vivo*. In this chapter, we describe the preparation and evaluation of plain and targeted lipoplexes, using targeting ligands, including epidermal growth factor and transferrin. Ligand-associated lipoplexes may be used to target DNA or other nucleic acid drugs to specific cells, particularly cancer cells that overexpress the receptors for the ligands. We provide examples of the enhancement of gene expression mediated by epidermal growth factor in murine and human oral squamous cell carcinoma cells, and human hepatoblastoma and rat colon adenocarcinoma cells. We also summarize the studies on the use of transferrin–lipoplexes for enhancing gene delivery to cervical carcinoma, murine colon carcinoma, and African green monkey kidney cells. We outline two animal models in which transferrin–lipoplexes

* Department of Biomedical Sciences, Arthur A. Dugoni School of Dentistry, University of the Pacific, San Francisco, California, USA
† Department of Pharmacy and Pharmaceutical Technology, School of Pharmacy, University of Navarra, Pamplona, Spain

Methods in Enzymology, Volume 509
ISSN 0076-6879, DOI: 10.1016/B978-0-12-391858-1.00018-6

have been used for antitumor therapy by delivering either the gene encoding interleukin-12 or a suicide gene: a CT26 murine colon carcinoma, and a syngeneic, orthotopic murine oral squamous cell carcinoma.

1. Introduction

Cationic liposome–DNA complexes, generally termed "lipoplexes," are convenient tools for the delivery of plasmid DNA to cultured cells as well as to various tissues *in vivo* (Düzgüneş and Felgner, 1993; Konopka *et al.*, 2004; Pedroso de Lima *et al.*, 2001; Simões *et al.*, 1999a; Tros de Ilarduya *et al.*, 2010). The introduction of therapeutic genes or oligonucleotides for the treatment of genetic and acquired diseases is a promising alternative to conventional drug therapies. Although viral vectors are generally believed to be more efficient in gene delivery to cells in culture and *in vivo*, this is, not always the case (S. Gebremedhin and N. Düzgüneş, unpublished data). Viral vectors have severe limitations, including their immunogenicity that may limit their use, especially when repeat administrations over a long period of time are required. The large-scale production of certain viral vectors may be problematic and in the case of retroviral vectors, potential random insertion into the genome of host cells may result in unwanted medical consequences such as the production of neoplastic cells.

Lipoplexes can carry relatively large pieces of DNA and protect the DNA from degradation. They are generally safer than viral vectors, and they can transfect a large variety of cell types. Unfortunately, lipoplexes also have many limitations. These include their low efficiency of transfection (i.e., the percentage of cells they can transfect), their toxicity at high concentrations, their limited ability to reach beyond the vasculature, and their strong interaction with negatively charged molecules in serum (Simões *et al.*, 1999a).

One approach to achieve cell-selective gene delivery is to associate targeting ligands with lipoplexes, either by covalent coupling of the targeting molecules to one of the lipid components of lipoplexes or by strong complexation of the ligand with the cationic lipids. In addition to facilitating the specific association of the targeted lipoplexes with cell surface receptors, such targeted lipoplexes might also induce the internalization of the complexes as well as destabilization of the endosome membrane following acidification. An additional advantage of the complexation of targeting ligands with lipoplexes is the reduction of the net positive surface charge of the lipoplexes, thereby reducing the interaction of the cationic lipids with serum components or interstitial fluid. This may then lower the probability that the lipoplexes would remain trapped at the site of injection or would bind nonspecifically to cells and tissues in the circulation.

Some of the targeting ligands used successfully to deliver genes include transferrin, folate, asialofetuin, and epidermal growth factor (EGF; Simões *et al.*, 1999a,b; Tros de Ilarduya and Düzgüneş, 2000; Tros de Ilarduya *et al.*, 2002; Zhao and Lee, 2004). Transferrin receptors are upregulated in many cancer cells and are internalized; and thus, can be used for the delivery of liposomes, dendrimers, and copolymers containing therapeutic agents into these cells (Daniels *et al.*, 2011; Kolhatkar *et al.*, 2011). Likewise, folate receptors are overexpressed in several malignancies, and liposomes with covalently attached folate can be targeted selectively to tumor cells and internalized via receptor-mediated endocytosis (Gosselin and Lee, 2002; Zhao and Lee, 2004). Asialofetuin is the natural ligand of the asialoglycoprotein receptor on mammalian hepatocytes, and may be used to target liposomes, recombinant lipoproteins, or polymers to these cells (Arangoa *et al.*, 2003; Motoyama *et al.*, 2011; Wu *et al.*, 2002).

EGF has been used in conjunction with cationic lipid–DNA complexes ("lipoplexes") to facilitate gene delivery to human lung carcinoma cells (Yanagihara *et al.*, 2000) as well as human hepatoblastoma and human cervical carcinoma cells (Buñuales *et al.*, 2011). EGF is a single-chain, 6 kDa polypeptide of 53 amino acids that binds with high affinity to the EGF receptor, and promotes the proliferation and differentiation of mesenchymal and epithelial cells. The EGF receptor is overexpressed in most epithelial malignancies, including head and neck squamous cell carcinoma (Hynes and Lane, 2005; Kalyankrishna and Grandis, 2006). Inhibition of EGF receptor activity by specific inhibitors, however, has only modest activity in suppressing tumor growth (Chen *et al.*, 2010).

2. Preparation of Protein–Cationic Lipid–DNA Ternary Complexes

2.1. Liposome preparation

A variety of cationic lipids are available for the preparation of liposomes. The most commonly used, biodegradable lipid is 1,2,-dioleoyl-3-(trimethylamonium) propane (DOTAP), available from Avanti Polar Lipids (Alabaster, AL, USA). Cationic liposomes can be composed of pure DOTAP or its mixtures with dioleoylphosphatidylethanolamine (DOPE) (usually at a 1:1 molar ratio) or cholesterol (usually at a 1:1 or 1:0.9 molar ratio). The lipids (usually 5–10 μmol cationic lipid), stored in chloroform at − 20 °C (either in sealed ampoules or under an argon atmosphere in Teflon-lined screwcap tubes), are transferred to a screwcap tube and dried in a rotary evaporator under vacuum until a uniform film is formed around the bottom of the tube. The tube is dried in high vacuum in a vacuum oven at room temperature. The film is then hydrated with deionized water or with

10% (w/v) glucose, 10 m*M* HEPES, pH 7.4, to a final lipid concentration of 6–10 m*M* DOTAP (or alternative cationic lipid), flushed with argon, and hand shaken or vortexed for 5–10 min. Brief sonication may be used to ensure that the lipid has been removed from the glass. The resulting multilamellar liposomes are extruded (5–21 times) through polycarbonate membranes of 50-nm pore diameter (available from Avestin, Toronto, Canada, or Avanti Polar Lipids), using a manual syringe extruder (Liposofast, Avestin, or Mini-Extruder, Avanti Polar Lipids). The resulting predominantly unilamellar liposomes are filter-sterilized by passage through 0.22-μm pore-size filters from Millipore (Bedford, MA, USA), in a bioguard hood. The liposomes are stored at 4 °C in glass tubes with Teflon-lined screwcaps, after flushing the tube for 5–10 s with a stream of argon or nitrogen. It is preferable to have a sterilizing filter between the gas tank and the outlet of the gas tube to avoid any contamination. The liposomes are generally used within 1 month of preparation.

Preformed liposomal transfection reagents may also be used for these experiments. For complexation with transferrin and EGF, we have used Metafectene Pro, a polycationic liposomal transfection reagent, containing a polyamino-lipid and DOPE, purchased from Biontex Laboratories GmbH (Munich, Germany). The liposomal reagents Metafectene (Konopka *et al.*, 2005) and Metafectene Easy are also available from Biontex.

2.2. Plasmids

A variety of plasmids can be used for these experiments. We have used extensively the plasmid pCMVluc (VR-1216, originally developed by Dr. P. Felgner at Vical, Inc., San Diego, CA, USA and kindly provided by Dr. Felgner) encoding luciferase under the control of the cytomegalovirus (CMV) promoter or pCMV.SPORT ß-gal (Life Technologies, Grand Island, NY, USA) encoding ß-galactosidase, again under the CMV promoter. Alternative plasmids include pCMVßlacZnls12co (Marker Gene Technologies, Inc., Eugene, OR, USA) expressing ß-galactosidase and gWiz-GFP (Aldevron, Fargo, ND, USA) expressing green fluorescent protein.

2.3. Lipoplex preparation

Ternary complexes are prepared by mixing the liposome suspension with 100 μl of 100 m*M* NaCl, 20 m*M* HEPES, pH 7.4, containing the appropriate amount of targeting ligand (or, for controls, with no ligand), to obtain the desired +/− (N/P) charge ratio. The lipid/DNA charge ratio is calculated as the molar ratio of DOTAP (or other monovalent cationic lipid), with one positive charge per molecule, to the phosphate of the nucleotide of DNA (average molecular weight of 330 Da). For lipids with multiple cationic charges, the charges per molecule have to be taken into consideration.

Some manufacturers, however, consider this as proprietary information rendering it impossible to obtain an accurate charge ratio. The mixture of liposomes and targeting ligand is incubated at room temperature for 15 min. Then, 100 μl of buffer containing 1 μg of the plasmid of choice is added, mixed gently, and the mixture is incubated for an additional 15 min at room temperature. For *in vitro* experiments, charge ratios of 0.5/1, 1/1, 2/1, and 4/1 or 5/1 (+/−) are used. Obviously, other ratios may also be used. Since the prepared complexes are larger than the pore size of filters used for sterilization, it is essential to prepare them in a bioguard hood or flow cabinet using sterile solutions. It is best to employ the complexes right after preparation, although the "maturation" of complexes by preincubation may improve transgene expression under certain circumstances (Tros de Ilarduya and Düzgüneş, 2000). Some studies have shown that the ternary complexes may be frozen in liquid N_2 and stored at −80 °C without losing activity (Faneca *et al.*, 2002).

For experiments utilizing transferrin, the initial buffer contains 16–32 μg of the ligand in 100 μl. In the case of EGF (recombinant murine or human EGF, available from eBioscience, San Diego, CA; Sigma-Aldrich, St. Louis, MO, USA; and Invitrogen/Life Technologies), 1–5 μg protein is utilized for this step. When producing EGF-lipoplexes using DOTAP/cholesterol liposomes (10 m*M* stock), the 100 μl HEPES-buffered 10% glucose solution (with EGF or without EGF as a control) is mixed with 0.6, 1.2, 2.5, or 6.3 μl of the liposome suspension for the 0.5/1, 1/1, 2/1, and 5/1 (+/−) complexes (Buñuales *et al.*, 2011).

For *in vivo* experiments, EGF-lipoplexes are prepared at a 5/1 charge ratio and contain 1 μg EGF per μg DNA. The final concentration of the DNA in the complex is 300 μg/ml, and 60 μg of DNA is injected in a volume of 200 μl.

3. Gene Delivery *In Vitro*

A number of cell types can be used for studies on targeted lipoplexes. Here, we will describe the cell lines that have been used in our laboratories for experiments involving ligand-mediated targeting of lipoplexes.

HSC-3, cells derived from oral squamous cell carcinoma (OSCC) of the tongue are provided by Dr. R. Kramer (UCSF). H-413, cells derived from OSCC of the buccal mucosa are obtained from Dr. R. Jordan (UCSF). Murine OSCC SCCVII, cells are obtained from Drs. D. Li and B. O'Malley (University of Pennsylvania). Human cervical carcinoma HeLa cells and fibroblast-like African green monkey kidney COS-7 cells are obtained from the American Type Culture Collection (ATCC; Manassas, VA, USA). CT26 undifferentiated murine colon carcinoma cells and HepG2 human hepatoblastoma cells are also provided by ATCC. DHDK12proB rat colon adenocarcinoma cells are obtained from Dr. D. García-Olmo (Hospital General de Albacete, Spain).

The cells are propagated in T-25 flasks in Dulbecco's modified Eagle's MEM medium (DMEM; Irvine Scientific, Santa Ana, CA, USA), supplemented with 10% (v/v) heat-inactivated fetal bovine serum (FBS; Sigma; UCSF Cell Culture Facility, San Francisco, CA; or other sources), penicillin (100 units/ml), streptomycin (100 μg/ml), and L-glutamine (4 m*M*) (DMEM/10). These reagents are available from Life Technologies or the UCSF Cell Culture Facility. CT26, HeLa, COS-7, HepG2, and DHDK12proB cells are incubated in DME-high glucose (DME-HG) medium with all the supplements. The cells are cultured at 37 °C under a 5% CO_2 atmosphere in appropriate cell culture incubators. The cells are passaged 1:3 or 1:6 twice a week.

For transfection, 2×10^5 cells (counted in a hemacytometer or in a Countess Automated Cell Counter; Life Technologies) are seeded in 1 ml of DMEM in 48-well culture plates 1 day before transfection and used at approximately 85% confluence. Cells are prewashed with serum-free DMEM medium and then covered with 0.4 ml of the same medium. Lipid/DNA or lipid/ligand/DNA complexes are added in a volume of 0.1 ml per well. The cells are incubated for 4 h at 37 °C and then 0.5 ml of medium containing 20% FBS is added to bring the final serum concentration to 10%.

Luciferase activity is usually assayed 48 h after transfection, using the Luciferase Assay System (Promega, Madison, WI, USA) and a luminometer. Naturally, luciferase expression can be assayed over a period of time to assess the time of optimal expression. A simple but convenient luminometer is the Turner Designs TD-20/20 instrument (Sunnyvale, CA, USA). Another luminometer is the Sirius-2 Berthold Detection Systems instrument (Pforzheim, Germany and Huntsville, AL, USA). The data are expressed as relative light units (RLU) per ml of cell lysate or as ng luciferase per mg protein. These values are designated as "transfection activity."

"Transfection efficiency" indicates the percentage of cells that measurably express the transfected gene. To visualize gene expression, plasmids expressing ß-galactosidase or green fluorescent protein may be used in the preparation of the targeted lipoplexes. When using, for example, pCMV. SPORT-ß-gal as the plasmid, following the transfection and subsequent incubation periods (usually 4 h + 48 h), the cells are washed with phosphate-buffered saline (PBS) and fixed in freshly prepared 2% formaldehyde and 0.2% glutaraldehyde for 1–2 h (Simões *et al.*, 2003). The cells are washed three times with PBS. They are then stained with an isotonic solution of 5′-bromo-4-chloro-3-indolyl-ß-D-galactopyranoside (X-Gal; Promega or Fermentas, Inc.). The cells are incubated for 24 h to enable full color development and then examined in an inverted phase contrast microscope.

Micrographs are obtained and are used for counting the percentage of transfected cells. The expressed ß-galactosidase hydrolyzes X-Gal into galactose and 4-chloro-3-brom-indigo, the latter forming a deep blue precipitate in the cells.

For the evaluation of green fluorescent protein expression, following the transfection and incubation periods (generally 4 h + 48 h), the cells are washed with PBS and then detached from the culture plate using 0.5 ml/well of Cell Dissociation Buffer (Life Technologies) or Cell Dissociation Solution (Millipore). The cells are mixed with DME-HG medium with FBS (1%) and propidium iodide (1 μg/ml; Sigma-Aldrich or Invitrogen). Propidium iodide is excluded from live cells but can enter dead cells and stains the DNA, which results in an enhancement of its fluorescence. Green fluorescent cells are counted in a flow cytometer (FACScan, BD, Franklin Lakes, NJ, USA; or equivalent flow cytometer) with excitation at 488-nm, emission around 530 nm for green fluorescent protein, and around 630 nm for propidium iodide (Simões *et al.*, 2003). The forward scatter signal is used to identify and gate the cells of interest or cell aggregates, and the propidium iodide signal is used to eliminate dead cells.

As an alternative to flow cytometry, fluorescence microscopy can be utilized to ascertain the percentage of cells expressing the florescent protein. In this case, individual cells are identified by counterstaining with Hoechst 33,342 fluorescent dye (Pierce, Thermo Scientific, Rockford, IL, USA; Molecular Probes, Life Technologies), which permeates the cell membrane of intact cells and binds the adenine–thymine regions in the minor groove of DNA. The dye emits blue fluorescence in the region 460–490 nm and is compatible with fluorescein and rhodamine fluorophores. Thus, the ratio of the number of green fluorescing cells as a result of green fluorescent protein expression to the number of blue-stained nuclei provides the percentage of cells that are transfected successfully, and hence, the transfection efficiency. Nevertheless, it is important to distinguish cell autofluorescence from fluorescence arising from green fluorescent protein. Thus, the fluorescence of untransfected cells should be used as a control.

4. Cell Viability Following Transfection

Cell viability assays are utilized to assess the potential toxic effects of targeted lipoplexes. These include assays for the release of lactate dehydrogenase or the reduction of various compounds, including MTT (3-(4,5-dimethylthiazol-2-yl)-2,5-diphenyltetrazolium bromide) (Sigma-Aldrich), XTT (2,3-bis(2-methoxy-4-nitro-5-sulfophenyl)-2H-tetrazolium-5-carboxanilide) (Sigma-Aldrich), and Alamar Blue (Promega and Life Technologies), whose active ingredient is resazurin

(7-hydroxy-3H-phenoxazin-3-one-10-oxide). The nontoxic resazurin enters cells and is reduced resorufin, which produces a bright red color (peak absorbance at 600 nm for the oxidized form and 570 nm for the reduced form) and fluorescence (emission at 590 nm, with excitation at 530–560 nm) (http://tools.invitrogen.com/content/sfs/manuals/PI-DAL1025-1100_TI%20alamarBlue%20Rev%201.1.pdf).

One milliliter of a 10% (v/v) solution of Alamar Blue in complete DME medium is added to the cells in the wells of a 48-well plate, usually about 3 h before the determination of transfection activity (which is usually performed 48 h after the 4-h transfection period). The cells are incubated with the dye for 2.5–4 h at 37 °C to allow for metabolic activity to convert the dye. The optimal time can be determined visually when there is a significant color change. Then, 200 μl of the supernatant from each well is transferred to the wells of a 96-well plate for ease of determination in a microplate reader. The absorbance at 570 nm and 600 nm is determined in a Molecular Devices (Sunnyvale, CA, USA) (or comparable) microplate reader. Cell viability is given by the following formula:

Viability (% of control cells) $= [(A_{570} - A_{600})$ of treated cells$] \times 100/[(A_{570} - A_{600})$ of control cells$]$

One of the advantages of the Alamar Blue assay is that the cells continue to be viable after the assay and can be returned to normal incubation conditions (Konopka *et al.*, 1996).

5. Enhancement of Transfection *In Vitro*

HepG2 and DHDK12proB cells are known to overexpress the EGF receptor. Transfection of these cells with EGF-lipoplexes at the 5/1 (+/−) charge ratio, with 1 μg DNA and 1 μg EGF resulted in a 3.6-fold increase in luciferase activity in HepG2 cells and 40-fold in DHDK12proB cells compared to plain lipoplexes (Buñuales *et al.*, 2011). The Alamar Blue assay indicated that both plain lipoplexes and EGF-lipoplexes at the different charge ratios had low cytotoxicity in HepG2 cells compared to untreated cells.

Murine OSCC (SCCVII) cells were transfected with either plain lipoplexes (using DOTAP:cholesterol liposomes) or lipoplexes complexed with recombinant murine EGF. Lipoplexes targeted to the EGF receptor facilitated a 14-fold enhancement of gene expression at 1 μg/ml EGF, and a 110-fold enhancement at 5 μg/ml (Table 18.1). By contrast, in nontumor derived NIH3T3 murine fibroblasts, gene expression was enhanced by only about fivefold at 5 μg/ml EGF. These observations indicate that significant specificity can be achieved in targeting reporter or therapeutic genes to oral cancer cells.

Table 18.1 Transfection of murine SCCVII and NIH3T3 cells by EGF-lipoplexes with DOTAP:cholesterol liposomes

Cells	EGF (μg/ml)	Transfection activity (RLU/ml)
SCCVII	0	6541
	1	93,553
	2.5	413,400
	5	717,933
NIH3T3	0	4272
	1	2578
	2.5	11,031
	5	23,000

Cells were transfected with 1 μg pCMV.luc complexed with murine EGF-complexed DOTAP:chol (2:1) liposomes at a 2/1 (+/−) charge ratio. (A. Streeter, N. Overlid, S. Gebremedhin, and N. Düzgüneş, unpublished data.)

Table 18.2 Transfection of oral squamous cell carcinoma cells by EGF-lipoplexes with Metafectene Pro liposomes

Cells	EGF (μg/ml)	Transfection activity (RLU/ml)
H-413	0	1009
	2	24,980
H-357	0	24,987
	2	114,180
HSC-3	0	50,087
	2	289,400

Cells were transfected with 1 μg pCMV.luc complexed with human EGF-complexed Metafectene Pro liposomes (1.5 μl). (S. Gebremedhin, M. Yee, J. Ouellette, and N. Düzgüneş, unpublished data).

The relatively novel liposomal transfection reagent Metafectene Pro can also be used for complexation with EGF for targeting DNA to various human OSCC cells. Different OSCC cell lines are highly variable in terms of their susceptibility to transfection. Complexation with human recombinant EGF with Metafectene Pro and subsequent complexation with DNA resulted in substantial enhancement of transfection activity (Table 18.2). When 2 μg/ml EGF was used, this enhancement was almost 25-fold in H-413 cells. For H-357 and HSC-3 cells, this enhancement was about five- and sixfold, respectively.

In CT26 cells, the use of transferrin–lipoplexes enhanced luciferase expression by a factor of 12 compared to nontargeted, plain lipoplexes, even in the presence of 60% serum (Tros de Ilarduya *et al.*, 2006), which is usually very inhibitory to transfection mediated by plain lipoplexes (Escriou *et al.*, 1998).

Transfection of HeLa cells with transferrin–lipoplexes composed of DOTAP:DOPE resulted in about a sixfold enhancement of transgene expression at the 1/1 (+/−) charge ratio, which produced the highest transfection activity (Simões *et al.*, 2003). Although the level of luciferase activity was lower at the 1/2 (+/−) charge ratio than at 1/1, the enhancement of gene expression by the use of transferrin–lipoplexes was in the range of 200, since the plain lipoplexes were essentially ineffective. In the case of COS-7 cells, transferrin–lipoplexes enhanced gene delivery by a factor of about 12 at the 1/1 (+/−) charge ratio, whereas at the 1/2 (+/−) charge ratio, the enhancement factor was about 600 (Simões *et al.*, 1998). The results at the 1/2 (+/−) charge ratio are intriguing, in that net negatively charged complexes are able to mediate gene transfer, contrary to the expectation that net positively charged lipoplexes are essential for this process. This observation indicates that transferrin that is associated with the lipoplexes is able to mediate binding to transferrin receptors, internalization, and destabilization of the endosomal membrane (Simões *et al.*, 1999b).

6. Targeted Lipoplexes *In Vivo*

We describe two animal models in which the antitumoral effect of therapeutic genes delivered in targeted lipoplexes is examined. In the first model, 5×10^5 CT26 murine colon carcinoma cells in 50 μl PBS are injected subcutaneously into the back of 6- to 8-week-old female Balb/C mice (Harlan Ibérica Laboratories, Barcelona, Spain; or equivalent source) (Tros de Ilarduya *et al.*, 2006). When the tumor size reaches 5–6 mm in diameter (about 10–20 days later), 60 μg DNA (pCMVluc or pCMVIL-12, expressing interleukin-12) in sterile saline or complexed with plain or transferrin–lipoplexes is administered by a single intratumoral injection using a 29-G insulin syringe. Lipoplexes are prepared as described above. Transferrin (32 μg/μg DNA) is complexed with DOTAP/cholesterol liposomes at a DOTAP/DNA charge ratio of 5/1 (+/−). The injection volume is 50 μl. Tumor growth is evaluated by measuring two perpendicular tumor diameters with a digital caliper. Mice are euthanized when the diameter of the untreated tumors reach 13 mm. For determination of cytokine levels, blood is collected by retro-orbital bleeding at 0, 6, and 24 h after injection of complexes or DNA, and centrifuged at $12{,}000 \times g$ for 10 min at 4 °C. Complete tumor regression was observed in 75% of the mice treated with targeted lipoplexes containing the plasmid encoding IL-12 (Tros de Ilarduya *et al.*, 2006). The survival rate of the mice was 88% at 23 days after the onset of therapy.

The second model is a syngeneic, orthotopic, and immunocompetent murine model of OSCC (Neves *et al.*, 2009; O'Malley *et al.*, 1996). Six- to eight-week-old female C3H/HeOuJ mice (Charles River Laboratories,

Barcelona, Spain) are anesthetized by intramuscular injection of chlorpromazine (2 mg/kg) and ketamine (100 mg/kg). They are then injected with a 30-G needle with 1×10^5 SCCVII cells in 50 μl PBS into the floor of the mouth. Five days later, the tumors are injected with 40 μg pCMVtk, expressing herpes simplex virus thymidine kinase (National Gene Vector Laboratory, University of Michigan, Ann Arbor, MI, USA), either plain or complexed with transferrin–DOTAP:cholesterol (1:1) liposomes, at a charge ratio of 3/2 (+/−), and with 32 μg transferrin per μg DNA. Four hours later, the animals are injected intraperitoneally, with 50 mg/kg ganciclovir daily for 8 days, to initiate suicide gene therapy via the conversion of ganciclovir to ganciclovir monophosphate by the viral thymidine kinase. The tumor size is measured every day by means of calipers. At the beginning of treatment, the average tumor volume is around 36 mm^3. Seven days after the initiation of treatment, the tumor volume decreased significantly compared to mice receiving PBS or naked pCMVtk (Neves *et al.*, 2009). Additional studies indicated the recruitment of NK cells and T-lymphocytes to the tumors as well as an increase in the levels of interleukin-2, interferon-γ, and the chemokine Rantes. Thus, targeted lipoplex-mediated delivery of suicide genes also have immune stimulatory effects that may contribute to tumor therapeutics.

7. Concluding Remarks

One of the advantages of lipoplexes in delivering nucleic acid drugs to cancer cells is that they are much more efficient in transfecting dividing cells than resting cells. Thus, lipoplexes delivering suicide genes have a built-in specificity for targeting rapidly dividing cancer cells. In additon, lipoplexes targeted to cell surface molecules overexpressed in cancer cells provide further selectivity in delivering nucleic acid drugs to cancer cells. It will be of interest to explore the use of targeted lipoplexes in delivering a combination of suicide genes and genes encoding immunostimulatory cytokines into tumors. Whether targeted lipoplexes, especially those that are net negatively charged to avoid substantial interaction with serum proteins and nontarget tissues, can reach neovascularized tumor tissue through both the enhanced permeation and retention effect and the targeting ligands will have antitumor effects after systemic or subcutaneous delivery, remains to be investigated.

ACKNOWLEDGMENTS

We thank all the members of our laboratories and other colleagues who have contributed to the development of the work described in this chapter.

REFERENCES

Arangoa, M. A., Düzgüneş, N., and Tros de Ilarduya, C. (2003). Increased receptor-mediated gene delivery to the liver by protamine-enhanced asialofetuin-lipoplexes. *Gene Ther.* **10,** 5–14.

Buñuales, M., Düzgüneş, N., Zalba, S., Garrido, M. J., and Tros de Ilarduya, C. (2011). Efficient gene delivery by EGF-lipoplexes *in vitro* and *in vivo*. *Nanomedicine* **6,** 89–98.

Chen, L. F., Cohen, E. E., and Grandis, J. R. (2010). New strategies in head and neck cancer: Understanding resistance to epidermal growth factor receptor inhibitors. *Clin. Cancer Res.* **16,** 2489–2495.

Daniels, T. R., Bernabeu, E., Rodríguez, J. A., Patel, S., Kozman, M., Chiappetta, D. A., Holler, E., Ljubimova, J. Y., Helguera, G., and Penichet, M. L. (2011). Transferrin receptors and the targeted delivery of therapeutic agents against cancer. *Biochim. Biophys. Acta.* (in press).

Düzgüneş, N., and Felgner, P. L. (1993). Intracellular delivery of nucleic acids and transcription factors by cationic liposomes. *Methods Enzymol.* **221,** 303–306.

Escriou, V., Ciolina, C., Lacroix, F., Byk, G., Scherman, D., and Wils, P. (1998). Cationic lipid-mediated gene transfer: Effect of serum on cellular uptake and intracellular fate of lipopolyamine/DNA complexes. *Biochim. Biophys. Acta* **1368,** 276–288.

Faneca, H., Simões, S., and Pedroso de Lima, M. C. (2002). Evaluation of lipid-based reagents to mediate intracellular gene delivery. *Biochim. Biophys. Acta* **1567,** 23–33.

Gosselin, M. A., and Lee, R. J. (2002). Folate receptor-targeted liposomes as vectors for therapeutic agents. *Biotechnol. Annu. Rev.* **8,** 103–131.

Hynes, N. E., and Lane, H. A. (2005). ERBB receptors and cancer: The complexity of targeted inhibitors. *Nat. Rev. Cancer* **5,** 341–354.

Kalyankrishna, S., and Grandis, J. R. (2006). Epidermal growth factor receptor biology in head and neck cancer. *J. Clin. Oncol.* **24,** 2666–2672.

Kolhatkar, R., Lote, A., and Khambati, H. (2011). Active tumor targeting of nanomaterials using folic acid, transferrin and integrin receptors. *Curr. Drug Discov. Technol.* **8,** 197–206.

Konopka, K., Pretzer, E., Felgner, P. L., and Düzgüneş, N. (1996). Human immunodeficiency virus type-1 (HIV-1) infection increases the sensitivity of macrophages and THP-1 cells to cytotoxicity by cationic liposomes. *Biochim. Biophys. Acta* **1312,** 186–196.

Konopka, K., Lee, A., Moser-Kim, N., Saghezchi, S., Kim, A., Lim, A., Suzara, V. V., and Düzgüneş, N. (2004). Gene transfer to human oral cancer cells via non-viral vectors and HSV-tk/ganciclovir-mediated cytotoxicity; Potential for suicide gene therapy. *Gene Ther. Mol. Biol.* **8,** 307–318.

Konopka, K., Fallah, B., Monzon-Duller, J., Overlid, N., and Düzgüneş, N. (2005). Serum-resistant gene transfer to oral cancer cells by Metafectene and GeneJammer: Application to HSV-tk/ganciclovir-mediated cytotoxicity. *Cell. Mol. Biol. Lett.* **10,** 455–470.

Motoyama, K., Nakashima, Y., Aramaki, Y., Hirayama, F., Uekama, K., and Arima, H. (2011). *In vitro* gene delivery mediated by asialofetuin-appended cationic liposomes associated with γ-cyclodextrin into hepatocytes. *J. Drug Deliv.* **2011,** 476137.

Neves, S. S., Faneca, H., Bertin, S., Konopka, K., Düzgüneş, N., Pierrefite-Carle, V., Simões, S. P., and Pedroso de Lima, M. C. (2009). Transferrin-lipoplex-mediated suicide gene therapy of oral squamous cell carcinoma in an immunocompetent murine model and mechanisms involved in the antitumoral response. *Cancer Gene Ther.* **16,** 91–101.

O'Malley, B. W., Cope, K. A., Chen, S. H., Li, D., Schwarta, M. R., and Woo, S. L. (1996). Combination gene therapy for oral cancer in a murine model. *Cancer Res.* **56,** 1737–1741.

Pedroso de Lima, M. C., Simões, S., Pires, P., Faneca, H., and Düzgüneş, N. (2001). Cationic lipid-DNA complexes in gene delivery: From biophysics to biological applications. *Adv. Drug Deliv. Rev.* **47,** 277–294.

Simões, S., Slepushkin, V., Gaspar, R., Pedroso de Lima, M. C., and Düzgüneş, N. (1998). Gene delivery by negatively charged ternary complexes of DNA, cationic liposomes and transferrin or fusigenic peptides. *Gene Ther.* **5,** 955–964.

Simões, S., Pires, P., Düzgüneş, N., and Pedroso de Lima, M. C. (1999a). Cationic liposomes as gene transfer vectors: Barriers to successful application in gene therapy. *Curr. Opin. Mol. Ther.* **1,** 147–157.

Simões, S., Slepushkin, V., Gaspar, R., Pedroso de Lima, M. C., and Düzgüneş, N. (1999b). Mechanisms of gene transfer mediated by lipoplexes associated with targeting ligands of pH-sensitive peptides. *Gene Ther.* **6,** 1798–1807.

Simões, S., Pires, P., Girão da Cruz, T., Düzgüneş, N., and Pedroso de Lima, M. C. (2003). Gene delivery by cationic liposome-DNA complexes containing transferrin or serum albumin. *Methods Enzymol.* **373,** 369–383.

Tros de Ilarduya, C., and Düzgüneş, N. (2000). Efficient gene transfer by transferrin lipoplexes in the presence of serum. *Biochim. Biophys. Acta* **1463,** 333–342.

Tros de Ilarduya, C., Arangoa, M. A., Moreno-Aliaga, M. J., and Düzgüneş, N. (2002). Enhanced gene delivery *in vitro* and *in vivo* by improved transferrin-lipoplexes. *Biochim. Biophys. Acta* **1561,** 209–221.

Tros de Ilarduya, C., Buñuales, M., Qian, C., and Düzgüneş, N. (2006). Antitumoral activity of transferrin-lipoplexes carrying the IL-12 gene in the treatment of colon cancer. *J. Drug Target.* **14,** 527–535.

Tros de Ilarduya, C., Sun, Y., and Düzgüneş, N. (2010). Gene delivery by lipoplexes and polyplexes. *Eur. J. Pharm. Sci.* **40,** 159–170.

Wu, J., Nantz, M. H., and Zern, M. A. (2002). Targeting hepatocytes for drug and gene delivery: Emerging novel approaches and applications. *Front. Biosci.* **7,** d717–d725.

Yanagihara, K., Cheng, H., and Cheng, P. W. (2000). Effects of epidermal growth factor, transferrin, and insulin on lipofection efficiency in human lung carcinoma cells. *Cancer Gene Ther.* **7,** 59–65.

Zhao, X. B., and Lee, R. J. (2004). Tumor-selective targeted delivery of genes and antisense oligodeoxyribonucleotides via the folate receptor. *Adv. Drug Deliv. Rev.* **56,** 1193–1204.

Author Index

Note: Page numbers followed by "*f*" indicate figures, and "*t*" indicate tables.

A

Aalinkeel, R., 41–60
Aarabi, S., 245
Abd El-Hady, S. S., 68–69
Abdin, M., 227
Abel, J., 232
Abo, S., 104–105
Abuharbeid, S., 341
Acharya, S., 63–67, 68–69, 73–75, 74*t*, 76, 79*f*, 80*f*
Acosta, N., 129
Adamo, R., 22–23
Adam, S. A., 314
Adams, R. L., 314
Addlagatta, A., 173–175
Adema, G. J., 146–147
Agger, E. M., 147
Agnihotri, S. A., 62–63, 73–75
Agrawal, S., 183
Aguilar, Z. P., 227
Ahmad, Z., 87–99
Ahmed, F. J., 227
Ahmet, F., 146–147
Ahsan, F., 151–152
Ahvenainen, R., 65–67
Aigner, A., 341
Ai, H., 263–276
Aitta, J., 75
Akaji, K., 316–318
Akashi, H., 340–341
Akbuga, J., 109, 112–113
Akinc, A., 341
Akin, D., 227
Akita, H., 278–279, 302–305, 307, 308–309, 315–318, 319–320
Aktas, Y., 131
Alam, M., 128–129, 145, 154
Albericio, F., 143–163
Albrecht, R. A., 128–129, 145, 154
Alcamí, J., 23, 24–25, 28, 33–34, 35–36
Alessandrini, A., 187–188
Ali, A., 341, 346
Alivisatos, A. P., 46, 148–149
Allaker, R. P., 87–99
Allan, J. S., 22
Allémann, E., 3
Allen, N. P., 314
Almeida, J. A., 288
Almodovar, J., 89
Almofti, M. R., 307
Alonso, M. J., 67–68
Altin, J. G., 146–147, 151–152
Altschuler, Y., 17–18, 226
Alves, I. D., 284–287
Alyautdin, R. N., 43
Amacker, M., 147
Amal, R., 264
Ambalavanan, N., 246, 252
Ambrosio, F., 245
Ambros, V., 107–108, 340
Amidi, M., 137–138
Amidon, G. L., 17–18
Amigo, R., 150
Amiji, M. M., 17–18, 67–68
Aminabhavi, T. M., 62–63, 73–75, 154
Amin, A. F., 43–44
Ammassari, A., 42–43
Ammoury, N., 3
Ananthanarayan, L., 171–172
Ances, B., 42–43
Andersen, M. O., 341
Andersen, P., 147
Andremont, A., 67–68
Andrew Wang, Y., 227
Andrich, M. P., 288
Andrieux, K., 67–68
Angel, C. E., 146–147
Angeli, V., 145–146
Angulo, J., 28, 29–31, 33, 35
Anhalt, M., 201
Anjos-Afonso, F., 146–147
Anthony, D. C., 27
Anton, M., 264
Antunes-Madeira, M. C., 288
Anumanthan, A., 180, 181–183
Apostolopoulos, V., 145, 146–147
Aquaro, S., 44–45
Aramaki, Y., 145–146, 357
Arangoa, M. A., 331, 357
Arani, R. B., 33–34
Arap, W., 227
Arar, K., 342
Araya, E., 150
Arbab, A. S., 207–208, 214–216, 217, 219–221

Arbiol, J., 148–149
Arcaro, K. F., 226
Arigita, C., 152
Arima, H., 357
Arking, D. E., 340–341
Arruda, T. M., 168–169
Arsianti, M., 264
Arya, G., 65–67
Arya, S. K., 169
Asakawa, T., 312–313
AshaRani, P. V., 227
Ashari, P., 217
Assodani, A., 227
Astronomo, R. D., 22–23
Attah-Poku, S., 134–135
Atyabi, F., 68–69
Aubry, S., 284–287
Audette, G. F., 168–169
Azab, B., 111–112

B

Babiuk, L. A., 134–135
Babu, V. R., 154
Bachem, A., 146–147
Bachmann, C., 304
Bachmann, M. F., 128–129, 144–145
Bachmeier, C., 17–18
Back, D., 49
Badawy, M. E., 65–67
Badve, S., 227
Bae, J., 341
Baert, J., 200
Bagga, S., 340
Baigude, H., 339–353
Bailey, R. E., 196–197
Bais, D., 112–113
Baker, E., 43–44
Bakhshi, R., 87–99
Bakkour, N., 293
Bak, M., 342
Bakowsky, U., 109
Bakthavatchalu, V., 227
Balasubramaniam, T. A., 165–194
Balavoine, F., 187
Baldrick, P., 129
Ballard, C. E., 106
Ballaun, C., 22
Bal, S. M., 151–152
Bangsaruntip, S., 184
Bankar, S. B., 171–172
Bantleon, R., 219–221
Bantz, K. C., 196–197
Baranek, T., 146–147
BarathManiKanth, S., 232
Bardhan, R., 227
Barenholz, Y., 288
Bar, H., 62–63
Barell, A. K., 188–190
Barin, F., 22
Barnett, B. P., 214–216
Barrett, J. F., 62
Barria, M. A., 227, 229*t*
Barrientos, A. G., 22–23, 24–26, 27
Bartel, B., 107–108
Bartel, D. P., 107–108, 340
Bartfai, T., 278–279
Bartlett, G. R., 281
Bartneck, M., 225–242
Basa, L. J., 22
Bashaw, L. A., 207–208
Bashir, R., 227
Baskerville, S., 342
Bass, B. L., 107–108
Bassik, M. C., 314–315
Basta, S., 144, 154
Bastus, N. G., 150
Bateman, A., 181–183
Bauer, D. J., 2
Bauer, G., 150
Bauer, K. D., 232–233
Bauer, M., 145
Bäuml, E., 32–33
Baumle, M., 225–226
Bawendi, M. G., 226
Beach, D., 107–108
Beck-Broichsitter, M., 3
Becker, A. L., 109
Becker, M. L., 227, 228–230
Bedner, E., 232–233
Bedoya, L. M., 23, 24–25, 28
Behlke, M. A., 340–341
Behm-Ansmant, I., 340
Behr, J. P., 264–265, 314, 328–329
Beijnen, J. H., 42–43, 44–45, 145–146
Belizaire, R., 145–146
Bellamkonda, R. V., 89
Belousov, V., 236–238
Bemelman, W. A., 107
Bencini, M., 4–5
Ben-Efraim, I., 314–315
Benhar, I., 62–63
Benie, A. J., 32–33
Benita, S., 3, 17–18, 226
Benoit, J. P., 8–9
Bentley, S. D., 181–183
Bento, D., 127–142
Bentolila, L. A., 44
Bergemann, C., 264
Berger, S., 28
Bergey, E. J., 43–44, 45–46
Bernabeu, E., 357
Bernstein, E., 107–108
Berthold, A., 129–130
Berti, F., 22–23

Bertin, S. L., 341, 364–365
Bertoldi, S., 88–89
Bertrand, J. R., 341
Besenbacher, F., 341
Bessa, J., 128–129
Bessou, G., 146–147
Bethell, D., 24–25
Bevaart, L., 152
Bevan, M. J., 146–147
Beyermann, M., 278–279
Beyrath, J., 151–152
Bhadada, S. V., 43–44
Bhakta, V., 254
Bharatwaj, B., 65–67
Bharti, A., 42–43
Bhaskaran, S., 229*t*
Bhaskar, S., 43–44
Bhatia, S. N., 244
Bhattacharya, S., 328–329
Bhattarai, N., 264–265
Bhatt, J. S., 43–44
Bienert, M., 278–279
Bierhals, K., 232
Bierman, M. J., 103
Bilbao, G., 328–329
Bilensoy, E., 131
Bindukumar, B., 50
Binley, J. M., 33–34, 34*f*
Birincioglu, M., 237
Bisazza, A., 1–19
Bischof, J. C., 229*t*
Bishop, D. K., 245
Bishop, N., 183
Blaas, D., 32–33
Blackwell, J. D., 214–216
Blakely, E. L., 318
Blakey, D., 328–329
Blanco-Prieto, M. J., 72–73
Blander, J. M., 144
Blokhin, A. V., 341
Blot, E., 341
Blow, D. M., 173
Blunk, T., 328–329
Boccaccini, A. R., 112–113
Bock, P., 49–50
Boddohi, S., 112–113
Bodmeier, R., 75, 134–135
Boesch, P., 318
Bolhassani, A., 147
Bolling-Sternevald, E., 67–68
Bond, D. R., 187
Bonetto, F., 157
Bongertz, V., 33–34
Bonifaz, L. C., 146–147
Bonnyay, D. P., 146–147
Bonoiu, A., 45–46
Boote, E. J., 229*t*
Borchard, G., 128–130, 137–138, 139
Borchert, H. H., 145–146
Border, W. A., 245 255*f*
Borges, O., 127–142
Borgna, L., 244
Borkow, G., 88
Bork, P., 340
Borm, P. J., 43–44, 88
Bornemann, J., 227
Botner, S., 102
Boudinot, P., 146–147
Bourgeois, D., 175
Bourin, M., 49
Boussif, O., 264–265, 328–329
Boutros, M., 340
Bouwstram, J. A., 151–152
Bowdish, K. S., 146–147
Bowersock, T., 134–135
Bracht, J., 340
Bradbury, M. W., 49
Bradford, M. M., 31, 150–151
Braeckmans, K., 195–224
Braich, R., 340–341, 342
Branca, M., 245
Branca, M. F., 245
Brancati, F. L., 176–177
Brandau, W., 226, 227, 229*t*, 230, 231
Brand, L., 288–289
Braun, S., 28
Bravman, T., 28–29
Braydich-Stolle, L. K., 196–197
Brayé, M., 151–152
Brehm, B. R., 219–221
Breitenbach, A., 139
Brenseke, B., 63–64
Breunig, M., 328–329
Breuzard, G., 314–315
Bridges, A. J., 148–149
Bright, F. V., 187
Brimnes, M. K., 146–147
Brinkmann, K., 147
Brisken, A. F., 328–329
Brisson, A. R., 207–208
Britton, W. J., 146–147, 151–152
Broadbent, R., 146–147
Broadrick, K. M., 112–113
Brodin, B., 145–146
Bromberg, L., 102–103
Bronner, V., 28–29
Brown, D. M., 246, 249
Brown, S., 181–183
Bruchez, M. P. Jr., 148–149
Brumfeld, V., 314–315
Brunner, S., 314
Brust, M., 24–25, 219–221
Buchacher, A., 22
Buchholz, V. R., 146–147
Budker, V., 314–315
Bule, M. V., 171–172

Bulte, J. W. M., 214–216
Bundgaard, T. J., 147
Buñuales, M., 357, 359, 362, 363
Burda, C., 45–46
Burgdorf, S., 146–147
Burge, C. B., 107–108
Burger, D. M., 42–43
Burger, H., 33–34
Burger, M. C., 226
Burgess, D. J., 3
Burgess, S., 328–329
Burkatovskaya, M., 65–67
Burke, D. S., 33–34
Burke, N. A., 112
Burkhart, J., 52–53
Burlingame, A., 314
Burton, D. R., 22–23, 33–34, 34*f*
Busch, D. H., 144, 154
Buschmann, I. R., 55–56
Buschow, S. I., 157
Bustin, S. A., 55–56
Byk, G., 363

C

Caiafa, P., 314
Cai, H., 45–46
Calabro, A., 256
Calarese, D. A., 22
Calin, G. A., 340–341
Caliò, R., 44–45
Callahan, P. J., 288
Camacho, A., 63–64
Cambillau, C., 175
Caminschi, I., 146–147
Camma, C., 103
Campanero, M. A., 72–73
Campbell, I. K., 146–147
Campbell, S. J., 27
Campoli-Richards, D. M., 2
Cañada, F. J., 27
Canada, J., 22–23
Cantley, W., 341
Cao, E., 43–44
Cao, H., 341, 346
Capaccioli, S., 233–235
Capel, P. J., 146
Caplan, A. I., 246, 256
Cardoso, A. L., 277–300
Cardoso, A. M. S., 277–300
Carlisi, B., 67–68
Carlyle, J. R., 146–147
Carman, C. V., 341
Carotta, S., 314
Carriere, F., 175
Carrington, J. C., 107–108
Carrino, D. A., 256
Carr, J. K., 33–34
Carthew, R. W., 340
Caruso, F., 109
Carvalho de Souza, A., 25–26
Casadevall, A., 62
Cassaday, R. D., 148–149
Castano, A. P., 65–67
Casteel, S. W., 229*t*
Castranova, V., 207–208, 226, 228
Caudy, A. A., 107–108
Cavalieri, F., 109
Cavalli, R., 1–19
Ceballos, G., 62–63
Cen, L., 208–211
Cerdeño-Tárraga, A.-M., 181–183
Cesbron, Y., 219–221
Chaki, S., 316–318
Chakraborty, S., 169
Chalkley, R. J., 314
Challis, G. L., 181–183
Chaloin, L., 278–279
Chambers, S. H., 232–233
Champion, C., 23, 28, 35
Chan, C. K., 314–315
Chanda, N., 229*t*
Chan, D. C., 318
Chandra, R., 341
Chandrasekhar, M., 229*t*
Chang, A. N., 342
Chang, C. J., 236–238
Chang, C. W., 45–46
Chang, M. W., 68–69
Chang, R. K., 70
Chang, T. C., 340–341
Chan, J. A., 342
Chan, S., 146–147
Chan, W. C. W., 196–197, 226, 229*t*
Chan, Y., 245
Chao, A., 104–105
Chappey, C., 33–34, 34*f*
Charalambous, A., 146–147
Chassaing, G., 278–279, 284–287
Chastain, M., 22–23
Chater, K. F., 181–183
Chatterjee, A. K., 88
Chaturvedi, A. K., 62
Chaudhary, A., 103
Chauhan, V. S., 148–149
Chavanpatil, M. D., 17–18
Chawda, R., 50
Chawla, J. S., 17–18
Chen, C. J., 17–18, 146–147
Chen, F., 264–265
Cheng, E., 88, 90
Cheng, H., 357
Cheng, P. W., 357
Cheng, Y. C., 31–32
Chen, H., 318
Chen, J., 188–190, 232–233, 341

Chen, J. L., 146–147
Chen, L. A., 288–289
Chen, L. F., 357
Chen, M., 227
Chen, R. J., 184, 188–190
Chen, S. H., 364–365
Chen, X., 45–46, 107–108, 263–276
Chen, Y. C., 214–216, 227, 229*t*
Chen, Y.-S., 148–149
Cheong, C., 146–147
Cheong, S. J., 264
Cheow, W. S., 68–69
Cheradame, H., 314–315
Cherry, R. J., 284–286
Chesebro, B., 33–34
Chiappelli, F., 42–43
Chiappetta, D. A., 357
Chin, M., 168–169
Chiodo, F., 21–40
Chithrani, B. D., 226, 229*t*
Chiu, Y.-L., 341, 346
Cho, C. K., 341
Cho, E. C., 229*t*
Choi, C. H., 226
Choi, H. S., 171, 226
Choi, K.-P., 173
Choi, M. R., 227
Choi, W., 165–194
Chomczynski, P., 55–56
Cho, M. J., 293
Chompoosor, A., 226
Chonn, A., 145–146
Choquenet, B., 112–113
Choquet, D., 227
Cho, T. J., 228
Chou, A., 184
Chou, C. C., 65–67
Chowers, Y., 107
Christensen, F. M., 227
Christensen, L. V., 340, 341
Christensen, M., 342
Chrzanowski, W., 88
Chuah, A. M., 112–113
Chuah, M. K., 200
Chu, C.-Y., 340, 341, 345–346
Chu, P. K., 88
Chu, Q., 341
Churchill, R., 229*t*
Ciancio, B., 42–43
Cingolani, A., 42–43
Ciolina, C., 363
Civra, A., 1–19
Clare, S. E., 227
Clark, G. J., 146–147
Clark, M. R., 26–27
Clausen, B. E., 146–147
Claussen, C. D., 219–221
Clavel, C., 23, 24–25, 28, 35
Clos, A., 227, 229*t*
Cohen, E. E., 357
Cohen, L., 293
Coligan, J. E., 22
Collard, F., 179
Collen, D., 200
Collins, D. S., 145–146
Collins, P., 2
Collis, G. E., 188–190
Colombel, J. F., 107
Columbo, P., 89, 90–91, 96–97
Colvin, V., 196–197
Commins, D., 42–43
Constantino, C. J., 112–113
Contreras, V., 146–147
Conway, B., 44–45
Cope, K. A., 364–365
Cordeiro-da-Silva, A., 129, 137–138
Cordobilla, B., 148–149, 150, 151–152
Coresh, J., 176–177
Cornelissen, I. L., 22, 35
Corot, C., 200–202
Correia, I., 284–287
Corti, A., 244
Cortie, M. B., 227
Costantini, D., 112–113
Costantino, P., 22–23
Cotten, M., 314
Cottone, M., 103
Cottrell, B., 304
Couarraze, G., 131
Couvreur, P., 67–68, 112–113, 131, 145–146
Craciun, L., 146–147
Cremer, K., 129–130
Crimeen-Irwin, B., 145
Crocker, A., 146–147
Croes, H. J., 157
Crommelin, D. J. A., 145, 152
Crosley, E. J., 341
Crouzier, T., 112–113
Crozat, K., 146–147
Cruz, L. J., 143–163
Cs-Szabo, G., 256
Cui, Z., 132
Cullis, P. R., 145–146
Cumberland, D. C., 328–329
Cummins, J., 245
Curiel, D. T., 328–329
Curnis, F., 244
Cutler, A., 67–68
Cygler, M., 173–175
Czubayko, F., 341

D

Da Cunha, H. N., 112–113
Dadabayev, A., 246
Dai, H., 184

Dailly, E., 49
Dai, T., 65–67
Dakappagari, N., 146–147
Dalby, A. B., 342
Dale, R. E., 288–289
Dalmasso, G., 102–103, 104*f*, 105*f*, 106, 109, 111–112, 111*f*, 115*f*, 116, 117–118, 119*f*, 120*f*
Dalpozzo, K., 147
Damgaard, C. K., 341
Danese, S., 107
Danev, R., 305, 308–309, 312–313, 315–316, 319–320
Dang, C. V., 340–341
Daniels, T. R, 357
Danishefsky, S. J., 22–23
Darguste, D. I., 146–147
da Rocha, S. R., 65–67
Darr, A. B., 256
Darzynkiewicz, Z., 232–233
Dash, C., 148–149
Dashevsky, A., 75
Dash, R., 111–112
Das, M., 63–65
Datta, M., 169, 175
David, C., 145
David, L., 112–113
Davidson, B. L., 278–279, 341
Davidson, M. C., 43–44
Davis, B. G., 27
Davis, M. E., 226
Davis, S. S., 75
Deasy, B. M., 245
de Boer, A. G., 43–44
Debotton, N., 17–18
De Brito, A. C., 112–113
Decker, J. M., 33–34
Deckmyn, H., 200
De Cuyper, M., 195–224
Deechongkit, S., 22
Deepak, V., 232
Defays, A., 146–147
de Gruijl, T. D., 145
De Guzman, R. C., 112–113
de Jonge, N., 229*t*
De Jong, W. H., 88, 226
de Kruijff, B., 286–287
de la Fuente, J. M., 22–23, 24–26, 27, 43–44
Delcayre, A., 147
Del Frari, B., 146–147
de Lima, M. C., 128–129, 278–279, 280–281, 282, 295
Dellacherie, E., 103–104, 112–113
Delrio, F. W., 228
DeLuca, A., 42–43
Demangel, C., 146–147, 151–152
Dembele, D., 146–147
Demeneix, B., 264–265, 328–329
De Meyer, S. F., 196–197, 200, 208–213, 209*f*
Demolombe, S., 314, 328–329
Denegri, M., 88–89
Deng, Y., 88–89
Den Haan, J. M. M., 146–147
Dennis, J. E., 246
Deo, Y. M., 146
Depollier, J., 278–279
Derdeyn, C. A., 33–34
Derossi, D., 278–279
DeRouchey, J., 329
De, S., 130–131
Desai, A., 109
Desai, M. P., 17–18
De Smedt, S. C., 195–224
de Sousa, A., 129–130, 137–138
Devine, D. V., 145–146
Devissaguet, J. Ph., 3, 67–68, 131
Devries, E. G., 104–105
de Vries, I. J., 144, 146–147, 157
de Wit, T. P., 146
Dey, S., 227
D'haens, G., 107
Dhand, C., 175
Dhar, S. K., 227
D'hoore, A., 107
Diamond, S. L., 314–315
Dickinson, L. E., 49, 88
Di Gianvincenzo, P., 21–40
Dignass, A., 107
Dilnawaz, F., 73–75, 74*t*, 76, 79*f*, 80*f*
Dinarvand, R., 68–69
Dinda, A. K., 227
Ding, H., 43–44, 45–46, 50
Ding, L., 341
Ding, Z., 151–152
Dios-Vieitez, M. C., 72–73
Dirmeier, U., 147
Di Santo, F., 44–45
Dispinseri, S., 33–34
Divita, G., 278–279, 280
Diwan, M., 154
Dizdaroglu, M., 237
Doelker, E., 3, 62
Doerks, T., 340
Dogan, A. L., 131
Do, L., 147
Domard, A., 112–113
Domingo, J. C., 143–163
Donahue, C. J., 232–233
Donalisio, M., 1–19
Dong, H., 341
Dong, M., 341
Donners, H., 33–34
Doores, K. J., 22–23
Doose, S., 44
Dorkin, J. R., 341
Dorner, S., 340

Dorsett, Y., 146–147, 342
Dos Santos, J. R. Jr., 112–113
Doud, M., 167
Dougan, G., 62
Douglas, J. F., 228–230
Dours-Zimmermann, M. T., 256
Dowdy, S. F., 278–279, 341
Dresselaers, T., 196–197
Dreyfuss, G., 107–108
Drezek, R., 196–197
Driessen, W. H., 151–152
Driver, S. E., 109
Drouvalakis, K. A., 184
Duarte, S., 278–279, 280–281, 282, 295
Duax, W., 173–175
Dudkin, V. Y., 22–23
Dudziak, D., 146–147
Dudzik, J., 168–169
Duesler, L., 246
Dunbar, P. R., 146–147
Duncanson, W. J., 154
Dunn, P. F., 91–93
Dunoyer, P., 342
Durand, A., 112–113
Dutertre, C. A., 146–147
Duttagupta, S. P., 88
Dutta, P. K., 65–67
Düzgüneş, N., 151–152, 181–183, 277–300, 327–338, 355–367
Dwek, R. A., 22
Dworetzky, S. I., 315

E

Eachus, R., 340
Ebbesen, T. W., 187
Eberle, A. N., 106
Ebner, S., 146–147
Ebstein, F., 146–147
Eddy, S. R., 107–108
Eden, H. S., 264
Edgington, T. S., 253
Edirisinghe, M., 89, 90–91, 96–97
Edwards, J. K., 227
Eguchi, A., 341
Ehrenberg, P. K., 33–34
Eiras, C., 112–113
Elaissari, A., 3
Elamanchili, P., 154
El-Andaloussi, S., 278–279
Elbashir, S. M., 107–108
Elder, A., 197
Elechiguerra, J. L., 63–64
Ellenbogen, R. G., 264–265
Ellis, R. J., 42–43
Elmen, J., 342
El-Nabarawi, M. A., 68–69
Elomaa, M., 329
El-Sayed, A., 310, 319–320
El-Sheikh, A., 253
Elson, H. E., 62
Eltringham-Smith, L. J., 254
Emelianov, S., 229*t*
Enatsu, M., 180
Endo, T., 304
Engbersen, J. F., 145–146
Engelbrecht, A. M., 233–235
Engelman, D. M., 288
Engering, A., 22
Eng, M., 177–179
Enouf, V., 147
Enríquez-Navas, P. M., 28, 29–31, 33, 35
Enting, R. H., 42–43
Ereifej, E. S., 112–113
Eremin, A. N., 172–173
Ernst, E., 62
Erogbogbo, F., 45–46
Esau, C. C., 342
Esbjorner, E. K., 279
Escande, D., 314, 328–329
Escriou, V., 363
Esendagli, G., 131
Esko, J. D., 252–253
Eslamian, M., 102
Esmaeili, F., 68–69
Essex, M., 22
Estrada, G., 43–44
Eulalio, A., 340
Everall, I., 42–43
Everse, L. A., 152
Ewe, K., 103
Eyquem, S., 342

F

Faas, S. J., 146–147
Fabian, M. R., 340
Facci, P., 187–188
Faham, A., 151–152
Fahmy, T. M., 154
Fallah, B., 358
Faneca, H., 278–279, 280–281, 282, 288, 295, 356, 358–359, 364–365
Fang, C., 264–265
Fanger, M. W., 146–147
Fanger, N. A., 146
Fang, G., 33–34
Farè, S., 88–89
Farook, U., 96–97
Fattal, E., 67–68
Faure, F., 146–147
Faust, S. M., 245
Fehr, C., 340
Fei, L., 228–230
Feizi, T., 22
Feldherr, C. M., 315

Felgner, P. L., 356, 362
Fellowes, V., 214–216
Felsenfeld, D. P., 227
Feng, M., 184
Feng, Z., 88–89
Fenouillet, E., 22
Fent, G. M., 229*t*
Fenyo, E. M., 33–34
Fernandez, A., 22–23, 24–26
Fernandez, B., 33–34, 35–36
Fernandez, S. F., 50
Fernig, D. G., 219–221
Ferrari, A., 214–216, 217, 218*f*, 219–221, 220*f*
Ferreira, L., 230
Ferreira, M. V., 230–231
Ferri, S., 177–179
Ferruti, P., 1–19
Fertig, S. J., 187
Fertsch, S., 226, 227, 229*t*
Fessi, H., 3, 131
Fetzer, C. A., 103
Feuillet, V., 146–147
Fiala, M., 49–50
Fifis, T., 145
Figa, M. A., 154
Figdor, C. G., 22, 35, 143–163
Filipek, S., 165–194
Filipowicz, W., 340
Fink, D. J., 246
Finnefrock, A. C., 22–23
Finn, M. G., 22–23
Finones, R. R., 214–216
Fiorese, C., 146–147
Fire, A., 109
Fischer, S., 144, 154
Fischler, M., 226, 230, 231
Fisher, K. D., 328–329
Fisher, P. B., 111–112
Flace, A., 145
Flacher, V., 146–147
Flammershein, H. J., 284–286
Flierl, A., 304
Florence, A. T., 139
Florence, R. L., 227
Flynn, B. J., 144
Focella, T. M., 227
Foged, C., 145–146, 152
Fokkink, R., 146–147, 154
Folini, M., 109
Fong, C. C., 230
Fontana, G., 67–68
Formigli, L., 233–235
Forrest, S. R., 185–186
Fortunati, E., 314
Foster, W., 245
Fournel, S., 151–152
Francesconi, M., 44–45
Frangioni, J. V., 226
Frankel, A. D., 278–279
Frank, J. A., 207–208, 214–216, 217, 219–221
Frazier, A. E., 304
Frechet, J. M., 146–147
Frederick, M. J., 183
Fredrickson, D. S., 188–190
Free, P., 219–221
Freytag, T., 75
Friedewald, W. T., 188–190
Frisch, B., 151–152
Froidevaux, S., 106
Frokjaer, S., 145–146, 152
Frormann, L., 201
Fujii, S., 146–147
Fujikura, K., 232–233
Fujimoto, M., 302–303, 316–318
Fujimura, R. K., 42–43
Fukasawa, M., 152
Fukuya, H., 177
Funatsu, J., 180
Funke-Kaiser, H., 55–56
Furtbauer, E., 314
Furukawa, R., 309, 319–322
Futaki, S., 278–279, 302–305, 309, 310, 312, 314–315, 316–320

G

Gabbay, J., 88
Gad, H. A., 68–69
Gaedtke, L., 329
Gaillard, P. J., 43–44
Galanth, C., 284–287
Gallagher, S., 137
Gallegos, C., 112–113
Gambhir, I. S., 63–64
Gambhir, S. S., 44
Gamvrellis, A., 145
Ganachaud, F., 3
Gander, B., 144, 154
Ganesan, P. G., 183
Gan, J., 148–149
Gansbacher, B., 264
Gan, Z., 146
Gao, F., 264, 271, 271*f*
Gao, S., 264
Gao, X.-G., 181–183, 196–197, 328–329
Garbuzenko, O., 264
García, L., 22–23, 327–338
Garcia-Perez, J., 33–34, 35–36
Garcion, E., 112–113
Garduno-Juarez, R., 173, 180
Garg, A., 102–103
Garnett, M. C., 75
Garrido, M. J., 357, 359, 362
Gaspar, R., 357, 364
Gasparri, A., 244
Gassull, M., 107

Gataiance, S., 254
Gautier, S., 112–113
Gavilanes, F., 280
Gebremedhin, S., 356, 363*t*
Geerkens, M., 225–226
Geertsma, R. E., 226
Geguan, P., 314–315
Geijtenbeek, T. B., 22, 35
Gems, D., 236–238
Gendelman, H. E., 49–50
Geng, X., 22–23
Gennis, R. B., 288–289
Gerace, L., 314–315
Gerard, B., 17–18
Gerber, A., 219–221
Gerecht, S., 88
Gerhardt, L. C., 112–113
Geuze, H. J., 145–146
Gewirtz, A. T., 107
Geyer, H., 22
Ghasemian, A., 173, 180
Ghazani, A. A., 226, 229*t*
Ghosh, G., 314–315
Ghosh, P. S., 196–197, 226
Ghosh, S., 88–89, 107
Giammona, G., 67–68
Giancola, M. L., 42–43
Giardiello, F. M., 104–105
Gieseler, R. K., 151–152
Gijzen, K., 146–147
Gilad, A. A., 214–216
Gilbert, B., 227
Gillespie, G. Y., 328–329
Gillet, P., 103–104, 112–113
Gillijns, V., 200
Giordano, R. J., 227
Giorgio, T. D., 314
Giralt, E., 150
Girão da Cruz, T., 361
Giunta, M., 103
Glish, G. L., 26–27
Gluckman, J. C., 22
Glud, S. Z., 341
Goasdoue, N., 284–287
Godbey, W. T., 264–265, 328–329
Goel, R., 229*t*
Goergen, D., 340
Gokhale, R., 3
Goksor, M., 279
Goldman, C. K., 328–329
Gole, A., 148–149
Gombotz, W. R., 130–131
Gomes, P., 278–279, 280–281, 282, 295
Gomollon, F., 107
Goncalves, C., 314–315
Gong, Q., 264, 270*f*, 271, 271*f*
Gonry, P., 112–113
Gonzalez-Ferreiro, M., 134–135
Gonzalez-Romero, D., 227, 229*t*
Goodenough, P., 181–183
Gooding, J. J., 184
Gopferich, A., 328–329
Gorczyca, W., 232–233
Gosselin, M. A., 357
Gou, J., 264
Gou, M., 102–103
Govender, T., 75
Goyal, B. R., 43–44
Graham, R. W., 314
Grandfils, C., 75
Grandis, J. R., 357
Graziano, R. F., 146
Grazú, V., 43–44
Greene, L. A., 214–216
Green, M., 278–279
Green, S. J., 26–27
Gref, R., 112–113
Griebel, P., 134–135
Griessinger, E., 146–147
Griffin, M. O., 62–63
Griffiths-Jones, S., 107–108
Grigoreva, I., 314–315
Grillo-Bosch, D., 150
Groettrup, M., 144, 154
Groopman, J. E., 22
Groscurth, P., 233–235
Grossin, L., 103–104, 112–113
Grudniak, A. M., 62–64
Grulke, E. A., 227
Grune, T., 207–208
Grunke, M., 233–235
Guenci, T., 44–45
Guillen, J., 288
Guiton, R., 146–147
Gulavita, S., 232–233
Gunn, J. W., 264–265
Guo, H., 340
Guo, L., 232–233
Guo, M., 146–147
Guo, W., 46
Gupta, A. K., 211–213
Gupta, M., 211–213
Gupta, P. K., 148–149
Gurtner, G. C., 245
Gurunathan, S., 232
Guttler, S., 146–147
Gutzeit, C., 207–208
Guyre, P. M., 146
Gu, Z., 264, 270*f*, 271

H

Haanen, J. B., 145–146
Hackley, V. A., 228
Haddadi, A., 154
Hadinoto, K., 68–69

Hagens, W. I., 226
Haglmuller, J., 150
Hagstrom, J. E., 314–315
Hahn, M. J., 151–152
Haid, B., 146–147
Hainfeld, J. F., 227
Halas, N. J., 227
Haley, B., 107–108
Hallbrink, M., 278–279
Halliwell, B., 236–238
Hallmann, R., 43–44
Hallock, K., 154
Hama, S., 278–279, 303–305, 309, 315, 316–318
Hamblin, M. R., 65–67
Hamdy, S., 154
Hamers, R. J., 103
Hamilton, J. A., 154
Hammond, S. M., 107–108
Hanauer, S. B., 107
Hande, M. P., 227
Hanes, J., 17–18
Han, G., 196–197
Han, J., 88–89, 340
Hankin, S., 227
Hank, J. A., 148–149
Han, M. S., 196–197, 341
Hannon, G. J., 107–108
Hansen, H. F., 342
Hansen, J. B., 107, 342
Hansen, O. C., 181–183, 341
Hao, R., 264
Harada, A., 109
Harada-Shiba, M., 109
Harashima, H., 278–279, 301–326, 328–329
Hardee, G. E., 75, 134–135
Harding, C. V., 145–146
Harding, K. G., 112
Haribalaganesh, R., 232
Harness, C. C., 154
Harper, D., 181–183
Harper, J. R., 245, 255*f*
Harris, D. E., 181–183
Harris, J. E., 26–27
Hart, D. N., 146–147
Hartung, E., 146–147
Harush-Frenkel, O., 17–18, 226
Hascall, V. C., 256
Hasegawa, Y., 329
Haseltine, W. A., 22
Haselton, F. R., 314
Hash, J. H., 62–63
Hassmann, J., 150
Hatanaka, M., 152
Hatzinikolaou, D. G., 181–183
Haugen, H. J., 88–89
Haugland, R. P., 288
Hawiger, D., 146–147
Hayakawa, T., 316–318
Hayashi, A., 145–146
Haynes, C. L., 196–197
Haynes, C. M., 236–238
Heath, A., 33–34
Hebbel, R. P., 200
Hedtjarn, M., 342
He, H., 264
Heidkamp, G. F., 146–147
Heijnen, I. A., 146
Heinegård, D., 245, 256
Heise, C. P., 328–329
Heitz, F., 278–279, 280
Helander, I. M., 65–67
Helguera, G., 357
Helm, I., 225–226
He, L. Z., 146–147
Hemminger, W. F., 284–286
Henderson, B. E., 104–105
Heng, B. C., 227
Henke, J. I., 264, 340
Hennink, W. E., 145–146, 152
Henn, V., 146–147
Henriques, A., 278–279, 280–281, 282, 295
Heras, A., 129
Hernández, R. M., 147
Herrmann, M., 233–235
He, S., 44, 46
Hester, S., 227
Heurtault, B., 151–152
Heyndrickx, L., 33–34
Hida, K., 17–18
Hidetaka Akita, 301–326
Higashi, T., 302–303, 316–318
Higashiyama, T., 232–233
Hijazi, K., 23, 28, 35
Hildebrand, A., 245
Hillaireau, H., 145–146
Himber, C., 342
Himmelreich, U., 196–197, 214–216, 217, 218*f*, 219–221, 220*f*
Hino, T., 75
Hiorth, M., 111–112
Hirayama, F., 357
Hirayama, N., 173
Hirling, H., 216–217
Hirn, S., 226, 227
Hirsh, P., 103
Hiruma, M., 177
Hirvonen, J., 75
Ho, A., 278–279
Hochstein, P., 237
Hodenius, M., 196–197
Hoek, E. M., 207–208, 226, 228
Hoekstra, D., 305
Hoetelmans, R. M., 42–43, 44–45
Hoffman, J. A., 251–252
Hoffman-Kim, D., 89
Hoffman, M., 76

Hoffman, R. M., 249
Hofhuis, F. M., 146
Hofmann, B., 52–53
Hohne, G. W., 284–286
Hollema, H., 104–105
Holler, E., 357
Holmes, D. E., 187
Holmes, H., 33–34
Holmgren, A., 236–238
Holschbach, C., 22
Holt, K., 63–64
Holtz, J., 340
Hombreiro Perez, M., 76
Hommes, D. W., 107
Hong, K., 328–329
Hoogeveen, A. T., 314
Hortensius, S., 151–152
Hosmalin, A., 146–147
Hosseini-Nasr, M., 68–69
Hosta, L., 148–149, 150, 151–152
Hosta-Rigau, L., 148–149
Hostetler, M. J., 26–27
House, A., 45–46
Hou, Y., 264
Hovgaard, M. B., 341
Howard, K. A., 341
Ho, Y. C., 111–112
Hsiao, J. K., 214–216
Hsieh, S. L., 22–23
Hsu, S. C., 214–216
Hsu, T. L., 22–23
Huang, C. Y., 22–23
Huang, D. M., 214–216
Huang, G. S., 148–149
Huang, H., 253, 264–265
Huang, J. L., 65–67
Huang, K., 45–46
Huang, L., 109, 311, 314, 328–329
Huang, S. K., 130
Huang, X., 259
Huang, Y., 146–147
Huard, J., 245
Hua, T., 128–129, 145, 154
Hubbell, J. A., 145–146
Hu, C., 112, 264–265
Hu, D., 88, 90
Hu, E., 219–221
Hu, F. X., 208–211
Hung, R. W., 154
Hung, Y.-C., 148–149
Hunsmann, G., 22
Hunter, E., 33–34
Hunter, S., 340
Huntzinger, E., 340
Hu, P.-A., 168–169
Hu, R., 45–46
Hurt, E. C., 314
Hu, S., 88–89
Hussain, F. N., 342
Hussain, N., 139
Hussain, S. M., 196–197
Hussell, T., 62
Hutchison, G. R., 227
Hutvagner, G., 107–108, 342
Hu, X., 232–233
Hu, Z., 103
Hwang, L. A., 146–147
Hwang, S., 171
Hyer, M. L., 259
Hynes, N. E., 357
Hyvonen, S., 75

I

Iacopetta, B., 259
Ichihara, J., 307
Ichikawa, S., 112–113
Idell, R. D., 107
Idoyaga, J., 146–147
Igartua, M., 147
Iida, M., 104–105
Iida, S., 316–318
Ikeda, C., 103
Ikeda, S., 104–105
Ilangovan, K., 181–183
Illert, W. E., 52–53
Illum, L., 75, 129
Illyes, E., 196–197, 208–211
Imagawa, N., 103
Imamoto, N., 304
Inaba, K., 146–147
Ingolia, N. T., 340
Inoue, K., 302–303
Inoue, T., 104–105
Ioji, T., 154
Isaya, G., 304
Ishida, T., 307
Ishizaki, T., 173
Ito, A., 103
Itoh, H., 104–105
Ito, R., 316–318
Itty Ipe, B., 226
Ivanov, A. A., 43
Iwasaki, S., 340
Izaurralde, E., 340

J

Jabbal-Gill, I., 128–129
Jackson, C., 304
Jacob, M., 216–217
Jahnen-Dechent, W., 225–242
Jain, V., 103
Jaitley, V., 139
Jakstadt, M., 207–208
James, E., 146–147
James, K. D., 181–183

James, M. B., 314
Jamieson, G., 146–147
Jana, S., 264
Jang, M. K., 68–69
Jan, M., 171
Jans, D. A., 314–315
Janssens, J., 67–68
Jaruga, P., 237
Jarver, P., 278–279
Järvinen, T. A. H., 243–261
Jassoy, C., 33–34
Jayamurugan, G., 109
Jayaraman, N., 109
Jeet, V., 146–147
Jell, G. M., 112–113
Jennings, G. T., 144–145
Jeong, H. J., 264
Jeong, J. H., 340, 341
Jeong, Y. I., 68–69, 103
Jeon, M., 171
Jerger, K., 102–103
Jerome, R., 75
Jesus, S., 127–142
Jiang, H., 264–265, 272–273, 272*f*, 273*f*
Jiang, X., 112
Jiao, X. A., 65–67
Ji, H., 341
Ji, J., 88
Jiménez-Barbero, J., 28
Jiménez, M., 27
Jin, F. L., 70
Jin, S., 43–44, 103, 214–216
Jiskoot, W., 151–152
Joffre, O. P., 146–147
Johnsen, M. G., 341
Johnson, M. B., 46
Johnson, P. A., 112–113
Johnsson, K. P., 148–149
Johnston, A. P., 109
Johnston, H. J., 227
Joliot, A. H., 278–279
Jolliet, P., 49
Jonas, L., 227
Jones, D. P., 107
Jones, I. M., 22
Jongbloed, S. L., 146–147
Joosten, B., 146–147
Joralemon, M. J., 227
Jordan, E. K., 207–208, 217
Jo, S., 340, 341
Josephson, L., 278–279
Joshi, P. B., 314
Joyce, J. G., 22–23
Juan, A. H., 342
Ju, H., 341
Junemann, C., 340
Junginger, H. E., 129–130, 137–138, 139
Jung, K.-E., 278–279, 341
Jung, T., 139, 207–208
Jurado, A. S., 277–300
Ju, X., 146–147

K

Kabanova, A., 22–23
Kabat, D., 33–34
Kabbaj, M., 103
Kai, M., 340–341
Kaiserling, E., 139
Kajiyama, N., 177
Kakiuchi, T., 145–146
Kakudo, T., 316–318
Kalala, W., 103
Kalay, H., 146–147
Kalden, J. R., 233–235
Kalish, H., 207–208
Kalishwaralal, K., 232
Kaltgrad, E., 22–23
Kalyanaraman, B., 236–238
Kalyanaraman, R., 22
Kalyankrishna, S., 357
Kamiya, H., 304, 307, 314–315, 316–318, 319–320
Kammerling, J. P., 25–26
Kamm, W., 139
Kam, N. W. S., 184
Kamphorst, A. O., 146–147
Kaneda, Y., 304
Kang, E. T., 208–211
Kang, S. H., 293
Kang, Y. J., 154
Kangueane, P., 42–43
Kannan, A. M., 168–169
Kannan, R., 229*t*
Kann, M., 315
Kanzaki, H., 307
Kappes, J. C., 33–34
Karan, H., 215
Karanikas, V., 146–147
Karim, A., 228–230
Karjalainen, M., 75
Karpilow, J., 342
Karp, J. M., 230
Karthikeyan, B., 232
Kashyap, M. P., 227
Kasimanickam, R., 63–64
Kassianos, A. J., 146–147
Kasturi, S. P., 128–129, 145, 154
Kataoka, K., 109, 278–279, 303–305, 309, 315
Katinger, H., 22, 33–34, 34*f*
Kato, N., 177
Katoono, R., 316–318
Katti, K. V., 229*t*
Kattumuri, V., 229*t*
Katz, J. L., 3
Kaushik, R. S., 134–135

Kauzlarich, S. M., 45–46
Kawahara, T., 94–96
Kawakami, T., 316–318
Kawashima, Y., 75, 103
Kean, T. J., 246
Keay, J. C., 46
Kedl, R. M., 144
Kedziorek, D. A., 214–216
Keene, A. M., 228
Kehlbach, R., 219–221
Kekos, D., 181–183
Keler, T., 146–147
Keller, A. M., 146–147
Keller, J. J., 104–105
Kelloff, G., 104–105
Kelly, C., 23, 28, 35
Kelly, J. W., 22
Kergueris, M. F., 49
Kersten, G. F., 152
Keskin, D. B., 154
KewalRamani, V. N., 22, 35
Khakoo, A. Y., 214–216
Khalil, I. A., 278–279, 303–305, 308–309, 310, 315–316, 318–320
Khambati, H., 357
Khamis, N., 227
Khan, M. A., 151–152, 157
Kharkevich, D. A., 43
Khdair, A., 17–18
Khoo, S., 49
Khuller, G. K., 68–69
Khvorova, A., 342
Kidane, A., 134–135
Kieler-Ferguson, H. M., 102–103
Kieser, H., 181–183
Kievit, F. M., 264–265
Kikuchi, H., 307, 318, 319–320
Kilby, J. M., 33–34
Kilk, K., 278–279
Killian, J. A., 286–287
Kim, A., 356
Kim, B., 134–135
Kim, D. G., 68–69
Kim, D. W., 167, 264
Kim, E. M., 264
Kim, J. H., 341
Kim, K. H., 102
Kim, K. M., 148–149
Kim, K. S., 49–50, 70
Kim, M. H., 278–279, 341
Kim, O. V., 91–93
Kim, S., 177–179, 229*t*
Kim, S. H., 70
Kim, S. I., 68–69
Kim, S. L., 264
Kim, S. W., 340, 341
Kim, T., 171
Kim, V. N., 340
Kim, W., 184
Kim, W. J., 340, 341
Kim, Y.-H., 340, 341
King, D. M., 148–149
Kinget, R., 103
Kinzig-Schippers, M., 76–77
Kipper, M. J., 89, 112–113
Kircher, M. F., 278–279
Kirpotin, D., 328–329
Kishida, H., 278–279, 303–305, 309, 315
Kissel, T., 3, 109, 139, 226, 229*t*
Kiss, V., 314–315
Kita, F., 307
Kitagawa, I. L., 112–113
Kitsoulis, S., 146–147
Kiwada, H., 328–329
Kjems, J., 341
Klaessig, F., 207–208, 226, 228
Klapproth, J. M., 107
Klauschenz, E., 278–279
Kleibeuker, J. H., 104–105
Klein, J. J., 341
Klibanov, A. M., 264–265
Kloetzel, P. M., 146–147
Klonoff, D. C., 166–167
Knight, S. W., 107–108
Knipe, D. M., 2
Kobayashi, H., 307, 318, 319–320
Kobayashi, I., 112–113
Ko, B. S., 214–216
Koch, C., 33–34
Koch, F., 146–147
Koeppe, I. R., 286–287
Kogan, M. J., 148–149, 150, 227, 229*t*
Kogerman, P., 278–279
Kogure, K., 278–279, 302–305, 307, 308–309, 310, 312, 315–320
Kohno, M., 180
Ko, J. J., 341
Kokkoli, E., 102–103
Kole, R., 293
Kolhatkar, R., 357
Kolinski, M., 168–169
Komatsu, M., 246, 249, 252
Kondo, M., 329
Koning, G. A., 152
Konings, F. A., 139
Konishi, Y., 75
Konopka, K., 356, 358, 362, 364–365
Koornstra, J. J., 104–105
Korber, B., 33–34, 34*f*
Koretsky, A. P., 196–197
Kornbluth, A., 107
Kosik, K. S., 342
Kostas, S. A., 109
Koup, R. A., 144
Kouri, J. B., 63–64
Koutsonanos, D., 128–129, 145, 154

Kovochich, M., 227
Koyama, Y., 177
Kozman, M., 357
Kraczkiewicz-Dowjat, A., 62–64
Krajewska, M., 259
Krajewski, S., 259
Kramer, M. G., 72–73
Krause, E., 278–279
Krauss, I. J., 22–23
Kretz-Rommel, A., 146–147
Kreuter, J., 43, 67–68, 129–130
Kreyling, W., 43–44
Kreyling, W. G., 226, 227, 229*t*
Krichevsky, A. M., 342
Krishnakumar, S., 65–67
Krishnan, K. J., 318
Krol, J., 340
Kruger, A., 264
Kruis, W., 107
Krusius, T., 254
Krutzfeldt, J., 340–341, 342
Krystek, P., 226
Kubo, K., 145–146
Kudo, A., 305, 308–309, 315–316, 319–320
Kudoh, M., 278–279, 303–305, 309, 315
Kudo, T., 104–105
Kugimiya, W., 180
Kuhmann, S. E., 33–34
Kumar, A., 132, 183
Kumar, B. N. P., 111–112
Kumar, P., 278–279, 341
Kumar, R., 45–46
Kumar, V., 227
Kundu, S. C., 111–112
Kunert, R., 22, 33–34, 34*f*
Kunisawa, J., 145–146
Kunou, N., 75
Kuo, P. L., 111–112
Kurek, A., 62–64
Kurihara, T., 304
Kuroiwa, T., 112–113
Kursa, M., 314
Kurts, C., 146–147
Kusuma, S., 88
Kwissa, M., 128–129, 145, 154
Kwon, D. S., 22, 35
Kwong, P. D., 22
Kwon, Y. J., 146–147
Kyriakouli, D. S., 318

L

Laakkonen, P., 251–252
Labarta, A., 150
Labhasetwar, V., 17–18, 65, 70–71, 75
Lacerda, S. H., 228–230
Lacroix, F., 363
Ladet, S., 112–113
Lahoud, M. H., 146–147
Lai, I. Y., 214–216
Lai, S. K., 17–18
Lai, W., 227
Lambris, J. D., 8–9
Lampleta, P., 329
Lamprecht, A., 76, 103
Landthaler, M., 342
Lane, H. A., 357
Lanford, R. E., 315
Lange, J. M., 42–43
Langel, U., 278–279
Langer, R., 230
Lan, M., 227
Lapasin, R., 112–113
Lario, P. I., 173
Larkin, M., 22
Laroui, H., 101–125
Larsson, N. G., 236–238
Larussa, D., 42–43
Lasagna-Reeves, C., 227, 229*t*
Lasic, D. D., 102–103
Lau, N. C., 107–108
Lauritsen, K., 67–68
Laus, R., 147
Lavasanifar, A., 154
Lavasani, M., 245
Lavelle, E., 128–129
Lavie, K., 28–29
Lavielle, S., 284–287
Lawrie, A., 328–329
Law, W. C., 41–60
Lazzell, A. L., 62
Leake, D., 342
Lebhardt, T., 3
Lebleu, B., 278–279, 280–281, 282, 294, 295
Lebre, F., 127–142
Lecellier, C. H., 342
Lee, A., 356
Lee, C. M., 264
Lee, D., 264–265
Lee, H. C., 68–69
Lee, H. H., 44, 46
Lee, H. W., 146–147
Lee, J. S., 146–147, 196–197
Lee, K., 341
Lee, L. K., 145–146
Lee, R. J., 311, 357
Lee, S. K., 263–276, 278–279, 341
Lee, S.-Y., 167, 171–172, 188–190
Lee, T. H., 22
Lee, Y. S., 340–341
Lehar, S. M., 146–147
Lehmann-Che, J., 342
Lehmann, K. E., 55–56
Lehr, C., 76
Leifert, A., 226, 227, 230, 231
Lemann, M., 107

Lemarchand, C., 112–113
Lembo, D., 1–19
Le Moine, A., 146–147
Lendeckel, W., 107–108
Lennox, K. A., 340–341
Lenti, L., 314
Leonard, M., 103–104, 112–113
Leonas, K. K., 70
Leong, K. W., 130
Letendre, S. L., 42–43
Leung, M. C., 264
Leung, R. K., 109
Le Verge, R., 67–68
Levin, C. S., 227
Levine, A. J., 42–43
Levine, A. S., 103
Levins, C. G., 341
Levitt, M. D., 103
Levy, H. V., 102
Levy, R. I., 188–190
Levy, R. J., 17–18
Lewinski, N., 196–197
Lewis, B. K., 207–208, 217
Lewis, D. L., 341
Lewis, R. V., 168–169
Lezoualc'h, F., 264–265
Lezoualch, F., 328–329
Liang, P. H., 22–23
Liang, W., 278–279
Licciardi, M., 67–68
Li, C. F., 65–67
Lichtenstein, G. R., 107
Li, D., 364–365
Liggett, S. B., 342
Lightowlers, R. N., 318
Li, H., 22–23, 341
Li, J. J., 44, 46, 145, 245, 264–265
Li, J. Y., 43–44
Li, L. S., 46, 102, 168–169
Lim, A., 356
Lim, L. P., 107–108
Lim, L. Y., 154
Lim, M., 264
Lim, S. T., 264
Li, N., 227, 236–238
Lincoln, P., 279
Lindblom, G., 286–287
Lindenberg, J. J., 145
Lindenstrø, T., 147
Lindgren, M., 278–279
Lindow, M., 342
Lindsay, J. O., 107
Lind-Thomsen, A., 342
Lin, E. Y., 50
Linke, S. P., 259
Lin, W.-H., 148–149
Liong, M., 227
Li, P., 165–194
Lipka, J., 226, 227, 229*t*
Li, S., 329
Litjens, M., 146–147
Littman, D. R., 22, 35
Liu, C., 253, 264
Liu, D., 230
Liu, G., 171, 263–276
Liu, H. M., 33–34, 70, 214–216
Liu, J., 184, 227
Liu, K., 146–147
Liu, L., 45–46
Liu, W., 226
Liu, X., 278–279, 341
Liu, X. F., 65–67
Li, W., 316–318
Li, X., 215, 232–233
Li, Y., 184, 245
Li, Z. Y., 229*t*
Ljubimova, J. Y., 357
Lloyd, K. O., 22
Lloyd, L. F., 173
Lobanok, A. G., 172–173
Loedige, I., 340
Loewenstein, P. M., 278–279
Lohani, M., 227
Lo, J. C., 146–147
Lombardo, J. R., 147
Lombardo, S., 168–169
Lomlim, L., 102–103
Londono, J. D., 26–27
Longaker, M. T., 245
Long, J., 102–103
Long, T. C., 227
Loo, J. S., 227
Loos, B., 233–235
Lopez, J., 146–147
Lopez-Ribot, J. L., 62
Lore, K., 144
Lorenzen, D. R., 145–146
Lorenz, H. M., 233–235
Lo, S. T., 146
Lote, A., 357
Lougheed, S. M., 145
Loughry, R. W., 104–105
Louie, A. Y., 45–46
Louie, G. P., 183
Lounis, B., 219–221
Loussouarn, G., 314, 328–329
Love, K. T., 341
Love, S. A., 196–197
Lovley, D. R., 187
Loweth, C. J., 148–149
Low Kah Mun, G., 227
Lowry, G. V., 227
Lowy, D. A., 187
Lubkin, A., 146–147
Luckert, P. H., 104–105
Ludtke, J. J., 314–315

Lu, G., 245
Lu, H., 341
Luhrmann, R., 107–108
Lui, P. C., 43–44
Lu, J., 271, 271*f*
Luke, G., 229*t*
Lukhtanov, E. A., 314–315
Lund, A. H., 340
Lungwitz, U. L., 328–329
Luo, F., 102–103
Luo, J., 88–89
Lusso, P., 33–34
Lutsiak, C. M., 154
Lutzmann, M., 314
Lv, F., 264

M

Macarthy, D. J., 226
Maccuspie, R. I., 228
MacEwan, M. R., 89
Machamer, C., 17–18
Macklin, M. D., 148–149
MacLachlan, I., 314
Macris, B. J., 181–183
Madeira, V. M., 284–286, 288, 290–291
Madhusudana, S. N., 109
Madler, L., 207–208, 226, 227, 228, 236–238
Maekawa, N., 173
Mages, H. W., 146–147
Magni, F., 244
Mahajan, S. D., 41–60
Maharramov, A., 65–67
Mahato, R. I., 328–329
Mahnke, K., 146–147
Mahon, K. P., 341
Mahvi, D. M., 148–149
Maier, O., 305
Maincent, P., 67–68, 76
Maitra, A., 43–44, 45–46, 227
Majer, Z., 196–197, 208–211
Makino, K., 226
Maksimenko, A., 341
Malhotra, B. D., 169, 175
Mallouh, V., 187
Malnati, M., 33–34
Mandal, M., 22–23, 111–112
Mandl, S. J., 147
Manfredi, A., 4–5
Mangham, A. N., 103
Manjunath, N., 278–279, 341
Manna, L., 46
Mann, D. M., 246, 252, 255*f*, 257–258
Manoharan, M., 340–341, 342
Manolova, V., 145
Mano, M., 278–279, 280–281, 282, 292, 295
Mansito, P. T., 329
Man, S. T., 17–18
Mansueto, S., 67–68
Mao, H. Q., 130
Maraskovsky, E., 146–147
Marenzi, S., 314
Marie, E., 112–113
Mark-Saltzman, W., 154
Marques, E. F., 288
Marquis, B. J., 196–197
Marquis, C. P., 264
Marquitan, G., 151–152
Marradi, M., 21–40
Marschinke, F., 207–208
Marshall, M., 107–108
Marshall, W. S., 342
Marteau, P., 107
Martinez-Ávila, O. M., 23, 24–25, 28, 29–30, 33, 35
Martinez, J., 107–108
Martin, K. W., 62
Martín-Lomas, M., 23, 27
Maruyama, H., 104–105
Maruyama, K., 316–318
Maruyama, R., 340–341
Marvel, J., 146–147
Massirer, K., 340
Masuda, T., 302–303, 305, 308–309, 315–316, 319–320
Mata, E., 147
Mathee, K., 167
Mather, J. P., 232–233
Matos, M. J., 288
Matshusita, K., 176–177
Matsuda, A., 304, 314–315
Matsuoka, Y., 304
Matsura, M., 329
Matsusaka, H., 103
Matsuura, M., 329
Matsuyama, K., 75
Mattapallil, J. J., 144
Mattie, D. M., 196–197
Matzke, M., 107–108
Maurer-Jones, M. A., 196–197
Mautner, V., 328–329
Ma, X., 45–46
Maxfield, F. R., 217
Maxwell, D. J., 196–197
Mayer, C., 150
Mayer, M., 28
Mayumi, T., 145–146
Mazumder, M. A., 112
McBride, J. L., 278–279, 341
McCarroll, J., 340–341, 342–343
McCutchan, F. E., 33–34
McCutchan, J. A., 42–43
McDonald, K. J., 146–147
McElhaney, L., 284–286
McElhaney, R. N., 284–286
McGinity, J. W., 75

McKenzie, I. F., 145
McLane, M. F., 22
McMurtry, I., 246, 252
Meade, B. R., 278–279
Medzhitov, R., 144
Mehrotra, G. K., 65–67
Meijering, E., 216–217
Meijer, S., 145
Meikas, A., 278 279
Meister, G., 342
Melamed, M. R., 232–233
Mello, C. C., 109, 342
Melzig, M., 278–279
Mendell, J. T., 340–341
Mendoza-Ramirez, J., 146–147
Men, K., 102–103
Mercatante, D. R., 293
Merchant, B., 148–149
Mergny, M. D., 264–265, 328–329
Merlin, D., 101–125
Mery, J., 278–279
Mesika, A., 314–315
Metelitsa, D. I., 172–173
Metsis, M., 278–279
Metz, K. M., 103
Meuse, C., 228–230
Meyer, B., 28, 31
Meyer, O., 328–329
Meyer, T. P., 52–53
Michael, N. L., 33–34
Michalet, X., 44, 45–46
Michetti, P., 107
Middel, J., 22, 35
Midoux, P., 314–315
Mi, F. L., 111–112
Mikhailova, R. V., 172–173
Mikos, A. G., 89, 264–265, 328–329
Miller, B. R., 207–208
Miller, D. W., 17–18
Miller, P. A., 62–63
Mimura, Y., 22
Minagar, A., 42–43
Minakawa, N., 304, 314–315
Minchin, R., 227
Minigo, G., 145
Minko, T., 264
Minoura, A., 305, 308–309, 315–316, 319–320
Mioskowski, C., 187
Miranda, O. R., 226
Mishima, T. D., 46
Mishina, Y., 278–279, 303–305, 309, 315
Misra, P., 226
Misra, R., 61–85
Miszka, D. G., 28–29
Mita, T., 312–313
Mitchel, J. A., 89
Mitra, M., 65–67
Mitsuhashi, K., 65–67
Mi, W., 146
Miyagishi, M., 302–303
Miyagishi, S., 312–313
Mizuguchi, H., 316–318
Mizukami, Y., 154
Mizuochi, T., 22, 152
Mohanty, C., 63–65
Mohler, J. L., 293
Mok, H., 264
Moller, W., 226
Moltedo, B., 146–147
Monnier, V. M., 179
Monteagudo, C., 62
Montenegro, J. M., 197
Montgomery, M. K., 109
Montgomery, R. L., 340–341
Monzon-Duller, J., 358
Moore, A., 232–233
Moore, J. P., 22
Moore, K., 112
Moore, N., 112–113
Moore, R., 232–233
Mora-Huertas, C. E., 3
Moran, T., 33–34
Moran, T. M., 146–147
Morein, S., 286–287
Moreira, J. N., 281, 283–284, 292, 295
Moreno-Aliaga, M. J., 331
Moreno, M. R., 288
Morgan, E. H., 43–44
Moriguchi, R., 278–279, 302–303, 305, 308–309, 312, 315–318, 319–320
Morishita, Y., 341
Morones, J. R., 63–64
Morris, M. C., 278–279, 280
Mortensen, N. J., 107
Mortimer, I., 314
Moser, C., 147
Moser-Kim, N., 356
Moser, R., 32–33
Motoyama, K., 357
Mottram, P. L., 145
Mourao-Sa, D., 146–147
Movassaghi, K., 146–147
Mo, X., 264
Mo, Y., 154
Mozumdar, S., 109
Mueller, M., 144, 154
Mui, B. L., 341
Mukerjee, S., 168–169
Mukerji, S., 88
Müller-Goymann, C. C., 3
Muller, R., 112–113
Mu, N., 264
Mundargi, R. C., 62–63, 73–75, 154
Mungan, N. A., 131
Munker, R., 340–341
Murdock, D., 304

Murdock, R. C., 196–197
Murooka, Y., 173
Murphy, E., 236
Murphy, M. P., 236–238
Murray, R. G., 62
Murri, R., 42–43
Murthy, N., 102–103, 128–129, 145, 154
Musameh, M., 187
Muster, T., 22
Mutwiri, G. K., 134–135
Myohoji, T., 304, 314–315

N

Naber, K. G., 76–77
Nachtwey, J. M., 147
Nagareya, N., 75
Nagasaki, T., 304, 314–315
Nagata, S., 103
Nagatsugi, F., 302–303
Nagayama, K., 305, 308–309, 312–313, 315–316, 319–320
Nah, J. W., 68–69
Na, H. S., 68–69
Nair, B. B., 41–60
Najid, A., 200–202
Nakagawa, S., 145–146
Nakai, A., 75
Nakai, M., 304
Nakajima, M., 112–113
Nakamura, T., 302–303, 316–318
Nakamura, Y., 302–303, 316–318
Nakanishi, M., 145–146
Nakanishi, T., 145–146
Nakase, I., 316–318
Nakashima, Y., 357
Naka, T., 302–303, 316–318
Nakata, M., 152
Nakaya, H. I., 128–129, 145, 154
Nakayama, Y., 304
Nandi, I., 102
Nango, M., 329
Nangrejo, M., 89, 90–91, 96–97
Nantz, M. H., 357
Naqvi, S., 227
Narasimhan, G., 173, 180
Narisawa, T., 104–105
Na, S. K., 68–69
Na, S. Y., 341
Nath, A., 42–43
Nativo, P., 219–221
Navarro, G., 102–103
Nayini, J., 104–105
Nel, A. E., 207–208, 226, 227, 228, 236–238
Nemet, I., 179
Neoh, K. G., 208–211
Netter, P., 103–104, 112–113
Neu, M., 109
Neuss, S., 226, 227, 230–231
Neutra, M. R., 151–152
Neves, S. S., 364–365
Newman, C. M., 328–329
Ng, K. W., 227
Nguyen, H. T. T., 102–103, 104*f*, 105*f*, 106, 109, 111–112, 111*f*, 115*f*, 116, 117–118, 119*f*, 120*f*
Nicol, F., 316–318
Nielsen, F. C., 340
Niepmann, M., 340
Nie, S., 196–197
Nieto, P. M., 30–31
Nihant, N., 75
Ni, J., 22–23
Nikol, S., 329
Nimesh, S., 341
Nimi, O., 173
Nimmerjahn, F., 146
Nir, S., 316–318
Nishikata, M., 75
Niu, G., 264
Niwa, K., 103
Niwa, M., 305
Niwa, T., 75
Noble, N. A., 245, 255*f*
Noble, R. L., 304
Nobre, L., 230
Nojima, M., 340–341
Nomura, N., 173
Norden, B., 279
Northoff, H., 219–221
Notcovich, A., 28–29
Nottet, H. S., 22, 35
Nouvel, C., 112–113
Novembre, F. J., 42–43
Novina, C. D., 109
Nozaki, M., 245
Ntziachristos, V., 43–44
Nuijen, B., 145–146
Nurmiaho-Lassila, E. L., 65–67
Nussenzweig, M. C., 146–147
Nuytten, N., 196–197, 208–213, 209*f*, 214–216, 217, 218*f*, 219–221, 220*f*
Nye, E., 146–147
Nyengaard, J. R., 341
Nystrom, T., 236–238

O

Oberdorster, G., 197
Oberle, V., 305
Ober, R. J., 146
Obertone, T. S., 107
O'Brien, J. J., 2
O'Brien, W. A., 33–34
Oehlke, J., 278–279
Offerhaus, J. A., 104–105

Officer, D. L., 188–190
Ogra, P., 128–129
Ohashi, W., 305
Ohkura, M., 103
Ohnishi, M., 307
Ohno, T., 103
Ohta, A., 312–313
Ohulchansky, T. Y., 44
Okada, Y., 304
Oka, M., 246, 252
Oktar, F., 87–99
Oku, N., 329
Oldenhuis, C. N., 104–105
Olenyik, A., 246
Oliveira, O. N. Jr., 112–113
Oliver, D. J., 238
Olmedo, I., 148–149, 227, 229*t*
Olson, A. H., 259
Olson, J. M., 264–265
Olson, W. C., 22–23
O'Malley, B. W., 364–365
Omnes, A., 67–68
O'Neil, C. P., 145–146
O'Neill, T., 146–147
O'Neill, W., 168–169
Ong, C. S., 146–147
Onitsuka, H., 104–105
Onnerfjord, P., 256
Ono, I., 103
Oohata, Y., 104–105
Oosterhuis, K., 145–146
Opitz, C., 146–147
Opolon, P., 341
Oresland, T., 107
Ori, H., 329
Orlandini, G. E., 233–235
Orlandini, S. Z., 233–235
Orlotti, N. I., 109
Orlova, M., 22–23
Orom, U. A., 340
Ortona, L., 42–43
Osai, Y., 103
Osemann, C., 225–226
Oshikawa, K., 328–329
Ouellette, J., 363*t*
Oussoren, C., 145
Overlid, N., 358, 363*t*
Oyewumi, M. O., 132

P

Pabo, C. O., 278–279
Paciotti, G. F., 229*t*
Paddon-Row, M. N., 184
Padro, D., 30–31
Paganinihill, A., 104–105
Pagano, R. E., 305
Page, T., 88
Pais, A. A., 288
Paiva, A., 278–279, 280–281, 282, 292, 295
Palm-Apergi, C., 341
Palmer, J. R., 104–105
Panda, J. J., 64–65
Pandey, R., 68–69
Pandit, S., 219–221
Panebra, A., 342
Pankow, J., 176–177
Panos, I., 129
Pant, A. B., 227
Pante, N., 315
Pantophlet, R., 22–23
Pan, Y., 225–242
Panyam, J., 17–18, 70–71, 75
Pan, Z. M., 65–67
Papahadjopoulos, D., 328–329
Papalia, G. A., 28–29
Papucci, L., 233–235
Parak, W. J., 197, 226, 227, 229*t*
Parathath, S., 342
Pareyn, I., 200
Parish, C. R., 146–147, 151–152
Park, C. G., 146–147
Park, E. H., 264
Park, E. J., 341
Parker, J., 227
Parkin, I. P., 88
Park, J. J., 228–230
Park, J. O., 264
Park, J. W., 328–329
Park, S. J., 70
Parolini, O., 88–89
Partal, P., 112–113
Partridge, L., 236–238
Parveen, S., 64–67
Parween, S., 148–149
Pasqualini, R., 227, 244–245, 248–249
Pasquinelli, A. E., 340
Passirani, C., 8–9, 112–113
Patarroyo, M. E., 147
Patel, H., 215
Patel, K. G., 147, 188–190
Patel, M. M., 43–44
Patel, P., 154
Patel, S., 232–233, 357
Patel, S. S., 314
Patel, V. R., 67–68
Pathak, A., 111–112
Pathak, S., 43–44
Patil, S. A., 62–63, 73–75
Patkaniowska, A., 107–108
Paulson, J. C., 22–23
Pawelczyk, E., 219–221
Pearson, J., 146–147
Peckys, D. B., 229*t*
Pecora, R., 132
Pedersen, J. A., 103

Pedraz, J. L., 147
Pedro Estrela, P., 168–169
Pedroso de Lima, M. C., 277–300, 356, 357, 358–359, 361, 364–365
Peer, D., 341
Pelisek, J., 329
Peltonen, L., 75
Penadés, S., 21–40
Peng, Q., 264–265
Peng, X., 46, 148–149
Penichet, M. L., 357
Pennington, D. J., 146–147
Penn, M., 246
Perdon, L. A., 151–152
Perez-Berna, A. J., 288
Perez-Olmeda, M., 33–34, 35–36
Pernelle, C., 278–279
Perni, S., 88
Perno, C. F., 44–45
Perrett, S., 228–230
Persidsky, Y., 49–50
Persson, D., 279
Peter, M. E., 233–235
Peters, D., 249
Petersen, S., 103
Peters, S., 227
Peters, T., 28, 31, 32–33
Petri, B., 112–113
Petromilli, C., 146–147
Petropoulos, C. J., 33–34, 34*f*
Peyramaure, S., 200–202
Pfanner, N., 304
Pfeiler, D., 304
Pham, Q. P., 89
Phillips, N. C., 103
Picart, C., 112–113
Piccirillo, C., 88
Pichon, C., 314–315
Pickl-Herk, B., 146
Pierard, G. E., 112–113
Pierrefite-Carle, V., 364–365
Pierre, P., 146–147
Pierschbacher, M. D., 245, 255*f*
Pietersz, G., 146–147
Pietersz, G. A., 145
Pilch, J., 249
Pilgrimm, H., 207–208
Pille, J. Y., 341
Pilobello, K., 187
Pinaud, F. F., 44
Pinheiro, T. J., 284–286, 290–291
Pintaske, J., 219–221
Pires, P., 356, 357, 361
Pirofski, L. A., 62
Pisanic, T. R., 214–216, 217, 218*f*, 219–221, 220*f*
Pishko, M. V., 109
Pissuwan, D., 227
Pitarresi, G., 67–68
Plank, C., 264, 316–318
Platt, E. J., 33–34
Plebanski, M., 145
Pletnev, V., 173–175
Pollard, H., 314, 328–329
Pollard, M., 104–105
Pooga, M., 278–279
Porkka, K., 251–252
Portegies, P., 42–43
Porter, M. D., 26–27
Pothayee, N., 63–64
Potter, M. A., 112
Poulin, L. F., 146–147
Prabhakaran, M., 185
Prabha, S., 70–71, 75
Prasad, B., 68–69
Prasad, P. N., 41–60
Prasad, S., 63–64
Prasad, V., 167
Prashant, C., 227
Pratten, J., 88
Prendergast, F. G., 288
Press, A. G., 103
Pretzer, E., 362
Price, J. C., 70
Prieto, M., 280
Prigent, P., 200–202
Prior, I. A., 219–221
Prisk, V., 245
Pristinski, D., 228–230
Pritchard, L. L., 341
Prochiantz, A., 278–279
Proietti, D., 22–23
Proietto, A. I., 146–147
Prokopovich, P., 88
Prusoff, W. H., 31–32
Pryor, W. A., 237
Puisieux, F., 3, 131
Pumera, S., 167, 169, 170*f*
Punt, C. J., 146–147
Puntes, V. F., 150, 197
Purtscher, M., 22
Puvvada, N., 111–112

Q

Qanungo, K. R., 179
Qian, C., 363
Qian, J., 44, 46
Qian, Z., 102–103
Qiao, C., 245
Qi, L., 112
Qin, F., 146–147
Qin, J., 341
Qin, L. L., 341
Quail, M. A., 181–183
Querbes, W., 341

Qu, F., 227
Quintanar-Guerrero, D., 3

R

Raasmaja, A., 329
Rabea, E. I., 65–67
Racie, T., 341
Racz, I., 67–68
Rad, A. M., 214–216
Rad-Malekshahi, M., 68–69
Rafati, S., 147
Raghuveer, M. S., 183
Rahbek, U. L., 341
Rahman, Q., 227
Rahmat, W., 184
Rajaonarivony, M., 131
Rajeev, K. G., 340–341, 342
Rajesh, Y. B., 109
Rajewsky, N., 340–341, 342
Rakhmilevich, A. L., 328–329
Ramakrishna, S., 168–169
Ramakrishna, V., 146–147
Ramanath, G., 183
Ramazanov, M. A., 65–67
Ramirez, J. T., 63–64
Ramkumarpandian, S., 232
Rana, T. M., 339–353
Randolph, G. J., 145–146
Ranganathan, P., 314–315
Rangaswamy, V., 154
Ranjan, A., 63–64
Ranucci, E., 1–19
Rao, C. V., 104–105
Rao, M., 148–149
Rapp, S., 52–53
Rappuoli, R., 22–23
Rasekh, M., 89
Rasmussen, L. M., 245
Ratner, L., 33–34
Ravetch, J. V., 146–147
Ravindran, R., 128–129, 145, 154
Ravindra, S., 62–63, 73–75
Raynal, I., 200–202
Raynaud, J., 112–113
Rayner, A. J., 342
Rayner, K. J., 342
Razansky, D., 43–44
Razzaq, M. Y., 201
Read, E. J., 214–216
Reale, A., 314
Rebuzzi, C., 200–202
Reddy, B. S., 104–105
Reddy, S. T., 145–146
Reed, J. C., 259
Reeve, A. K., 318
Rehwinkel, J., 340
Reichmann, H., 304
Reich, Z., 314–315
Reimers, C. E., 187
Reinhard, B. M., 110–111
Reinherz, E. L., 154
Reinhold, B., 154
Reip, P., 88, 90
Reis e Sousa, C., 144, 146–147
Remy, J. S., 314, 328–329
Ren, G., 88, 90
Ren, G. G., 87 99
Renugopalakrishnan, V., 165–194
Reppen, T. W., 341
Reshetnyak, Y. K., 288
Resina, S., 278–279, 280–281, 282, 294, 295
Rexach, M., 314
Reynolds, J. L., 41–60
Rhee, S. G., 236–238
Rhoades, J., 65–67
Richard, C., 187
Rich, R. L., 28
Riffle, J. S., 63–64
Rigden, D. J., 219–221
Rigotty, J., 104–105
Rijcken, F. E., 104–105
Rivas, I. P., 151–152
Rivera Gil, P., 197, 227
Rivera, M., 146–147
Riviere, K., 102–103
Rizzitelli, A., 146–147
Rizzo, M. G., 42–43
Roa, W., 232–233
Robbins, S., 146–147
Robertson, B., 342
Robertson, D. J., 229*t*
Robinson, D., 130–131
Robinson, J. P., 227
Robinson, L., 49
Roblot-Treupel, L., 67–68
Rodriguez, A., 146–147
Rodriguez, H., 237
Rodríguez, J. A., 357
Roederer, M., 144
Roesch, P. L., 341
Rogers, N. C., 146–147
Rogers, R. A., 112–113
Rohrer, D., 146–147
Roh, S. H., 68–69
Roizman, B., 2
Rojas, T. C., 22–23, 24–26
Rojo, J., 22–23, 146–147
Rolland, A., 328–329
Roller, S., 65–67
Romani, N., 146–147
Romaris, M., 245
Romeijn, S. G., 137–138
Ron, D., 236–238
Rong, J., 259
Rosa, S. M., 288

Rosenberg, L., 104–105
Rosen, C. A., 22
Rosenkrands, I., 147
Rosselli, M., 103
Rossi, M., 88–89
Rossin, R., 227
Ross, R. K., 104–105
Rotello, V. M., 196–197, 226
Rother, R. P., 146–147
Roth, S., 288–289
Rountree, R. B., 147
Roux, K. H., 22
Roy, A., 167
Roy, I., 41–60
Roy, K., 130
Roy, S., 167, 168–169, 171
Rozema, D. B., 341
Rozentur, E., 226
Ruau, D., 227
Rudd, P. M., 22
Rueda, F., 143–163
Ruedl, C., 227
Ruoslahti, E., 243–261
Ruparelia, J. P., 88
Ruponen, M., 329
Ruthaven, N. A., 284–286
Ruvkun, G., 107–108
Ruz, N., 72–73
Ryan, F., 87–99
Rytting, E., 3

S

Sacchi, N., 55–56
Saag, M. S., 33–34
Sabel, B. A., 43
Saboktakin, M. R., 65–67
Sacchi, A., 244
Sachar, D. B., 107
Sadagopa Ramanujam, V. M., 227, 229*t*
Sado, P., 67–68
Saeland, E., 146–147
Safaiyan, S., 147
Sagan, S., 284–287
Sagawa, I., 307
Saghezchi, S., 356
Saha, K., 226
Sah, D. W. Y., 341
Sahoo, S. K., 61–85
Sahu, A., 8–9
Saib, A., 342
Sailor, M. J., 244
Sainkar, S. R., 148–149
Sakai, Y., 177
Sakakibara, R., 152
Sakakibara, S., 75
Sakaue, R., 177
Salaita, K., 102–103
Salama, A., 146–147
Saleh, N., 227
Salio, M., 146–147
Salminen, M. O., 33–34
Saltik, M., 314
Salva, E., 109
Samadi, N., 68–69
Sama, P., 227
Samia, A. C., 45–46
Samim, M., 227
Sampson, N., 173
Samstein, R. M., 154
Samuel, J., 154
San Biagio, P. L., 67–68
Sanchez, M., 134–135
Sanchez-Navarro, M., 146–147
Sanchez-Palomino, S., 33–34, 35–36
Sancho, D., 146–147
Sandborn, W. J., 107
Sanders-Buell, E., 33–34
Sanders, R. W., 22
Sande, S. A., 111–112
Sandhu, A. P., 341
Sandy, J. D., 256
Sano, M., 104–105
Santos-Ferreira, T., 295–296
Saphire, E. O., 22
Saramunee, K., 102–103
Saravolac, E. G., 314
Sarisozen, C., 131
Sarkar, D., 111–112
Sarria, J. C., 216–217
Sasada, T., 154
Sasaki, K., 278–279, 302–303, 312
Sasaki, S., 302–303
Sasaki, Y., 103, 340–341
Sastry, M., 148–149
Sato, A., 103
Satoh, M., 104–105
Sato, M., 103
Sato, N., 103
Saudan, P., 145
Sauer, T., 314
Saulnier, P., 8–9
Sauzieres, J., 67–68
Savigni, D. L., 43–44
Saville, S. P., 62
Savla, R., 264
Sawada, J., 246, 252
Sawitowski, T., 225–226
Sazani, P., 293
Scanlan, C. N., 22
Scaria, P. V., 316–318
Schaefer, M., 146–147
Schafer, R., 219–221
Schaffler, M., 226, 227
Schalch, H. G., 148–149
Schalkhammer, T., 150

Schapira, A. H., 318
Schatz, G., 304
Schefe, J. H., 55–56
Scheller, A., 278–279
Scheper, R. J., 145
Scher, E. C., 46
Scherer, F., 264
Scherman, D., 264–265, 328–329, 363
Schiavone, N., 233–235
Schiffner, L., 22
Schiffrin, D. J., 24–25
Schillaci, D., 67–68
Schillinger, U., 264
Schlager, J. J., 196–197
Schleh, C., 226, 227, 229*t*
Schlosser, E., 144, 154
Schmid, G., 225–226, 227, 229*t*, 230, 231
Schmid, G. N., 227
Schmiedel, J., 304
Schmitz, A., 341
Schmitz, J., 52–53
Schneider, J., 22
Scholte, B. J., 314
Scholzen, A., 145
Schoonmaker, P. L., 147
Schrag, J. D., 173–175
Schrand, A. M., 196–197
Schroeder, U., 43
Schroer, T. A., 17–18
Schuber, F., 151–152
Schubert, M. A., 3
Schugens, C., 75
Schultz, P. G., 148–149, 187
Schulz, O., 146–147
Schumacher, T. N., 145–146
Schuttler, C. G., 340
Schwarta, M. R., 364–365
Schwartz, A. G., 89
Schwartz, S. A., 41–60, 103
Schwarz, D. S., 107–108
Schwarze, S. R., 278–279
Schwarz, K., 145
Scolaro, M. J., 151–152
Scott, P. G., 256
Sebestyen, M. G., 314–315
Seder, R. A., 144
See, V., 219–221
Seibel, M., 304
Seibel, P., 304
Seleem, M. N., 63–64
Sellars, M., 146–147
Selvin, E., 176–177
Semba, T., 304
Semmler-Behnke, M., 226, 227, 229*t*
Semple, S. C., 341
Senden, T., 314–315
Sen, M., 131
Seo, D. H., 68–69
Seo, H., 65–67
Sepehrizadeh, Z., 173, 180
Serada, S., 302–303, 316–318
Serres, S., 27
Setty, S. B., 62–63, 73–75
Sevanian, A., 237
Seymour, L. W., 328–329
Sezer, A. D., 112–113
Sfakianos, J. N., 33–34
Shafran, S. D., 44–45
Shaheen, S. M., 316–318
Shaheen, U., 219–221
Shah, N., 229*t*
Shanahan, T. C., 50
Shankar, P., 278–279, 341
Shao, Y., 264–265, 272–273, 272*f*, 273*f*
Shapiro, E. M., 196–197
Shapiro, S., 104–105
Shapshak, P., 42–43
Sharma, A., 68–69
Sharma, S., 68–69
Sharma, U., 89
Sharp, F. R., 146–147
Sharp, P. A., 107–108, 109
Shastri, N., 146–147
Shaw, A., 232–233
Shaw, G. M., 33–34
Shaw, W., 328–329
Sheahan, M., 103
Sheedy, F. J., 342
Sheetz, M. P., 227
Sheffield, W. P., 254
Sheikpranbabu, S., 232
Shekarriz, M., 102
Shekhar, M. P., 17–18
Shen, F., 112
Shen, W., 245
Shibata, N., 103
Shi, F., 328–329
Shi, H., 102–103, 227
Shikata, K., 152
Shimamoto, N., 167
Shimaoka, M., 341
Shimizu, S., 307
Shimizu, Y., 152
Shim, M., 184
Shimada, K., 173
Shimokado, K., 109
Shin, H., 167
Shinitzky, M., 288
Shinkawa, H., 173
Shinohara, Y., 307, 318, 319–320, 328–329
Shin Teh, J., 146–147
Shirai, K., 94–96
Shishko, Zh. F., 172–173
Shiver, J. W., 22–23
Shockman, G. D., 62
Shuai, X., 271, 271*f*

Shubayev, V. I., 214–216
Shulaev, V., 238
Sibson, N. R., 27
Siebenbrodt, I., 145–146
Silahtaroglu, A., 342
Silva, G. A., 43–44
Simard, M. J., 342
Simeoni, E., 145–146
Simões, S. P., 278–279, 280–281, 282, 292, 294, 295–296, 356, 357, 358–359, 361, 364–365
Simón, L., 148–149, 150, 151–152
Simon, U., 226, 227, 230, 231
Singer, E. J., 42–43
Singhal, R. S., 171–172
Singh, M., 63–64
Singh, R., 103
Singh, S. K., 22–23, 63–64, 146–147
Sinnott, J. T., 42–43
Sintov, A. C., 102
Siomi, M. C., 340
Sips, A. J., 226
Sitaraman, S. V., 101–125
Sitterberg, J., 109
Six, J. L., 112–113
Skewis, L. R., 110–111
Skoien, T., 111–112
Skountzou, I., 128–129, 145, 154
Skrabalak, S. E., 229*t*
Skrtic, S., 196–197
Slatkin, D. N., 227
Slepushkin, V., 357, 364
Slot, J. W., 145–146
Sluijter, B. J., 145
Smagghe, G., 65–67
Smilowitz, H. M., 227
Smith, J., 137
Smith, L. C., 328–329
Smith, L. H., 259
Smith, N., 328–329
Smith, P. J., 232–233
Soares, H., 146–147
Sode, K., 177–179
Soderholm, J., 107
Sodroski, J., 22
Sodroski, J. G., 22
Soenen, S. J., 195–224
Sohn, M. H., 264
Solanki, P. R., 175
Solís, D., 27
Solomon, J., 22
Soman, C., 148–149
Somasundaran, P., 168–169, 207–208, 226, 228
Somboonwit, C., 42–43
Sommerfeld, P., 43
Sonavane, G., 226
Sondel, P. M., 148–149, 328–329
Sonenberg, N., 340
Song, B., 264, 270*f*, 271, 271*f*
Song, H. C., 22–23
Song, J., 102–103
Song, Y., 340
Sontheimer, E. J., 340
Sorgel, F., 76–77
Sorgi, F. L., 328–329
Soria, J., 341
Soroceanu, L., 328–329
Sorrell, J. M., 256
Soto, C., 227, 229*t*
Sparidans, R. W., 44–45
Spellman, M. W., 22
Sperling, R. A., 226, 227, 229*t*
Spiller, D. G., 219–221
Squadrito, G. L., 237
Srinivasan, K., 22
Srinivasan, S., 107
Sriranganathan, N., 63–64
Srirangarajan, S., 62–63, 73–75
Srivastava, R. K., 227
Stamm, S., 293
Stanfield, R. L., 22
Stange, E. F., 107
Stanley, J. K., 227
Stanton-Maxey, K. J., 227
Stark, A., 340
St Clair, D. K., 227
Stebbing, D., 341
Stecher, H. A. III., 187
Steed, P., 62
Steenbergen, C., 236
Steffes, M. W., 176–177
Steiner, P., 216–217
Steinman, R. M., 146–147
Steitz, J. A., 340
Sternberg, B., 328–329
Steurbaut,W., 65–67
Stevens, C. V., 65–67
Stewart, M., 314–315
Stewart, S., 128–129, 145, 154
Stewart, T. E., 252–253
Stiegler, G., 33–34, 34*f*
Stins, M., 49–50
Stoeger, T., 43–44
Stoffel, M., 340–341, 342
Stoitzner, P., 146–147
Stokes, J. J., 26–27
Stolley, P. D., 104–105
Stolnik, S., 75
Stoltz, J. F., 103–104, 112–113
Stone, V., 227
Storch, J., 288
Storm, G., 145–146, 152
Storset, A. K., 146–147
Stougaard, P., 181–183
Stover, H. D., 112
Straarup, E. M., 342
Streeter, A., 363*t*

Streng-Ouwehand, I., 146–147
Stricker, M., 340
Stride, E., 89, 90–91, 96–97
Stroh, A., 207–208
Struck, D. K., 305
Stuart, M. C., 146–147
Sturgis, J., 227
Suarez, Y., 342
Subramanian, A., 314–315
Subr, V., 328–329
Sugiura, Y., 305
Sugiyama, M., 329
Suhara, T., 302–303
Sui, R., 88–89
Su, J., 340–341, 343
Sullivan, N., 22
Sullivan, S. M., 151–152
Sun, B., 245
Sun, C., 196–197
Sundarapandiyan, K., 146
Sundaresan, G., 44
Sundblad, A., 145–146, 152
Sung, H. W., 111–112
Sun, H. X., 147
Sun, J., 264, 270*f*, 271
Sun, L., 227
Sun, S., 264, 270*f*, 271
Sun, X., 227, 232–233
Sun, Y., 88–89, 356
Su, W. H., 65–67
Suzara, V. V., 181–183, 356
Suzuki, H., 340–341
Suzuki, R., 302–303
Suzuki, T., 232–233, 305
Swain, P. M., 340–341, 342–343
Swartz, C., 227
Swartz, J. R., 147
Swartz, M. A., 145–146
Swiegers, G. F., 188–190
Swierczewska, M., 264
Swihart, M. T., 45–46
Sydlik, U., 232
Sykes, D. E., 50
Szoka, F. C. Jr., 102–103, 316–318

T

Tabatabaie, R. M., 65–67
Tachibana, T., 304, 314–315
Tachikawa, H., 184
Tacken, P. J., 143–163
Taira, K., 302–303
Tait, B., 146–147
Tajuba, J., 227
Takabe, K., 177
Takada, K., 103
Takada, T., 103
Takahashi, T., 104–105
Takamisawa, I., 109
Takata, K., 232–233
Takatsuka, S., 340–341
Takayama, S., 316–318
Takaya, T., 103
Takenaka, S., 226, 227, 229*t*
Takeuchi, H., 75, 103
Takeuchi, Y., 94–96
Takihara, 173
Talley, N. J., 67–68
Tamagaki, S., 304, 314–315
Tam, P., 314
Tamura, Y., 103
Tanaka, E., 226
Tanaka, K., 145–146
Tanaka, S., 305
Tancrede, C., 67–68
Tan, E. C., 227
Tang, G., 187, 264–265
Tang, N., 278–279
Tang, Y., 214, 245
Tani, A., 233–235
Tanibe, H., 65–67
Tanimoto, M., 304, 314–315
Tannert, A., 146–147
Tanzi, M. C., 88–89
Tao, Z.Y., 187
Taratula, O., 264
Tatke, S. S., 185
Tavares, J., 137
Taylor, R. W., 318
Taylor, W. H., 252–253
Tazi, J., 293
Tegos, G. P., 65–67
Tehan, E.C., 187
Temel, R. E., 342
Tempestini, A., 233–235
Tender, L. M., 187
Teodosio, C., 280, 292
Terada, H., 307
Terao, K., 307
Terrones, M., 196–197
Tertil, M., 314–315
Teyssie, P., 75
Thanavala, Y., 128–129
Thankappan, U. P., 109
Thavasi, V., 168–169
Theis, C., 131
Theiss, A. L., 102–103, 107, 109, 116, 118, 120*f*
Thibault, C., 146–147
Thomann, J. S., 151–152
Thomas, A., 112
Thomas, L., 49
Thomas, M., 264–265
Thomas-Tikhonenko, A., 340–341
Thompson, M., 207–208, 226, 228
Thomson, N. R., 181–183
Thoren, P. E., 279

Thornalley, P. J., 236–238
Thynne, G., 146–147
Tian, F., 43–44
Tian, J., 264
Tillman, L., 75, 134–135
Tingey, A., 181–183
Tischler, A. S., 214–216
Tiwari, N., 103
Tjia, J., 49
Tokino, T., 340–341
Tokumo, K., 104–105
Tomarchio, V., 67–68
Tomari, Y., 340
Tomoda, K., 226
Tongchusak, S., 154
Tong, Y., 340
Tontini, M., 22–23
Too, C. O., 188–190
Torchilin, V. P., 278–279
Torensma, R., 22, 35, 144, 146–147, 152
Torres Suarez, A. I., 151–152
Touma, M., 154
Trabulo, S., 277–300
Tramposch, K. M., 44, 46
Travis, S. P., 107
Trevisan, A., 112–113
Triller, A., 278–279
Tripathi, S., 65–67
Tripp, C. H., 146–147
Tritton, T. R., 62–63
Trkola, A., 22
Tros de Ilarduya, C., 102–103, 327–338, 355–367
Trotta, F., 4–5
Trotter, P. J., 288
Trumpfheller, C., 146–147
Tsai, D. H., 228
Tsai, G. J., 65–67
Tsay, J. M., 44
Tseng, M. T., 227
Tseng, W. C., 314
Tseng, Y. C., 109
Tsubokawa, N., 94–96
Tsuchiya, S., 145–146
Tsugawa, W., 177–179
Tsutsumi, Y., 145–146
Tu, C., 45–46
Tuder, R. M., 227
Tuppen, H. A., 318
Turiel, A., 150
Turnbull, D., 318
Turnbull, D. M., 318
Turner, J. L., 227
Tuschl, T., 107–108, 340–341, 342
Twardzik, D. R., 245
Tycko, B., 217
Tyler, R. D. Jr., 63–64
Tyner, K. M., 228

U

Ubrich, N., 76
Uchida, T., 75
Udit, A. K., 22–23
Ueda, K., 305
Uekama, K., 357
Ueno, M., 278–279, 302–305, 309, 312, 315
Ulbrich, K., 328–329
Ulrich, S., 43
Unanue, E. R., 145–146
Unfried, K., 232
Unger, T., 55–56
Unger, W. W., 146–147
Unser, M., 216–217
Urakami, T., 246, 252
Urayama, A., 227, 229*t*
Urban-Klein, B., 341
Urbiola, K., 327–338
Urlaub, H., 107–108
Urtti, A., 329
Utz, P. J., 184

V

Vachet, R. W., 26–27
Vakil, N., 67–68
Valdes-Sueiras, M., 42–43
Valenzuela, S. M., 227
Valiyaveettil, S., 227
Valkna, A., 278–279
Van Assche, G., 107
van Broekhoven, C. L., 146–147, 151–152
Vandana, M., 65–67
Van den Berg, J. H., 145–146
van den Broek, M., 314
Vandendriessche, T., 200
van den Hout, M. F., 145
Van Den Mooter, G., 103
van den Tol, M. P., 145
van der Aa, L. J., 145–146
Vanderkooi, J. M., 288
van der Lubben, I. M., 129, 139
Van Der Sluis, T., 104–105
van der Vlies, A. J., 145–146
van der Werf, S., 147
van de Ven, R., 145
Vandevord, P. J., 112–113
van de Winkel, J. G., 146, 152
van Dommelen, S. L., 146–147
van Duijnhoven, G. C., 22, 35
van Gils, J. M., 342
Vanhoorelbeke, K., 200
van Kasteren, S. I., 27
van Kooyk, Y., 22, 35, 146–147
van Leeuwen, P. A., 145
Vannier, J. P., 341
van Rooij, E., 340–341
van Vliet, S. G., 22, 35

van Vugt, M. J., 152
Vargas-Reus, M. A., 87–99
Vasudevan, S., 340
Vathy, L. A., 45–46
Vaughan, H., 146–147
Vauthier, C., 131
Vedala, H., 167, 168–169, 171
Veiseh, O., 264–265
Velegol, D., 207–208, 226, 228
Venkatesan, P., 111–112
Ventre, E., 146–147
Venturi, M., 22
Verbeek, J. S., 146
Vercauteren, D., 196–197, 208–211
Vergara, L., 227, 229*t*
Verger, R., 175
Verhoef, J. C., 129–130, 139
Verma, C. S., 173, 180
Verma, S., 3
Vermeulen, A., 342
Veronesi, B., 227
Vervoort, L., 103
Vestweber, D., 304
Vidal, P., 278–279
Vijay-Kumar, M., 107
Villalain, J., 288
Villarreal, F. J., 62–63
Villinger, F., 128–129, 145, 154
Vingsbo-Lundberg, C., 147
Visaria, R., 229*t*
Visosky, A., 33–34
Visser, C. C., 43–44
Viswanathan, S., 165–194
Vitale, L. A., 146–147
Vitale, S. A., 3
Vivier, E., 146–147
Vocero-Akbani, A., 278–279
Voinnet, O., 342
Voll, R., 233–235
Volovitch, M., 278–279
Vonarbourg, A., 8–9
Vremec, D., 146–147
Vrielink, A., 173
Vugt, M. J., 146
Vulink, A. J., 146–147
Vu Manh, T. P., 146–147

W

Waddell, W. R., 104–105
Wadley, R. B., 146–147
Wagenknecht, L., 176–177
Wagenlehner, F. M., 76–77
Wagner, E., 109, 314, 329
Wagner, J., 246
Wakefield, D. H., 341
Waksman, S. A., 62
Walczak, P., 214–216
Walker, G. F., 329
Walker, M., 24–25
Wallace, D. C., 304, 318
Wallace, G. C., 188–190
Walter, E., 17–18
Walter, H., 150
Walzer, T., 146–147
Wang, B., 106
Wang, J., 22–23, 171, 187
Wang, L. V., 103, 229*t*
Wang, L. X., 22–23
Wang, M., 33–34, 34*f*
Wang, Q., 264–265
Wang, S. K., 22–23
Wang, W. C., 342
Wang, X. T., 3, 146–147
Wang, Y. A., 46, 264
Wang, Y. Y., 43–44
Wang, Z., 168–169, 263–276
Wanjie, S., 146
Wan, Y., 102–103
Ward, C. M., 328–329
Ward, E. S., 146
Warner, T. F., 148–149
Warren, J. D., 22–23
Warshauer, M. E., 104–105
Watson, J. V., 232–233
Wawrzak, Z., 173–175
Way, D., 49–50
Webster, P., 226
Wee, S., 130–131
Wehrly, K., 33–34
Weidenfeller, B., 201
Weidner, S., 151–152
Weidner, W., 76–77
Wei, G., 94–96
Wei, H., 227
Weinand, M., 49–50
Weinauer, F., 52–53
Weinstein, E. G., 107–108
Weiser, B., 33–34
Weissenberg, A., 232
Weissleder, R., 278–279
Weissman, J. S., 340
Weiss, S., 44
Weiss, W. R., 144
Wei, X. W., 33–34, 173, 180, 187
Welch, M. J., 227
Wen, F., 226, 230, 231
Wenk, A., 226, 227, 229*t*
Wentzel, E. A., 340–341
Werth, S., 341
West, K. M., 340–341
Whitehead, K. A., 341
White, M. R., 219–221
Whitley, R. J., 2
Whitmore, M., 329
Whittaker, P. A., 109

Whittum-Hudson, J. A., 65–67
Whitworth, C. W., 70
Whyman, R., 24–25
Wiedemann, N., 304
Wiesner, B., 278–279
Wignall, G. D., 26–27
Wihart, M. T., 45–46
Wijers, M., 157
Wilde, F., 304
Wilke, M., 314
Wilken, N., 304
Wille-Reece, U., 144
Williams, T., 293
Wilson, D. S., 102–103
Wilson, I. A., 22
Wilson, L. B., 217
Wilson, M., 88
Wilson, T. E., 148–149
Wils, P., 363
Wingate, J. E., 26–27
Wischke, C., 145–146
Wiseman, R. L., 236–238
Wishnick, M., 62–63
Wiskirchen, J., 219–221
Withers-Martinez, C., 175
Witschi, C., 62
Witte, M. H., 49–50
Wiwattanapatapee, R., 102–103
Woith, W., 233–235
Wolburg, H., 219–221
Wolff, J. A., 314–315
Wolska, K. I., 62–64
Wong, C. H., 22–23
Wong, D. T., 169–171
Wong, J. Y., 154
Wong, S. C., 341
Woodle, M. C., 328–329
Wood, S. C., 245
Wooley, K. L., 227
Woo, S. L., 364–365
Wormald, M. R., 22
Wormley, F. L. Jr., 62
Wrin, T., 33–34, 34*f*
Wu, A. M., 44
Wu, B., 264
Wu, C. H., 109
Wu, D., 146–147
Wu, F., 45–46
Wu, G. Y., 109
Wu, H., 278–279, 341
Wu, J., 357
Wu, K. K., 264–265, 328–329
Wu, L., 65–67, 146–147
Wu, P., 227
Wu, X., 33–34
Wyman, T. B., 316–318

X

Xhauflaire-Uhoda, E., 112–113
Xia, C., 264, 271, 271*f*
Xiang, M., 102–103
Xiang, S. D., 145
Xiao, J. X., 139
Xiao, X., 245
Xia, T., 207–208, 226, 227, 228, 236–238
Xia, Y., 89, 229*t*
Xie, J., 89, 264
Xie, Y., 147
Xing, J. Z., 232–233
Xing, P. X., 146–147
Xing, R., 264
Xiong, S., 227
Xiong, Y., 232–233
Xu, G., 50
Xu, H., 146–147, 227
Xu, J., 227
Xu, S., 109
Xu, W., 278–279
Xu, Z., 112, 264

Y

Yacaman, M. J., 63–64
Yacoby, I., 62–63
Yamada, M., 278–279, 303–305, 309, 315
Yamada, Y., 301–326
Yamaguchi, Y., 245, 255*f*, 257–258
Yamaguti, M., 305, 308–309, 315–316, 319–320
Yamamoto, A., 112
Yamamoto, E., 340–341
Yamamoto, H., 103
Yamamoto, K., 307
Yamamoto, N., 152
Yamamoto, T., 245, 255*f*, 307, 318, 319–320
Yamano, H. O., 340–341
Yamashita, A., 316–318
Yamashita, K., 307, 318, 319–320
Yamashita, M., 173
Yamashita, T., 103
Yamauchi, K., 109
Yamauchi, T., 94–96
Yamazaki, S., 146–147
Yamazaki, Y., 329
Yanagihara, K., 357
Yan, F., 341
Yang, C. S., 214–216, 340–341, 342–343
Yang, H. S., 227, 264–265, 272–273, 272*f*, 273*f*
Yang, M., 230, 249
Yang, N. S., 328–329
Yang, S. F., 340

Yang, T. C., 65–67
Yang, X., 229*t*
Yang, Z. S., 187
Yan, Q., 88
Yan, Y. T., 102–103, 104*f*, 105*f*, 106, 109, 111–112, 111*f*, 115*f*, 116, 117–118, 119*f*, 120*f*
Yao, M., 214–216
Yasenko, M. I., 172–173
Yasuzaki, Y., 309, 319–322
Yaworski, E., 341
Yazdi, M. T., 173, 180
Yee, M., 363*t*
Yee, V. C., 179
Yeh, J. I., 227
Ye, L., 45–46
Yeom, J. H., 341
Ye, Y., 227
Ye, Y. P., 147
Yi, C., 230
Yin, Y. X., 65–67
Yockman, J. W., 340, 341
Yocum, G. T., 214–216
Yokel, R. A., 227
Yoneda, Y., 304
Yong, K., 45–46
Yong, K. T., 41–60
Yoo, E.-H., 167, 171–172, 188–190
Yoshida, H., 177
Yoshida, K., 75
Yoshida, N., 177
Yoshiizumi, M., 180
Yoshikawa, Y., 103
Yotsumoto, S., 145–146
Young, D., 62
Young, R. G., 246
Youssef, M., 67–68
Yu, D., 340–341
Yu, H., 106, 264–265
Yui, N., 316–318
Yu, K. O., 196–197
Yurimoto, H., 177
Yuste, E., 28, 29–30, 33, 35
Yu, X., 109

Z

Zabala, M., 72–73
Zachariah, M. R., 228
Zaffaroni, N., 109
Zahoor, A., 68–69
Zaisserer, J., 52–53
Zalba, S., 357, 359, 362
Zalipsky, S., 154
Zamore, P. D., 107–108, 342
Zampa, M. F., 112–113
Zanella, M., 227
Zang, E., 104–105
Zanta, M. A., 264–265, 328–329
Zardo, G., 314
Zauber, A. G., 104–105
Zdrahala, I. J., 88–89
Zdrahala, R. J., 88–89
Zehnter, I., 52–53
Zelenay, S., 146–147
Zelikin, A. N., 109
Zelphati, O., 316–318
Zeps, N., 259
Zern, M. A., 357
Zhang, C., 278–279
Zhang, F., 227
Zhang, G., 65–67, 314–315
Zhang, H., 89, 90 91, 96–97
Zhang, H. F., 146–147
Zhang, H. R., 96–97
Zhang, J., 102–103, 168–169, 179, 184, 230
Zhang, L., 187
Zhang, M., 196–197, 264–265
Zhang, Q., 229*t*
Zhang, W., 88
Zhang, X. M., 45–46, 232–233, 264–265, 272–273, 272*f*, 273*f*
Zhang, Y., 88
Zhang, Z., 33–34, 154
Zhao, G. C., 187
Zhao, J., 88
Zhao, M., 278–279
Zhao, W., 45–46
Zhao, X. B., 102–103, 227, 271, 271*f*, 357
Zhao, Y., 70
Zheng, J., 340
Zhong, C.-J., 26–27
Zhong, Z., 264–265
Zhou, F., 151–152
Zhou, J., 88–89
Zhou, S. S., 88
Zhou, X., 88–89
Zhu, A. P., 65–67
Zhu, H., 176–177
Zhu, J., 245
Zhuo, R., 264–265
Zhu, P., 22
Zhu, Z. J., 226
Ziegler, U., 233–235
Zimmer, C., 207–208
Zimmer, J. P., 226
Zimmermann, D. R., 256
Zimmermann, J., 145–146
Zink, J. I., 227
Zohar, M., 314–315
Zolla-Pazner, S., 33–34, 34*f*
Zoonens, M., 288

Zorko, M., 278–279
Zou, G., 22–23
Zou, X., 112
Zuckerman, J. E., 226
Zucolotto, V., 112–113
Zuidema, J., 145
Zuleger, C. L., 148–149
Zurbriggen, R., 147
Zwart, N., 104–105
Zwick, M. B., 22, 33–34, 34*f*

Subject Index

Note: Page numbers followed by "*f*" indicate figures, and "*t*" indicate tables.

A

Acyclovir-loaded and fluorescent nanoparticles
  β-CD-PACM inclusion complexes. *See* (β-Cyclodextrin-poly(4-acryloylmorpholine) monoconjugate (β-CD-PACM) nanoparticles)
  characterization, β-CD-PACM nanoparticles. *See* (β-Cyclodextrin-poly(4-acryloylmorpholine) monoconjugate (β-CD-PACM) nanoparticles)
  NPs formation, 4
  physical stability, 8
  preparation, 6
  Pyris program, 5
  quantitative determination, 6
  radical polymerization, 4–5
  sterilization, 8
  structure, 4–5
  thermal analysis, 5
AFM. *See* Atomic force microscopy (AFM)
Alkyl-PEI2k-IOs, gene delivery
  air, 269
  cytotoxicity, 264–265
  gene delivery nanovectors, 264
  hydrophobic SPIO nanoparticles, 270*f*, 271
  materials
    equipment, 265–266
    reagents, 265
  methods
    cell transfection, IO–siRNA complexes, 268
    cytotoxicity assay, 268–269
    fabrication, 266–267
    MRI study, 269
    SPIO-siRNA complex formation and property, 267–268
  molecular weight, 264–265
  N/P ratios, 269
  optimal incubation period, 269
  siRNA complexes
    fluc gene expression, 272–273, 274*f*
    heparin, 272, 273*f*
    physical characterization, 272, 272*f*
    T2-weighted images, 272*f*, 273
  size data, DLS and AFM, 269
  TEM and AFM analyses, 271, 271*f*
  washing, PBS, 269
Amprenavir
  HPLC
    conditions, 49
    measures, 49
    SPE cartridge, 49
  levels, 49
  QD-Amprenavir-Tf Nanoplex
    BBB transverse, 48–49
    bioconjugation use, 48–49
    carbodiimide chemistry, 48–49
    control, 48
    formation, 48–49
    monoclonal antibody, 48
Antifibrotic compounds
  cell culture experiments
    CAR phage and peptide, 252–253, 253*f*
    CHO-K, 252–253
    heparin, 253
  description, 244
  in vivo phage display
    CAR, 246
    clones, 250–251
    CX7C-peptide library, 248–249, 248*f*
    ex vivo screen, 249
    generation library, 246–247, 248*f*
    mammalian expression system, 246, 247*f*
    PBS and BSA, 249
    PCR, 251
    rescue and amplification, 249–250
    tissue samples, 249
  immunohistochemistry, 251–252
  multifunctional fusion proteins, 254–256
  peptide and homing experiments, 251
  phage, 244–245
  TGF-β1, 245
Antimicrobial properties, electrically formed elastomeric polyurethane
  bacterial testing, 91
  control and function, polymer-active agent, 89
  direct deposition, 97
  diverse applications, 88
  drug delivery carriers, 88
  electrospinning, 89
  electrospun fibers, 97
  fabricating method, 89
  fiber, film and pattern generation
    "breathable" and "interaction,", 91–93
    control, 91–93
    CuO, 91–93, 92*f*

Antimicrobial properties, electrically formed elastomeric polyurethane (*cont.*)
deposition distance, 91–93
elastic deformation, 91–93
electrical writing method, 93–94
electrospinning process, 90–93
individual solutions, 90–91
maintain structures, 91–93
materials, 90
metal and metal oxide particle system, 88
microscopy and EDX, 91
MRSA. *See* (MRSA and antimicrobial properties)
polyurethanes, 88–89
preparation, solution, 90
PU–CuO composite mats, 89–90
structural analysis, fiber, 94, 95*f*
Antivirals drug, nanoparticles
acyclovir administration, 2
acyclovir-loaded and fluorescent particles. *See* (Acyclovir-loaded and fluorescent nanoparticles)
advantages and control, 2–3
assessment
acyclovir concentration, 15–17
coumarin 6 β-CD-PACM, 17–18
HSV-1 infection and treatment, 14
inhibition, HSV-1 infection, 15
virus titration, 14–15
β-CD-PACM, 4
biocompatibility assays
activation, 11
cell viability, 9–10
EpiVaginal Tissue Model, 12–14
hemolytic properties, 10–11
intracellular ROS, 11–12
recognition and elimination, 8–9
safety concerns, 8–9
toxicity, 8–9
classification, 3
delivery system, 2–3
formation and sizes, 3
HSV-1 and HSV-2, 2
molecular mechanism, 3
nanoprecipitation, 3
polymer and encapsulation efficiency, 4
preparation, 3
properties, 3
therapeutic index, 2–3
Atomic force microscopy (AFM)
hydrophilic nanocomposites, 271, 271*f*
size data, 269

B

BBB. *See* Blood–brain barrier (BBB)
B-cell lymphoma 2 (BCL2), 233–235
β-Cyclodextrin-poly(4-acryloylmorpholine) monoconjugate (β-CD-PACM) nanoparticles
characterization
coumarin 6, 7–8
drug loading and encapsulation efficiency, 7
in vitro release kinetics, 8
morphology, 7
size and distribution, 6–7
surface charge, 7
inclusion complexes
coumarin, 6
DSC and FTIR spectroscopy, 5
Edwards Modulyo gree-drier, 5
HPLC method, 5
KBr pellet, 5
Modulyo freeze-drier, 6
skin irritation
controls, 13
cytokine IL-1α release, 13–14
cytotoxicity, 12
evaluation, skin irritation potential, 12
incubation, 12–13
LDH release, 13
measures, LDH cytotoxicity, 13
MTT ET-50, 12–13
"non irritant" compound, 14
quantification, cell viability, 13
skin irritation potential, 12
Biocompatibility assays
activation
human serum, 11
ISO procedure, 11
measurement, ELISA systems, 11
cell viability. *See* (Cell viability)
EpiVaginal Tissue Model. *See* (EpiVaginal Tissue Model)
hemolytic properties, 10–11
intracellular ROS
ECFH conversion, 12
fluorescence analysis, 12
generation, 11
recognition and elimination, 8–9
safety concerns, 8–9
toxicity, 8–9
Biomaterials
biosensors, 89
characteristics, 91–93
electrospinning, 89
medicine, 94–96
polymer-active agent function, 89
polysaccharide hydrogels, 114, 114*t*
polyurethanes, 88–89
technologies, 89
tissue-engineering constructs, 88–89
Blood–brain barrier (BBB)
description, 43
human BBB, *in vitro* model

cultures, BMVECs and NHAs, 50
formation and function, 50
Persidsky model, 49–50
preparation, 49–50
measurement, TEER, 50
Bovine serum albumin (BSA), 249
BSA. *See* Bovine serum albumin (BSA)

C

Cationic lipid. *See* Protein-cationic lipid-DNA ternary complexes
Cell death
apoptosis *vs.* necrosis, 225–242
caspases and BCL2, 233–235
TUNEL, 233–235
Cell-penetrating peptides (CPPs). *See* CPP-based system
Cell viability
antiviral activity, acyclovir and b-CD-PACM, 10, 16*f*
assessment
cell staining, 205–207
LDH assay, 203–205
cell line formulation, 9
endothelial cells test, 10
lipoplexes
advantages, Alamar Blue, 362
assays, 361–362
control, 362
transfection activity, 362
MTS test, 9–10
optical characteristics, nanoparticles, 10
Vero cells, 9–10
Cell viability tests, MTT, 231
Chinese hamster ovary cells (CHO-K), 252–253
Chitosan-based nanoparticles
antigen adsorption
particle's loading capacity, 134–135, 136*f*
polymeric particles, 134–135
antigen bioactivity evaluation, 137
cell viability
MTT assay, 138–139
spleen cell suspensions, 138
characterization
SEM, 133–134
size measurement, 132–133
zeta potential titration, 133
in vitro release study, 135–137
Peyer's patches, 139–140
preparation
chitosan/PCL particles, 131
precipitation/coacervation method, 129
vaccination, 128–129
CHO-K. *See* Chinese hamster ovary cells (CHO-K)
Cholesterol esterase (ChE)
Pseudomonas fluorescen, 173–175
X-ray structure, 176*f*
Cholesterol oxidase (ChoA)
catalysis, 174*f*
cloning, 173
description, 173
flowchart, CES/COX production by *Pichia pastoris*, 182*f*
X-ray structure, 175*f*
Confocal laser scanning microscopy (CLSM), 320–321
CPP-based system
biological activity
goals, 292
pre-mRNA splicing pattern, 293
splicing correction assay, 293–294
$S4_{13}PV$ and reverse NLS peptides, 292
transfection efficiency, 292, 292*f*
composition, 296–297
cytotoxicity, measurement
activation mechanisms, 295
apoptosis rate, 295–296
caspase-3/7 activity, 296
unspecific, 295
physical characterization
DSC, 284–288
fluorescence anisotropy assay, 288–291
lipid interactions, importance, 283–284
preparation
amino acid analysis, 280
binary complexes, 281–282
cationic liposomes, 281
complexation, 282–283
noncovalent association, 280
plasmid DNA, 282
sequences, 280, 280*t*
ternary complexes, 282
Cytotoxicity assay, 268–269

D

DCs. *See* Dendritic cells (DCs)
Dendritic cells (DCs)
active targeting, 146–147
AuNPs, 148–150
description, 144
liposome based vaccines
advantage, 152
Fc fragment, 153–154, 153*f*
peptide encapsulation, 152–153
phosphatidylserine, 152
PLGA, 154
use, 151–152
passive targeting, 144–146
pDNA, 315–316
role, 315
vaccines, 147–148
Differential scanning calorimetry (DSC). *See* CPP-based system
DLS. *See* Dynamic light scattering (DLS)

DNA/PEI condensation assays, 331
"Double-gavage" method, 113
Doxycycline-loaded nanoparticles
advantages, 65
bacterial infection, 62
bioadhesive drug delivery systems, 63–64
chitosan
antimicrobial and antifungal activity, 65–67
Chlamydophila pneumoniae, 65–67
features, 65–67
intranasal immunization, 65–67
preparation and side, 65–67
production, 65–67
properties, 65–67
colloidal particles, 64–65
development, antibiotic resistance, 62
difficulties, 62–63
encapsulation, hydrophilic drug, 81
gram-positive and negative bacteria, 62
investigation, 81
nanocarriers deliver, 63–64
oxytetracycline, 62–63
PCL. *See* (PCL)
Poly(lactide-*co*-glycolides), 68–69
polyalkylcyanoacrylate polymers, 67–68
polyketides classification, 62–63
polymeric, 65
protection, 64–65
role, nanotechnology, 63–64
tetracyclines, 62–63
treatment, 62
using different polymers, 64–65
vital cellular process, 62
Dual function (DF)-MITO-Porter
mitochondrial
delivery, DNase I, 321–322
diseases, 318
genome targeting and construction, 318–320, 319*f*
targeting activity, evaluation, 320–321, 321*f*
Dynamic light scattering (DLS)
IO–siRNA complexes, 267
size data, 269
SPIO nanocrystals, 266

E

EGF, 357
Electrospinning
process, 90–93
structures, 89
Elemental analysis (EDX), 91
EpiVaginal Tissue Model
controls, 13
cytokine IL-1α release, 13–14
cytotoxicity, 12
evaluation, skin irritation potential, 12
incubation, 12–13
LDH release, 13
measures, LDH cytotoxicity, 13
MTT ET-50, 12–13
"non irritant" compound, 14
quantification, cell viability, 13
skin irritation potential, 12

F

Fluorescence anisotropy
adequate interpretation, 289
DPH, 288
fluorescent probes, 288
I_{II} and $I_{\perp}$ components, use, 288–289
lipid bilayers, 291
lipid phase transition, 290–291
liposome preparation, 289, 290–291
r value, 289
$S4_{13}PV$ peptide, 291
temperature range, 290–291, 290*f*
Fluorescence resonance energy transfer (FRET)
fusion activity, 308–309
mitochondrial membrane fusion assay
analysis, 307
isolation, rat liver, 307
nuclear membrane fusion assay
analysis, 306
isolation, nucleus from HeLa cells, 306
screening assay, liposome preparation, 305
FRET. *See* Fluorescence resonance energy transfer (FRET)
Fructosyl valine amino oxidase
catalysis, 178*f*
crystal structure, 179
HbA1C, 177
X-ray structure, 179*f*

G

Gastrointestinal drug delivery
anti-inflammatory compound, encapsulated drug
advantages, 106
clinical trials, 104–105
complexation, 105–106
effects, inflammation and colon cancer, 104–105
lower dose and side effects, 105–106
NSAIDs, 104–105
peptides, 106
plasmids, 109–110
proteins, 107
siRNA, 107–109
biomaterial choice
chitosan, 112
extraction, alginate, 112
hydrogel formation, 111–112
KPV-loaded NPs, 111–112, 111*f*
colon cancer, 102

degradation, stomach acidic PH, 102
digestive tract, 119–120
GI tract, 103
hydrogels, 103
material, drug application
bioactive drug, 104
double emulsion/evaporation, solvent technique, 103–104
hydrophilic drug, 103–104
synthesis, NPs, 103–104, 105*f*
ultrasound, 103–104
nanomedicine, 102–103
NP matrices and vectorization, 103
Nps loaded NSAIDs
Ab covered NPs, 110–111
characteristics, 110
PVA covered NPs, 110
surface engineering types, 110
NP therapy, 119–120
oral vaccines, 119–120
polymer NPs, 118–119
polysaccharides, 119
specificity, efficiency and modulation, 102–103
systemic treatment, 103, 104*f*
targeting NPs
alginate and chitosant formation, 113
biomaterial, 114
dextran-FITC NPs, 113, 115*f*
"double-gavage" method, 113
hydrogel. *See* (Hydrogel preparation)
hydrogel formation, 112–113
macrohydrogel, double-gavage method, 113
polymers, 112–113
polysaccharide charge and characteristics, 112–113
purposes, double-gavage method, 113
techniques, 102
Gene delivery
alkyl-PEI2k-IOs. *See* (Alkyl-PEI2k-IOs, gene delivery)
in vitro
cell derivation, 359–360
cell lines, 359
green fluorescent protein evaluation, 361
luciferase activity, 360
reagents and cell culture, 360
transfection, 360–361
Genetic nanomedicine
advantages, lipoplexes, 356, 365
cell derivation, 359–360
cell lines, 359
cell viability. *See* (Cell viability)
DNA degradation, 356
EGF, 357
enhancement transfection *in vitro*
Alamar Blue assay, 362
EGF receptor, 362
HeLa cells and DOTAP:DOPE, 364
HepG2 and DHDK 12proB, 362
human OSCC cells, 363, 363*t*
luciferase expression, CT25, 363
murine SCCVII and NIH3T3 cells, 362, 363*t*
green fluorescent protein evaluation, 361
ligands, 356, 357
limitation, 356
lipoplexes *in vivo*. *See* (Protein-cationic lipid-DNA ternary complexes)
luciferase activity, 360
protein–cationic lipid–DNA ternary complexes. *See* (Protein-cationic lipid-DNA ternary complexes)
reagents and cell culture, 360
therapeutic genes, 356
transfection, 360–361
viral vectors, 356
2G12/GNPs interaction, STD-NMR
calculation, dissociation constant, 32–33
competitive titration, Te-50 GNP, 31, 32*f*
decreases, STD signals, 31
deuteration, 31
direct and indirect interaction, 30–31
inhibition experiments, 31
monovalent oligomannoside displacement, 30–31
titration, 2G12-oligomannosides, 31
Glucose oxidase (GOx)
β-d-glucose to gluconic acid, 171–172, 172*f*
description, 171–172
X-ray structure, 172–173, 172*f*
2G12-mediated HIV-1 neutralization and *mano*-GNPs effect
determination, 33–34
human T cells
cell culture and preparation, PBMCs, 36
DC-SIGN-mediates interaction, 35
DC-SIGN-receptors, 35
trans-infection assay, Raji DC-SIGN+ cells, 36
trans-infection experiments, 35–36
inhibition
Luciferase Assay System, 34–35
measures, 34–35
reproducible, 35
TZM-bl cells, 34–35
Te-10 and Te-50, 33
Gold nanoparticles (AuNPs)
aggregation state, fluids, 227–228
cell-based nanotoxicity
cell cycle arrest and proliferation inhibition, 232–233
cell viability tests, MTT, 231
death, 233–236
DNA content measurement, 233
measurement, 230

Gold nanoparticles (AuNPs) (*cont.*)
morphology, 230–231
oxidative stress, 236–239
cell-based toxicity tests, 227
dosage and quantification, 227–228, 229*t*
Fc fragment preactivation, 151
Fc receptors, 149, 149*f*
mice, 149
nanotechnology, 225–226
peptide
antigen, 148–149, 148*f*
conjugation, 150–151
preparation, 150
toxicity, 226, 227

H

Hank's Buffered Salt Solution (HBSS), 306
HeLa cells isolation, 306
Hydrogel preparation
biomaterial encapsulation delivery, 117*t*
chelation solution, 114*t*
hydrogel solution, 116*t*
NPs
vs. colon, 117–118
determination, colon, 118
DSS-induced colitis parameters, 118, 119*f*
FACS, 118
inflammation activity, 118
TNFa siRNA/polyethylenimine (PEI), 118, 120*f*
sodium chloride solution, 114*t*
synthesis, NPs, 116*t*

I

Interfering nanoparticles (iNOPs)
design and creation, 342–343
miRNA delivery, 341
IONPs. *See* Iron oxide nanoparticles (IONPs)
Iron oxide nanoparticles (IONPs)
cell culture
endothelial cells, 200
murine neural progenitor cells, 199
PC12 rat phechromocytoma cells, 199
cellular iron
cell morphology, 198
functionality assay, 198
pH-dependent degradation, 198–199
ROS, 198
cell viability assessment
LDH assay, 203–205
staining, 205–207
degradation
intracellular, 219–221
pH-dependent, 217–219
description, 196–197
functionality assay
assessment, 214–216
neurite length and number, 216–217
PC12 cells, 214–216, 215*f*
morphology investigation, 211–214
proliferation, 208–211
ROS induction, 207–208

L

Lactate dehydrogenase (LDH) assay
magnetoliposome, 203–205, 203*f*
use, 203–205
Lipases
description, 175
triglyceride determination, 176*f*
X-ray structure, 175, 177*f*
Lipoplexes
in vivo, targeted lipoplexes
antitumoral effect, therapeutic genes, 364
evaluation, tumor growth, 364
OSCC, 364–365
transferrin, 364
preparation
EGF, 359
maturation, 358–359
ratio, lipid/DNA, 358–359
ternary complexes, 358–359
transferrin, 359
Lipopolyplexes (LPP)
characterization, complexes
DNA condensation, 332, 332*f*
molecular weight and PEI, 333, 334*t*
particle size, 332–333, 333*t*
zeta potential, 332–333, 333*t*
experimental procedures
cell culture, 330
DNA/PEI condensation assays, 331
in vitro transfection activity, 331–332
materials, 329–330
particle size and zeta potential measurement, 331
preparation, 330
gene therapy, 328
in vitro transfection activity measurements
HepG2 cell, 334–335, 334*f*
IL-12 gene expression, 335–336, 336*f*
polyplexes, 335, 335*f*
PEI, 328–329
principle, 329
viral and nonviral vectors, 328

M

Manno-glyconanoparticles and HIV
biorecognition processes, 22–23
characterization
destructive technique, 27
^{1}H-NMR and elemental analysis, 26–27, 27*f*

organic ligands, 25–26
TEM analysis, 26
thermo spectra-tech, 25–26
by UV/Vis and TEM, 25–26
development, biofunctional NPs and biomedicine, 22
diverse types, 24–25
eradication, 22
evaluation
biosensor and MRI, 28
direct binding, GNPs and gp120-2G12, 28–29
interaction, 2G12/GNPs. *See* (2G12/GNPs interaction, STD-NMR)
macromolecule-ligand interaction, 28
SPR technology and STD-NMR spectroscopy, 28
Te-10 and Te-15 GNPs, 29–30
2G12, 23
glucose conjugate, 24–25
glycoprotein gp120, 22
gold glyconanoparticles
direct gold salt reduction, 23
direct synthesis, 24–25
ligand exchange, 23
interaction, DC-SIGN, 22
oligomannosides, 24–25
preparation
direct formation, 25
H-nuclear magnetic resonance, 25
$NaBH_4$, 25
thiol-ending glycoconjugate, 25
presentation, gp120, 22–23
structure, 22
validation and HIV-1 neutralization
cell culture and preparation, PBMCs, 36
DC-SIGN mediate interaction, 35
determination, 2G12 concentration, 33–34
human T cells, 35
inhibition, 34–35
Te-10 and Te-50 GNPs, 33–35
trans-infection assay, 36
Microfluidics, protein-CNT sensors, 169
MicroRNAs (miRNAs) silencing
anti-miR and therapies, 340–341
biogenesis, 340
delivery, inhibitors
amino groups and functions, 342–343
anti-miRs, 342
iNOPs-7, 342–343, 344*f*
nanoparticles–siRNA complex and cellular entry, 342–343, 343*f*
siRNA localization, 343, 345*f*
description, 340
efficiency determination
anti-miRs, 344–346
dual luciferase assay, 345–346
northern blotting, 344
quantitative PCR, 345
in vitro transfection
cell preparation, 347
complete medium, 347–348
material preparation, 346–347
serum-free, 347
in vivo
delivery, anit-miR-122, 348–349
determination, delivery efficiency, 349–350
formulation, iNOP-7/anit-miR-122, 348
oligonucleotides, 346
regulation, 350
role, 340
siRNA delivery, 341, 342
MITO-Porter. *See* Dual function (DF)-MITO-Porter
MRSA and antimicrobial properties
ANOVA, 94–96
biomaterial, 94–96
contact-based inhibition, 96–97
elastomeric structures, 96–97, 97*f*
fibrous film coatings, 94–96
PU polymer, 94–96
survival, EMRSA 16, 94–96, 96*f*
two-tier mechanism, 96–97
MTT assay, 138–139
Multifunctional envelope-type nano device (MEND)
DF-MITO-Porter
intracellular observation, 320–321
mitochondrial delivery, DNase I, 321–322
mitochondrial genome targeting and construction, 318–320
FRET
fusion activity, 308–309
mitochondrial membrane fusion assay, 307
nuclear membrane fusion assay, 305–306
screening assay, liposome preparation, 305
NLS peptide, 304
nuclear gene delivery
nuclear membrane, 314–315
pDNA to DC, 315–316
T-MEND function, 316–318
pDNA, 304
problems, original and innovative strategy, 304–305
programmed packaged concept and construction, 302–304
tetra-lamellar MEND (T-MEND)
characteristics, 311–313
development, 309
fusogenic lipid composition, 310
preparation, 310–311
Multifunctional fusion proteins
baculovirus expression, 254–256, 255*f*

Multifunctional fusion proteins (*cont.*)
cloning, 254
function
cell proliferation and binding assays, 257–258, 257*f*
histology, 258
quantitative analysis, 259
wound healing model, 258
mammalian expression, 254, 255*f*
recombinant decorins characterization, 256

N

Nanoparticle-drug delivery, brain
Amprenavir
activity, 44
anti-HIV-1 activity, 44–45
concentrations, 44–45
mature HIV-1 virions production, 44–45
zidovudine and lamivudine, 44–45
analysis, 45
antiretroviral therapies, 42–43
ART regimens, 43
ARV drugs, 43
BBB, 43
evolution, anitretorviral containing nanoplex
BBB function, 49
cell viability measurement, MTT, 51
HIV-1, 54
measurement, TEER, 50
monocyte isolation, PBMCs, 52–53
transfer, Tf-QD-Amprenavir NPs, 51–52
HIV-1 and CNS, 42–43
nanotechnology, 43
QDs. *See* (Quantum dots (QDs))
quantum dots (QDs), 43–44
Tf and TfRs, 43–44
therapeutic advantage, 43
NLS. *See* Nuclear localization signal (NLS)
NPC. *See* Nuclear pore complex (NPC)
Nuclear gene delivery
nuclear membrane, 314–315
pDNA to DC, 315–316
T-MEND function
cholesterol, 316–318
KALA peptide, 316–318
luciferase activity, 316–318, 317*f*
pH environment, 316–318
transfection activity, 316–318
Nuclear localization signal (NLS)
drawback, 315
mechanism, 314–315
peptide, 304
Nuclear pore complex (NPC), 314, 315
Nucleic acid delivery. *See* CPP-based system

O

Oral squamous cell carcinoma (OSCC)
DNA targeting, 363
gene delivery, 359–360
lipoplexes, 364–365
transfection, 362
Oxidative stress
measurement, 238–239
ROS, 236–238

P

PBS. *See* Phosphate-buffered saline (PBS)
PCL. *See* Poly-ε-caprolactone (PCL)
PCL nanoparticles and doxycycline-loaded PLGA
aliphatic polyester family, 70
antibacterial activity
doxycycline testing, 76
vs. E. coli, 76
determination, MIC and MBC
concentration, drug, 76–77
control, bacterial growth, 76–77
by microdilution method, 76–77
value, 76–77
drug release behavior, 70
formulation and degradation, 70
improvement, encapsulation efficiency, 70
microcapsules, 70
and native doxycycline effect
activity, 78–81
in bacterial culture, 80*f*
bacterial life cycle, 78–81
drug concentration, 78–81
E. coli growth, 78–81
growth kinetics, bacteria, 78–81
MHB and void nanoparticles, 78–81
preparation, PLGA
different formulation, size and encapsulation efficiency, 73–75, 74*t*
distribution, 71–72
doxycycline aqueous solution, 70–71
drug loading, 76
estimation, 72–73
hydrophilic drug, 70–71
improvement, doxycycline encapsulation efficiency, 75
measuring, drug concentration, 72–73
osmotic gradient, 75
parameters, 71–72
polarity organic solvent, 75
role, polymer ratio and solvent, 75
separation, 72–73
structure, 70–71, 72*f*
stability, 77–78
pDNA. *See* plasmid DNA (pDNA)
PEI. *See* Polyethylenimine (PEI)
Peyer's patches, 139–140
Phosphate-buffered saline (PBS), 249

Plasmid DNA (pDNA)
advantages, 302–303
coating, 309
consecutive steps, 302–303
DCs, 315–316
effect, 314
encapsulation, 312
MEND, 303*f*
nuclear delivery, 316*f*
size, 314
T-MEND, 310*f*
Poly(lactic-co-glycolic acid) (PLGA)
antigen quantification, NPs, 156
conjugating antibodies, NPs, 157
NP preparation, 155–156, 155*f*
quantification, TLR-L, 156–157
use, 154
Poly-ε-caprolactone (PCL), 131
Polyethylenimine (PEI)
alkyl-PEI2k-IOs. *See* (Alkyl-PEI2k-IOs, gene delivery)
use, 328–329
Polyurethane. *See* Antimicrobial properties, electrically formed elastomeric polyurethane
Programmed packaging, 302
Protein–carbon nanotube (CNT) sensors
amperometric sensor, blood analytes detection
blood cholesterol, 171
four-channel biosensor, 169, 170*f*
surface modification, 171
biosensors, 169
characterization, SWCNT enzyme adduct, 186
concept
bio-inspired systems, 168
bio–nano interface, 168–169
description, 167
experimental protocol
amperometric response, 188–190
chain anchors, 188–190
HDL, 188–190
protein and transducer attachment, 189*f*
technical problems, 190
fabrication, 188
fermentation process control, 181
gene structure and purification, overexpressed protein
column chromatography, 181
encoding, 181–183
expression vector, 183*f*
health care, 166
mediators
criteria, 186
description, 186
microfluidics, 169
nanoscience, 190–191
protein probes. *See* (Protein probes, serum detection and monitoring)
proteins immobilization
self-assembled monolayers, 185–186
on SWCNT, 183–185
prototype
functions, 167–168
microscale prototype lab-on-a-chip, 168
nanotube growth, 168
signal detection, 188
surface characterization, 187–188
Protein-cationic lipid-DNA ternary complexes
in vivo, targeted lipoplexes
antitumoral effect, therapeutic genes, 364
evaluation, tumor growth, 364
OSCC, 364 365
transferrin, 364
lipoplex preparation
EGF, 359
maturation, 358–359
ratio, lipid/DNA, 358–359
ternary complexes, 358–359
transferrin, 359
liposome preparation
biodegradable lipid use, 357–358
DOTAP, 357–358
extraction, storage and period, 357–358
transfection reagents, 358
plasmids, 358
Protein probes, serum detection and monitoring
ChE, 173–175
ChoA, 173
fructosyl valine amino oxidase, 176–179
GOx, 171–173
lipase, 175
protein engineering and molecular biology
ChoE, 180
construction, expression vectors, 180–181
electrochemical responsiveness, 180
thermal properties, *Rhizopus niveus* lipase (RNL) activity, 180
Proteins immobilization
self-assembled monolayers, 185–186
on SWCNT
electronic coupling, 184
four-step process, 183
pET28 system-step process, 184*f*
two-nanosecond molecular dynamics simulation, 185
Protein transduction domains (PTDs), 278–279
PTDs. *See* Protein transduction domains (PTDs)

Q

Quantum dots (QDs)
advantages, organic dyes, 45–46
aquesous dispersion
carboxyl functional, 46–47
ligand exchange, 46–47

Quantum dots (QDs) (*cont.*)
description, 45–46
double-shelled CdSe/CdS/ZnS
graded shell, 46
nucleation, 46
purification, 46
size, 46
synthesis, organic media, 46
dynamic light scattering, 47
generations, 45–46
high-resolution transmission electron microscopy, 47
imaging modalities, 45–46
physical properties, 47
spectrophotometry and fluorometry, 47
structure, 45–46
surface coatings, 45–46

R

Reactive oxygen species (ROS)
induction, 198, 207–208
intracellular
ECFH conversion, 12
fluorescence analysis, 12
generation, 11
ROS. *See* Reactive oxygen species (ROS)

S

Single wall carbon nanotube (SWCNT)
characterization, enzyme adduct, 186
proteins immobilization on
electronic coupling, 184
four-step process, 183
pET28 system-step process, 184*f*
two-nanosecond molecular dynamics simulation, 185
Small interfering RNA (siRNA), 341, 342
Splice-switching oligonucleotides (SSOs), 282, 293

T

TEM. *See* Transmission electron microscope (TEM)
Tetra-lamellar MEND (T-MEND)
characteristics
discontinuous sucrose density, 312
phase-contrast transmission electron microscopy, 312–313
physiochemical properties, 311–312
development, 309
function. *See* (Nuclear gene delivery)
fusogenic lipid composition, 310
nuclear delivery, 315–316
preparation, 310–311, 310*f*
TGF-β1. *See* Transforming growth factor-β1 (TGF-β1)
Transferase dUTP nick end labeling (TUNEL), 233–235
Transforming growth factor-β1 (TGF-β1), 245
Transmission electron microscope (TEM)
hydrophilic nanocomposites, 271
SPIO nanocrystals, 266
TUNEL. *See* Transferase dUTPnick end labeling (TUNEL)

W

Wound healing model, 258

Z

Zeta potential titration
colloidal system, 133
measurements, nanoparticle, 133, 134*f*
pH values, 133

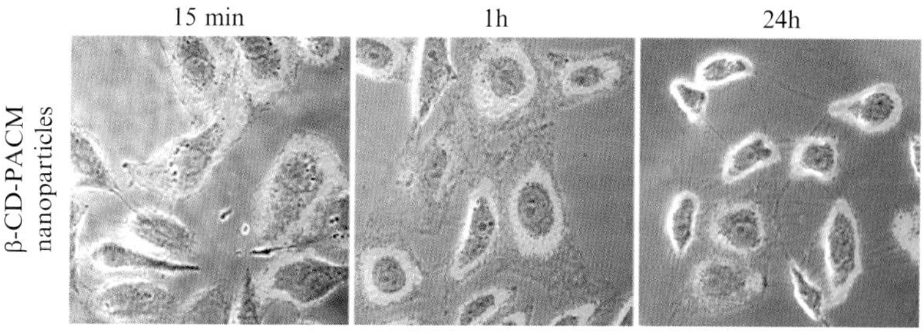

Roberta Cavalli *et al.*, Figure 1.3 Cell uptake of β-CD-PACM nanoparticles. Vero cells are incubated with the compound for the times indicated and then analyzed by confocal laser scanning microscopy without fixation. Merged phase-contrast and immunofluorescence images are shown. β-CD-PACM nanoparticles appear to accumulate in a perinuclear compartment 1 h post-exposure.

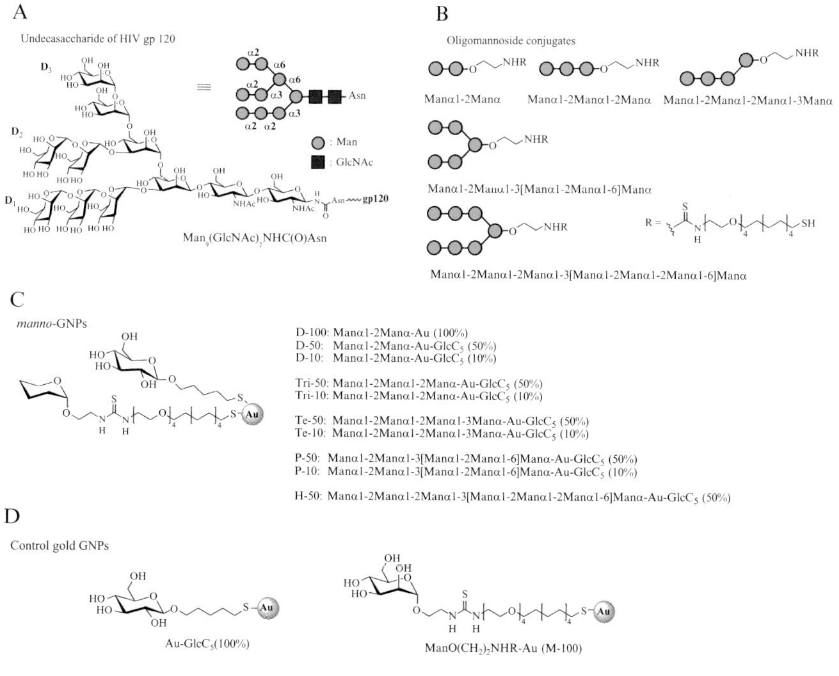

Paolo Di Gianvincenzo *et al.*, Figure 2.1 (A) gp120 *N*-glycan undecasaccharide; (B) thiol-ending oligomannosides conjugates used for *manno*-GNP preparation; (C) *manno*-GNPs. D, T, Te, P, and H stand for di- tri-, tetra-, penta-, and heptamannoside conjugates, respectively; the numbers indicate the percentages of oligomannosides on GNP, the rest being the 5-(mercapto)pentyl β-D-glucopyranoside (GlcC$_5$) component; (D) control GNPs bearing glucose and mannose conjugates.

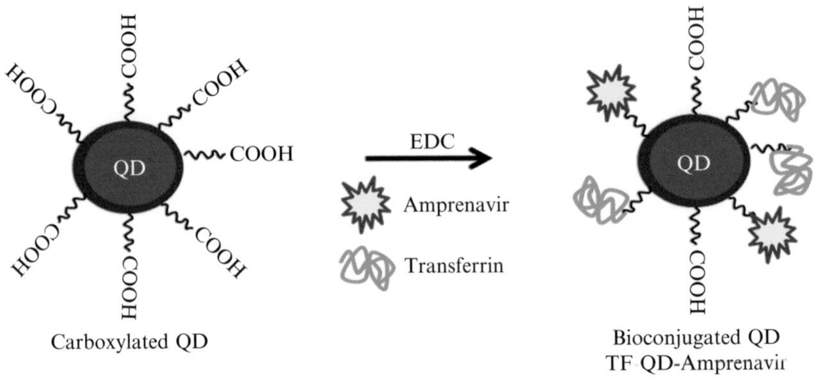

Supriya D. Mahajan *et al.*, Scheme 3.1 Formation of Tf-QD-Amprenavir nanoplex.

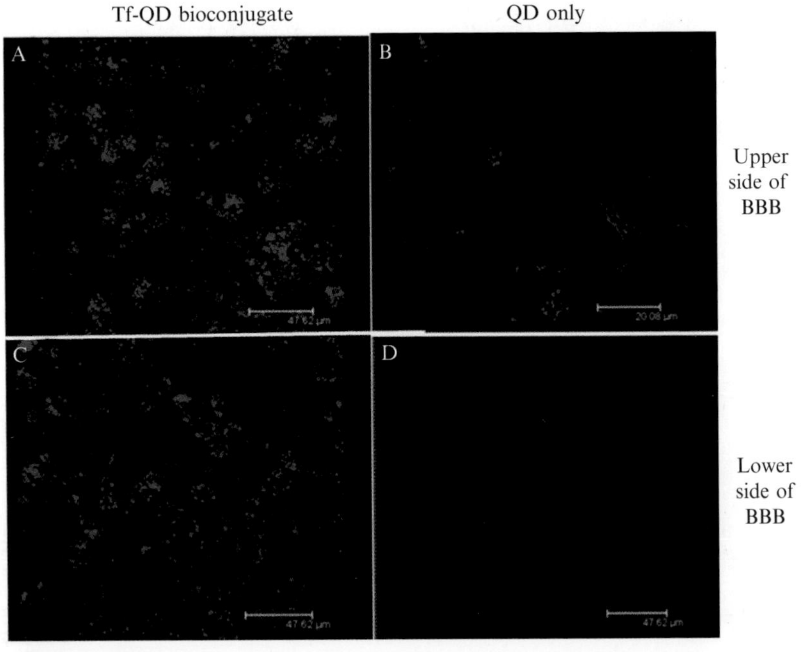

Supriya D. Mahajan *et al.*, Figure 3.2 Transmigration of nanoplexes across the BBB. Confocal Images of the BBB, (A) upper side of the *in vitro* BBB model following treatment with (a) QD-Tf and (b) QD alone. (B) Lower side of the *in vitro* BBB model following treatment with (c) QD-Tf and (d) QD alone.

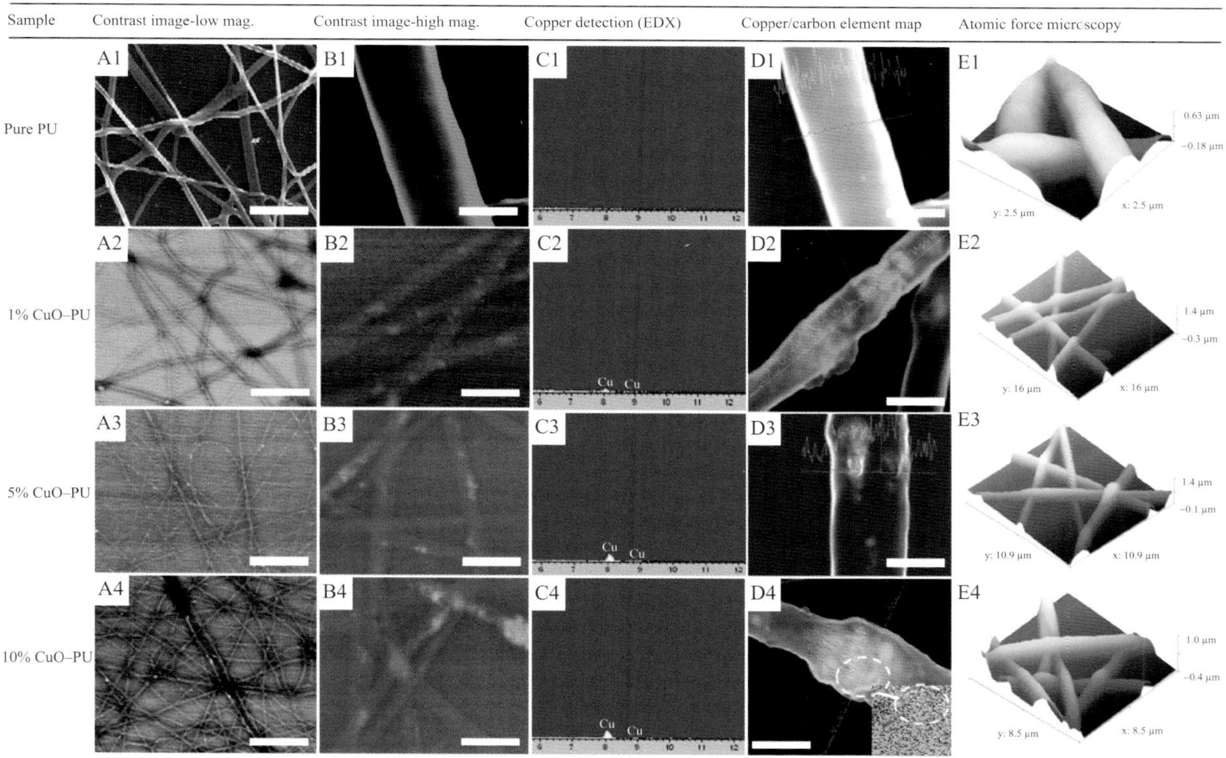

Z. Ahmad *et al.*, Figure 5.2 Fiber analysis on various electrospun compositions. Showing (A1–A4) low magnification contrast imaging—electron microscopy, (B1–B4) high magnification contrast imaging—electron microscopy, (C1–C4) element analysis, (D1–D4) element mapping, and (E1–E4) atomic force microscopy. (Scale bars: A1 = 10 μm, B1 = 1 μm, D1 = 1 μm, A2 = 10 μm, B2 = 5 μm, D2 = 1 μm, A3 = 10 μm, B3 = 5 μm, D3 = 1 μm, A4 = 10 μm, B4 = 5 μm, D4 = 1 μm.)

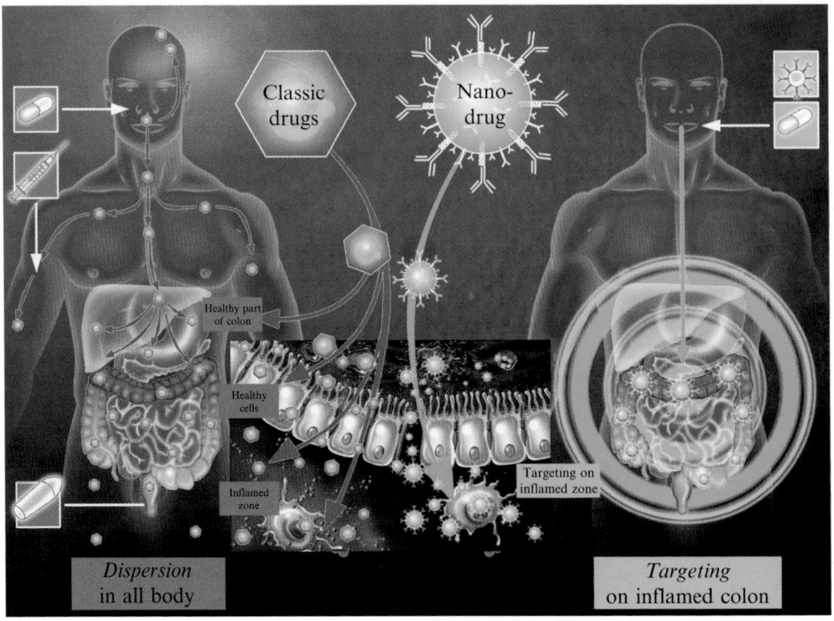

Hamed Laroui *et al.*, Figure 6.1 Illustration representing the different localization of drug in targeted strategy compared with systemic treatment. For oral intake or intravenous injection of the classical drug, the bioactive component is distributed throughout the body without any distinctions between healthy and inflamed tissue. Enema strategy can only target the distal part of colon. In targeting strategy, nanoparticles (NPs) are covered with an antibody whose ligands are overexpressed in inflamed areas. The NPs accumulate and the drug is released in the specific area (from Laroui *et al.*, 2010b).

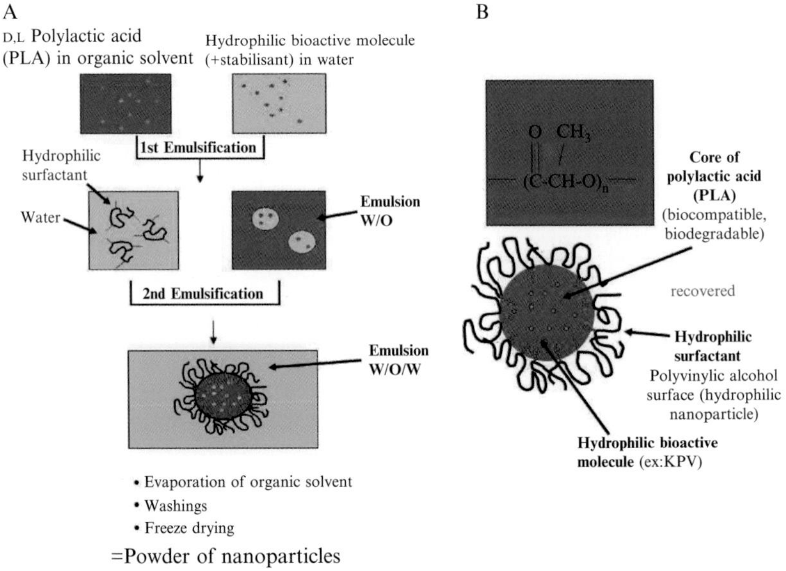

Hamed Laroui *et al.*, Figure 6.2 (A) Schematic process of NP synthesis. A hydrophilic drug is encapsulated by double emulsion water in oil in water (W/O/W). (B) Schematic representation of PLA NPs loaded with KPV, a hydrophilic drug, and coated with PVA (from Laroui *et al.*, 2010a).

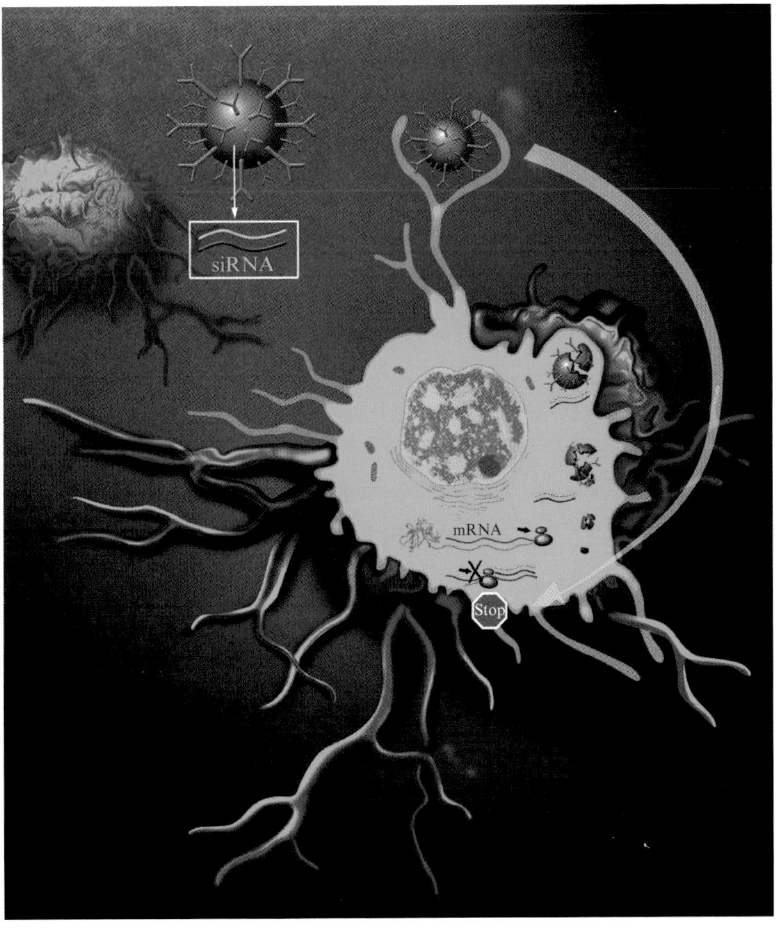

Hamed Laroui *et al.*, Figure 6.3 Schematic of the delivery of siRNA-loaded nanoparticles to a macrophage to stop the translation of mRNA and decrease the level of the corresponding protein.

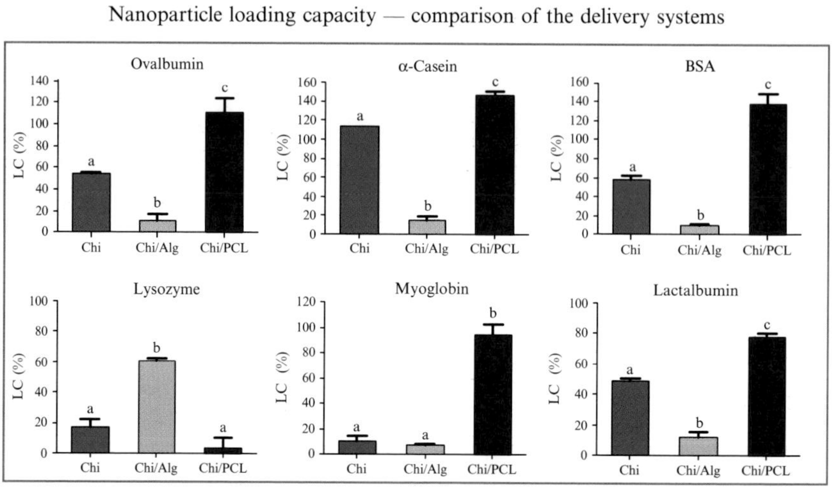

	Chi	Chi/Alg	Chi/PCL
Size (nm)	592.0 ± 25.6	3613.9 ± 1655.6	169.0 ± 7.4
Zeta potential (mV)	+ 5.50 ± 0.55	−15.74 ± 2.17	−8.18 ± 1.77

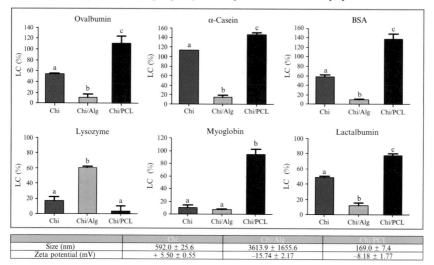

Filipa Lebre *et al.*, Figure 7.3 Comparison of the particle's loading capacity (LC) using different proteins. The results illustrated on this figure were obtained during adsorption studies performed with buffer phosphate, pH 7.1 ± 0.2 with six proteins with different isoelectric points. Chi/PCL nanoparticles have a significantly higher LC when compared with Chi and Chi/Alg particles for all the proteins except lysozyme. The inclusion of a hydrophobic polymer like PCL into chitosan particles allows the establishment of hydrophobic interactions between proteins and particles, which explains, at least in part, the result. In contrast, Chi/Alg particles have the lowest LC for almost all the proteins. The adsorption of proteins with low IEP (<7.0) is below 20% and about 60% for lysozyme (IEP ~ 11.0). This last result can be explained by the negative charge of particles in PB 7.4 which favors electrostatic interactions only with the positively charged proteins. The table below the chart illustrates the size and zeta potential of chitosan-based delivery systems suspended in phosphate buffer, pH 7.4 (mean ± SD, $n = 3$), during adsorption studies.

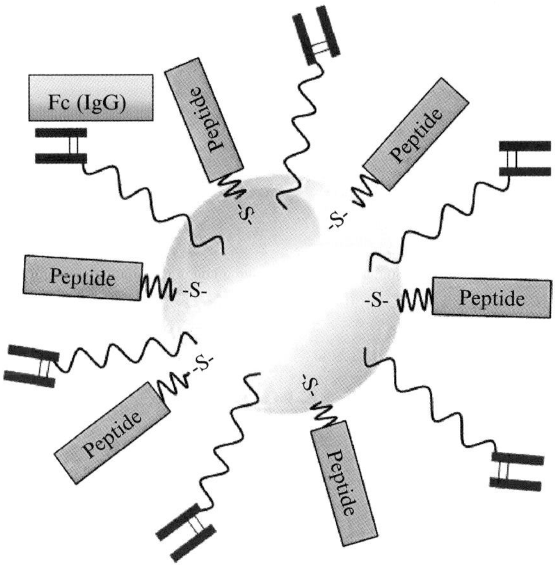

Luis J. Cruz *et al.*, Figure 8.1 AuNPs carrying peptide antigen and targeted to FcR using the Fc fragment of IgG.

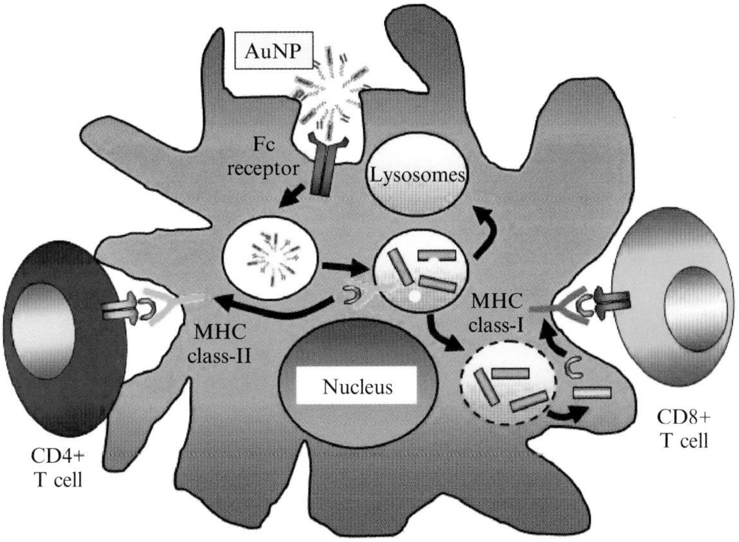

Luis J. Cruz *et al.*, Figure 8.2 AuNPs targeted to Fc receptors on DCs are phagocytosed and processed by endosomal and lysosomal acidification and enzymes. Subsequently, they are presented in the context of MHC class II. Endosomal scape allows the processing and cross-presentation of these particles by MHC class I.

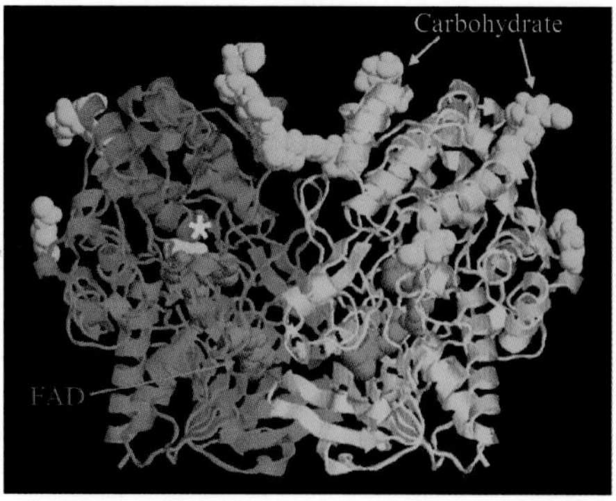

Sowmya Viswanathan *et al.*, Figure 9.4 X-ray structure of glucose oxidase, PDB code 1gpe.

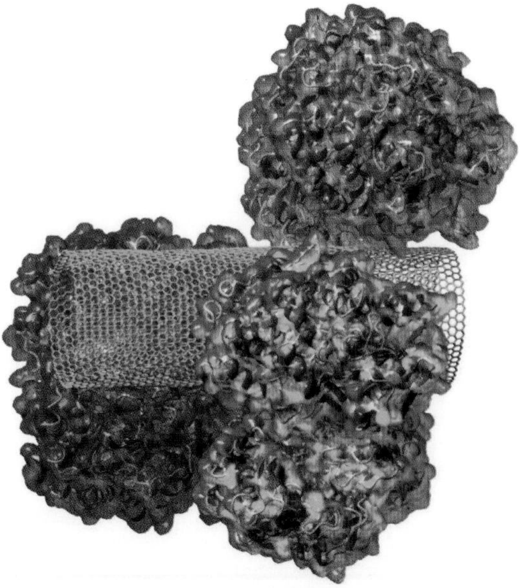

Sowmya Viswanathan *et al.*, Figure 9.14 A snapshot from 20 ns molecular dynamics simulation of glucose oxidase.

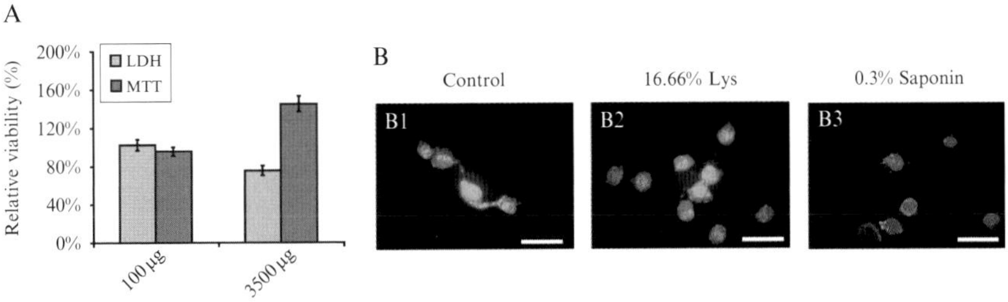

Stefaan J. Soenen *et al.*, Figure 10.1 Assessing cell viability of magnetoliposome (ML)-treated 3T3 fibroblasts. (A) Results of an MTT and an LDH assay for 3T3 fibroblasts incubated with 3% cationic lipid-containing MLs at 100 and 3500 μg Fe/ml for 24 h. Values are given relative to those of untreated control cells. The error bars indicated are mean ± SEM ($n = 10$). (B) Representative images depicting the viability of untreated NIH 3T3 cells (B1) or cells incubated for 24 h with 16% (B2) cationic lipid-containing MLs at 100 μg Fe/ml, as determined by calcein AM (green; live cells) and ethidium homodimer (red; damaged cells) treatment; blue color indicates DAPI nuclear staining. B3 shows control cells treated with 0.3% saponin; scale bars: 75 μm. See Soenen and De Cuyper (2009) for more details. Reproduced with permission from Soenen and De Cuyper (2009), © Wiley VCH.

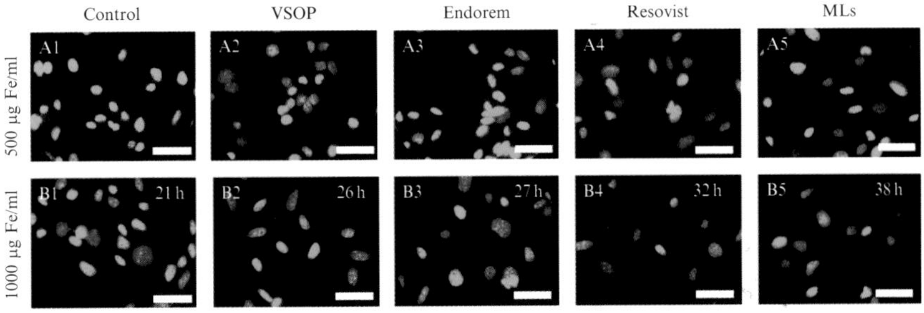

Stefaan J. Soenen *et al.*, Figure 10.2 (A,B) Cellular proliferation of C17.2 NPCs as assessed by EdU-staining (green) for cells not exposed to any particles (Control) or cells incubated with the indicated particles (VSOP: citrate-coated 4 nm diameter IONPs; Endorem: dextran-coated IONPs with hydrodynamic diameter between 60 and 150 nm; Resovist: carboxydextran-coated IONPs with hydrodynamic diameter of about 75 nm; MLs: lipid-coated 14 nm diameter IONPs) at 500 μg Fe/ml (A) and 1000 μg Fe/ml (B) at 3 days postparticle incubation; nuclei of non-proliferative cells are colored blue (DAPI). Scale bars: 75 μm. Reproduced with permission from Soenen *et al.* (2010b), © Wiley VCH.

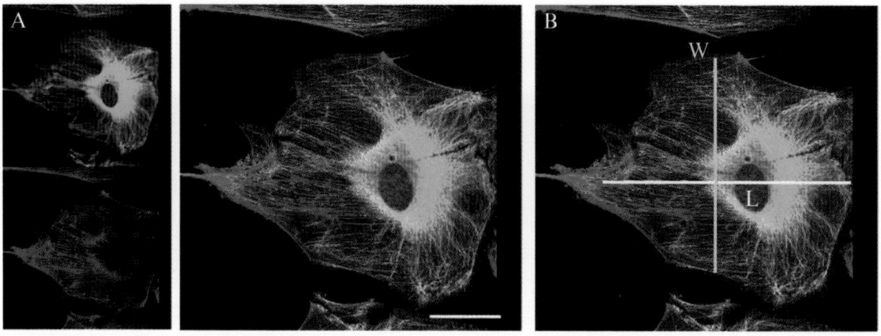

Stefaan J. Soenen *et al.*, Figure 10.3 (A) Representative confocal image of a hBOEC depicting α-tubulin (green), F-actin (red), and a merged image showing α-tubulin, F-actin, and DAPI nuclear counterstaining (blue). Scale bar: 50 μm. (B) For measuring cell polarity, the cellular length (white bar) and width (light blue bar) must be determined. The ratio of cellular width over length then gives information on the cellular spreading. Please note that cytoplasmic extensions should be excluded for a proper determination of cellular length and width.

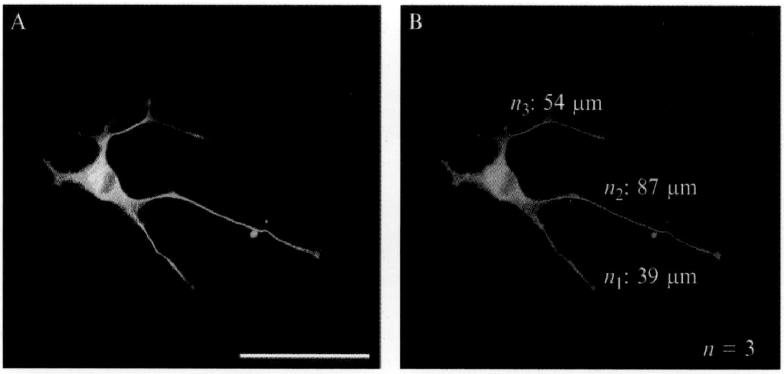

Stefaan J. Soenen *et al.*, Figure 10.4 (A) Representative confocal image of a PC12 cell depicting both α-tubulin (green) and F-actin (red) staining. Scale bar: 50 μm. (B) The results obtained from calculating neurite lengths and number of neurites using the NeuriteJ plugin for ImageJ. A grayscale version of the previous image is shown along with the calculated trajectories for the neurites, their individual lengths, and the total number of neurites extending from this cell.

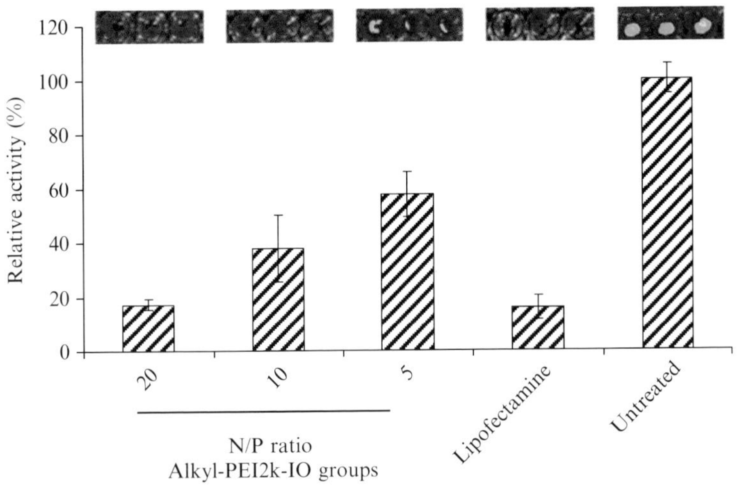

Gang Liu *et al.*, Figure 13.5 Inhibition of fluc gene expression by *N*-alkyl-PEI2k-IOs/siRNA (siRNA = 6 pmol) at various N/P ratios.

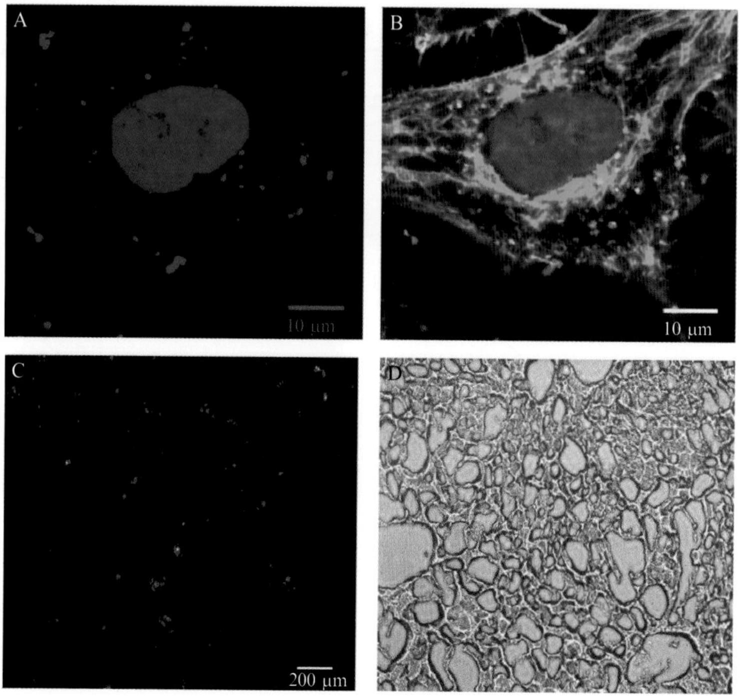

Huricha Baigude and Tariq M. Rana, Figure 17.3 Localization of siRNA. (A) Cells were transfected with iNOP-7 containing Cy3-labeled siRNA. Localization of the duplex siRNA after 24 h was monitored by confocal microscopy. Overlay images of siRNA and nuclear DNA (4′,6-diamidino-2-phenylindole [DAPI] stained) are shown. (B) Cells transfected with iNOP-7/Cy3 siRNA complex were stained fluorescein labeled Phalloidin to show cell outlines. (C) Fluorescence microscopy analysis of siRNA distribution *in vivo*. Histology slide of liver shows the distribution of siRNA in liver 4 h after intravenous administration of iNOP-7 was injected. Scale bar: 100 μm. (D) Bright field microscopic image of (C).